No. 1150
$29.95

HANDBOOK OF MODERN ANALYTICAL INSTRUMENTS

BY RAGHBIR SINGH KHANDPUR

FIRST EDITION

FIRST PRINTING

Printed in the United States of America

Library of Congress Cataloging in Publication Data

Khandpur, Raghbir Singh, 1942-
Handbook of modern analytical instruments.

Bibliography: p.
Includes index.
1. Chemical apparatus. I. Title.
QD53.K63 543'.07 81-9133
ISBN 0-8306-1150-9 AACR2

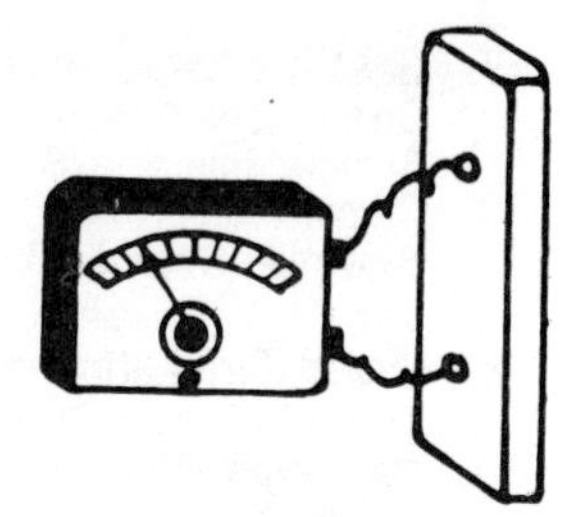

Contents

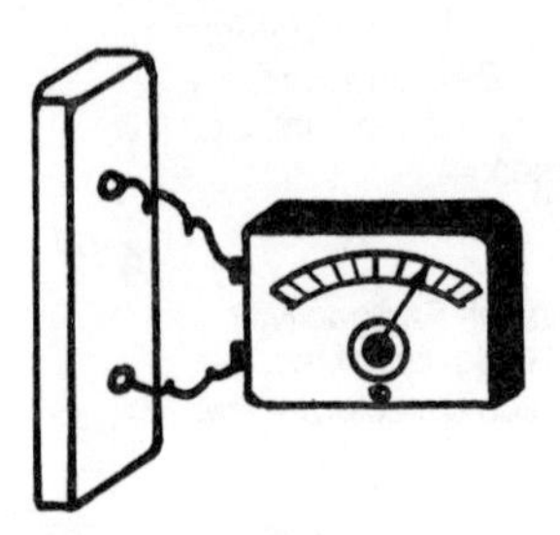

Preface

Instruments used for analysis constitute the largest number of instruments in use today. Their range is spectacular and baffling in variety. It is difficult to imagine a field of activity where analytical instruments are not required and used. They are used in food processing laboratories, oil refineries, drug and pharmaceutical industry, chemical industry, environmental pollution monitoring and control and, above all, in hospitals for routine clinical analysis and research. Modern industry has grown so fast, and quality control of manufactured products has become so important a problem, that the few skilled analytical wet chemists that are available cannot meet this challenge. Moreover, large volumes of data are needed everyday in many walks of life; in the hospitals, diagnosis of illness is dependent to a large degree on the blood tests. These functions are elegantly carried out by analytical instruments.

This book has been designed to cater to a wide variety of readers. The users of analytical instruments will find the text useful as they will be able to appreciate the principle of operation and basic building blocks of the instruments they work on everyday. An attempt has been made to balance the highly technical details of the instruments with descriptive and lucid explanations of only the necessary information. With the wide-spread use of analytical instruments, it is essential to have qualified and sufficiently knowledgeable service and maintenance engineers. Besides having a basic knowledge about the principle of operation, it is important for them to know the details of commercial instruments from different manufacturers. A concise description of instruments of each type from leading manufacturers has, therefore, been provided. The area of analytical instrumentation involves a multi-disciplinary approach, with electronic and optics forming the major disciplines. The scintillating progress in the field of integrated circuits technology has fulfilled many a dream. The present day instruments have attachments for computing results from graphic recorders, which were hitherto done manually with great difficulty. This has helped to extend the application and range of instruments and, therefore, several new approaches have been successfully attempted.

A stage is fast approaching when repair and servicing will be carried out only at the printed circuit card level without having to repair the instrument at the component level. Work of this nature necessitates the knowledge of the worker about various building blocks of an instrument, and this has been the pervading spirit in the presentation of the text. Thus, the persons entrusted with the task of looking after the instruments are likely to find the book highly useful. A similar advantage would be felt by the students of chemistry, physics, chemical engineering, etc.

The book starts with the explanation of basic concepts in electronics and covers various types of display systems and laboratory recorders in the first chapter. Integrated circuits and operational amplifiers and their use in analytical instrumentation have been illustrated to make the reader acquainted with these devices.

Colorimetry and spectrophotometry have become the central techniques of analytical chemistry. Although the optical designs of these instruments were almost perfected quite early, and hence precision measurement of the wavelength of spectral lines was possible in the 1930s, measurements of intensity were relatively crude. Photographic methods were used for this purpose. The development of phototubes resulted in the construction of precision spectrophotometers for the first time. Spectrophotometry, particularly in the ultraviolet region, offered a sensitive, precise and non-destructive method of analysis of biologically important substances such as proteins and nucleic acids; the technique when later extended to the visible region of the spectrum led directly to the explosion of analytical methodology, characterized by the use of an ever decreasing amount of biological materials and multi-sample analytical methods. The use of the double beam principle helped in the development of infrared spectroscopy, where direct recording of spectra is now routine. Similarly, flame photometers and atomic absorption spectrophotometers have become almost indispensable tools in clinical and research laboratories. Various types of spectrophotometers are included in the first part of the book.

Next is a study of the chromatography which offers a unique method of separation of closely similar substances. Various methods like liquid-liquid partition, gas-liquid partition, ion-exchange, molecular sieve, etc., have all provided a basis for the same. Gas-liquid chromatography is perhaps the most widely used than any other single analytical method for small to medium sized molecules.

However, a liquid mobile phase possesses certain advantages over a gaseous mobile phase since it can contribute to the separation achieved through the specificity of its interaction with the solutes. This advantage has resulted in the development of high pressure, high performance liquid chromatography, using specialized column packing of very small size. Two chapters are devoted to the different types of chromatographic instruments.

Electrophoresis, used by Tiseluis in 1937 as a medium scale method for the separation of proteins, and which has now developed into an ultramicro method with the introduction of zone electrophoresis, is covered in the next chapter. Further improvements took place with the use of synthetic gel material polyacrylamide and led to the most generally used method for the characterization of proteins and nucleic acids. The continuous improvements in staining and optical visualization methods have led to this technique being applicable to the microgram level or even less.

Perhaps the most sophisticated ultra-micro scale analytical instrument yet developed is the combination of the gas chromatograph with the mass spectrometer, and the same has been sell illustrated in the book in the chapter on mass spectrometry. Extremely compact versions of this instrument were used in the space missions.

The most important spectroscopic techniques for structural determination are the nuclear magnetic resonance and electron spin resonance spectrometry. These topics are covered in the next two chapters. Fourier transform NMR and spin decouplers are also concisely dealt with.

Introduction of newer electronic devices has resulted in the obtaining of accurate, repetitive readings to be performed with speed and minimal error. The lengthy warmup periods, very common in circuits using thermionic valves, have been eliminated with the availability of solid state devices. For example, in a digital pH meter with printer, the electronics is much faster than measurement is possible as electrodes must be rinsed between measurements, and they take time to respond when immersed in solutions. Electrochemical instruments have not only been considerably improved but also their operations are automated. These instruments can be connected to a strip chart recorders whereby complete titration curves can be plotted. While the chapter on electrochemical instruments covers various types of instruments like conductivity meters and polarographs, the chapter on pH meters includes the latest in electronic circuitry, ion-selective electrodes and chemically-sensitive semiconductor devices. An exhaustive treatment has been given to industrial gas analyzers and blood gas analyzers.

The awareness and concern about the deteriorating environment is increasing the world over. It is necessary to monitor changes taking place in the quality of the environment for initiating efforts to accomplish the environmental pollution monitoring program. Many of the analytical tools used elsewhere in other areas of applicaitons could be profitably utilized for air and water pollution monitoring purposes. A special chapter on this subject illustrates various techniques employed for monitoring different pollutants in air and water.

Qualitative and quantitative analysis has been greatly facilitated by the increasing availability of isotopically labeled compounds. The measurement of these has required the development of increasingly complex instrumentation, usually capable of measuring and printing out the radioactive counts from 100 or more samples wholly automatically. Multichannel counters, scanners and gamma camera are representative of the highly sophisticated equipment and have richly deserved their place in the book.

In hospital laboratory equipment, great advances have been made in the instrumental handling of large numbers of samples and in the data processing. Ergonomic design of the instrument panel layout and incorporation of internal electronic automatic control circuits have enabled equipment performing complex functions to be operated by laboratory staff relatively unskilled in instrumental techniques. Analysis procedures have been automated, with the result that a very large number of samples can be handled per hour for a multiplicity of tests. The instruments like autoanalyzers have brought about tremendous procedural and conceptual innovations. A complete description of the automated analysis system has been provided in a separate chapter.

The modern laboratory has a large number of analytical instruments churning out information. Mechanical procedures for handling the huge amount of data are imperative. The marriage of computers and instruments is offered as a way of easing the burden on the scientists, as well as optimizing the performance of the analytical instruments. It is now possible to apply computer systems to all the major analytical instrument procedures. Many of the leading instrument manufacturers are developing and producing systems for use in the laboratory, both for data acquisition and for control purposes. The small general purpose computer, dedicated to a specific laboratory function, has been virtually responsible for a revolution in the methodology, quantity, economy and quality of experiments performed. It is possible to buy dedicated computer systems for gas chromatography, NMR, high resolution mass spectrometry and emission spectroscopy, etc. The fundamentals of computers and computer directed analyzers constitute the last chapter of the book.

Fortunately, many times instruments required for different applications work on the same principle for doing the same or similar functions. For example, a carbon dioxide monitor for respiratory gas analysis works on the principle of absorption of infrared radiation and so does the instrument used for measuring carbon dioxide content in the emission gases from industrial effluents. The method of sample collection may, however, vary in each application. Keeping this in mine, this volume not only attempts to explain the core of the instruments but also illustrates their utility separately for specific applications.

The book has a sufficient degree of comprehensiveness and depth to give the reader the most important information without requiring him to delve in more specialized books on the subject. It is the hope of the author that the material in the book would be found useful by a large number of readers working in different disciplines.

Thanks go to Professor Harsh Vardhan, Director, Central Scientific Instruments Organization, Chandigarh, for according permission for publication of this book. Credit is due to Mr. P.R. Kalia and Mr. Venkatanathan who transformed the original notes and drafts into a form which made it possible for the author to publish the same. Mr. Makhan Singh is to be commended for his skill and an almost inexhaustible patience in helping me in completing the diagrams. I would also like to thank my other colleagues of CSIO who have helped me directly or indirectly in the preparation of the book. Particular mention should be made of Mr. J.K. Chhabra and Mr. Rajinder Singh.

The author feels indebted to various manufacturers of analytical instruments who supplied valuable information on their products along with some interesting photographs which have given a charming look to this publication. Special mention can be made of magnificent assistance from Beckman Instruments, USA; Varian, USA; Schoeffel Instrument Corp., USA, Pye Unicam, U.K., Arthur H. Thomas Co., USA; Carl Zeiss, Jena; Corning, USA; Technicon, USA; Baird Atomic, USA, and Gilford, USA.

Finally I am grateful to my wife Ramesh Khandpur and my children Vimal, Gurdial and Popila who bore the brunt of uncalled for neglect during the preparation of the manuscript.

Raghbir Singh Khandpur

Chapter 1

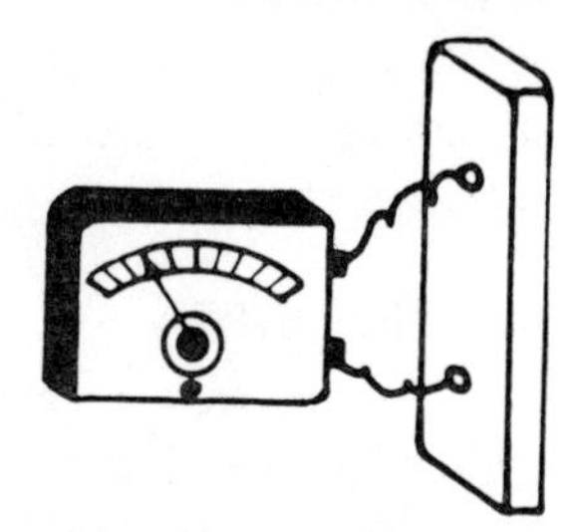

Basic Components of Analytical Instruments

Analytical instruments are used to provide information about the composition of a sample of matter. They are employed, in some instances, to obtain qualitative information about the presence or absence of one or more components of the sample whereas in other instances quantitative data are sought from them. In the broadest sense, any analytical instrument (Fig. 1-1) would consist of the following four basic units:

■ A *chemical information source* which generates a set of signals containing necessary information. The signal may be generated from the sample itself. For example, the yellow radiation emitted by heated sodium atoms constitutes the source of the signal in a flame photometer.

■ A *transducer,* which converts the signal to one of a different nature. Because of the familiar advantages of electric and electronic methods of measurement, it is the usual practice to convert into electric quantities all non-electric phenomena associated with the analysis of a sample. For example, the photocell and the photomultiplier tube are transducers that convert radiant energy into electrical signals.

■ The *signal conditioner* converts the output of the transducer into an electrical quantity suitable for operation of the display system. Signal conditioners may vary in complexity from a simple resistance network or impedance matching device to multi-stage amplifiers and other complex electronic circuitry. They help in increasing the sensitivity of instruments by amplification of the original signal or its transduced form.

- A *display system* which provides a visible representation of the quantity as a displacement on a scale, on the chart of a recorder or on the screen of a cathode ray tube.

Thus, the instrument can be conveniently considered in terms of flow of information where operation of all parts is essentially simultaneous.

BASIC DEFINITIONS

The progress in instrumental methods of analysis has closely paralleled developments in the field of electronics because the generation, transduction, amplification and display of signals can be conveniently accomplished with electronic circuitry. All electronic circuits are constructed with the help of some basic components and circuit blocks. Some of the important system components are described in the following sections.

Potential Difference

Potential difference between two points is the amount of work required to transfer a unit quantity of electricity from one point to another. It usually refers to the difference of potential between two points which may be electrically isolated. A potential difference may exist between them because of a difference in electrostatic charge.

Voltage

Voltage is the difference in electric pressure between two points. Between any two points, the voltage difference and the potential difference are the same and are measured in volts. Both these terms are used interchangeably.

The practical unit of voltage or potential difference is the *volt.* A potential difference of one volt will cause a current of one ampere in a resistance of one ohm. Signal voltages encountered in analytical instruments are usually much smaller, of the order of 1/1000 volt. Larger units are the *kilovolts* (1KV = 1000 volts) encountered in cathode ray tube power supplies.

Electric Current

Electric current is the rate of electric charge transfer from one point in a conductor to another. It arises from a drift of electrons

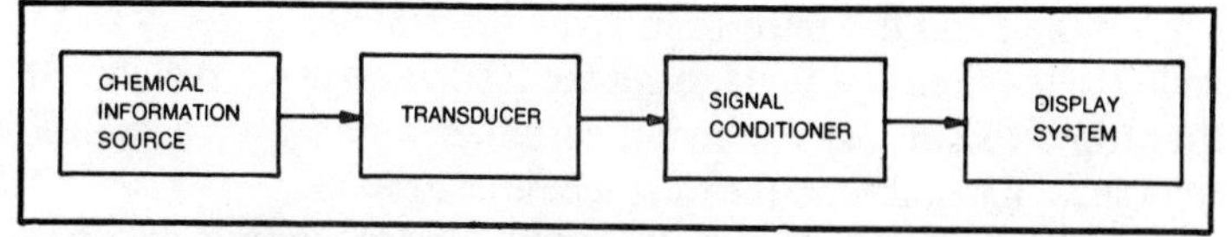

Fig. 1-1. Elements of an analytical instrument.

from the atoms of the conductor in one direction on connection of the voltage source.

The practical unit of current is ampere (A). It is equal to a flow of 6×10^{18} electrons per second past a given point on the conductor. A constant current of one ampere flowing for one second will deposit .001118 grams of silver.

Charge

The integral of current with respect to time is known as *electric charge.* It is equal to the amount of electricity carried past a given point on a wire in one second when the current flowing is one ampere.

The practical unit of electric charge is the *coulomb.* It is defined as the charge on 6.24×10^{18} electrons or the charge transferred in one second by a current of one ampere.

Resistance

Resistance is the opposition to the flow of current offered by a conductor, device or circuit. It is related to current as follows:

resistance = voltage/current (Ohm's Law)

The resistance is expressed in ohms (abbreviated Ω). The value of resistance of a metal is temperature dependent and it increases with the rise in temperature. The most commonly encountered resistors in electronic circuitry are *carbon resistors.* Their value is either printed in numbers or it is put in the form of color coded bands around the body. Each number from 0 to 9 has been assigned a color according to the following chart (Fig. 1-2).

Black	*0*	*Blue*	*6*
Brown	*1*	*Violet*	*7*
Red	*2*	*Grey*	*8*
Orange	*3*	*White*	*9*
Yellow	*4*	*Gold*	*±5% tolerance*
Green	*5*	*Silver*	*±10% tolerance*

The first band closest to the end of the resistor represents the first digit of the resistance value. The second band gives the second digit and the third band gives the number of zeros to be added to the first two digits to get the total value of the resistor. If there is a fourth band, it indicates either a ± 5% or a ± 10% tolerance. If it is absent, the tolerance is ±20%.

High stability resistors are made by depositing a layer of "cracked" carbon on a ceramic rod and completely insulating the resistive element so that it can withstand arduous operating conditions. Still superior in stability and for low noise applications are metal oxide film resistors.

Some resistors are made to give an adjustable or variable resistance. They are three terminal devices in which the central terminal is connected to the movable contact or *wiper.* A variable resistance is obtained between the wiper arm and any one end of the resistor track. Such a device is called a *rheostat.*

If several resistances (R_1, R_2, R_3, R_4) are connected in series, then total value (R_T) is given by

$$R_T = R_1 + R_2 + R_3 + R_4$$

When these resistances are connected in parallel, the value R_T is given by

$$\frac{1}{R_T} = \frac{1}{R_1} + \frac{1}{R_2} + \frac{1}{R_3} + \frac{1}{R_4}$$

Kirchoff's Laws

Although the most important law in electrical measurements is the Ohm's Law which defines the relationship between current, potential and resistance in a circuit, the rigorous solution of all the currents and voltages of an extensive electronic circuit is only possible by the application of Kirchoff's two laws: The sum of the currents at any single point of an electrical circuit is zero and the sum of the voltages around a closed electrical circuit is zero. Fortunately, this is normally unnecessary since the current involved in transferring the signal from stage to stage is small compared with the current required to place the individual stages of a circuit in operation. Note that dc measurements are generally made by using moving coil type galvanometers. For higher sensitivities, light spot galvanometers are preferred.

Power

Power is defined as amount of work done per unit time. In an electrical circuit, the unit of work is *joules.* Therefore, power is expressed as joules per second or *watts.*

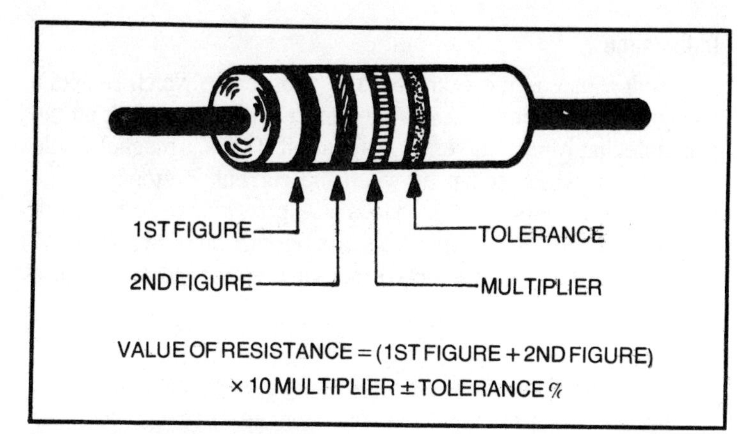

Fig. 1-2. Color code of resistors.

If a current of I amperes is caused to flow through a resistance of R ohms, then the electrical power dissipated in the resistance is given by the expression:

$$P = I^2R = \frac{V^2}{R} = VI$$

Each type of resistor is designed to work up to a maximum permissible wattage and it is important not to exceed this rating.

Capacitor

A *capacitor* (also called condenser) consists of two conductors separated by a dielectric or an insulator. The dielectric can be paper, mica, ceramic or plastic film or foil. High value capacitors are usually of electrolyte type. They are made of a metal foil with a surface that has an anodic formation of metal oxide film. The anodized foil is in an electrolytic solution. The oxide film is the dielectric between the metal and the solution. The high value of capacity of electrolytic capacitors in a small space is due to the presence of a very thin dielectric layer. An electrical charge can be placed on the plates by connecting a voltage source to the capacitor.

Capacitance is measured in *farads*. A capacitor has a capacitance of one farad when one coulomb charges it to one volt. The farad is too large a unit. Usual sub-units used are microfarad (10^{-6}) and the picofarad (10^{-12}).

A trimmer capacitor is a variable capacitor. Spacing between the metal plates which are separated by a dielectric can be adjusted to give a variable capacitance.

Inductance

Inductance is the characteristic of a device which resists a change in the current through the device. Inductors work on the principle that when a current flows in a coil of wire, a magnetic field is produced which collapses when the current is stopped. The collapsing magnetic field produces an electromotive force which tries to maintain the current. When the coil current is switched, the induced emf would be produced in such a direction, so as to oppose the buildup of the current.

$$e = -L\,di/dt$$

The unit of inductance is a *henry*. An inductance of one henry will induce a counter emf of one volt when the current through it is changing at the rate of one ampere per sec. Inductances of several henries are used in power supplies as smoothing chokes, whereas smaller values (in the milli or microhenry ranges) are used in audio and radiofrequency circuits.

Alternating Current Circuits

An alternating current is one whose instantaneous values undergo a definite cycle of changes at regular intervals of time. A sinewave signal is a special type of alternating signal with its voltage or current varying sinusoidally (with simple harmonic motion) with time. It is characterized by the equation:

$$i = I_m \operatorname{Sin} 2\pi ft$$

where I_m represents the maximum or peak current, i is the instantaneous current at time t and f is the frequency in cycles per second.

Frequency is numerically equal to the reciprocal of the time of one cycle. It is usually expressed as cycles per second or *hertz*. The mains power supply has a frequency of 60 Hz or in some countries 50 Hz.

A second sine wave might have the same frequency as the first but differ in the times at which the current starts from zero. Such waves are said to be out of phase. The difference in phase of two signals is measured either in degrees or in a fraction of a complete angular rotation. Figure 1-3 shows two sine waves with a phase difference of $\pi/2$ or 90°.

A value characteristic of sinusoidal waves needed in power calculations is the root-mean-square value. This value is given by:

$$I_{rms} = 0.707\, I_m$$

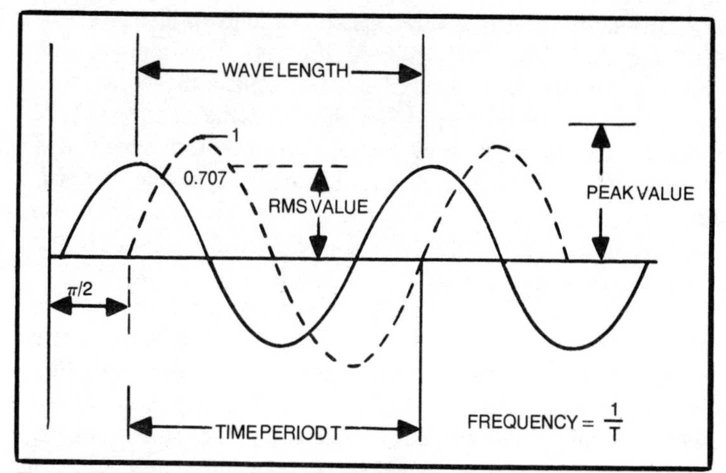

Fig. 1-3. Two sine waves with 90° phase difference.

When an ac signal source is connected to a pure resistance, the voltage at any instant across the resistance is governed by Ohm's Law, and the current and voltage are said to be in phase. Obstruction to the flow of current in an ac circuit is normally characterized as *impedance* which is a complex quantity and is often a function of frequency.

Transformers are used in ac circuits to change the voltage level and consequently the current level of ac power, ideally, with very little power loss. They are invariably present in the operated instruments.

THE ELECTRON AND ELECTRON TUBES

All the three forms of matter, namely solids, liquids and gases consist of minute particles called molecules which, in turn, are composed of atoms. Atoms have a positively charged nucleus around which revolve tiny charges of negative electricity known as *electrons*. It has been estimated that the mass of an electron is approximately 9.1×10^{-31} kilograms and it carries a negative charge of -1.6×10^{-19} coulombs.

The movement of electrons can be accelerated by supplying additional energy to them. This action can be conveniently done by using heat energy. If the temperature of a metal is gradually raised, there is an increase in the velocity of electrons in the metal and they even tend to break away from the surface of the metal. This action which is speeded up when the metal is heated up in vacuum is used in electron tubes to produce the necessary electron supply.

An electron tube basically consists of a source which produces electrons, designated as the *cathode* (Fig. 1-4). Also, there may be one or more additional electrodes which control these electrons and an electrode called the plate which collects the electrons. All of these electrodes are in an evacuated envelope. This envelope is usually made up of glass, but in special cases it may be of metal, ceramic or a combination of these materials. The property of the electron tube to control the movement of electrons supplied by the cathode makes it an exceedingly sensitive and accurate device.

When the cathode is heated, electrons leave the cathode surface and form an invisible cloud around it. Any positive electric potential within the evacuated envelope is a strong attraction to the electrons. The positive electric potential is supplied by the electrode called *anode* which is located within the same envelope. The vacuum inside the envelope or the tube permits free movement of the electrons and prevents damage to the hot electron emitting surface of the cathode.

The cathodes are classified based on the method by which they are heated to emit electrons. A directly heated cathode is a wire heated by the direct flow of electric current through it. The material of the wire used may be *tungsten*, thoriated tungsten or coated with alkaline earth acids. An indirectly heated cathode consists of a thin metal sleeve coated with electron-emitting material and is heated by a separate tungsten resistance wire placed inside the sleeve but electrically insulated from it. Directly heated cathodes require less power for heating. They are, therefore, preferred for tubes designed to operate on battery power. An indirectly heated tube reduces introduction of hum from the ac power line. It also allows close spacing between the cathode and the other electrodes in the tube which improves the tube performance as an amplifier.

Diode

The simplest form of electron tube consisting of two electrodes, a cathode and an anode is called a *diode* (Fig. 1-4A). In a diode the anode (plate) is connected to positive electrical potential with respect to cathode. The electrons emitted from the cathode move towards the anode under the influence of the positive plate potential and find a return path through the plate and the supply circuit. This flow of electrons constitutes what is known as the *plate current.*

If a negative potential is applied to the plate, the electrons emitted from the cathode are repelled back to the surface of the

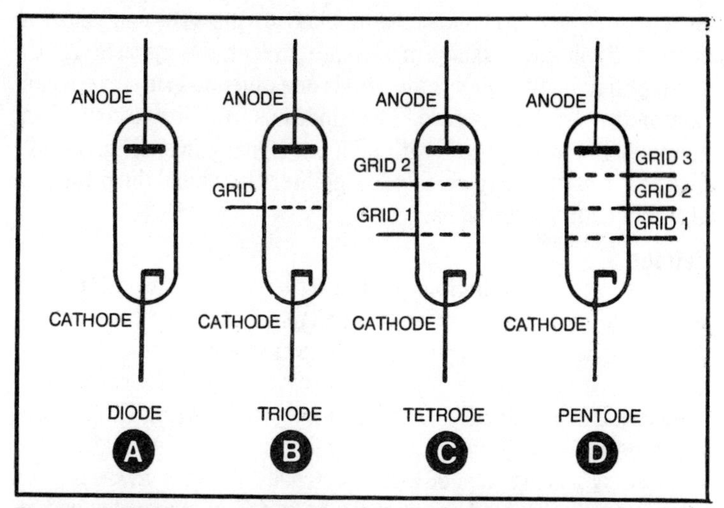

Fig. 1-4. Various types of electron tubes. (A) Diode. (B) Triode. (C) Tetrode. (D) Pentode.

cathode and no plate current flows. If an alternating voltage like the ac mains supply is applied to the plate, the flow of current shall take place only when the plate is positive with respect to cathode. A diode thus offers an excellent method of achieving a unidirectional current from alternating current. This process is called rectification and the diode is called a *rectifier.* Since almost all the vacuum tube circuits require direct current for their operation, a diode, therefore, becomes an essential and integral component of all electronic instruments.

Triode

In order to control the flow of plate current in the tube, the introduction of a third electrode is necessary. This additional electrode is called the *grid* and is placed between the cathode and the anode. The grid usually consists of relatively fine wire wound on two support rods and extending almost the length of the cathode. When the tube is used as an amplifier, a small negative dc potential is generally applied to the grid. Under this condition, the grid does not draw appreciable current. The flow of electrons from cathode to the plate then becomes a function of the negative voltage present on the grid. The grid being closer to the cathode is more effective than the plate. If the grid is made more negative, the plate current decreases and vice versa. When a signal is applied on the grid, the plate current varies in accordance with the signal.

Because a small voltage signal applied to the grid can cause a comparatively large change in the plate current, the signal is said to be amplified by the tube. If the grid is negative, no grid current can flow and power input is zero. The triode has an ability to control an appreciable output power with no input signal power (Fig. 1-4B). Figure 1-5 shows a typical triode amplifier circuit and the principle of amplification of the signal.

Tetrode

One of the limitations of the *triode* is the inter-electrode capacitance existing between grid and plate, plate and cathode and grid cathode. These capacitances produce undesirable coupling between various electrodes, particularly at high frequencies and then cause instability and unsatisfactory performance. The capacitance between grid and plate, the most significant of all, can be reduced by placing an additional electrode called the *screen grid* in the tube. With this addition, the tube has four electrodes and is accordingly called a *tetrode* (Fig. 1-4C). The screen grid is mounted between control grid and the plate and acts as an electrostatic shield between them, thus reducing the grid to plate capacitance. The screen grid is operated at a positive voltage usually less than the plate voltage. The presence of the screen grid has the desirable effect that it is possible to obtain high amplification without plate to grid feedback and resultant instability.

Pentode

In all electron tubes, electrons striking the plate at high speeds dislodge some electrons from its surface. This emission of electrons from the electrode other than the cathode is called *secondary emission.* In the case of screen grid tubes, the proximity of the positive screen grid offers a strong attraction to these secondary electrons and results in reduction of plate current and consequently the useful plate voltage swing.

This undesirable effect can be minimized when a fifth electrode is placed within the tube between the screen grid and plate. This fifth electrode is known as the *suppressor grid* (Fig. 1-4D) and is usually connected to the cathode. Because of its negative potential with respect to the plate, the suppressor grid retards the flight of secondary electrons and diverts them back to the plate. The five-electrode tube is called *pentode*.

Electrometer

The term *electrometer* refers to dc voltmeters having a very high input impedance. They are used to measure the emf of a

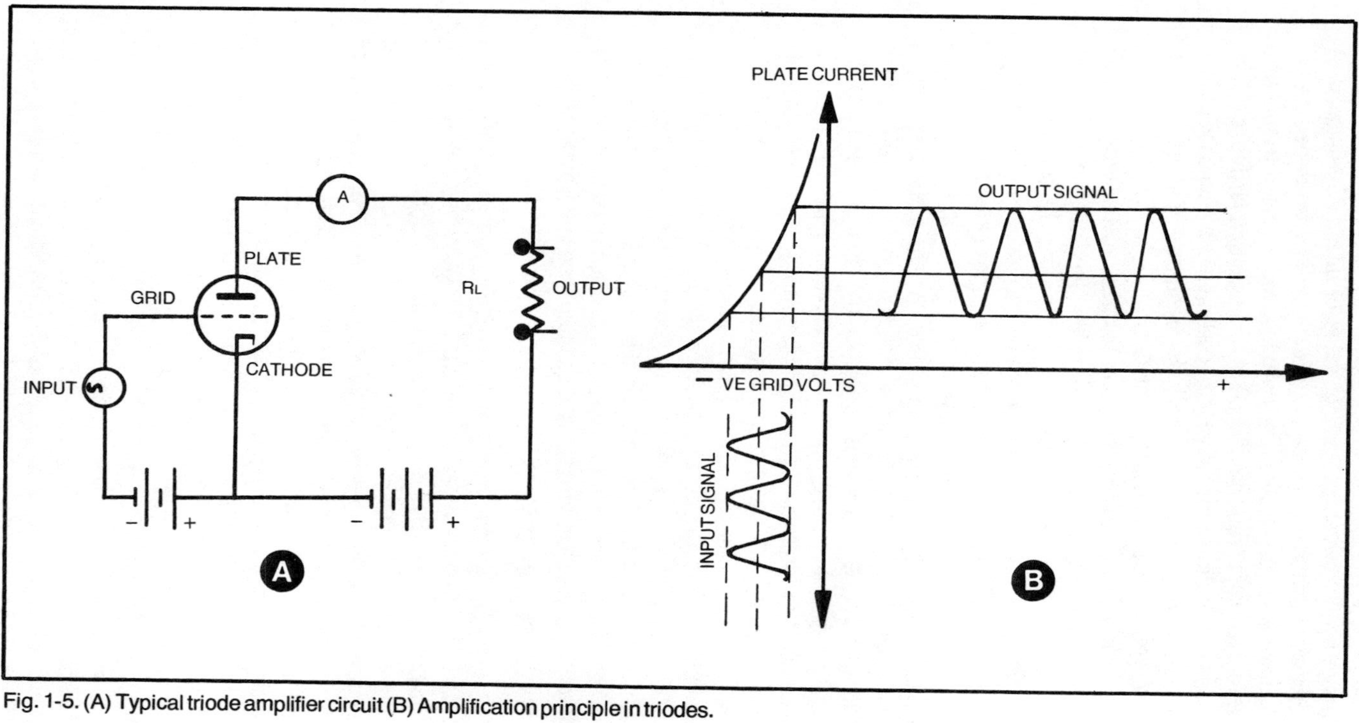

Fig. 1-5. (A) Typical triode amplifier circuit (B) Amplification principle in triodes.

voltage source which has a high value of internal resistance. A typical application is the measurement of the emf developed at a pH sensitive glass electrode. They are also widely used to measure the ionization current developed in an ionization chamber by the presence of radioactivity in which a current of 10^{-12} amps is passed through a high resistance and the resulting voltage drop is measured by means of an electrometer.

In order to attain a high input resistance, special electrometer tubes have been designed. The glass envelope of an electrometer tube is treated to reduce leakage currents along the surfaces. The cathode is operated at low temperature to minimize radiant heating of the grid. This reduces thermionic emission of electrons from the grid and also minimizes photoelectric emission resulting from cathode light striking the grid wires. A low plate potential is employed to reduce ionization of residual gas in the tube by the plate current. If this is not done, positive ions attracted to the negative grid result in unwanted grid current. With these precautions, grid resistors as high as 10^{12} ohms are feasible and currents of the order of 10^{-12} amp can be measured.

However, these valves have almost been displaced by insulated gate field effect transistors (IG FET). In addition to its small size, IGFET provides the required degree of stability and capability of a millivolt sensitivity. It is used in the input circuit as a source follower followed by operational amplifiers connected for appropriate gain to give an output suitable for display devices.

SEMICONDUCTOR DEVICES

Semiconductor devices are considerably smaller in size than the vacuum tubes but can perform their equivalent functions with many advantages. Besides their small size and light weight, they have no filaments or heaters for electron emission and therefore require no heating power or warmup time. They are solid in construction, extremely rugged and are free from microphonics and therefore can withstand severe environmental conditions. Because of these advantages, semiconductor devices have almost displaced the vacuum tube from many applications in which it was formerly pre-eminent. The emphasis in this book will therefore be on circuits making use of semiconductor devices in discrete and integrated forms.

Unlike other electron devices, which depend for their functioning on the flow of electrons through a vacuum or a gas, semiconductor devices make use of the flow of current in a solid. These solids, called the semiconductors, have poorer conductivity than a conductor but better conductivity than an insulator. These elements belong to the fourth group of the periodic table. The materials most often used in semiconductor devices are *germanium* and *silicon.* The atoms of these elements have a crystal structure and the outer or valence electrons of individual atoms are tightly bound to the electrons of adjacent atoms in electron pair bonds or the covalent bonds. In order to obtain free electrons from such structures, very small amounts of other elements having different atomic structure can be added as impurities. These impurities can thus modify and control the basic electrical characteristics of the semiconductor materials.

When the impurity atom has one more valence electron than the semiconductor atom, this extra electron can not be bound in the electron pair and is held loosely by the atom. This free electron requires very little excitation to break away from the atom and the material becomes a better conductor. The presence of this extra free electron makes the material have a negative charge and the resulting material is called *N-type.*

A different effect is produced when the impurity added is the atom having one less valence electron than the semiconductor atom. This means that one of the bonds in the resulting crystal structure cannot be completed because the impurity atom lacks one valence electron. A vacancy or hole thus exists in the lattice structure which encourages the flow of electrons in the semiconductor material consequently increasing its conductivity. The vacancy or hole in the crystal structure is considered to have a positive electrical charge because it represents the absence of an electron. Semiconductor material which contains these holes or positive charges is called *P-type* material.

P-N Junction

When N-type and P-type materials are joined together (Fig. 1-6A), a very important phenomenon takes place at the junction of the two materials. The free electrons from the N-type material diffuse across the junction and recombine with holes in the P-type material. This is called *diffusion current.*

Let us consider the condition when an external battery is connected across a P-N junction (Fig. 1-6B) with its positive terminal connected to the P-type material and the negative terminal to the N-type material. Under these circumstances, the junction is said to be forward-biased as the polarity of the external battery facilitates the movement of the charge carriers across the

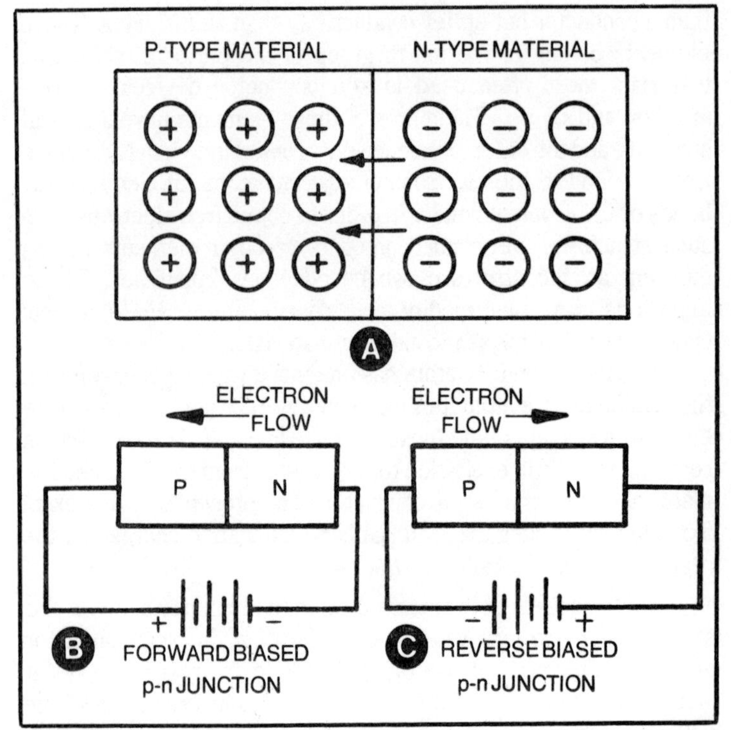

Fig. 1-6. (A) P-N junction. (B) Forward biased P-N junction. (C) Reverse biased P-N junction.

junction and the junction offers a very low resistance to the current flow. The P-N junction under this condition is called *forward-biased.*

If the polarity of the externally applied battery is reversed (Fig. 1-6C), the charge carriers, under the influence of the external battery, move towards the adjacent sides of the junction and form a barrier layer with the result that the electron flow in the external circuit is very low. The P-N junction under this condition is said to be *reverse-biased.*

Semiconductor Diode

The simplest type of semiconductor device is the diode which is basically a P-N junction (Fig. 1-7A). The N-type material which serves as the negative electrode is called *cathode* and the P-type material which serves as the positive electrode is referred to as the *anode*. The arrow symbol shows the direction of the flow of conventional current which is opposite to the flow of electrons.

Just like the vacuum tube diode, the semiconductor diode conducts current more easily in one direction than in the other, so that it is an effective rectifying device. The semiconductor diodes are available in a wide range of current capabilities suitable practically for almost all applications.

Transistor

When a second junction is added to a semiconductor diode, the resulting device is called a *transistor*. This device is capable of providing power or voltage amplification. The three regions of the transistor are called the *emitter,* the *base* and the *collector.* In normal operation (Fig. 1-7B) the emitter-base junction is biased in the forward direction and the collector-base junction in the reverse direction.

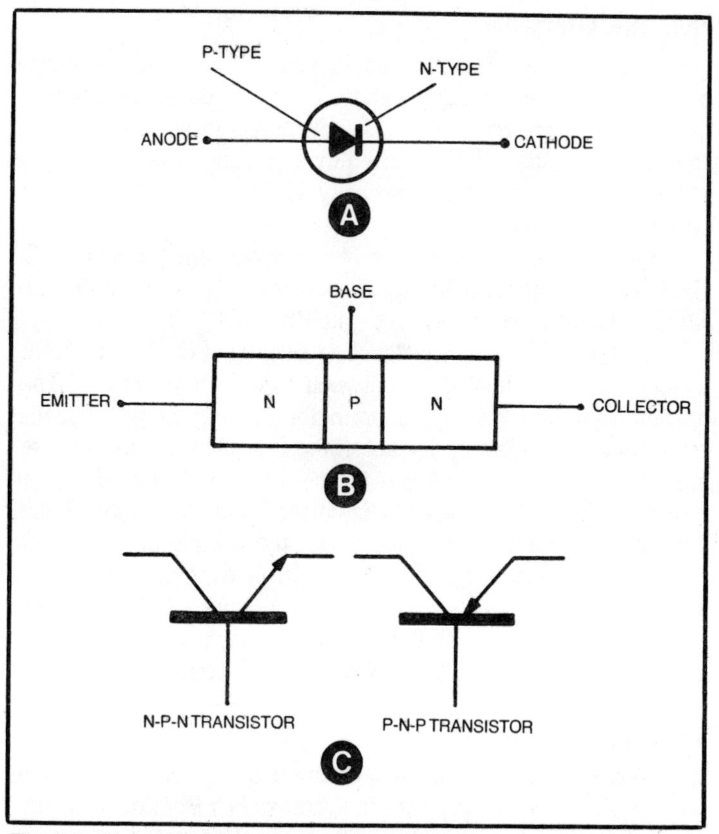

Fig. 1-7. (A) Semiconductor diode (B) Functional diagram of a transistor (C) Symbols used for transistors.

Depending upon the direction of current flow, different symbols (Fig. 1-7C) are used for N-P-N and P-N-P transistors. The first two letters of the N-P-N and P-N-P designations indicate the respective polarities of the voltages applied to the emitter and the collector in normal operation. In an N-P-N transistor, the emitter is made negative with respect to both the collector and the base, and the collector is made positive with respect to both the emitter and the base. In a P-N-P transistor, the emitter is made positive with respect to both the collector and the base, and the collector is made negative with respect to both emitter and the base.

The transistor can be used for a wide variety of control functions including amplification, oscillation and frequency conversion. Other semiconductor devices are zener diode, silicon-controlled rectifier, unijunction transistor, etc.

Field-Effect Transistor

The field-effect transistor is basically a three terminal semiconductor device having characteristics similar to that of a pentode vacuum tube. It is a voltage-operated device and, therefore, instead of being biased by a current, it is biased by a voltage. No input current normally flows and hence its input resistance is virtually infinite.

Field-effect transistors are of two types: the junction FET (JFET) and the metal-oxide semiconductor FET (MOSFET). The MOSFET is also called insulated-gate FET (IGFET).

In the JFET, a thin conducting channel (Fig. 1-8) of finite conductance is established between two P-N junctions. The current from the source to the drain, for a given voltage, depends on the dimensions of the channel. If the P-N junctions are reverse biased by applying a voltage to the gate, a depletion region containing no mobile carriers is formed and the width of the conducting channel is reduced. Thus, the magnitude of current between source and the drain can be controlled by the reverse bias applied to the gate electrode. This provides a means of controlling the amplification in the FET. Depending upon the type of material of the channel, the FET may be N-channel or P-channel.

MOSFET

In the metal-oxide semiconductor FET (MOSFET), a thin layer of silicon oxide insulates the gate contact from the channel. There are two types of MOSFET: *depletion* (normally on at zero bias) and *enhancement* (normally off at zero bias). Figure 1-9 shows

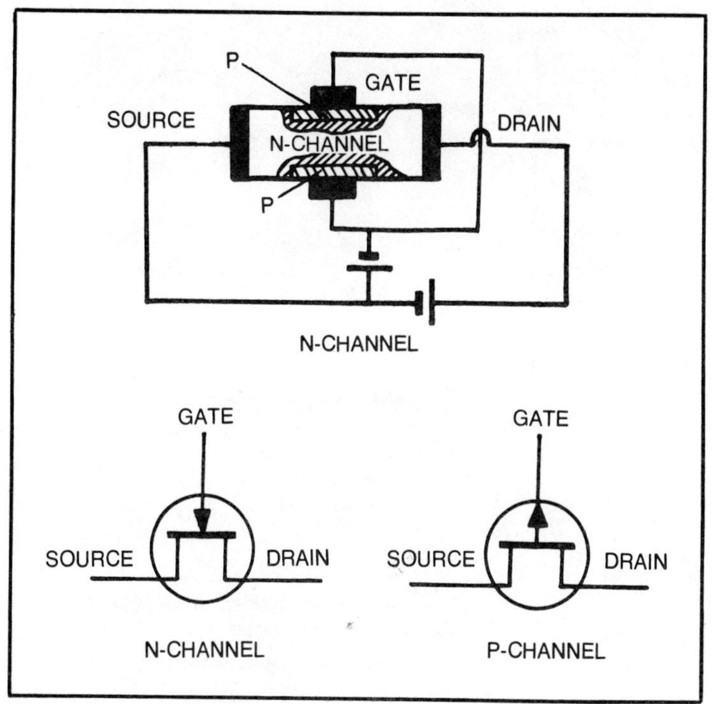

Fig. 1-8. Field-effect transistor.

depletion type N and P channels and enhancement type N and P channels. Also, there is a fourth lead. This is the substrate or body of the MOSFET and is usually connected to the source or ground. The substrate is not always shown in an illustration of a MOSFET and for all intensive purposes, can be disregarded in considering the FET operation. We can identify depletion MOSFET by the solid channel and enhancement FET by the broken channel.

The P-channel enhancement type transistor is mostly used. The MOSFET exhibits an extremely high input resistance which may be in the range of 10^{12} to 10^{15} ohm. Unlike the junction FET, the MOS maintains a high input resistance without regard to the magnitude or polarity of the input gate voltage.

The gate insulating material in a MOSFET is typically below 2×10^{-5}cm thick. Consequently, gate to substrate voltage of the order of 50 volts or so will cause breakdown of the insulation and the MOSFET is ruined. Any static charge on a person's finger or some tool he is using can destroy a MOSFET. To prevent destruction, most MOSFET are supplied by the manufacturers with their leads twisted so that the gate is shorted to the substrate.

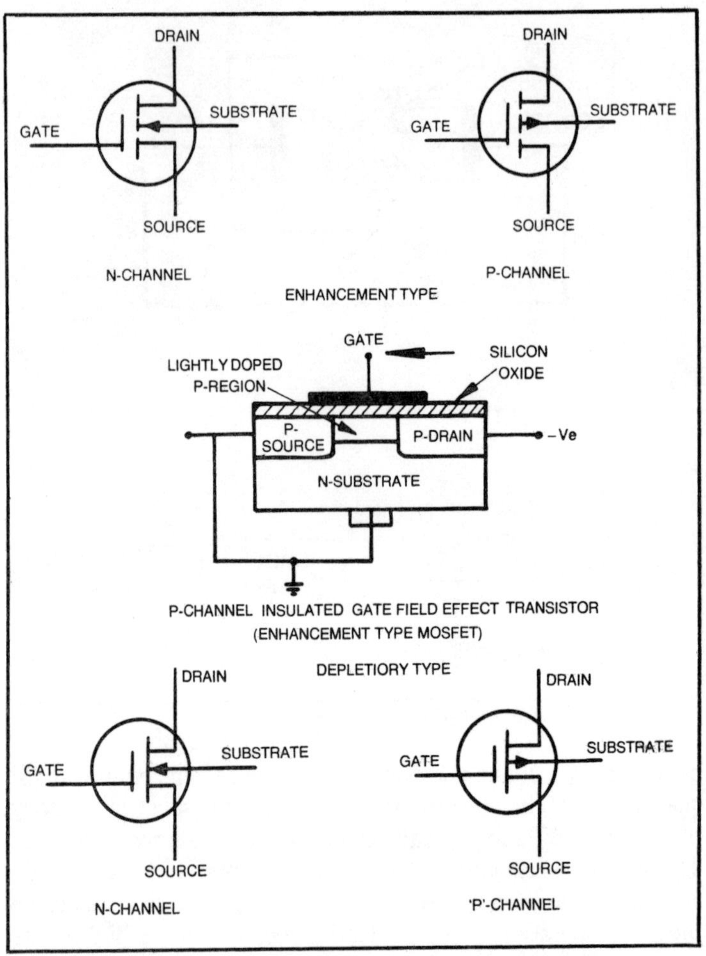

Fig. 1-9. MOS-FET (metal-oxide insulated gate field-effect transistor).

Shorting leads should be clipped to the gate when soldering or installing MOSFET.

INTEGRATED CIRCUITS

During the past decade, the electronics industry saw a new development with the invention of integrated circuits technology. Integrated circuits eliminate the use of individual electronic parts, such as resistors, capacitors and transistors, etc., as the building blocks of electronic circuits. In their place, we have tiny chips of semiconductor material whose functions are not those of single components but of scores of transistors, resistors and capacitors and other electronic elements, which are all interconnected to perform the function of a complete and complex circuit. So, instead of the usual semiconductor devices, resistances and capacitances, complete multistage amplifiers, complex flip-flops and dozens of other functional circuits have become the basic components of electronic equipment.

Circuits requiring close matching and tracking of components can be more advantageously adapted for integrated circuits. Also, it is less expensive to fabricate integrated transistors than passive elements such as resistors and capacitors. Large values of resistances and capacitors are to be avoided as they are not economical. Also, the circuit design has to be such that it should be non-critical as to component tolerances. For example, 20% tolerances of resistances and capacitances shall have much lower cost than 10% tolerance values. Another severe limitation of integrated technology for linear applications is the non-availability of inductances as integrative circuit elements. So, either the inductances are to be eliminated while designing a particular integrated circuit or a hybrid circuit would have to be used. In digital electronics, standard circuits are available for virtually every application.

OPERATIONAL AMPLIFIERS

An *operational amplifier* is a high gain amplifier originally intended to be used for doing mathematical operations. But the versatility of this amplifier owing to the provision for external selectable feedback has made it possible for adaptation to many applications in the analytical instruments field. It is used for the construction of ac and dc amplifiers, active filters, phase inverters, multivibrators, comparators, etc. The operational amplifiers are available in the integrated form and thus simplify the design of equipment by offering a high quality amplifier in one package and result in considerable size reduction. They are popularly known as *op amps.*

The block diagram of a typical operational amplifier is shown in Fig. 1-10. The input stage is a conventional differential amplifier

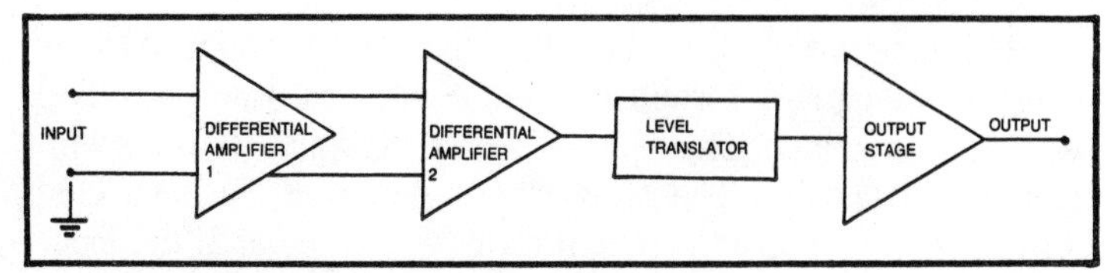

Fig. 1-10. Block diagram of an operational amplifier.

with a constant current source placed in the emitters of the two transistors. It is desirable to have a high gain in this section so that any imbalances or imperfections in the succeeding stages have little or no effect on the output signal.

The signal output from the first stage is fed differentially to the second stage differential amplifier. Because common mode rejection capability is not as stringent in this stage, this stage does not require a constant current source in the emitters. Normally, the second stage is needed only to provide some additional gain. Its input resistance should be high enough to prevent excessive loading of the first stage. Therefore, an emitter follower circuit is often employed.

The output of the second stage is normally taken to be single ended. Here, the amplified signal is associated with a certain amount of dc voltage at its output. Some means of level translation is thus necessary between the second and the final stages. By eliminating the dc level at the final stage, the output voltage will vary about a zero reference level, thus preventing any undesired dc current in the load and also enhancing the permissible output voltage swing.

An ideal operational amplifier has the following characteristics:

Voltage gain	$A = \infty$
Bandwidth	$BW = \infty$
Input impedance	$= \infty$
Output impedance	$= 0$

Typically, an operational amplifier has two input terminals and one output terminal. It is basically a differential amplifier and is normally used with external feedback networks that determine the function performed. Several examples of the use of operational amplifiers appear at different places in the text. Some of the basic application circuits are given below:

Basic Inverting Circuit

In the circuit of Fig. 1-11A, the input signal is applied to the inverting (negative) terminal and the non-inverting (positive) terminal is grounded. The input voltage E_1 is applied in series with input resistance R_1. The feedback resistance R_2 is connected between the input and the output terminals. At point 1, the input current is equal to the feedback current. Therefore, we can write.

$$\frac{E_1 - E_i}{R_1} + \frac{E_0 - E_i}{R_2} = 0$$

If the voltage gain of the amplifier is A, then $E_0/E_i = A$

$$\therefore E_i = E_0/A$$

$$\frac{E_1}{R_1} - \frac{E_0}{A} \cdot \frac{1}{R_1} + \frac{E_0}{R_2} - \frac{E_0}{A} \cdot \frac{1}{R_2} = 0$$

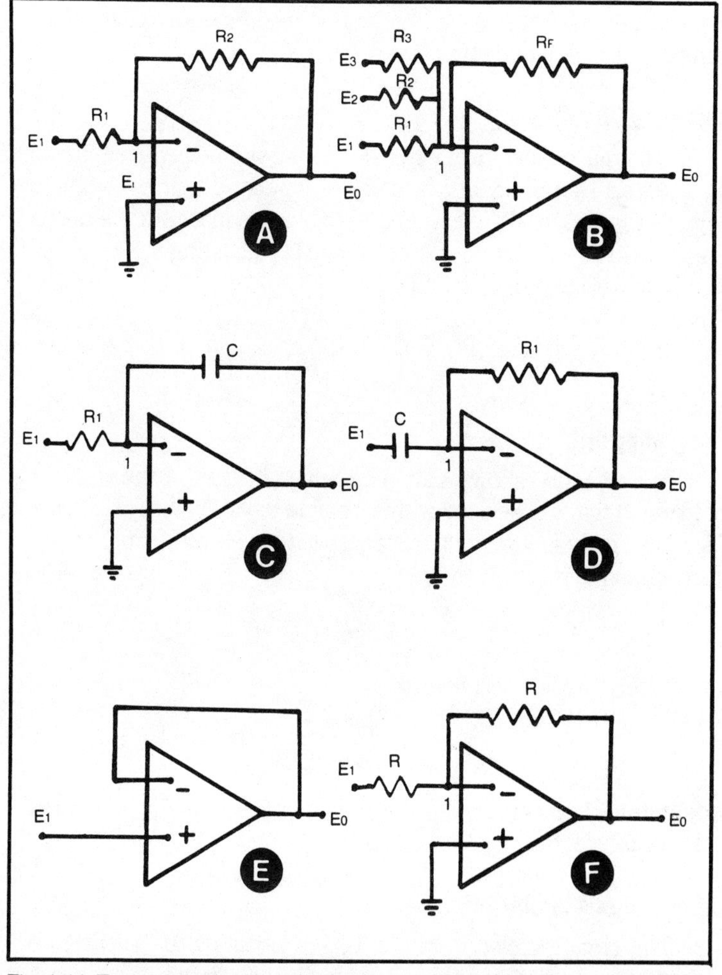

Fig. 1-11. Typical application circuits using operational amplifiers. (A) Inverting circuit. (B) Summing circuit. (C) Integrating circuit. (D) Differentiating circuit. (E) Voltage follower. (F) Unity gain voltage inverter.

$$E_0 = -\frac{E_1 (R_2/R_1)}{1 - \left[\frac{1}{A}\right]\left[1 + \frac{R_2}{R_1}\right]}$$

Since $A >> (1 + R_2/R_1)$

$$\therefore E_0 = -\frac{R_2}{R_1} \cdot E_1$$

Hence $E_0/E_1 = -R_2/R_1$

This shows that the gain of the circuit is independent of the voltage gain of the amplifier. Also, the gain is unaffected by changes in temperature, device parameters or frequency.

Summing Circuit

A useful and practical extension to the simple inverter circuit is obtained by providing a number of inputs so that the voltage signals can be added to each other. The various input signals can be summed in different proportions by suitably adjusting the values of the input resistors (Fig. 1-11B).

$$E_0 = -R_F \left(\frac{E_1}{R_1} + \frac{E_2}{R_2} + \frac{E_3}{R_3}\right)$$

Integrating Circuit

Figure 1-11C shows a simple running integrator (no reset or hold logic) that can be used within a stable feedback loop. Here the feedback path is provided by a capacitor. Hence, sum of the currents at point 1 is given by

$$\frac{E_1}{R_1} + C\frac{dE_0}{d_t} = 0$$

Integrating with respect to time,

$$E_0 = -\frac{1}{R_1 C} \int E_1 d_t$$

Integrating circuits are required for modeling dynamic systems and solving differential equations.

Differentiating Circuit

The circuit shown in Fig. 1-11D is arranged to generate an output voltage which is proportional to the differential with respect to time of the input voltage. The circuit can be analyzed by considering sum of the currents at point 1.

$$C.\ \frac{dE_1}{d_t} + \frac{E_0}{R} = 0$$

$$E_0 = -\ RC \times \frac{dE_1}{d_t}$$

This shows that the output voltage is proportional to the derivative of the input. Differentiating circuits are susceptible to noise and instability and need to be used with care. They are usually followed by a filter to limit the effective bandwidth.

Voltage Follower

If the output of the operational amplifier is connected back to the input at the inverting end while the input is given at the non-inverting terminal, the circuit functions as a *voltage follower* (Fig. 1-11E). In this case, the voltage gain of the amplifier is almost unity and the output voltage follows changes in the input voltage. The circuit provides a very high input impedance and low output impedance. Therefore, this configuration is ideal for isolating and driving other circuits. It is a more refined version of the emitter follower.

Unity Gain Inverter

A *unity gain voltage inverter* is formed by using identical resistances in the input and feedback paths (Fig. 1-11F). Inverters are used wherever sign changes are necessary. They are also used simply to lower the impedance level or raise the power level of a signal. The circuit is capable of functioning well over a wide range of signal levels, frequencies and impedances.

DIGITAL TECHNIQUES

In analytical instrumentation, most circuitry is concerned with the amplification and processing of signals which are available in an analog form. However, the introduction of compact digital computers have made possible the digital manipulation of the analog signals after they have been converted into a digital form. To do this, an analog to digital converter is employed which basically samples the analog signal at a predetermined rate to get the digital equivalents. Using combinations of pulse circuits such as logic gates and flip-flops, it is possible to carry out arithmetical manipulations upon the series of digital values. Logic circuits are used to test whether a predetermined set of conditions has been obeyed and can route the path of signals accordingly. Flip-flops are used to store information in the form of noughts and ones corresponding to a transistor being cut off or switched on respectively. They can be cascaded to get an electronic counter. The simplest counter is a *ripple* counter in which the output of one flip-flop drives the input of the next flip-flop, the flip-flops being all connected in series. It is basically a binary counter as each time a clock pulse comes in, the count advances to the next binary number. If "n" flip-flops are cascaded together, we get 2^n states. For example, for four flip-flops, there are 2^4 or 16 states.

Generally, we prefer to have a *decimal* or *decade counter* instead of a binary counter. A decade counter uses a base of 10, i.e., it has 10 distinct states. It is made by using feedback around a four flip-flop binary counter. The idea is to make the binary counter skip through six out of its 16 states. It is also possible to precondition the logic inputs to each flip-flop in order to omit certain states. Logic circuits can decode the binary coded decimal (BCD) number stored in a decade-counter unit so that we can directly read out the decimal count. Measuring frequency and voltage are two important applications of electronic counters.

The availability of integrated circuit logic modules at low costs has led to a universal use of digital techniques. These modules are used in large quantities in digital computers because of their small size, low power consumption and high reliability. The small size of MOS elements has resulted in *large scale integration* (LSI) in which thousands of elements are created on a single chip. The digital ICs are available in a big range as simple logic gates, memory units and even as complete data processing units like microprocessors.

Digital Displays

Digital displays present the values of the measured quantities in numerical form. Instruments with such a facility are directly readable and slight changes in the parameter being measured are easily discernible in such displays as compared to their analog counterparts. Digital measuring instruments were originally developed for an accurate measurement of frequency and time. Because of their higher resolution, accuracy and ruggedness, they are preferred for display over conventional moving coil indicating meters.

The outputs from most of the transducers are analog signals. For digital operations which may include computations and display, these analog signals are converted to a digital representation.

Several types of analog to digital converters have been developed. One of the most popular and widely used circuits for this purpose is the *dual slope integration* method. Referring to Fig. 1-12, at time t, (at the beginning of the measurement cycle), the capacitor C is fully discharged. The unknown input voltage is then applied to the integrator so that the capacitor C begins to charge at a rate proportional to V_{in}. The charging continues until a predetermined number of clock pulses have been counted. At this instant t_2, the control logic switches the integrator input to V_{ref}, which is a known voltage with a polarity opposite to that of V_{in}. Capacitor C then discharges at a rate determined by V_{ref}. As this voltage is greater than the voltage to be measured, the charge on the capacitor decreases more rapidly than it is built up. The counter (Fig. 1-13) gets reset at time t_2 and again it counts clock pulses. It continues to do so until the comparator indicates that the integrator output has returned to zero. This is at time t_3 and at this instant, the control logic switches the input of the integrator to zero volts and also commands the counter to store the count. The count stored in the counter is then directly proportional to the input voltage because the time taken for capacitor C_1 to discharge is proportional to the charge acquired which, in turn, is proportional to the input voltage. The number stored in the counter is then displayed to give a direct reading of the parameter being monitored.

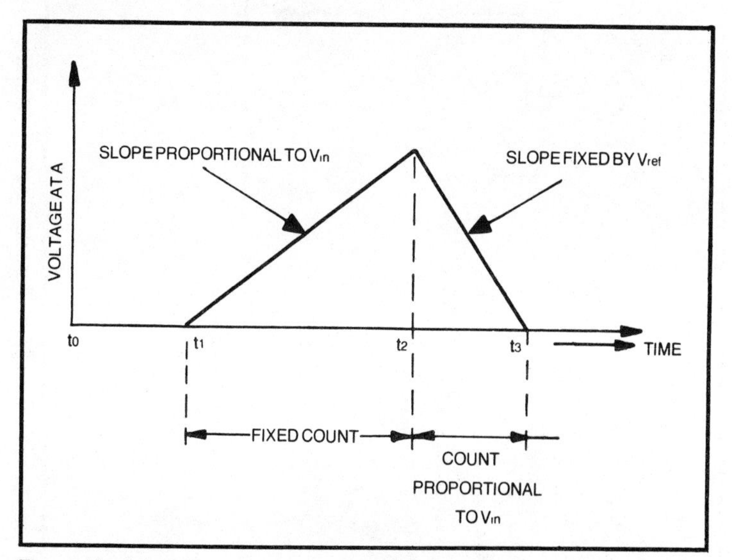

Fig. 1-12. Principle of dual-slope integration method of analog-to digital conversion.

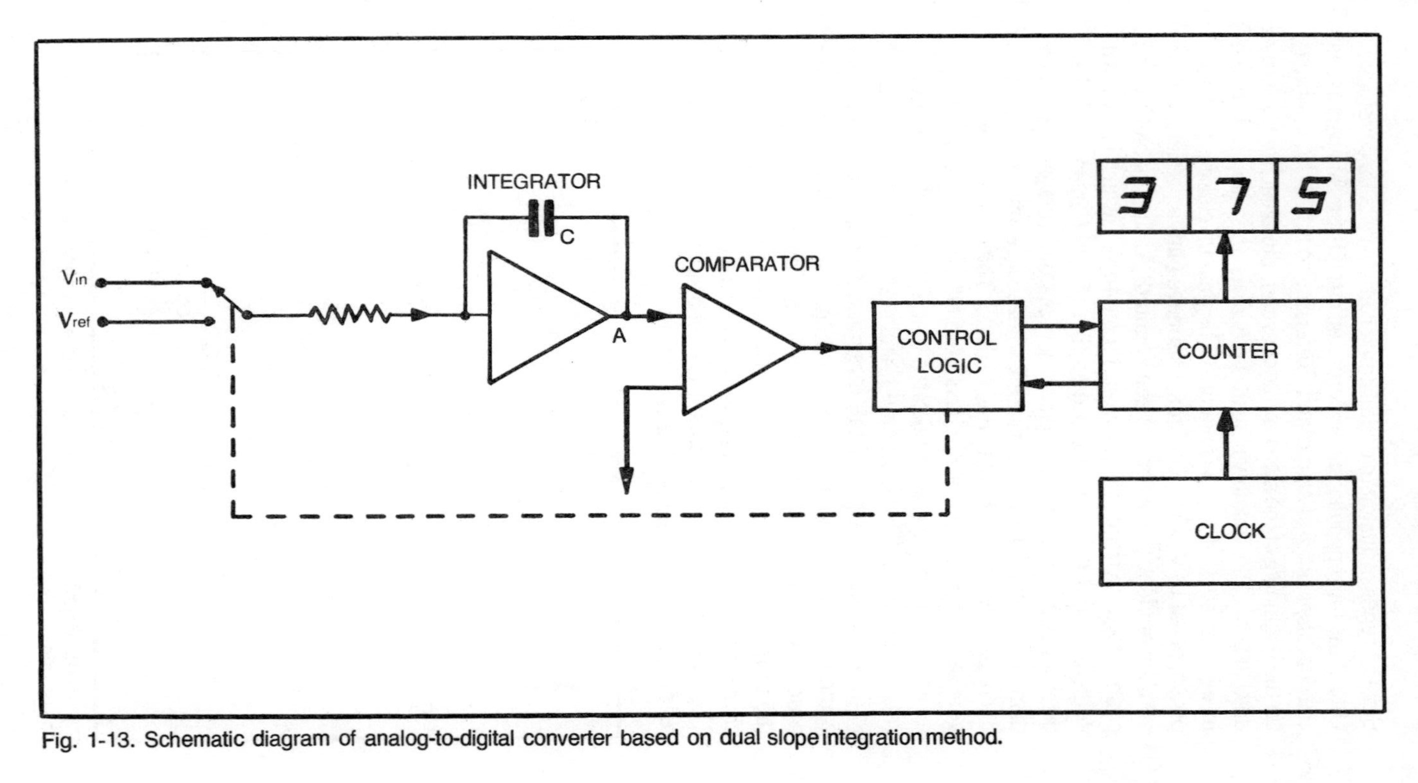

Fig. 1-13. Schematic diagram of analog-to-digital converter based on dual slope integration method.

Fig. 1-14. Seven segment display.

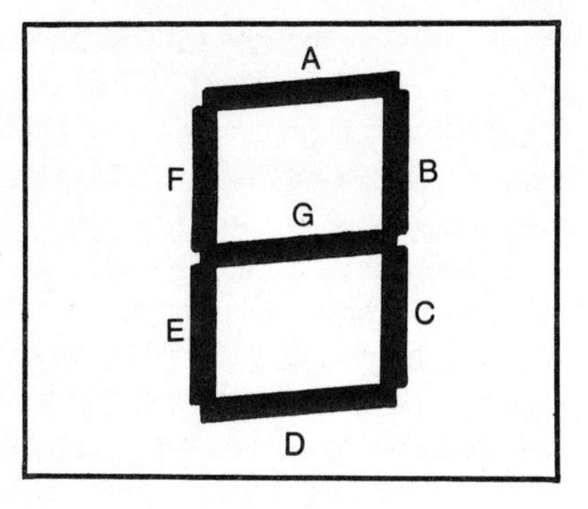

Display Devices

Different types of devices are available for display in numerical form. The oldest in the series is the gas-discharge glow tube, popularly known as the *nixie.* This tube has 10 cathodes and one anode. Each of the cathodes is shaped like the characters 1, 2, 3 . . . 9, 0 and stacked one behind the other in a glass envelope. The nixie tube works on the principle that discharge takes place between the cathode with the lowest potential and the anode, thus causing only one of the characters to glow at a time. The nixie tube requires a higher supply voltage, usually of the order of 120 volts or above for operation.

The seven segment display is an arrangement which has become quite popular in digital displays. The numerical indicator consists of seven bars (Fig. 1-14) positioned in such a way that the required figure can be displayed by selection of the appropriate bars. For example, when the decimal digit 4 has to be shown, bars f, g, b and c are illuminated. A special decoder is required for driving such an indicator from a BCD code. The seven segments can be illuminated by incandescent lamps, neons and directly heated filaments.

The principle of seven segment display has also been used in vacuum fluorescent tubes. Basically, this tube consists of a vacuum tube with seven anodes and a filament. The anodes which are each coated with a phosphor are arranged in the standard planar seven segment format. The filament, which is very fine, lies in front of the anodes and is heated to just below incandescence so that it remains invisible. A positive voltage between 12 volts and 25 volts is applied to the anodes to make them glow. Vacuum fluorescent displays are available in a soft bluish green color.

Light emitting diodes are used in small sized displays. These semiconductor diodes are made of gallium arsenide phosphide and are directly compatible with 5 volt supplies typically encountered in digital circuitry. The LEDs are very rugged and can withstand large variations in temperature. The LEDs are arranged in seven segment format for display and are available in deep red, green and yellow colors.

Liquid crystal displays consist of two parallel glass plates, with conductive coatings on their inner surfaces and a drop of liquid crystal material sandwiched between them. The front plate of the display is coated with a conductive transparent material, usually indium oxide, in a seven segment pattern. The back plate may have either a transparent or reflective coating depending upon the mode of display. There are two types of liquid crystal displays. In the dynamic scattering type, a normally clear material becomes turbid when a current flows through it. In the field effect type, the material either does or does not rotate the plane of polarization or the polarized light passing through it, depending upon whether or not a voltage is applied across it. This type requires very low current for its operation.

Digital Panel Meter

A/D converters are commercially available in the form of a single monolithic chip. They even include features like auto-zero and auto-polarity, so that they can be readily used for constructing digital panel meters. The three digit A/D converter LD 130 from Siliconix is a typical example. This converter employs a *quantized feedback* conversion technique which is characterized by a single phase digitization interval in which a digital control system feeds back quantized units of charge in response to the sampled state of an analog comparator. These quanta of charge balance the charge being supplied to the integrator by the analog voltage. The magnitude of the quantized charge being fed back and its sign arise from the fact that the control logic has two up/down duty cycles available during the *measure* interval. The *auto-zero* interval provides a means to null out the offset voltages of the amplifiers used in the LD 130. In this interval, the input buffer amplifier is switched to ground, the count-correcting operation is established and the integrator output is brought back to analog ground, the comparator threshold. The two fundamental intervals of the sampling period, the auto-zero and measure intervals, are provided by internal clock. The up/down BCD counter increments by each clock pulse when the u/d logic is up and decrements by each clock pulse when the logic is down. The contents of the BCD counters are loaded into the internal latches and externally decoded. Connecting a reference voltage and light emitting diodes externally, a three digit panel meter can be constructed. The detailed circuit of the A/D converter used as a digital panel meter is shown in Fig. 1-15.

LABORATORY RECORDERS

Every modern laboratory possess recorders of different varieties for recording various types of analog signals available from a variety of sources and even from analytical instruments. Recorders are available in a large range of sizes, speeds, sensitivities and prices. No doubt the various curves which are now traced automatically with a recorder can be plotted manually from point by point measurements. The procedure is not only time consuming but may cause valid bits of information to be overlooked entirely, simply because the points were taken too far apart. Also, artifacts that might not be observable at all in point by point observations will often be readily identifiable on a recording.

The most elementary electronic recording system consists of three important components—namely the *signal conditioner,* the *writing part* and the *chart drive mechanism.* The signal conditioner converts the input signal to a suitable level for operation of the writing part. The writing part provides a visible graphic representation of the quantity of interest. The signal conditioners usually consist of a preamplifier and the main amplifier. Both these amplifiers have to satisfy specific operating requirements such as input impedance, gain and frequency response characteristics. The writing part may be the direct writing galvanometric type of null balancing potentiometric type.

Direct Coupled Amplifiers

In most of the requirements in the analytical instrumentation field, *direct coupled amplifiers* are required. These amplifiers give an excellent frequency response but they tend to drift. The drift is a slow change of output having no relation with the input signal applied to the amplifier. There are several reasons for the presence of drift in the dc amplifiers, but the most common cause is the slow changes of component values with temperature. Since the frequency response of the RC coupled amplifier does not extend all the way down to dc, it does not drift. Arrangements are worked out so that the advantages of both types of coupling can be obtained in one amplifier. Typically, all stages except one are direct coupled. The one RC coupled stage prevents the drifting of the output.

Chopper Amplifiers

The *chopper amplifier* is a useful device in the field of analytical instrumentation as it gives a good solution to the problem of achieving adequate low frequency response while avoiding the drift problem inherent in direct coupled amplifiers. This type of amplifier makes use of a chopping device which converts a slowly varying direct current to an alternating form with an amplitude proportional to the input direct current and with phase dependent on the polarity of the original signal. The alternating voltage is then amplified by a conventional ac amplifier whose output is rectified back to get amplified direct current. A chopper amplifier is an excellent device for signals of narrow-band width and reduces the drift problem to zero.

There are two types of choppers, *mechanical* and *transistor*. A mechanical chopper is simply an electronic switch driven by an alternating current. It is so designed that the flux saturates a magnetic circuit such that a switching operation occurs only near the cycle zero points. For mechanical choppers, 50 Hz mains frequency is usually used as the chopping frequency. Choppers which operate at higher frequencies, say about 400 Hz, are also available.

The use of a transistor as a chopper increases the possible rate of switching and, therefore, can be useful for signals of wider bandwidths. Several typical transistor chopper circuit configurations are available in the literature. The action of the transistor chopper is based on its low saturation resistance for on mode and high resistance for off mode. The transistor can be used as a signal chopper at rates conveniently up to 100 kHz. Greater stability against drift can be achieved in the chopper stabilized dc amplifier shown in Fig. 1-16.

Since drift is a low frequency phenomenon, the signal can be applied simultaneously to a chopper modulated ac amplifier of the narrow bandwidth, and through an ac coupling to a differential output dc amplifier of wide bandwidth. The low frequency components are given high gain over the restricted bandwidth, and the drift output is fed back for drift stabilization. For higher frequency components, the gain is provided by the dc amplifier. In this arrangement, medium and higher frequencies pass through the dc amplifier with the gain, say, A_1. The drift and very low frequencies signals pass through both amplifiers in series, with the gain $A_1 A_2$. The drift factor is reduced by A_1/A_2 and it is possible to obtain drift figures of only a few microvolts.

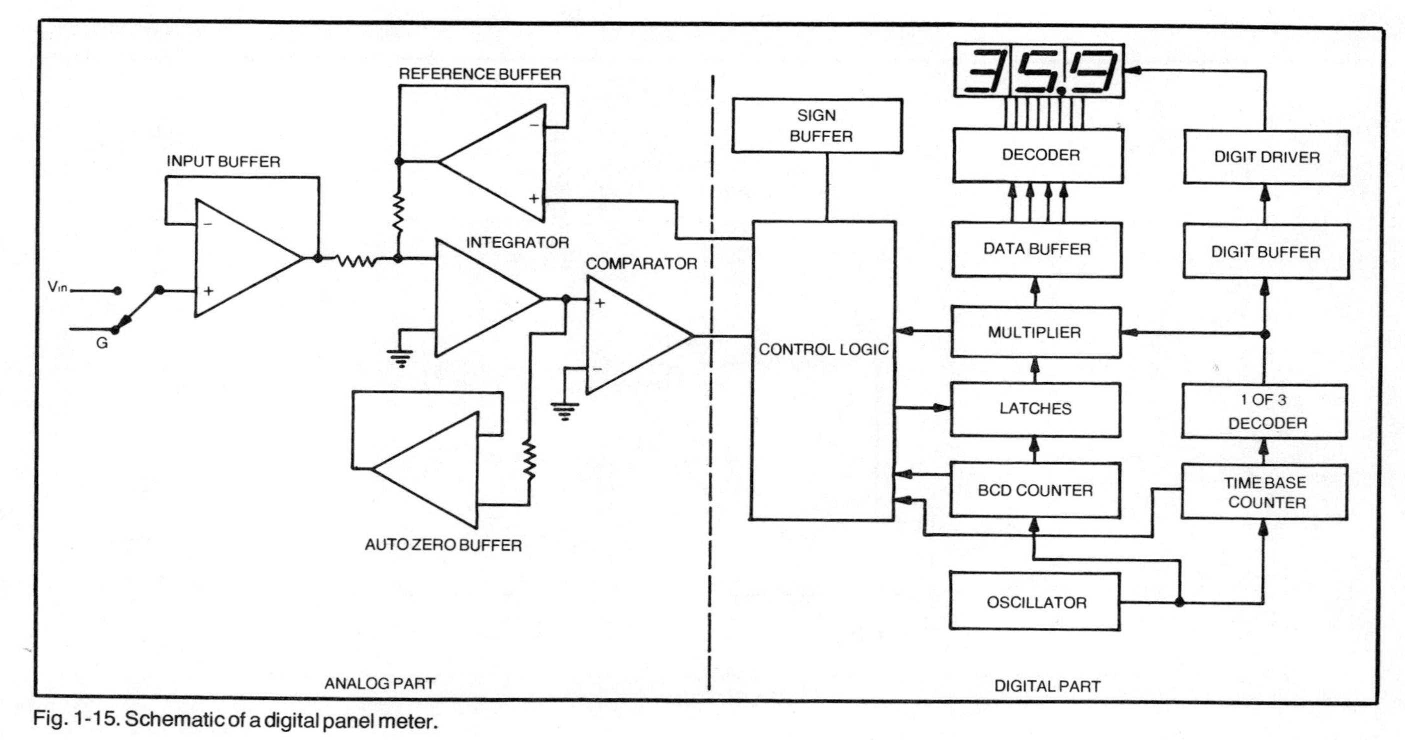

Fig. 1-15. Schematic of a digital panel meter.

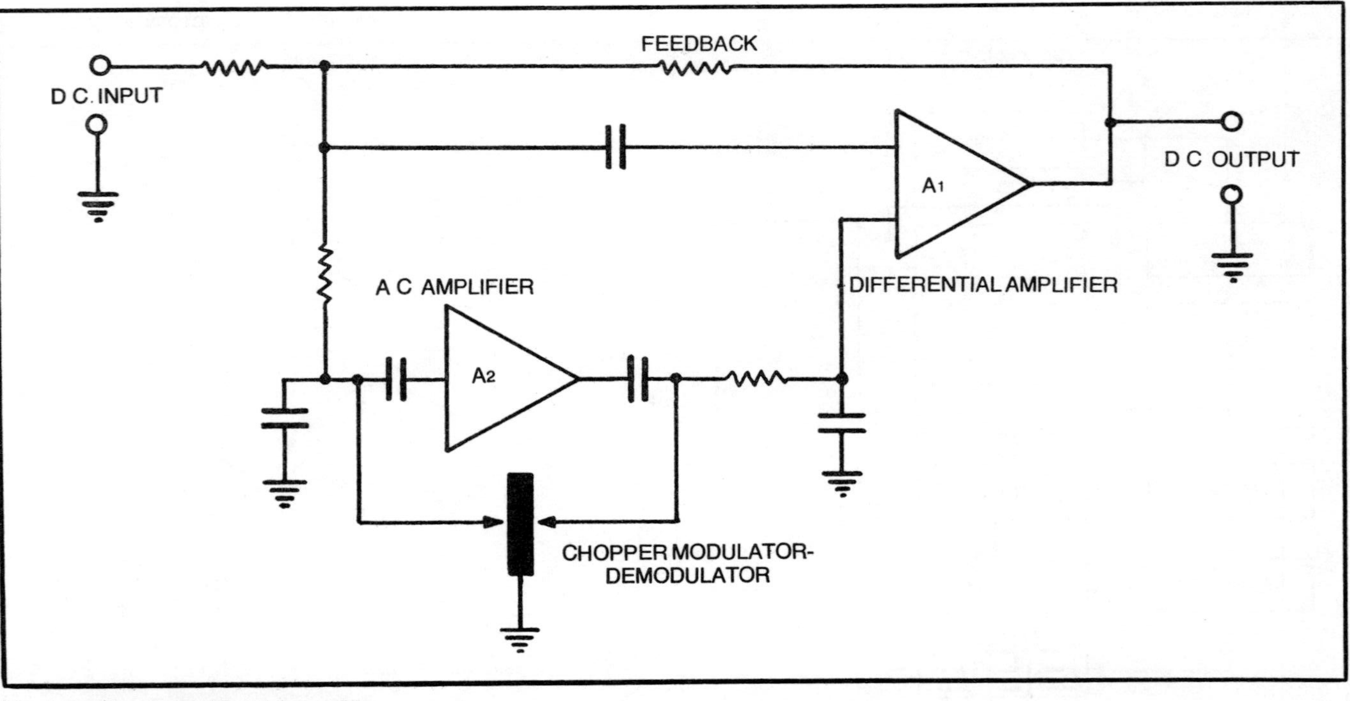

Fig. 1-16. Chopper stabilized dc amplifier.

Chopper amplifiers find application in analytical instrumentation in the amplification of small dc signals of a few microvolts. Such order of amplitudes are obtainable in atomic absorption spectrophotometers, pH meters, etc. The frequency response of a chopper amplifier depends upon the value of the chopping frequency. The input impedance can be made high by using a sub-miniature electrometer tube or insulated gate field effect transistor at the input stage.

Direct Writing Recorders

In the most commonly used direct writing recorders, a galvanometer activates the writing arm called the *pen* or the *stylus*. The mechanism is a modified form of the *D'Arsonval meter movement.* This arrangement owes its popularity to its versatility combined with reasonable ruggedness, accuracy and simplicity.

A coil of thin wire, wound on a rectangular aluminum frame is mounted in the air space between the poles of a permanent magnet (Fig. 1-17). Hardened-steel pivots attached to the coil frame fit into jeweled bearings so that the coil rotates with a minimum of friction. Most often, the pivot and jewel is being replaced by a tautband system. A lightweight pen is attached to the coil. Springs attached

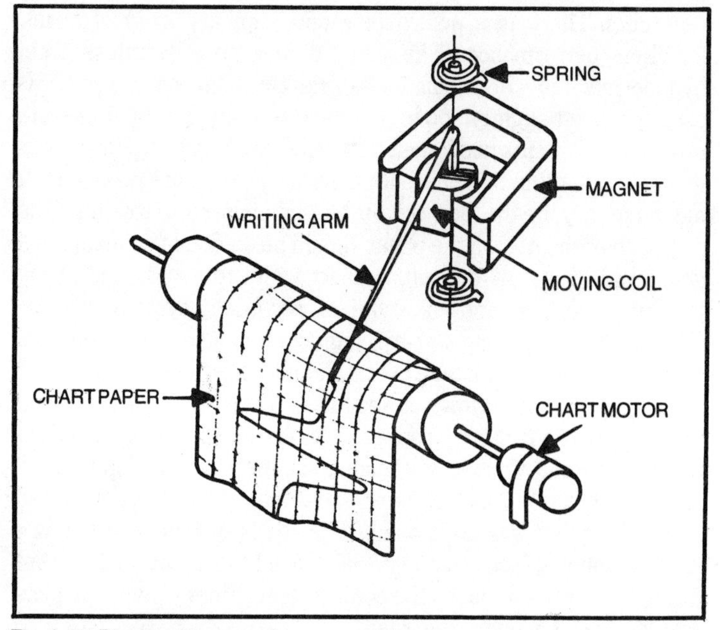

Fig. 1-17. Direct writing recorder.

to the frame return the pen and coil always to a fixed reference point.

When current flows through the coil, a magnetic field is developed which interacts with the magnetic field of the permanent magnet. It causes the coil to change its angular position, as in an electric motor. The direction of rotation depends upon the direction of flow of current in the coil. The magnitude of pen deflection is proportional to the current flowing through the coil. As mentioned earlier, the stylus can have an ink tip, or it can have a tip that is the contact for an electrosensitive, pressure sensitive or heat sensitive paper. If a writing arm of fixed length is used, the ordinate will be curved. In order to convert the curvilinear motion of the writing tip into rectilinear motion, various writing mechanisms have been devised to change the effective length of the writing arm as it moves across the recording chart.

Tautband instruments are preferred over pivot and jewel type instruments because they have the advantages of increased electrical sensitivity, elimination of friction, better repeatability and increased service life.

Writing Methods

Out of the several writing methods available, the *ink recording* method is widely used in slow speed as well as high speed recorders. The writing pens depend upon capillary action of the ink for their performance. The pen tip may be a stainless steel hypodermic tube or a small glass nozzle. The ink reservoir is usually placed slightly above the plane of writing to facilitate the flow of ink. The ink used must flow out from the pen tip in an even manner so that the trace is continuous and no gaps are produced. It must not dry in the pen feed-tube and thus stop writing. The pressure of the pen on the paper should be sufficient to maintain a good contact across the entire chart width to avoid gaps in the tracing. The pen and ink type of recording system has the advantage of producing a sharp, high contrast and easily readable trace. But the disadvantage is that the ink reservoir must be filled frequently and the hollow pen and tube ink must be cleared of dried ink that blocks the narrow passageway. Also, except in the hands of the most careful worker, ink tends to be messy, easily spilled and often spattered. To overcome this problem, the inkless or heated stylus recorder was developed. The stylus of this recorder is a narrow ribbon of resistive material shaped to a point, and current through the ribbon causes the point to heat. Fiber tip writing pens are also used in modern recorders.

Most of the portable recorders use the heated stylus writing system, where recording paper moves over a steel knife edge, kept at a right angle to the paper motion, and the stylus moves along the knife edge. The hot stylus burns off the white cellulose covering of the heat sensitive paper, exposing the black under-surface of the paper, thus forming the trace. The current for heating the stylus is usually obtained from a separate winding of the main transformer. A *rheostat* is included in series with the stylus so that the current and consequently the temperature of the stylus can be varied. By means of these adjustments, a sharply defined trace can be produced.

Paper Drive

The usual paper drive is by a synchronous motor and a gear box. The speed of the paper through the recorder is determined by the gear ratio. If it is desired to change the speed of the paper, one or more gears must be changed.

Paper speed is an important consideration for several reasons. If the paper moves too slowly, the recorder signal variations will be "bunched up" and will be difficult to read and interpret. If the paper moves too fast, the recorded waveform will be so spread out that large lengths of paper will be required to record the variations of signal. This will not only be uneconomical in terms of paper consumption, but it also makes the task of reading the waveforms more difficult.

A constant speed is the basic requirement of the paper drive because the recorded events are time-correlated. The frequency components of the recorded waveform can be determined if it is known how far the paper moved past the pen position as the record was being taken.

Most of the recorders contain an additional timing mechanism that prints series of small dots on the edge of the paper as it moves through the recorder. This time marker produces one mark each second or at some other convenient time interval. The speed of the paper in centimeters per second in such cases need not be known as the paper chart is already marked off in units of time. This feature is particularly useful in recorders having more than one speed. Even when the speed is changed, the marker will put the marks at the same time interval.

Potentiometric Recorder

For recording of low frequency phenomenon, strip chart recorders based on the potentiometric null-balance principle are generally used. The operating principle of a simple potentiometer is shown in Fig. 1-18.

A slide wire AB is supplied with a constant current from a stabilized power supply. The slide wire is constructed from a length of resistance wire of high stability and uniform cross section such that the resistance per unit length is constant. The unknown dc voltage is fed between the moving contact C and one end A of the slide wire. The moving contact is adjusted so that the current flowing through the galvanometer placed across AC is zero. At that moment, the unknown input voltage is proportional to the length of the wire AC. In practice, the slide wire is calibrated in terms of span voltage, the typical spans being 100 mV, 10 mV or 1 mV.

For converting the simple potentiometer into a recording unit, the moving contact of the slide wire is made to carry a pen, which writes on a calibrated chart, moving underneath it. For obtaining the null-balance a self-balancing type potentiometer is generally used. The balancing of the input unknown voltage against the reference voltage is achieved by using a servo-system. The potential difference between the sliding contact C and the input dc voltage is given to a chopper type dc amplifier in place of a galvanometer. The chopper is driven at the mains frequency and converts this voltage difference into a square wave signal. This is amplified by the servo-amplifier and then applied to the control winding of a servo motor. The servo motor is a two phase motor, whose second winding is supplied with 50 Hz mains supply that works as a reference phase winding. The motor is mechanically coupled to the sliding contact. The motor turns to move the pen and, simultaneously, varies the voltage of the sliding contact such that the potential difference between the input voltage and a reference voltage is zero. The circuit operates in such a manner that the motor moves in one direction if the voltage across AC is greater and in the opposite direction if it is less than the input voltage. The servo motor is shaft coupled to a tacho-generator which provides the necessary damping to the servo motor. It slows down as it approaches balance position and thus minimizes the overshoot.

The servo motor generally used to drive the pen in the self-balancing potentiometric recorders is the ac two phase induction motor. The motor has two separate stator windings which are physically perpendicular to each other. The out-of-phase alternating currents in the two stator windings produce a rotating magnetic field. This rotating magnetic field induces a voltage in the

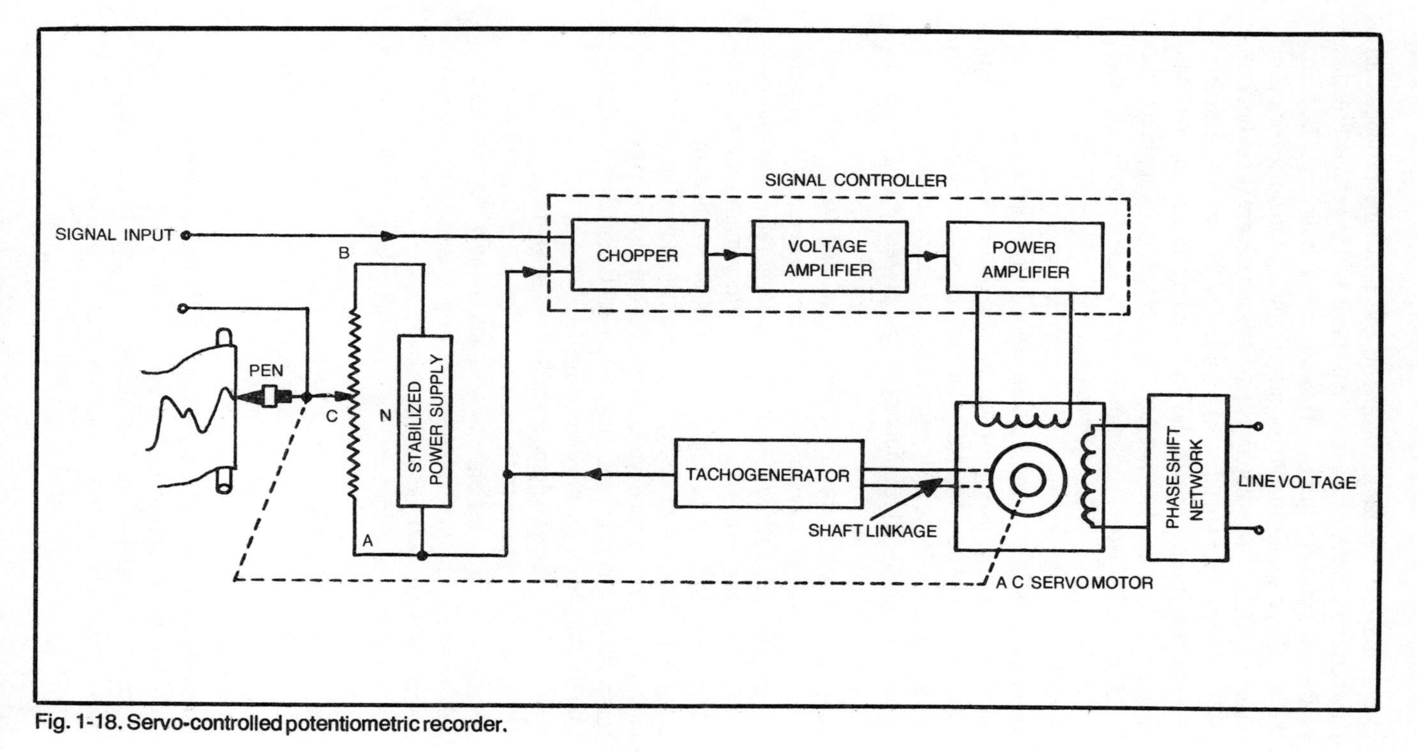

Fig. 1-18. Servo-controlled potentiometric recorder.

rotor, and the resulting current in the armature produces an interacting field which makes the rotor turn in the same direction as the rotating magnetic field. To produce a rotating magnetic field, the ac voltage applied to the one stator winding should be 90° out of phase with the voltage applied to the other winding. It can be done either in the power amplifier which supplies the control winding or in a phase-shift network for the line winding. For no input signal, obviously, the rotor does not turn. When a voltage is applied to the input, the rotor would turn slowly and its speed would increase with the magnitude of the input voltage.

The tacho-generator used to provide damping by electrical feedback is usually a two-phase device with a squirrel cage rotor. One phase is supplied from the ac line voltage to induce eddy currents in the rotor. When the rotor moves, it causes a shift in the space alignment of the eddy current field so that voltage is induced in the other phase. This induced voltage is directly proportional to the rotor velocity.

In older recorders, the reference voltage was usually provided by a Weston standard cell. It required frequent adjustment to validate the calibration. In some more recent recorders, mercury cells have been used. This type of battery maintains its rated voltage throughout its life. Calibration is needed much less frequently. In modern designs, all batteries are eliminated and constancy of reference voltage is assured through the use of temperature-compensated zener diodes. Calibration is seldom required after initial factory adjustment.

The slide wire could be a straight resistive material or it can be made of resistive plastic material. The latter is quite popular because it is highly resistant to wear and is easily cleaned. Also, it has infinite resolution as compared to a wound resistance wire where the contactor touches successive turns.

Most servo motors are conventional rotary motors and are connected to the slide wire and pen through mechanical linkages. To simplify the design, linear motors have been developed, in which the prime mover is in a straight line rather than circular. However, being expensive, it has not been widely accepted.

The chart is driven by a constant speed motor to provide a time axis. Therefore, the input signal is plotted against time. The recorders of this type are called T-Y recorders. If the chart is made to move according to another variable, then the pen would move under the control of the second variable in the x-direction. Such type of recorders are called X-Y recorders.

THE OSCILLOSCOPE

Among the indicating and display devices, the cathode ray oscilloscope occupies an important place in analytical instrumentation. Like a pen recorder, the oscilloscope presents a two dimensional graphical display. In recorders, a mechanical device does the writing while in the oscilloscope an electronic beam does the same job with speed and elegance unthinkable in other types of display devices.

A block diagram showing the various stages required for a complete oscilloscope and location of various important controls is shown in Fig. 1-19. It will be seen that in effect there are two separate sections in the oscilloscope. These are the horizontal and vertical sections. Each section has its own amplifiers, controls and deflection plates and so on. Basically, whatever is done on the vertical section does not affect the horizontal section and vice versa. Working together, they will display the incoming signal on the face of the cathode ray tube.

A cathode ray oscilloscope consists of the following main components: cathode ray tube, horizontal deflection circuit, vertical deflection system and power supplies.

Cathode Ray Tube/Picture Tube

The *cathode ray tube* is an electron device that produces a visual display of electrical phenomena. Its chief advantage is that it produces visual representation directly with extremely high speeds. This is made possible because of the very high velocity with which the electrons can move.

The fundamental components of a cathode ray tube are the *electron gun,* an arrangement for producing and focusing the electron beam; the *deflecting device,* a means for deflecting the electron beam either electrostatically or electromagnetically; and the fluoroscent screen, upon which the electron beam is focused to form a fine spot and which emits light when bombarded by electrons. The essential parts of a cathode ray tube are shown in Fig. 1-20.

The electron gun consists of a cathode for generation of electrons, a control electrode for varying the electron current density, an accelerating electrode for attracting the electrons and an arrangement for focusing the electron beam.

The cathode is a cylinder of nickel with *barium oxide* on the cap at one end. The cathode is indirectly heated by tungsten heater wire which is insulated from the cathode. Surrounding the cathode

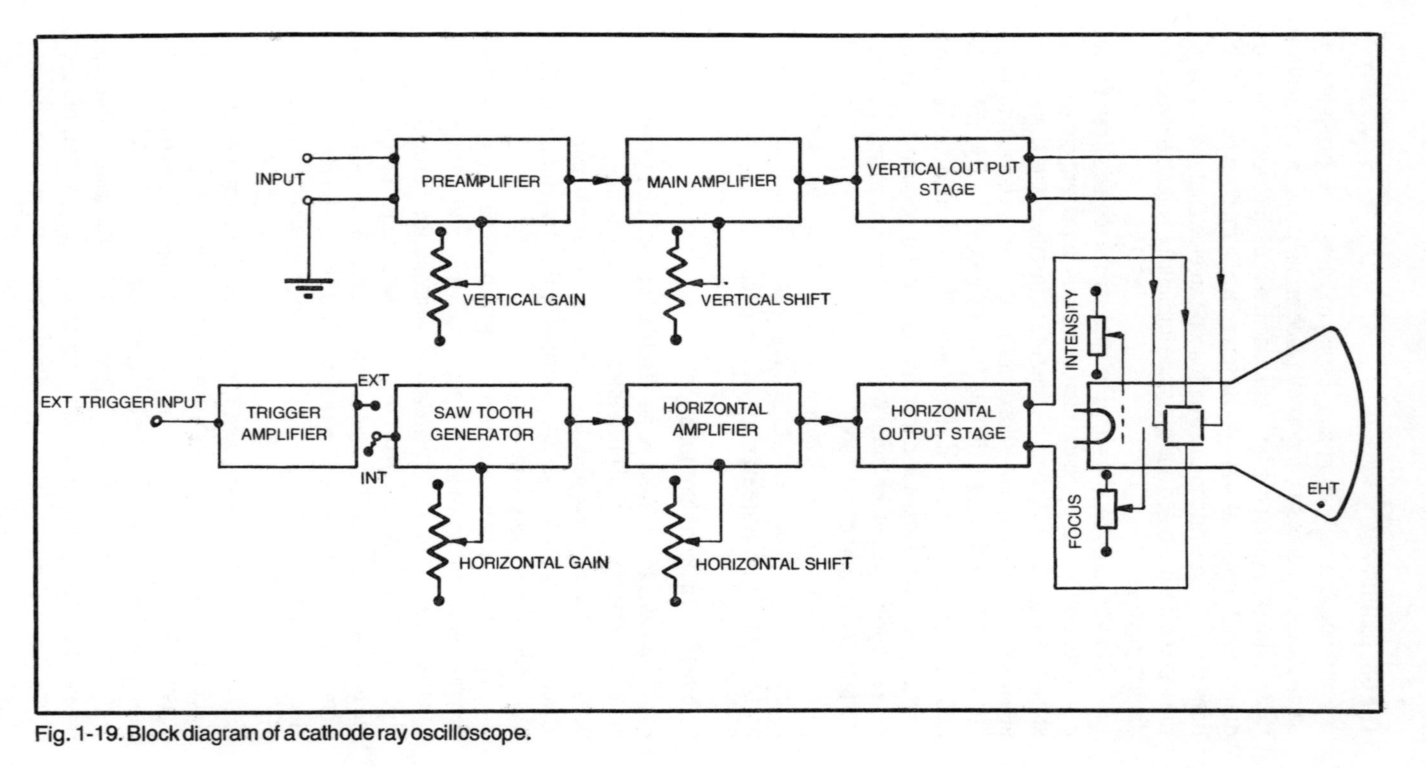

Fig. 1-19. Block diagram of a cathode ray oscilloscope.

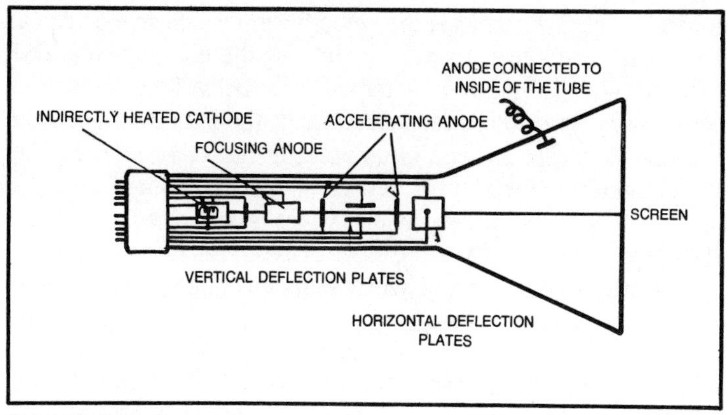

Fig. 1-20. Cathode ray tube.

at a distance of a fraction or a mm is a heat shield in the form of a nickel cylinder, which projects beyond the end of the cathode. This shield helps to concentrate the emitted electrons in the form of a beam. The control electrode over the heat shield functions as a control grid in the CRT. It is also cylindrical in shape and has a properly spaced aperture with respect to the cathode. This electrode is operated at some low positive or negative potential relative to the cathode, the polarity being dependent upon the design details.

Next to the control electrode is the first anode, often called the *accelerating electrode,* which is kept at a moderately high positive potential relative to the cathode. This electrode is provided with an aperture in the control grid side to permit the electron beam to enter this electrode and one or more additional apertures on the sides to remove strongly diverging electrons from the beam.

The electron beam from the accelerating electrode has a tendency to spread because of the mutual repulsion between the electrons. Hence, some focusing device is required to bring the beam to a sharp focus at the screen. In the electrostatic system of focusing, the final two electrodes of the electron gun are the first anode and the second or final electrode. They form an electron lens system for focusing the beam into a fine spot on the fluorescent screen. Both these anodes are kept positive with respect to the cathode. The focusing is achieved by varying the potential in the first anode while maintaining the accelerating electrode and the second electrode at a fixed potential. The optimum adjustment is that which produces the smallest spot on the screen. The final anode potential determines the velocity with which an electron in the beam travels to the screen on leaving the electron gun. The positive voltage on this electrode is usually kept anywhere between 1000 to 4000 volts with respect to the cathode in a general purpose oscilloscope. The glass walls of the tube between the second anode and the screen are usually coated with *aquadag,* a conducting layer of carbon particles which is connected to the final electrode. The intensity of the electron beam can be controlled by the potential of the control electrode. Its function is similar to the grid in an ordinary vacuum tube. As the electrode is made more negative, fewer electrons are drawn from the space charge adjacent to the cathode and result in giving a less intense spot on the screen.

The astigmatism control sets the mean voltage of each pair of deflection plates to make it as near as possible to that of the gun's final electrode. If this is not so, an aberration of the image is produced on the screen due to the formation on a low power cylindrical lens between the deflection plates and the final anode. The presence of astigmatism is established if the spot cannot be simultaneously focused in both the horizontal and vertical planes.

The beam emerging from the electron gun may be deflected either electrostatically or magnetically. In electrostatic deflection, two pairs of deflecting plates at right angles to each other are provided. In this arrangement, the position of the spot can be located by controlling the voltages on these plates. The deflection of the spot is directly proportional to the deflecting voltages on the plates. The deflection sensitivity is usually between 0.1 and 0.3 mm/volt. Electrostatic deflection is common in cathode ray oscilloscopes where screen widths are not very large. Magnetic deflection is preferred in case of tubes with big screen sizes because it would require large deflecting voltages in case the electrostatic method is employed.

In order to obtain a bright spot on the screen and at the same time achieve a reasonably good deflection sensitivity, it is desirable to accelerate the electrons constituting the electron beam. This arrangement of post deflection acceleration is obtained by the use of a ring of conducting material called the intensifier electode, inside the tube near the fluorescent screen, and is operated at a potential several times that of the second electrode.

The cathode ray screen is composed of material, called phosphors, that will emit light when bombarded by electrons. There are several types of phosphors which differ in color, persistence and efficiency. The commonest is *willemite* which produces the familiar greenish trace of general purpose cathode ray tubes and the special purpose large screen monitor tubes. For making measurements, the flat face tube is usually preferred.

Vertical Deflection System

The vertical amplifier usually has a high input impedance preamplifier in the input stage. The preamplifier may be single-ended or differential type. The output of the vertical amplifier is directly fed to the vertical plates of the CRT. The Y gain is controlled in the preamplifier stage and the Y shift is controlled in the main amplifier stage.

Horizontal Deflection System

For most applications of the oscilloscope, the light spot is made to move horizontally across the face of the tube at a uniform rate. The trace thus obtained provides a linear time scale against which an unknown voltage can be plotted by applying it across the vertical deflection plates. The electron beam can be swept across the screen at a uniform rate by applying a sawtooth waveform to the horizontal plates. The sawtooth waveform is often called *sweep*. The oscillator circuit capable of generating the sweep is called the *sweep generator*. Good linearity of the sweep is a major requirement in order to obtain a reliable time base. At the end of the sweep, the electron beam is deflected back to the starting point by lowering the sweep voltage back to zero. The finite time required to change the sweep voltage back to the starting value can produce a low intensity visible trace, known as the flyback. The flyback can be eliminated or blanked out by applying a negative pulse to the grid of the cathode ray tube during the period of flyback.

The sweeps are calibrated in terms of a direct unit of time for a given distance of spot travel across the screen, hence the term *time base*. A transparent scale with vertical and horizontal lines spaced one centimeter apart usually is fitted against the face of the CRT. This scale allows time and amplitude to be read directly.

To obtain an accurate time axis and make time and frequency measurements possible, it is necessary to apply to the X deflection plates a linearly increasing voltage proportional to time. Such a sawtooth voltage can be obtained by charging a capacitor by a constant current source.

Operational amplifiers are used to generate a linear sweep. By using separate amplifiers for the integration and the voltage comparison, waveform control and greater precision is achieved. Figure 1-21 shows one such circuit. The amplifier A_2 in conjunction with R_1 and C acts as an integrator with a feedback comparator that switches the voltage to be integrated. The comparator A_1 forces a rapid negative integration and by making R_1 much smaller than R_2, the fall time of the ramp can be made much less than the rise time. The diode D disconnects the comparator during the positive integration period.

The waveform parameters are largely controlled by three potentiometers. To set the amplitude of the ramp, the comparator trip points are adjusted by means of R_3. The ramp frequency is varied by adjusting the integration resistor R_2. By varying the midpoint of comparator trip points with R_4, the waveform is offset by any desired voltage.

A certain amount of minimum fall time is imposed by the limited currents available from the amplifier outputs to discharge the capacitor. To avoid this limit, the comparator can be used to drive an FET RESET switch connected across C.

Synchronized and Triggered Operation

The waveshape can be displayed continuously on the CRT screen only if the measured voltage is continuous and repetitive. To produce a stable display, the repetitive displays must be superimposed, and for that, the sweep generator must oscillate at exactly the same frequency as the repetition rate of the signal or some submultiple thereof. The sweep generator is synchronized to the signal frequency by feeding the input waveform to the sweep generator in such a way that it controls the repetition rate. This type of sweep is called a synchronized sweep (Fig. 1-22).

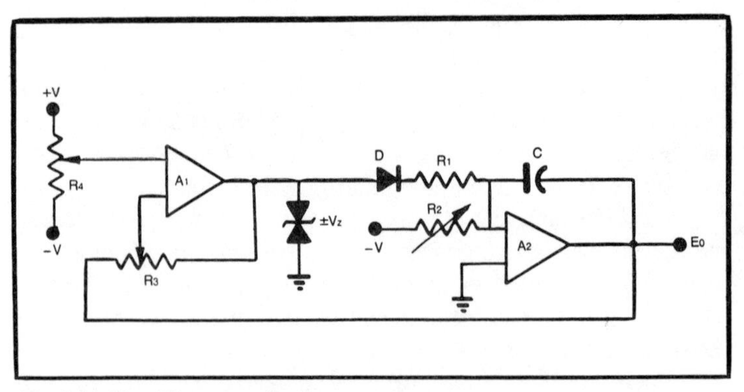

Fig. 1-21. Circuit for generating linear sweep.

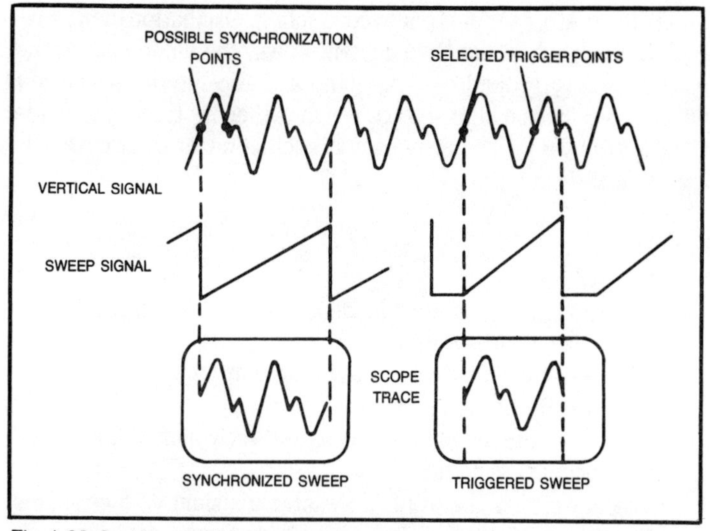

Fig. 1-22. Synchronized and triggered operation.

For measurement of time and repetition rate, a triggered sweep is desirable. With this type of sweep, the particular voltage and slope of the incoming waveform is selected with which sweep is to be triggered. When this point in the waveform occurs, the electron beam moves horizontally across the CRT at the desired rate. When the scan is complete, the sweep circuit is receptive to the next trigger point in the wave train. It is possible to trigger the sawtooth generator by a signal derived from the vertical amplifier. This is called internal mode. The generator can also be triggered from an external source. The sweep when calibrated in time per division can be used for time measurement.

Power Supply

In the solid state instruments, supply voltages for circuit operation are generally obtained from a dc-dc converter operating at audio-frequencies. The supply is highly stabilized to make the system independent of supply variations over a wide range.

The importance of a dc-dc converter lies in stepping up dc voltage of 6 or 12 volts to several thousand volts for operation of the cathode ray tube. The circuit shown in Fig. 1-23 is typical of a dc-dc converter. It consists of an overdriven, push-pull, transformer coupled, transistor oscillator. The circuit is self oscillating and produces a square wave current in the transformer windings. When the supply is switched on, current through resistance R_1

biases transistors Q_1 and Q_2 into conduction. Oscillation then starts rapidly. As soon as the base current flows, the diode clamps the base return to ground. A large step-up ratio for the secondary winding produces a high voltage which is then rectified to provide EHT for the CRT. The frequency of oscillation for square wave is approximately given as:

$$f = \frac{V}{4BNA} \times 10^8$$

where

B is the saturation flux density of the transformer core, in lines/cm^2,

A is the core cross-sectional area, in cm^2,

V is the dc supply voltage,

N is the number of turns on each half of the primary winding.

The capacitor C_1 prevents excessive transient voltage spikes on the collector of transistors Q_1 and Q_2, as they switch from one state to the other. The output voltage may be increased by increasing the transformer turns-ratio or by adding voltage multiplier circuits. For example, double voltage may be obtained by replacing the voltage doubler circuit with a quadrupler circuit (Figs. 1-23A and 1-23B). The transformer is the most important item in a dc-dc converter. It should have close coupling, low leakage reactance and a core which saturates sharply. The core used is generally of the toroidal form with which very close coupling can be achieved. The frequency of oscillations is selected for optimum operation of the power transistors and the transformer. The low limit of frequency is generally determined by the allowable size of the magnetic components. The upper frequency in most cases is limited by the frequency cut-off characteristics of the transistors and the core alloys of the magnetic material used in the transformer.

Most commonly used frequencies are from 5000 Hz to 16000 Hz. The low voltage supplies are taken from separate secondary windings and rectifier networks.

SHIELDING AND GROUNDING

Low level signals which are usually encountered in analytical instruments are quite sensitive to external contaminations. This is especially troublesome where the signal source impedance is very high. The spurious signal or noise is an unwanted signal caused by the stray capacitance, inductance or resistance which accidentally

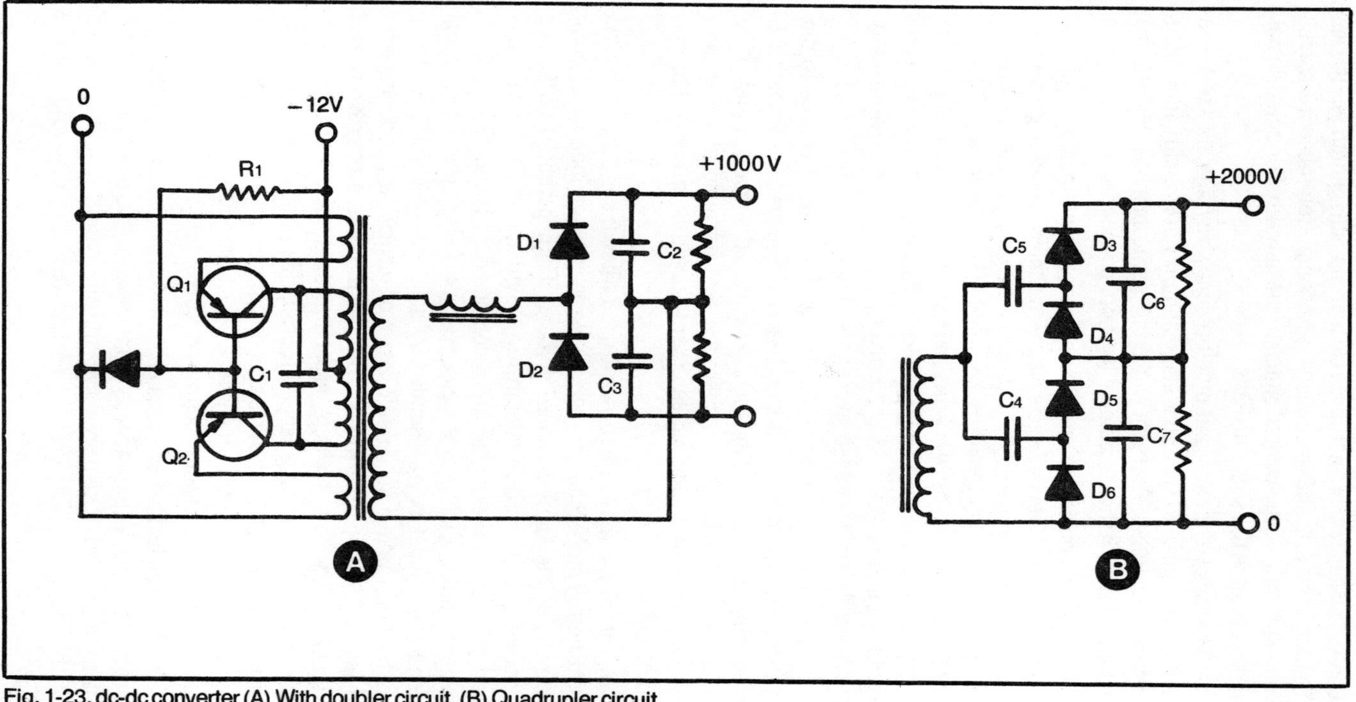

Fig. 1-23. dc-dc converter (A) With doubler circuit. (B) Quadrupler circuit.

couple various parts of the circuit or its surroundings. It can produce errors in measurements and completely obscure useful data. The ratio of the wanted signal to the unwanted or noise signal is expressed as the signal-to-noise ratio.

The most common and omnipresent stray signals are those derived from the 50 Hz line voltage and they are readily identified on the recordings. They are caused by numerous reasons but are picked up more if the instrument has poor connections.

A major consideration in combating stray signals in all low level measurements and recording systems is properly grounding the circuit. Its primary function is to assure that electronic enclosures and chassis are maintained at zero potential. Modern laboratories have a third copper conductor in all electrical circuits which is non-current carrying and is connected to the electric power ground or the cold water mains pipeline. This will usually provide a satisfactory system ground.

Where it is not practical to connect the signal source to the system ground, then it is imperative that a second low impedance grounding point be established. It is called signal ground. It is generally undesirable to connect the signal ground to the system ground. Moreover, the signal circuit should be grounded at one point and at only one point.

Interference is sometimes caused when the ground current is returned by more than one path. Two separate grounds are seldom, if ever, at the same absolute voltage. Their potential difference creates an unwanted current in series with one of the signal leads and causes a noise signal to be combined with useful signal. To prevent noise pickup from electrostatic fields, low level signal conductors must be surrounded by an effective shield. A woven metal braid under an outside layer of insulation is adequate for many applications. However, for transmission of microvolt-level signals, very low leakage capacitance is essential. Specially designed cables having lapped foil shields plus a low resistance drain wire in place of the braided wire shield are used for this purpose. This design reduces leakage capacity from about 0.1 $\mu\mu$f per foot to 0.01 $\mu\mu$f per foot for a typical cable.

The signal cable shield is grounded at the signal source. This prevents signal-cable capacity from shunting the amplifier's impedance to ground. It also preserves the high common mode rejection of the amplifier. The shield is connected to the low side of the signal source.

To sum up:

■ Every low level recording or display system should have a stable system ground and a good signal ground.

■ The signal cable shield should not be attached to more than one ground, and this ground should be at the signal source.

■ More than one accidental or intentional ground on either the signal circuit or the signal cable shield will produce excessive electrical noise in any low level circuit. Therefore, the signal circuit should be grounded at only one point and never at more than one point.

■ Always ground a floating signal circuit and the signal cable shield only at the signal source.

Chapter 2

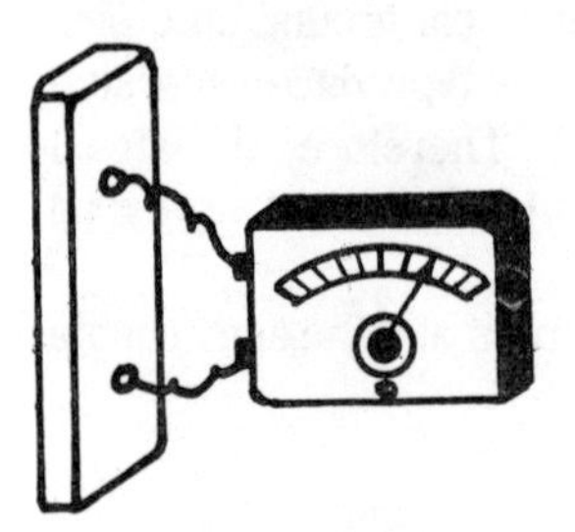

Colorimeters and Spectrophotometers

The most important of all the instrumental methods of analysis are the methods based on the absorption of electromagnetic radiation in the visible, ultraviolet and infrared ranges. According to the *quantum theory*, the energy states of an atom or molecule are defined and any change from one state to another would, therefore, require a definite amount of energy. If this energy is supplied from an external source of radiation, the exact quantity of energy required to bring about a change from one given state to another will be provided by photons of one particular frequency, which may thus be selectively absorbed. The study of the frequencies of the photons which are absorbed would thus indicate a lot about the nature of the material. Also, the number of photons absorbed may provide information about the number of atoms or molecules of the material present in a particular state. It provides us with a method to have qualitative and quantitative analysis of a substance.

Molecules possess three types of internal energy—*electronic, vibrational* and *rotational*. When a molecule absorbs radiant energy, it can increase its internal energy in a variety of ways. The various molecular energy states are quantized and the amount of energy necessary to cause any change in any one of the energy states would generally correspond to specific regions of the electromagnetic spectrum. Electronic transitions correspond to the ultraviolet and visible regions, vibrational transitions to the near infrared and infrared regions and rotational transitions to the infrared and far-infrared regions. The method based on the absorption of radiation of a substance is known as *absorption spectroscopy*. The main advantages of spectrometric methods are speed, sensitivity to very small amounts and a relatively simple operational methodology. The time required for the actual measurement is very short and most of the analysis time, in fact, goes into preparation of the samples. Absorption spectroscopy has a tremendously wide range of analytical applications and is proving extremely useful for analysis even at trace levels.

ELECTROMAGNETIC RADIATION

Electromagnetic radiation is a type of energy that is transmitted through space at a speed of approximately 3×10^{10} cm/sec. Such a radiation does not require a medium for propagation and can readily travel through vacuum. Electromagnetic radiation may be considered as discrete packets of energy called *photons*. The relation between the energy of a photon and the frequency of its propagation is given by

$$E = h\nu$$

where

E = energy in ergs
ν = frequency in cycles per sec.
h = Planck's constant (6.6256×10^{-27} ergs - sec)

Electromagnetic energy, for some purposes, can be more conveniently considered as a continuous wave motion in the form of an alternating electric field in space. The electric field produces magentic field at right angles to its own direction. These fields, in turn, are mutually perpendicular to the direction of propagation. If λ is the wavelength (interval between successive maxima or minima of the wave), then

$$C = \nu\lambda$$

where C is the velocity of propagation of radiant energy in vacuum and ν is the frequency in cycles per second. The practical units employed to express wavelength are:

nm	nanometers	
or		$= 10^{-7}$ cm
mμ	millimicron	
μm	micrometer	
or		$= 10^{-4}$ cm
μ	micron	
A^0	the Angstrom	$= .1\ m\mu = 10^{-8}$ cm

Wave number is defined as the number of waves per centimeter.

Figure 2-1 shows the various regions in the electromagnetic spectrum which are normally used in spectroscopic work. Visible light represents only a very small portion of the electromagnetic spectrum and generally covers a range from 380 to 780 m μ. The ultraviolet region extends from 185 mμ to the visible. Shorter wavelengths lie in the far-ultraviolet region which overlaps the soft X-ray part of the spectrum. The infrared region covers wavelengths above the visible range.

INTERACTION OF RADIATION WITH MATTER

When a beam of radiant energy strikes the surface of a substance, the radiation interacts with the atoms and molecules of the substance. The radiation may be transmitted, absorbed, scattered or reflected, or it can excite fluorescence depending upon the properties of the substance. The interaction, however, does not involve permanent transfer of energy.

The velocity at which radiation is propagated through a medium is less than its velocity in vacuum. It depends upon the kind and concentration of atoms, ions or molecules present in the medium. Figure 2-2 shows various possibilities which might result when a beam of radiation strikes a substance. The radiation may be transmitted with little absorption taking place, and therefore, without much energy loss. The direction of propagation of the beam may be altered by reflection, refraction, diffraction or scattering. The radiant energy may be absorbed, in part or entirely, by the substance.

In absorption spectrophotometry, we are usually concerned with absorption and transmission. Generally, the conditions under which the sample is examined are selected to keep reflection and scattering to a minimum.

Absorption spectrophotometry is based on the principle that the amount of absorption that occurs is dependent on the number of molecules present in the absorbing material. Therefore, the intensity of the radiation leaving the substance may be used as an indication of the concentration of the material.

Let us suppose P_0 is the incident radiant energy and P is the energy which is transmitted. The ratio of the radiant power transmitted by a sample to the radiant power incident on the sample is known as the transmittance.

$$\text{Transmittance } T = P/P_0$$

$$\% \text{ Transmittance} = (P/P_0) \times 100$$

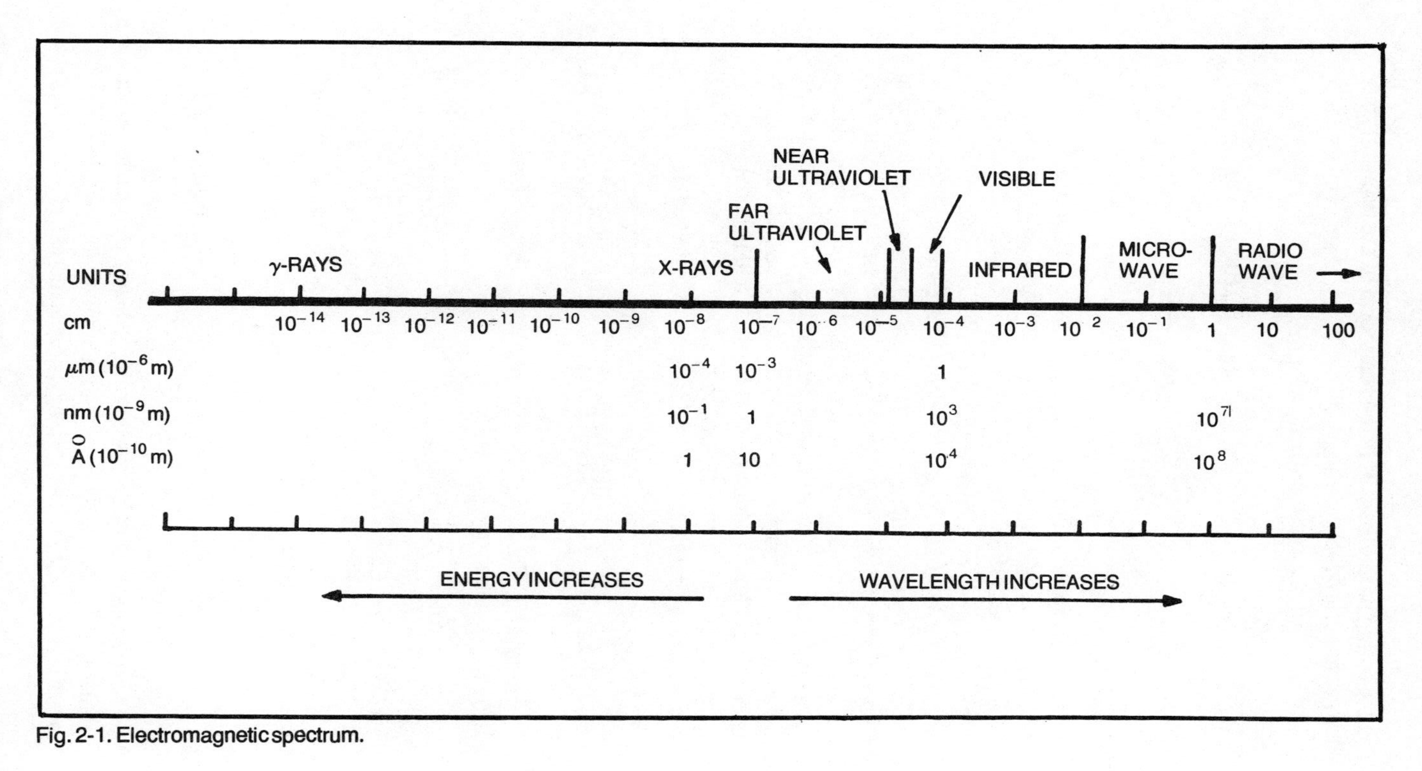

Fig. 2-1. Electromagnetic spectrum.

The logarithm to the base 10 of the reciprocal of the transmittance is known as *absorbance.*

$$\text{Absorbance} = \log_{10}\frac{(1)}{T}$$

$$= \log_{10}(P_0/P)$$

$$\text{Optical density} = \log_{10}\frac{(100)}{T}$$

THE BEER-LAMBERT LAW

The relationship between energy absorption and concentration is of great importance to analysts. It was proposed by Lambert (1760) and Beer (1852) that the amount of monochromatic radiant energy absorbed or transmitted by a solution is an exponential function of the concentration of the absorbing substance present in the path of radiant energy. This means that successive equal thickness of a homogeneous absorbing medium will reduce the intensity by successive equal fractions and, therefore, radiant energy will diminish in geometric or exponential progression. In other words, if a particular thickness absorbs half the radiant energy the thickness which follows the first and is equal to it will not absorb the entire second half, but only the half of this half and will consequently reduce it to one quarter. Expressed mathematically

$$-\frac{dP}{P} = K.db \qquad \textbf{Equation 2-1}$$

Integrating Equation (1), changing logarithm to base 10 and putting $P = P_0$ when $b = 0$, we obtain

$$2.303 \log\frac{P_0}{P} = K \times b \qquad \textbf{Equation 2-2}$$

This equation shows that the radiant power of the unabsorbed light decreases exponentially as the thickness of the absorbing medium increases arithmetically.

$$\therefore \quad P = P_0 \times e^{-Kb} \qquad \textbf{Equation 2-3}$$

$$= P_0 \times 10^{-0.434\,Kb}$$

Also, a decrease in radiant power of a beam of parallel monochromatic radiation takes place in a similar manner when the concentration of the light absorbing material increases.

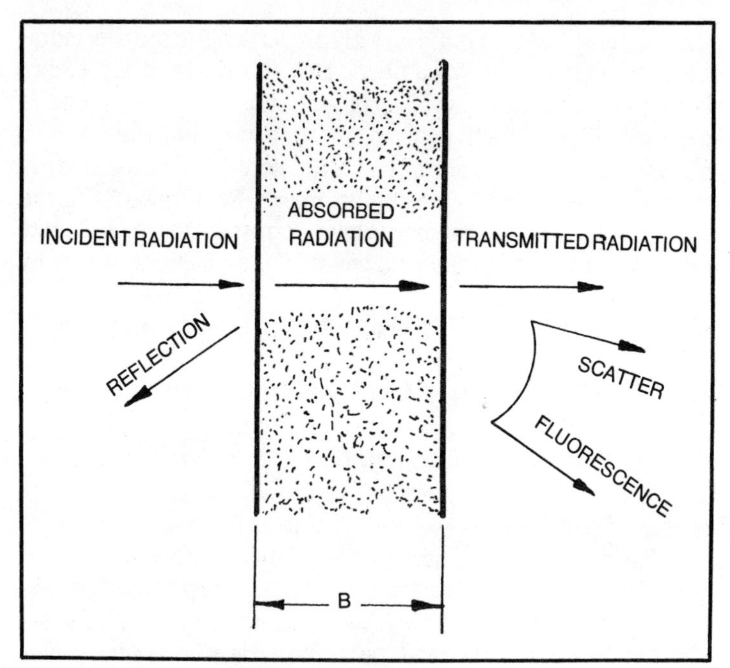

Fig. 2-2. Interaction of radiation with matter.

$$\therefore \quad 2.303 \log \frac{P_0}{P} = K'c \qquad \textbf{Equation 2-4}$$

Equations 2 and 4 can be combined and written as

$$\log \frac{P_0}{P} = a\,b\,c = \log \frac{1}{T} \qquad \textbf{Equation 2-5}$$

$$\text{Absorbance } A = a\,b\,c$$

where a is the absorptivity which is constant depending upon the wavelength of the radiation and nature of the absorbing material.

b is thickness of the absorbing material.

c is concentration in gram per liter.

The equation for absorbance thus describes the relationship between absorbance, thickness and concentration. Absorptivity may be referred to as *specific extinction* and absorbance as *optical density*.

Absorbance is the property of a sample, whereas absorptivity is the property of a substance and is a constant. When the absorbance of a compound is directly proportional to the concentrations, the compound follows Beer's Law. Therefore, if absorbance is plotted graphically against concentrations, a straight line is obtained. A graph derived from transmittance data will not be a straight line unless transmittance (or percent transmission) is plotted on the log-axis of a semi-log paper. The constant "a" in the Beer-Lambert Law must have units corresponding to the units used for concentration and path length of the sample. It is usually written as

$E^{1\%}_{1\,cm}$ = The value of "a" for a 1% sample concentration of 1 cm thickness.

τ = The molar extinction coefficient, i.e., the value of "a" for a sample concentration of 1 gm molecule per liter and 1 cm thickness.

The relationship between transmittance and absorbance as marked on the scales of indicating meters is shown in Fig. 2-3.

Beer's Law only describes the relationship between absorbance, thickness and concentration. It does not imply that these are the only factors that affect absorbance. The direct, linear relationship between absorbance and concentration is used as a fundamental test of a system's conformity to the combined laws. A straight line passing through the origin indicates conformity to the Beer's Law. Discrepancies are usually found when the absorbing solute dissociates or associates in solution.

Beer's Law is derived by assuming that the beam of radiation is monochromatic. However, in all photometers and most of the spectrophotometers, a finite bandwidth of frequencies is always present. The wider the bandwidth of radiation passed by the filters or dispersing device, the greater is the deviation of a system from adherence to Beer's Law.

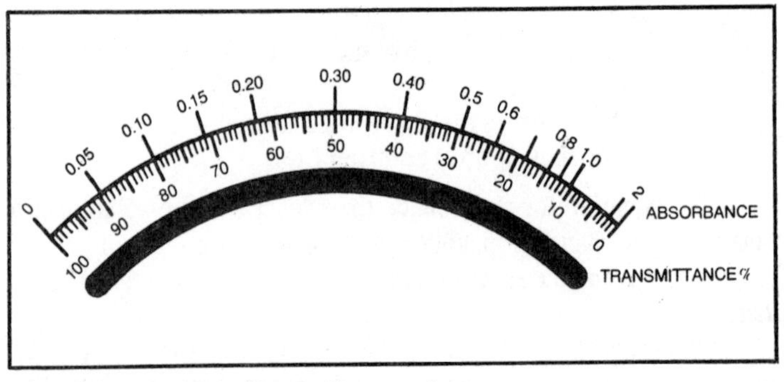

Fig. 2-3. Absorbance and transmittance scale.

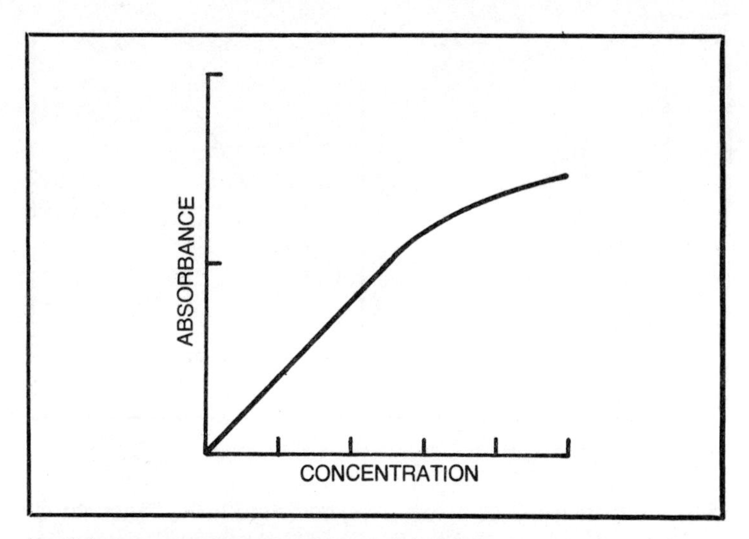

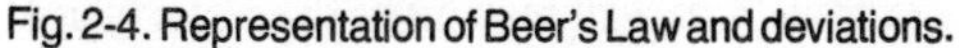
Fig. 2-4. Representation of Beer's Law and deviations.

Beer's Law is a limiting law and should be expected to apply only at low concentrations. It has been observed that the deviation becomes evident at higher concentrations (Fig. 2-4) on an absorbance versus concentration plot when the curve bends towards the concentration axis.

QUANTITATIVE ANALYSIS

The most usually employed quantitative method consists of comparing the extent of absorption or transmittance of radiant energy, at a particular wavelength, by a solution of the test material and a series of standard solutions. It can be done with visual color comparators, photometers or spectrophotometers.

Choice of Wavelength

The selection of a suitable wavelength in the spectrum for quantitative analysis of a sample can be made during the course of preparing the calibration curve for the unknown material. The calibration curve is plotted to show the absorbance values of a series of standards of known concentration and the concentrations of actual samples can then be read directly (Fig. 2-5). A series of standard solutions is prepared along with a blank. Using one filter at a time, calibration curves are plotted in terms of absorbance versus concentration. The filter which provides closest adherence to linearity, over the widest absorbance interval, and which yields the largest slope with a zero intercept, will constitute the best

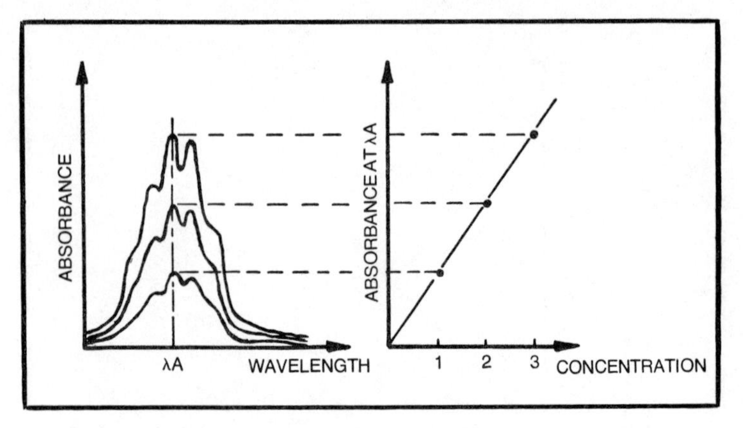

Fig. 2-5. Absorption spectra and Beer-Lambert calibration.

choice for analysis. In a spectrophotometer, the wavelength of maximum absorbance is readily ascertained from the absorbance wavelength curve for the material.

Simultaneous Spectrophotometric Determination

It is often found in practice that when there are several components which absorb radiation of the same wavelength, their absorbances add together and it would no longer be true that the absorbance of the sample is proportional to the concentration of one component. If there is no reaction or interaction between the different solutes, the absorbances are additive of all the components at a given frequency. The Beer-Lambert Law can then be written as

$$A = a_1bc_1 + a_2bc_2 + a_3bc_3 + \dots\dots\dots\dots + a_nbc_n$$

where the subscripts refer to the respective components. In determining the various values for a, if the cell thickness b is held constant, the n b may be included in a. Hence, this is an equation with n c's when n compounds are present. In n such equations were determined from values of the absorbance at n different frequencies, n simultaneous linear equations would result which may be solved to find the required concentrations. Certainly, the procedure is difficult, and in the presence of many components, one would really need a computer to solve the equations. However, much of the classical work in the analysis of petroleum fractions was carried out in this way. The method is not preferred if there are more than two or three components absorbing radiation of the same wavelength.

A simplified and more common method is to convert the component under analysis, by adding a chemical reagent which specifically reacts with it to form a highly absorbing compound. The addition of this reagent to the mixture would result in change of the wavelength of the absorption maxima so that there is no longer interference among the compònents. The analysis then becomes very simple.

ABSORPTION INSTRUMENTS

Figure 2-6 shows an arrangement of components of an absorption instrument. The essential components are:

—A source of radiant energy, which may be a tungsten lamp, Xenon-mercury arc, hydrogen or deutriùm discharge lamp, etc.

—Filtering arrangement for selection of a narrow band of radiant energy. It could be a single wavelength absorption filter, interference filter, a prism or a diffraction grating.

—An optical system for producing a parallel beam of filtered light for passage through an absorption cell (cuvette). The system may include lenses, mirrors, slits, diaphragm, etc.

—A detecting system for measurement of unabsorbed radiant energy which could be human eye, barrier-layer cell, phototube or photo-multiplier tube.

—A readout system for display, which may be an indicating meter or numerical display.

Generally, the components are selected appropriate to their intended use. Figure 2-7 shows optical characteristics of various optical components and their range of suitability in the electromagnetic spectrum.

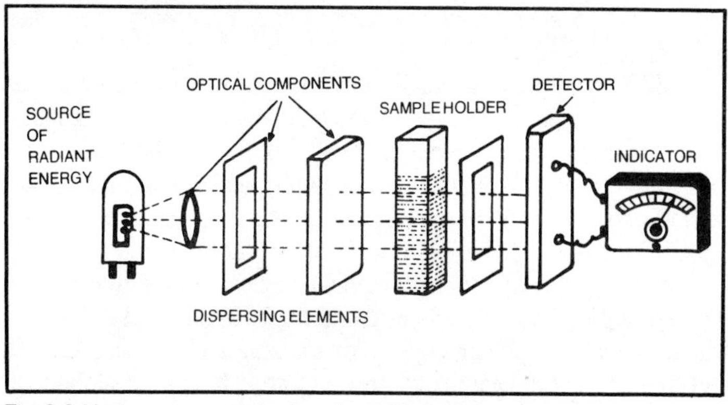

Fig. 2-6. Various components of an absorption instrument.

Radiation Sources

The function of the radiation source is to provide sufficient intensity of light suitable for making a measurement. The most common and convenient source of light is the *tungsten lamp*. This lamp consists of a tungsten filament enclosed in a glass envelope. It is cheap, intense and reliable. Major portion of the energy emitted by a tungsten lamp is in the visible region and only about 15 to 20% is in the infrared region. When using a tungsten lamp, it is desirable to use a heat absorbing filter between the lamp and the sample holder to absorb most of the infrared radiation without seriously diminishing energy at the desired wavelength. For work in the ultraviolet region, a hydrogen or deutrium discharge lamp is used. In these lamps, the envelope material of the lamp puts a limit on the smallest wavelength which can be transmitted. For example, quartz is suitable only up to 200 mμ and fused silica up to 185 mμ. The radiation from the discharge lamps is concentrated into narrow wavelength regions of emission lines. Practically, there is no emission beyond 400 mμ in these lamps. For this reason, spectrophotometers for both the visible and ultraviolet regions always have two light sources, which can be manually selected for appropriate work.

For work in the infrared region, a tungsten lamp may be used. However, due to high absorption of the glass envelope and the presence of unwanted emission in the visible range, tungsten lamps are not preferred. In such cases, nernst filaments or other sources of similar type are preferred. They are operated at a lower temperature and still radiate sufficient energy.

For fluorescent work, an intense beam of ultraviolet light is required. This requirement is met by a xenon arc or a mercury vapor lamp. A cooling arrangement is very necessary when this type of lamp is used.

Mercury lamps are usually run direct from the ac power line via a series ballast choke. This method gives some inherent lamp power stabilization and automatically provides the necessary ionizing voltage. The ballast choke is physically small and a fast warmup to the lamp operating temperature is obtained. Moss (1977), however, points out that ac operation gives a strong modulation to the visible flux and thus may cause amplitude noise within the pass band of the spectrometers. He describes a circuit which provides a power supply for ionizing and running 125 W high-pressure mercury argon arc discharge lamps on stabilized direct current.

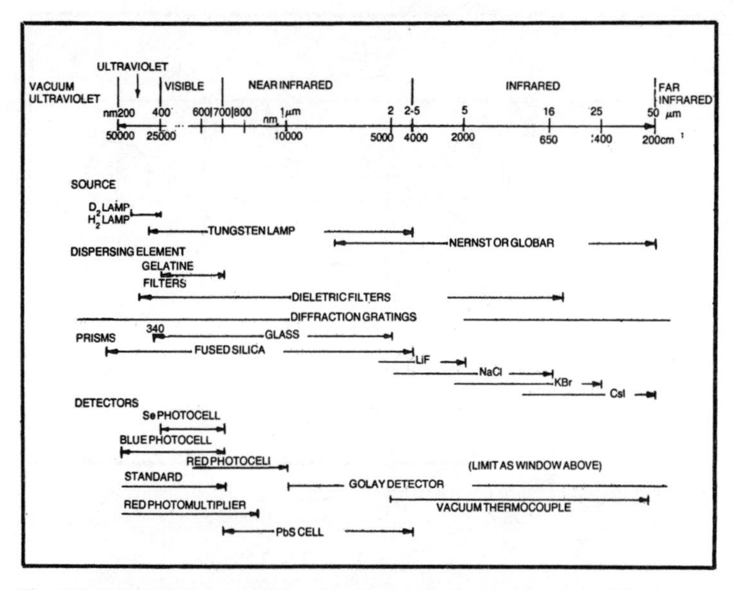

Fig. 2-7. Spectral characteristics of various optical components and their range of suitability in the electromagnetic spectrum.

The latest development in light sources is the introduction of the tungsten-halogen light source which has a higher intensity output than the normal tungsten lamp used in colorimetry and spectrophotometry. It also has a larger life and does not suffer from blackening of the bulb glass envelope. In the ultraviolet region of the spectrum, the deutrium lamp has superseded the hydrogen discharge lamp as a ultraviolet source. The radiation sources should be highly stable and preferably emit out a continuous spectrum.

Lamp Regulator

A simplified circuit diagram of a lamp regulator that delivers constant power to a tungsten lamp is shown in Fig. 2-8. A part of the voltage across the lamp is summed with a voltage proportional to the current through the lamp. This total voltage V_f is compared with a reference voltage V_{ref} by operational amplifier A_1. A_1 controls the voltage at the emitter of transistor Q, such that V_f will always be equal to V_{ref}.

$$V_f = V_{ref} = V_1 + V_2$$

The voltage V_f can be written in terms of lamp current (I_L) and filament resistance R_f

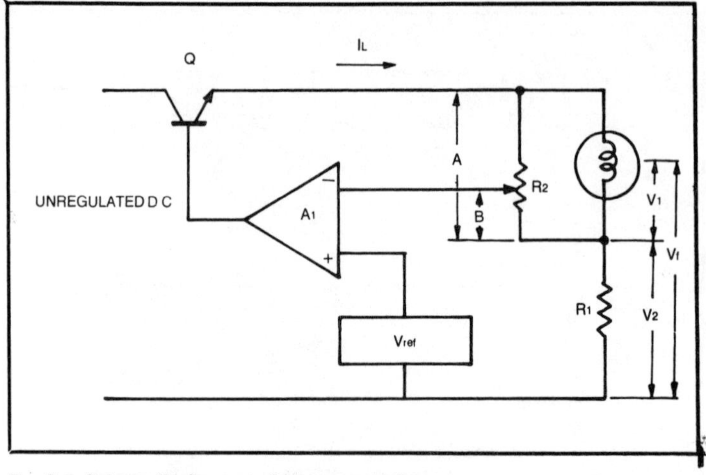

Fig. 2-8. Schematic diagram of a lamp regulator.

$$V_f = \frac{b}{a} \times I_L \times R_f + I_L \times R_1$$

where b/a is the ratio corresponding to the setting of potentiometer R_2. R_1 is the current sensing resistor. The ratio b/a can be selected so that constant power is delivered for a particular type of lamp. The power delivered to the lamp can be varied by adjusting the reference voltage.

Optical Filters

A filter may be considered as any transparent medium which by its structure, composition or color enables isolation of radiation of a particular wavelength. For this purpose, ideal filters should be monochromatic, i.e., they must isolate radiation of only one wavelength. A filter must meet the following two requirements—high transmittance at the desired wavelength and low transmittance at other wavelengths.

However, in practice, the filters transmit a broad region of the spectrum. Referring to Fig. 2-9, they are characterized by the relative light transmission at the maximum of the curve Tλ, the width of the spectral region transmitted (the half-width—the range of wavelength, between the two points on the transmission curve at which the transmission value equals ½ Tλ) and T_{res} (the residual value of the transmission in the remaining part of the spectrum). The ideal filter would have the highest value of Tλ and the lowest values for the transmission half-width and T_{res}.

Absorption Filters

The *absorption* type optical filters usually consist of color media: color glasses, colored films (gelatin etc.) and solutions of the color substances. This type of filter has a wide spectral

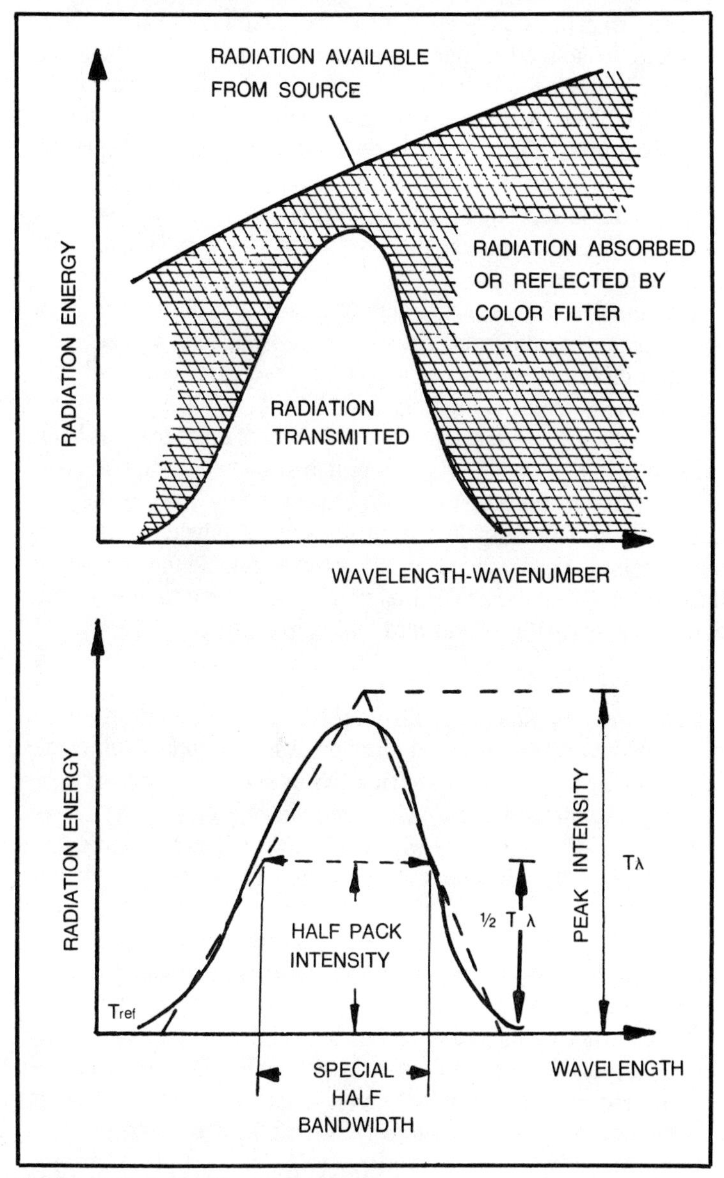

Fig. 2-9. Optical properties of a light filter.

bandwidth which may be 40 to 50 μ in width at one half the maximum transmittance. Their efficiency of transmission is very poor and is of the order of 5 to 25%.

It is possible to obtain more selective light filters from colored media by increasing their thickness two, three or more times. Here the transmission of the filter at the maximum for light of the wavelength isolated is decreased, but there is a simultaneous increase in the selectivity. By using this technique, it is theoretically possible to achieve a very good selectivity, but the fall in transmission efficiency would have to be compensated by suitable amplification of the photocurrent. As absorption type filters do not provide a high degree of monochromaticity required for isolating complex systems, their use is restricted to only very simple types of photometers.

Composite filters, consisting of sets of unit filters, are often used. In the combination, one set consists of long wavelength, sharp cutoff filters and the other of short wavelength cut-off filters. Combinations are available from about 360 mμ to 700 mμ.

The glass filter consists of a solid sheet of glass that has been colored with a pigment, which is either dissolved or dispersed in glass, whereas the gelatin filter consists of a layer of gelatin impregnated with suitable organic dyes and sandwiched between two sheets of glass. Gelatin filters are not suitable for use over a long period. With the absorption of heat, they tend to deteriorate due to changes in the gelatin and bleaching of the dye.

Interference Filters

Interference filters usually consist of two semitransparent layers of silver, deposited on glass by evaporation in vacuum and separated by a layer of dielectric (ZnS or MgF_2). In this arrangement, the semitransparent layers are held very close. The spacer layer is made of substance which is of low refractive index. The thickness of the dielectric layer determines the wavelength transmitted. Figure 2-10 shows the path of light rays through an interference filter. Some part of light that is transmitted by the first film is reflected by the second film and again reflected on the inner face of the first film.

Constructive interference between different pairs in superposed light rays occurs only when the path difference is exactly one wavelength or some multiple thereof. The relationship expressing a maximum for the transmission of a spectral band is given by

$$m\lambda = 2\,d(n)\,\text{Sin}\,\theta$$

when light is incident normally, $\sin\theta = 1$

$$\text{therefore } m\lambda = 2d(n)$$

where d is the thickness of the dielectric spacer, whose refractive index is n. The multiple of frequencies harmonically related to the wavelength of the first order rays is the order (m) of the interference.

Interference filters allow a much narrower band of wavelengths to pass and are similar to monochromators in selectivity. They are simpler and less expensive. However, as the selectivity increases, the transmittance decreases. The transmittance of these filters varies between 15 to 60% with spectral bandwidths of 10 to 15 mμ.

One type in interference filters is the continuous wedge filter which permits a continuous selection of different wavelengths. The continuity of an interference filter is achieved by using a spacer film of graded thickness between the two semitransparent layers of silver. They usually have a working interval of 400 to 700 mμ and a dispersion of 5.5 mμ per mm. Transmittance is usually not more than 35%. With less transmittance, the sensitivity gets lower

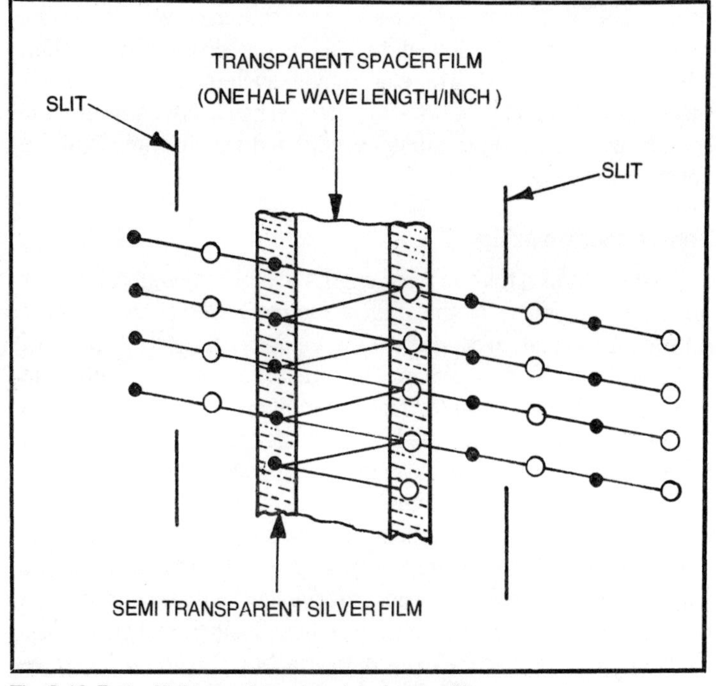

Fig. 2-10. Path of light rays through an interference filter.

which may be compensated by using electronic amplifiers after the photodetectors.

For efficient transmission, multilayer transmission filters are often used. They are characterized by a band-pass width of 8 mμ or less and a peak transmittance of 60-95%. Interference filters can be used with high intensity light sources, since they remove unwanted radiation by transmission and reflection, rather than by absorption. Palmer (1971) explains a monochromator using graded interference filters with a half-width band-pass of 7.2 mμ. The arrangement makes use of two graded filters and provides a flux of monochromatic light not easily obtainable with other types of monochromators of comparable dimensions.

MONOCHROMATORS

Monochromators are the optical systems which provide better isolation of spectral energy than the optical filters and are, therefore, preferred where it is required to isolate narrow bands of radiant energy. Monochromators usually incorporate a small glass of quartz prism, or a diffraction grating system, as dispersing media. The radiation from a light source is passed either directly or by means of a lens or mirror into the narrow slit of the monochromator and allowed to fall on the dispersing medium where it gets isolated. The efficiency of such monochromators is much better than that of filters and spectral half bandwidths of 1 mμ or less are obtainable in the ultraviolet and visible regions of the spectrum.

Prism Monochromators

Isolation of different wavelengths in a prism monochromator depends upon the fact that the refractive index of materials is different for radiation of different wavelengths. If a parallel beam of radiation falls on a prism, the radiation of two different wavelengths will be bent through different angles. The greater the difference between these angles, the easier it is to isolate the two wavelengths. This becomes an important consideration for selection of material for the prisms, because only those materials are selected whose refractive index changes sharply with wavelength.

Figure 2-11 shows the use of a prism as a monochromator. Light from the source S is made into a parallel beam and made to fall on a prism after it is passed through entrance slit S_1 and mirror M_1. The entrance slit is at the focus of mirror M_1. The prism disperses the light, and photons of different wavelengths are deflected at

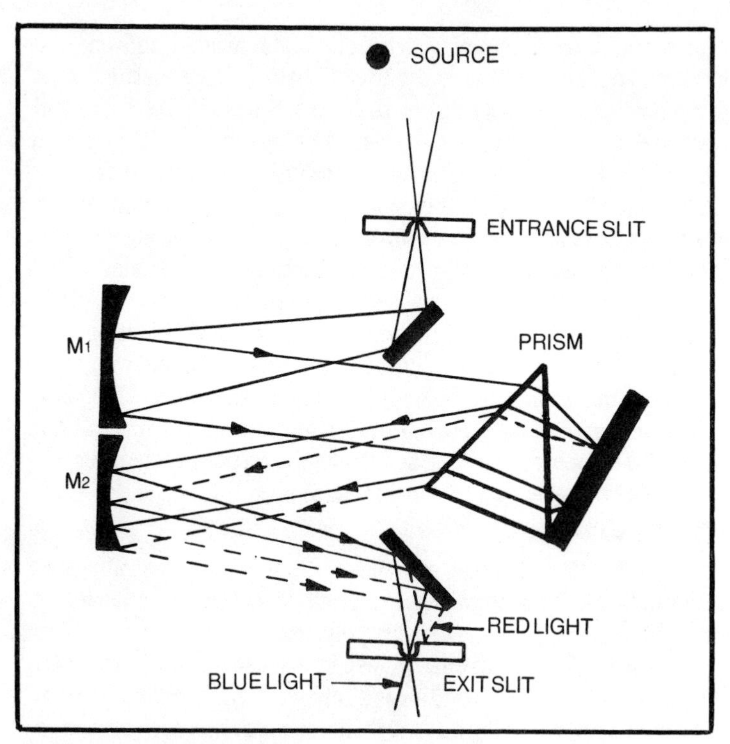

Fig. 2-11. Prism monochromator.

different angles. If the dispersed beam is again refocused, the focal point for photons of one wavelength will be displaced from that for photons of a different wavelength. The light of any one wavelength can be selected by moving a slit across the focal plane. The required wavelength passes through the slit; the other wavelengths are blocked. The optical arrangement used in practice may differ from that illustrated in this figure, but the principle is the same. In most of the cases, the prism is moved to shift the spectrum across the exit slit, rather than the slit being moved across the focal plane. The same collimating mirror is used for both M_1 and M_2 to save costs of high grade optical components. The instrument manuals can be consulted to know the alternative systems.

Prisms may be made of glass or quartz. The glass prisms are suitable for radiations essentially in the visible range, whereas the quartz prism can cover the ultraviolet spectrum also. It is found that the dispersion given by glass is about three times that of quartz. However, quartz shows the property of double refraction.

Therefore, two pieces of quartz, one right handed and one left handed are taken and cemented back to back in the construction of 60° prism (cornu mounting) or the energy must be reflected and returned through a single 30° prism so that it passes through the prism in both directions (*Littrow mounting*). The two surfaces of the prism must be carefully polished and optically flat. Prism spectrometers are usually expensive because of exacting requirements and difficulty in getting quartz of suitable dimensions.

There are several ways of selecting a particular wavelength in prism monochromators. It may be chosen by local selection with movable slits, or by local selection with fixed slits behind which are placed as many photosensitive elements as there are slits. The selection can also be achieved by prism rotation in which all the lines of the spectrum are passed through a fixed slit one after the other. The wavelength scale in this case is non-linear.

Diffraction Gratings

Monochromators may also make use of diffraction gratings as dispersing medium. A *diffraction grating* consists of a series of parallel grooves, ruled on a highly polished reflecting surface. The grooves are ruled at extremely close intervals as 15,000 or 30,000 lines per inch. When the grating is put into a parallel radiation beam so that one surface of the grating is illuminated, this surface acts as a very narrow mirror. The reflected radiation from this grooved mirror overlaps the radiation from neighboring grooves (Fig. 2-12). The waves would, therefore, interfere with each other. On the other hand, it could be that the wavelength of radiation is such that the separation of the grooves in the direction of the radiation is a whole number of wavelengths. Then, the waves would be in-phase and the radiation would be reflected undisturbed. When this is not a whole number of wavelengths, there would be destructive interference and the waves would cancel out, and no radiation would be reflected. By changing the angle at which the radiation strikes the grating, it is possible to alter the wavelength reflected.

The expression relating the wavelength of the radiation and the angle (θ) at which it is reflected is given by

$$m\mu = 2d \, \text{Sin}\theta$$

where d is the distance separating the grooves and is known as the grating constant and m is the order of interference

when m =1, the spectrum is known as first order and with m =2, the spectrum is known as second order.

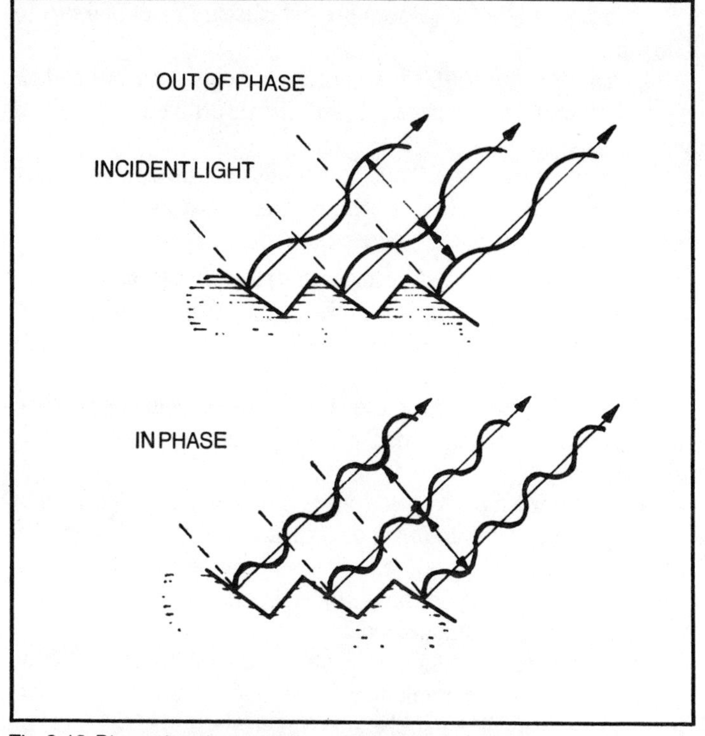

Fig. 2-12. Dispersion phenomenon in diffraction gratings.

The resolving power of a grating is determined by the product mN, where N is the total number of grooves or lines on the grating. Higher dispersion in the first order is possible when there are a larger number of lines. When compared with prisms, the gratings provide much higher resolving powers and can be used in all spectral regions. Gratings would reflect, at any given angle, radiation of wavelength λ and also $\lambda/2$, $\lambda/3$, etc. This unwanted radiation must be removed; otherwise, it will appear as stray light.

OPTICAL COMPONENTS

Several different types of optical components are used in the construction of analytical instruments based on the radiation absorption principle. They could be windows, mirrors and simple condensers. The material used in the construction of these components is a critical factor and depends largely on the range of wavelength of interest. Normally, the absorbance of any material should be less than 0.2 at the wavelength of use.

—Ordinary silicate glasses are satisfactory from 350 mμ to 3000 mμ.

—From 300 mμ to 350 mμ, special corex glass can be used.

—Below 300 mμ, quartz or fused silica is utilized, the limit for quartz is 210 mμ.

—From 180 mμ to 210 mμ, fused silica can be used, provided the monochromator is flushed with nitrogen or argon to eliminate absorption by atmospheric oxygen.

Reflections from glass surfaces are reduced by coating these with magnesium fluoride which is one quarter wavelength in optical thickness. With this, scattering errors are also greatly reduced.

With a view to reduce the beam size or render the beam parallel, condensers are used. These condensers operate as simple microscopes.

To minimize light losses, lenses are sometimes replaced by front-surfaced mirrors to focus or collimate light beam in absorption instruments. Mirrors are aluminized on their front surfaces. With the use of mirrors, chromatic aberrations and other imperfections of the lenses are minimized.

Beam splitters are used in double beam instruments. These are made by giving a suitable multilayer coating on an optical flat. The two beams must retain the spectral properties of the incident beam. Half-silvered mirrors are often used for splitting the beam. However, they absorb some of the light in the thin metallic coating. Beam splitting can also be achieved by using a prismatic mirror or stack of thin horizontal glass plates, silvered on their edges and alternatively oriented to the incident beam.

PHOTOSENSITIVE DETECTORS

After isolation of radiation of a particular wavelength in a filter or a monochromator, it is essential to have a quantitative measure of their intensities. This is done by causing the radiation to fall on a photosensitive element, in which the light energy is converted into electrical energy. The electric current produced by this element can be measured with a sensitive galvanometer directly, or after suitable amplification.

Any type of photosensitive detector may be used for detection and measurement of radiant energy, provided it has a linear response in the spectral band of interest, and has a sensitivity good enough for the particular application. There are two types of photoelectric cells—*photovoltaic* cells and *photoemissive* cells.

Photovoltaic or Barrier Layer Cells

Photovoltaic or barrier layer cells usually consist of a semiconducting substance, which is generally selenium deposited on a metal base, which may be iron and which acts as one of the electrodes. The semiconducting substance is covered with a thin layer of silver or gold deposited by cathodic deposition in vacuum. This layer acts as a collecting electrode. Figure 2-13 shows the construction of the barrier layer cell. When radiant energy falls upon the semiconductor surface, it excites the electrons at the silver selenium interface. The electrons are thus released and collected at the collector electrode.

The cell is enclosed in a housing of insulating material and covered with a sheet of glass. The two electrodes are connected to two terminals for connecting the cell with other parts of the electrical circuit.

Photovoltaic cells are very robust in construction, need no external electrical supply, and produce a photocurrent sometimes stronger than other photosensitive elements. Typical photocurrents produced by these cells is as high as 120 μ amp per lumen. At constant temperature, the current set up in the cell usually shows a linear relationship with the incident light intensity. Selenium photocells have very low internal resistance and, therefore, it is difficult to amplify the current they produce by dc amplifiers. The currents are usually measured directly by connecting the terminals of the cell to a very sensitive galvanometer.

Selenium cells are sensitive to almost the entire range of wavelengths of the spectrum. However, their sensitivity is greater within the visible spectrum and highest in the zones near to the

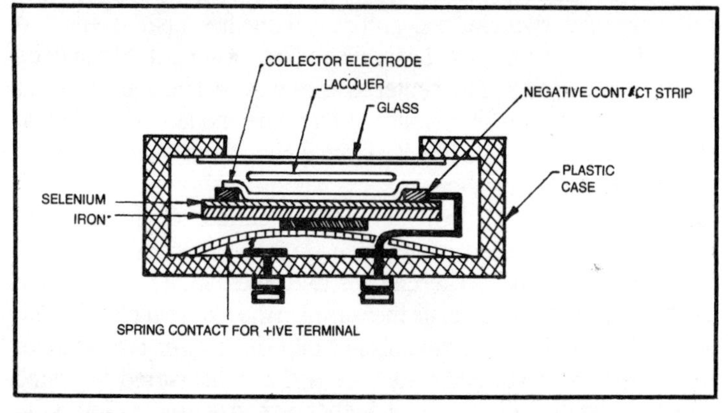

Fig. 2-13. Construction of a barrier layer cell.

yellow wavelengths. Figure 2-14 shows the spectral response of the selenium photocell and the human eye.

Selenium cells have a high temperature coefficient and, therefore, it is very necessary to allow the instrument to warm up before the readings are commenced. They also show fatigue effects. When illuminated, the photocurrent rises to a value several percent above the equilibrium value and then falls off gradually. When connected in the optical path of the light rays, care should be taken to block all external light and to see that only the light from the source reaches the cell. Apart from selenium, photocells may be made of some other materials. The spectral sensitivity is different for different types of cells and should be chosen in accordance with the wavelength of the radiation to be measured.

Selenium cells are not suitable for operations in instruments where the levels of illumination change rapidly because they fail to respond immediately to those changes. They are thus not suitable where mechanical choppers are used to interrupt light 15-60 times a second.

Photoemissive Cells

Photoemissive cells are of three types— *high vacuum photocells, gas-filled photocells* and photo *multiplier* tubes. All of these types differ from selenium cells in that they require an external power supply to provide a sufficient potential difference between the electrodes to facilitate the flow of electrons generated at the photosensitive cathode surface. Also, amplifier circuits are invariably employed for the amplification of this current.

High Vacuum Photoemissive Cells. The vacuum photocell consists of two electrodes. The cathode has a photosensitive layer of metallic cesium deposited on a base of silver oxide and the anode is either an axially centered wire or a rectangular wire that frames the cathode. The construction of the anode is such that no shadow falls on the cathode. The two electrodes are sealed within an evacuated glass envelope.

When a beam of light falls on the surface of the cathode, electrons are released from it which are drawn towards the anode, which is maintained at a certain positive potential. This gives rise to photocurrent which can be measured in the external circuit. The spectral response of a photoemissive tube depends on the nature of the substance coating the cathode and can be varied by using different metals, or by variation in the method of preparation of the

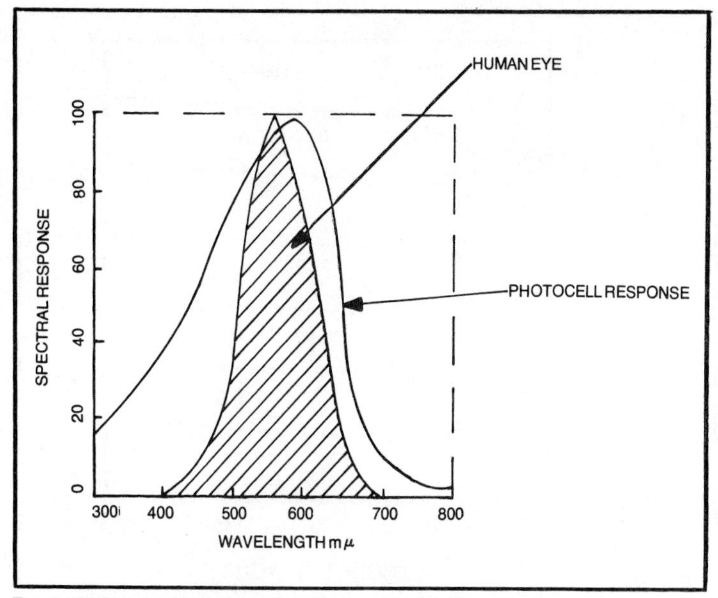

Fig. 2-14. Spectral response of a selenium photocell and the human eye.

cathode surface. Cesium-silver oxide cells are sensitive to the near infrared wavelengths. Similarly, potassium-silver oxide and cesium-antimony cells have maxima of sensitivity in the visible and ultraviolet regions. The spectral response also depends partly on the transparency to different wavelengths of the medium to be traversed by the light before reaching the cathode. For example, the sensitivity of the cell in the ultraviolet region is limited by the transparency of the wall of the envelope. For this region, the use of quartz material can be avoided by using a fluorescent material like *sodium salicylate*, which when applied to the outside of the photocell transforms the ultraviolet into visible radiations.

Figure 2-15 shows current-voltage characteristics of vacuum photoemissive tubes at different levels of light flux. They show that as the voltage is increased, the point is reached where all the photoelectrons are swept to the anode, as soon as they are released, and result in saturation photocurrent. It is not desirable to apply very high voltages, as they would result in excessive dark current without any gain in response.

Figure 2-16 shows a typical circuit configuration usually employed with photoemissive tubes. Large values of phototube load resistor are employed to increase the sensitivity up to the practical limit. Load resistances as high as 10,000 megohms have

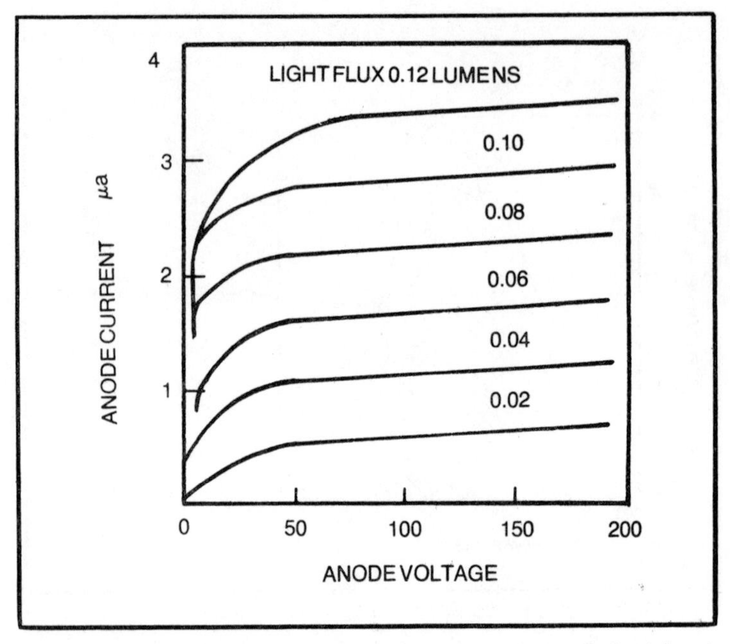

Fig. 2-15. Current voltage characteristics of vacuum photoemissive tube at different levels of light flux.

been used. This, however, almost puts a limit as further increase of sensitivity induces difficulties in the form of noise, non-linearity and slow response. At these high values of load resistors, it is very essential to shield the circuit from moisture and electrostatic effects. Therefore, special type of electrometer tubes, carefully shielded and with grid cap input, are employed in the first stage of the amplifier.

Gas-Filled Photoemissive Cells. This type of cell contains small quantities of inert gas like argon whose molecules can be ionized when the electrons present in the cell possess sufficient energy. The presence of the small quantities of this gas prevent the phenomenon of saturation current when higher potential differences are applied between the cathode and anode. Due to repeated collisions of electrons in the gas-filled tubes, the photoelectric current produced is greater even at low potentials.

Photomultiplier Tubes. Photomultiplier tubes are used as detectors when it is required to detect very weak light intensities. The tube consists of a photosensitive cathode and has multiple cascade stages of electron amplification in order to achieve a large amplification of primary photocurrent within the envelope of the phototube itself. The electrons generated at the photocathode are attracted by the first electrode, called *dynode*, which gives out secondary electrons. There may be 9-16 dynodes. The dynode consists of a plate of material coated with a substance having a small force of attraction for the escaping electrons. Each impinging electron dislodges secondary electrons from the dynode. Under the influence of positive potential, these electrons are accelerated to the second dynode and so on. This process is repeated at the successive dynodes which are operated at voltages increasing in steps of 50-100 volts. These electrons are finally collected at the collector electrode.

The sensitivity of the photomultiplier tube can be varied by regulating the voltage of the first amplifying stage. Because of the relatively small potential difference between every two electrodes, the response is linear. The output of the photomultiplier tube is limited to an anode current of a few milliamperes. Consequently, only low intensity radiant energy can be measured without causing any appreciable heating effect on the electrode surface. They can measure light intensities about 10^7 times weaker than those measurable with an ordinary phototube. For this reason, they should be carefully shielded from stray light. The tube is fairly fast in response, to the extent that they are used in scintillation counters where light pulses as brief as 10^{-9} seconds duration are

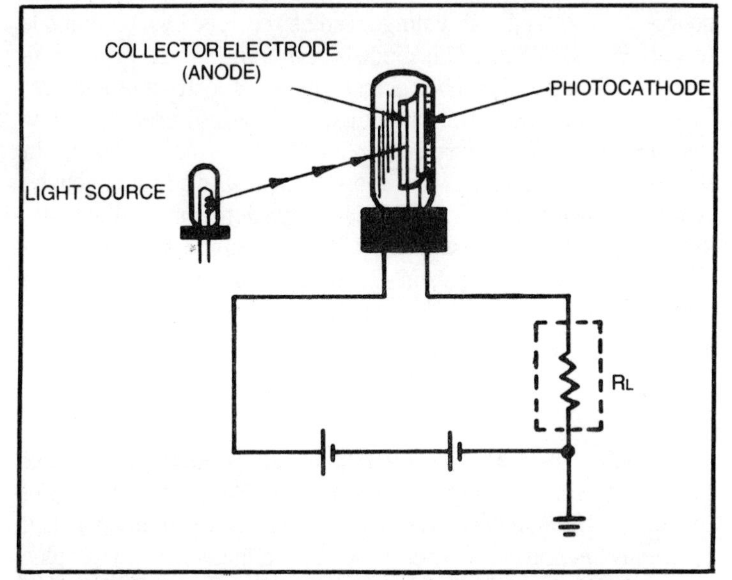

Fig.2-16. Typical circuit configuration employed with photoemissive tubes.

encountered. A direct current power supply is required to operate a photomultiplier, the stability of which must be at least one order of magnitude better than the desired precision of measurement. For example, to attain precision of 1%, fluctuation of the stabilized voltage must not exceed 0.1%.

Fatigue and saturation can occur at high illumination levels. The devices are sensitive to electromagnetic interference, and they are also more costly than other photoelectric sensors. Photomultipliers are not uniformly sensitive over the whole spectrum and, in practice, manufacturers incorporate units best suited for which the instrument is designed. In the case of spectrophotometers, the photomultipliers normally supplied cover the range of 185 to 650 nm. For measurements at longer wavelengths, special red-sensitive tubes are offered. They cover a spectral range from 185 to 850 nm, but are noticeably less sensitive at wavelengths below 450 nm than the standard photomultipliers.

Photomultiplier tubes may be damaged if excessive current is drawn from the final anode. Since accidental overload may easily occur in a laboratory and tubes are expensive to replace, it is advisable to adopt some means of protection from overloads. Gough (1977) explains a circuit which automatically cuts off the EHT supply to a photomultiplier tube, if accidental overload of the tube should occur. The EHT once cut off has to be reset manually. The photomultiplier, which is large and expensive and requires a source of stabilized high voltage, can be replaced by a photodiode (e.g. H.P. 5082-4220). This diode is usable within a spectral range of 0.4-1.05 μm, in a number of instruments (spectrophotometers, flame photometers). The photodiode can be powered from low voltage source. The signal is amplified by a low noise op amp, in the feedback of which a twin T resistor is utilized advantageously. A diode array is used as a multi-wavelength detector for simultaneous absorbance measurements in rapid scanning spectrometers (Horlick, 1976). Yotsuyanagi (1976) describes an instrument of this type using a linear diode array as a replacement of the exit slit and photomultiplier tube.

SAMPLE HOLDERS

Liquids may be contained in a cell or cuvette made of transparent material such as silica, glass or Perspex. The faces of these cells through which the radiation passes are highly polished to keep reflection and scatter losses to a minimum. Solid samples are generally unsuitable for direct spectrophotometry. It is usual to dissolve the solid in a transparent liquid. Gases may be contained in cells, which are sealed or stoppered to make them air-tight. Sample holder is generally inserted somewhere in the interval between the light source and the detector.

For the majority of analyses, a 10 mm path length rectangular cell is usually satisfactory. For the far-ultraviolet region below 210 mμ, a 10 mm path length rectangular cell made from special silica that has better transmission characteristics at shorter wavelengths than does the standard cell is recommended. However, some samples, such as turbid or densely-colored solutions may absorb so strongly that shorter path lengths are necessary for the sample to transmit sufficiently. Rectangular liquid cells are commercially available in both 5 mm and 1 mm path lengths, while the standard 10 mm path length cell can be reduced to 1 mm pathlength by using a silica spacer. These shorter path length cells have lesser volumes, of the order of 0.43 ml for a 1 mm path length, which are necessary for some studies.

In analyses, where only minimal volumes of liquid samples are practical, microcells which have volumes as small as 50 microliters can be employed. Most of the rectangular liquid cells have caps and, for analyses of extremely volatile liquids, some of the cells have ground-glass stoppers to prevent escape of vapor. Studies of dilute or weakly absorbing liquid samples, or of samples where trace components must be detected, require a cell with a long path length. For such applications, a 50 cm path length with about 300 ml volume cell is employed.

Cylindrical liquid cells offer higher volume to path length ratios than do rectangular cells, being available in path lengths of 20, 50 and 100 mm and in volumes of 4, 8, 20 and 40 ml respectively. Similar to these cylindrical cells are the demountable cells that have silica windows which are easily removable. This demountable feature is especially useful for containment of samples that are difficult to remove and clean from conventional cylindrical cells. Demountable cells are equipped with ground-glass stoppers.

Gas cells are available in both rectangular and cylindrical configurations. Both configurations incorporate glass stopcocks. Two stopcocks are installed on the cylindrical cells to permit connecting it into a flow system for dynamic measurements. Rectangular gas cells usually have pathlengths of 2 mm and 10 mm, while cylindrical gas cells have a pathlength of 100 mm.

Alexander, et. al., (1974) describes the design and operation of a cell for making precise absorbance measurements of dilute aqueous solutions over the temperature range 25-250°C. This cell overcomes the major problem which stems from the reactivity of hot aqueous solution when measurements are made at high temperatures.

Erickson and Surles (1976) discuss the importance of sample handling techniques specifically, and of instrumental and operational parameters, in general, on accuracy and precision. Chalzell (1977) discusses the construction of a cell that can withstand pressures up to 25×10^5 Pa and temperatures up to 230°C to measure the ultraviolet and visible spectra of numerous materials.

The measurement of absorption spectra of liquids at low temperatures presents practical difficulties like misting of cell windows and accommodation of attachment for accurate control of the sample temperature in the small space normally available in the cell compartment. Coe and Slaney (1969) illustrate an attachment which enables spectra of liquids to be measured over the normal range of the instrument and at moderately low temperatures (down to at least −96°C) with standard cells. Problems of misting are eliminated by immersing the cells in liquid and placing long silica rods in the light path. The complete attachment is inserted between the monochromator and the photocell compartment in place of the cell compartment generally attached in the instrument.

Another low temperature cell for absorption spectroscopy in the ultraviolet and visible region is described by Aurich (1969). This cell provides the temperature range from room temperature to liquid nitrogen with an accuracy to 0.1°C.

COLORIMETERS

A colorimetric method in its simplest form uses only the human eye as a measuring instrument. This involves comparison, by visual means, of the color of an unknown solution with the color produced by a single standard or a series of standards. The comparison is made by obtaining a match between the color of the unknown and that of a particular standard by comparison with a series of standards prepared in a similar manner to the unknown. Errors of 5 to 20% are not uncommon because of the relative inability of the eye to compare light intensities.

In the earlier days, visual methods were commonly employed for all colorimetric measurements, but now photoelectric methods have largely replaced them and are used almost exclusively for quantitative colorimetric measurements. These methods are more precise and eliminate the necessity of preparing a series of standards every time a series of unknowns is run.

Strictly speaking, a colorimetric determination is one that involves visual measurement of color; and a method employing photoelectric measurement is referred as a photometric or spectrophotometric method. However, usually any method involving measurement of color in the visual region of the electromagnetic spectrum (400-700 mμ) is referred as the colorimetric method.

In a colorimeter, the sample is normally a liquid. The sample compartment of a colorimeter is provided with a holder to contain the cuvette in which the liquid is examined. Usually, this holder is mounted on a slide with positions for at least two cuvettes so that sample and reference values can be measured one after the other. Usually, the reference cuvettes are measured first and a shutter is moved into or out of the light beam until the microammeter gives a full scale deflection (100% T scale reading). The sample is then moved into the beam and the light passing through it is measured as a percentage to the reference value.

$$\frac{\text{Sample}}{\text{concentration}} = \frac{\text{Standard}}{\text{concentration}} \times \frac{\text{Sample reading}}{\text{Reference reading}}$$

Colorimeters are extremely simple in construction and operation. They are used for a great deal of analytical work where high accuracy is not required. The disadvantage is that a range of filters is required to cover different wavelength regions. Also, the spectral bandwidth of these filters is large in comparison with that of the absorption band being measured.

Single-Beam Filter Photometers

Figure 2-17 illustrates the basic components of a filter photometer. The source of light is a tungsten-filament lamp which is held in a reflector and throws light on the sample holder through a filter. The filter may be either of absorption or interference type. The sample holder is a cuvette with parallel walls or may be a test tube. The light, after passing through the sample holder, falls on the surface or the photocell. The output of the cell is measured either on the light-spot galvanometer or microammeter.

The lamps must be energized from a highly stabilized dc source or by the output of a constant voltage transformer. To operate the instrument, the following steps are taken:

■ With the photocell darkened, the meter is adjusted mechanically to read zero.

■ The blank or a pure solvent or a reference solution is inserted in the path of the light beam and the incident light intensity

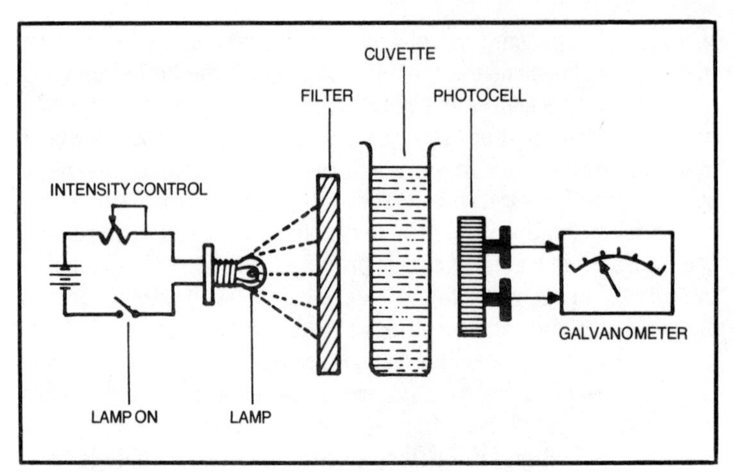

Fig. 2-17. Basic components of a filter photometer.

is regulated. This can be done, either by adjusting the rheostat in series with the lamp, rotating the photocell about an axis perpendicular to the light beam, or adjusting a diaphragm in the light beam. With the help of anyone of these adjustments, the meter reading is brought to 100 scale divisions.

■ Solutions of both standards and unknowns are inserted in place of the blank and the reading of the specimen relative to the blank is recorded. The meter scale is calibrated in linear transmittance units (0-100%).

Such types of instruments are easy to operate and are also inexpensive. Errors of 1 to 5% are quite common but are nevertheless acceptable in many applications.

Double-Beam Filter Photometer

In double-beam filter photometers, two photocells are normally employed. The two photocells are connected to two potentiometers, P_1 and P_2 in Fig. 2-18. Each of the two potentiometers is a low resistance and is wound linearly. Light from the source lamp is made to pass through the filter F and is then divided into two parts, one part passing through the solution in the cuvette before falling on the measuring photocell, and the other part passing directly on to the reference photocell. The galvanometer G receives opposing currents through it. The potentiometer P_1 is graduated in transmittance and absorbance.

The operation of the instrument is very simple. With the lamp off, the galvanometer zero is adjusted mechanically. The potentiometer P_1 is set to T = 1 or A = 0. Then with the lamp on, the blank solution is placed in the light path of the measuring cell. Potentiometer P_2 is adjusted until galvanometer G reads zero. The solution to be analyzed is then substituted for the blank and P_1 is adjusted until the current through the galvanometer is zero, the setting of P_2 remaining unchanged. The absorbance or transmittance can then be read directly on the scale of potentiometer P_1. This principle is employed in the electrophotometer manufactured by Fischer Scientific Company.

In two cell photometers, the errors resulting from the fluctuations of the lamp intensity are minimized. The scale of potentiometer P_1 (transmittance scale) can be made much larger in size than the scale of the meter in single cell instruments.

Different manufacturers adopt different means for restoring the balance of current through the galvanometer. In the Klett-Summerson photoelectric colorimeter, it is done by adjusting the intensity of the reference light beam by means of a diaphragm (Fig. 2-19). A Lumetron photoelectric colorimeter makes use of the rotation of the reference photocell about an axis perpendicular to the light beam through an angle of 90°, plus a series of fixed apertures for coarse adjustment. These adjustments remain unchanged while standards and unknowns are inserted. The potentiometer in series with the reference photocell is adjusted to obtain the scale reading.

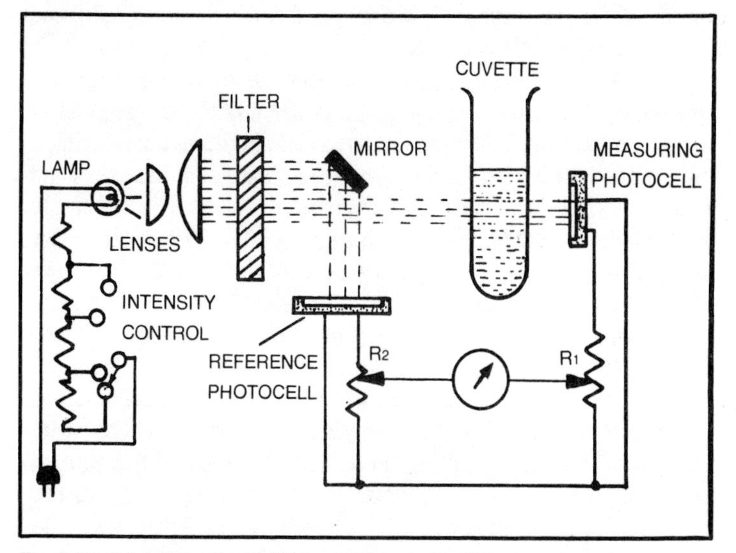

Fig. 2-18. Schematic of a double beam filter photometer.

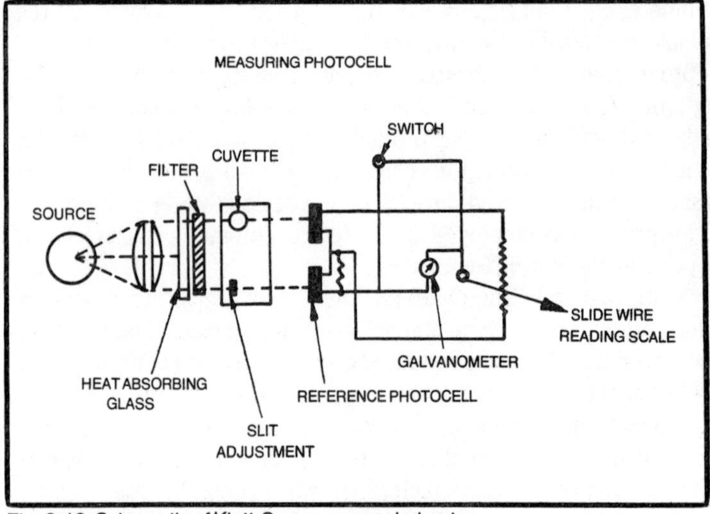

Fig. 2-19. Schematic of Klett-Summerson colorimeter.

Multichannel Photometer

An increasing number of chemical analyses is carried out in the laboratories of industry and hospitals and, in most of these, the final measurement is performed by a photometer. Obviously, it is possible to increase the capacity of the laboratory by using photometers which have a large measuring capacity. One of the limitations for rapid analyses is the speed at which the samples can be transferred in the light path.

In a multichannel photometer, instead of introducing one sample at a time into a single light path, a batch of samples is introduced and measurements carried out simultaneously using a multiplicity of fiberoptic light paths (Fig. 2-20) and detectors and then scanned electronically instead of mechanically. The 24 sample cuvettes are arranged in a rack in a three key eight matrix. The twenty-fifth channel serves as a reference beam and eliminates possible source and detector drifts. The time required to place the cuvette rack into the measuring position corresponds to the amount of time necessary to put one sample into a sample changer.

The light source is a 50 W tungsten-halogen lamp driven from a precisely controlled voltage source. The light is chopped by means of a mechanical rotating chopper. A lens focuses the light on the end of a bundle of fiber optic elements. The output of the detectors is amplified and displayed on a digital voltmeter. The whole operation is synchronized with digital logic circuits.

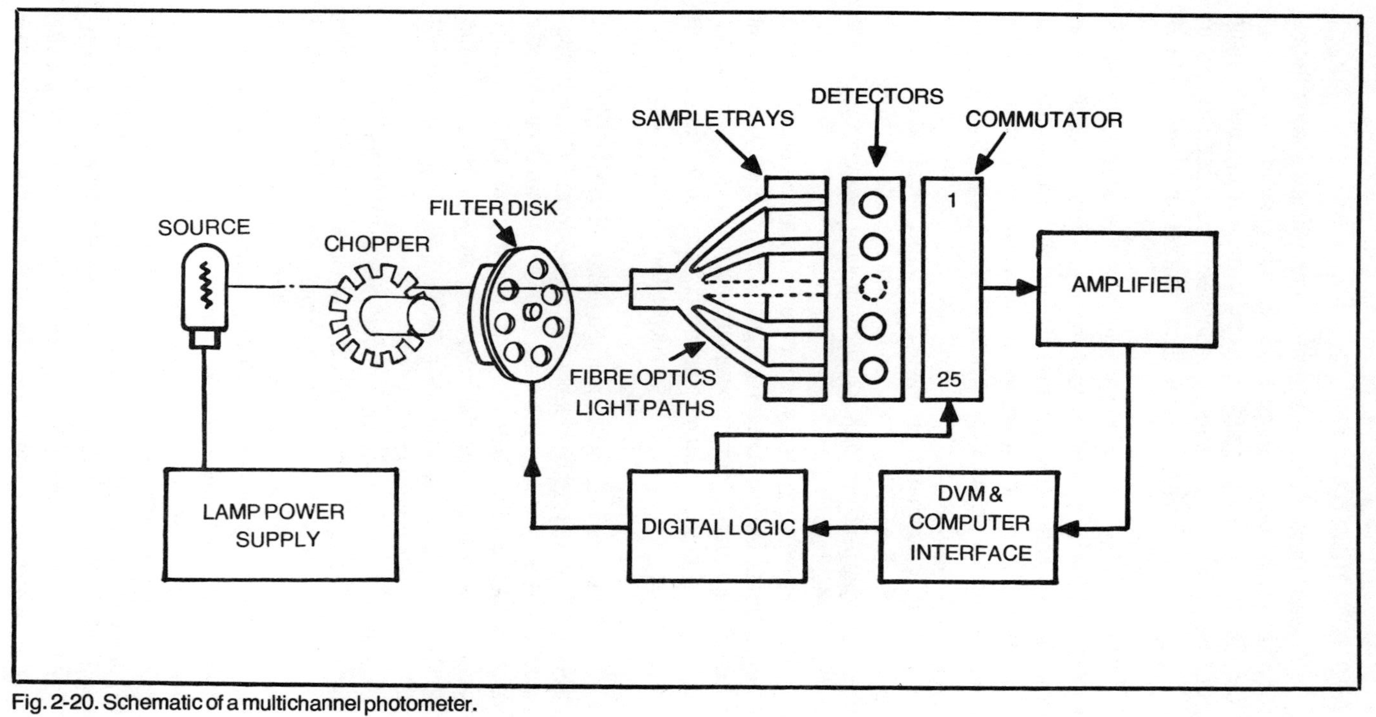

Fig. 2-20. Schematic of a multichannel photometer.

SPECTROPHOTOMETERS

A *spectrophotometer* is an instrument which isolates monochromatic radiation in a more efficient and versatile manner than color filters used in filter photometers. In these instruments, light from the source is made into a parallel beam and passed to a prism or diffraction grating where the light of different wavelengths is dispersed at different angles.

The amount of light reaching the detector of a spectrophotometer is generally much smaller (Fig. 2-21) than that available for a colorimeter because of the small spectral bandwidth. Therefore, a more sensitive detector is required. A photomultiplier or vacuum photocell is generally employed. The electrical signal from the photoelectric detector can be measured by using a sensitive microammeter. However, it is difficult and expensive to manufacture a meter of the required range and accuracy. To overcome this problem, either of the following two approaches are generally adopted.

The detector signal may be measured by means of an accurate potentiometric bridge. A reverse signal is controlled by a precision potentiometer until a sensitive galvanometer shows that it exactly balances the detector signal and no current flows through the galvanometer. This principle is adopted in Beckman Model DU single beam spectrophotometer.

The detector current is amplified electronically and displayed directly on an indicating meter or in digital form. These instruments have the advantage in speed of measurement. As in the case of colorimeters, the instrument is adjusted to give a 100% transmission reading with the reference sample in the path of the light beam. The sample is then moved into the beam and the percentage transmission is observed.

Modern commercial instruments are usually double beam, digital reading and/or recording instruments which can provide absorbance, concentration, percent transmission and differential absorbance readings. It is also possible to make reaction rate studies. They can be used to include specialized techniques, such as automatic sampling and batch sampling with the addition of certain accessories. The measurements can be made, generally, with light at wavelengths from 340 to 700 nanometers and from 190 to 700 nanometers with deutrium source. In the variable slit type of instruments, the slit can be made to vary from 0.05 mm to 2.0 mm. The wavelength accuracy is ± 0.5 nm. The recorders are usually single channel, strip-chart potentiometric recorders. They are

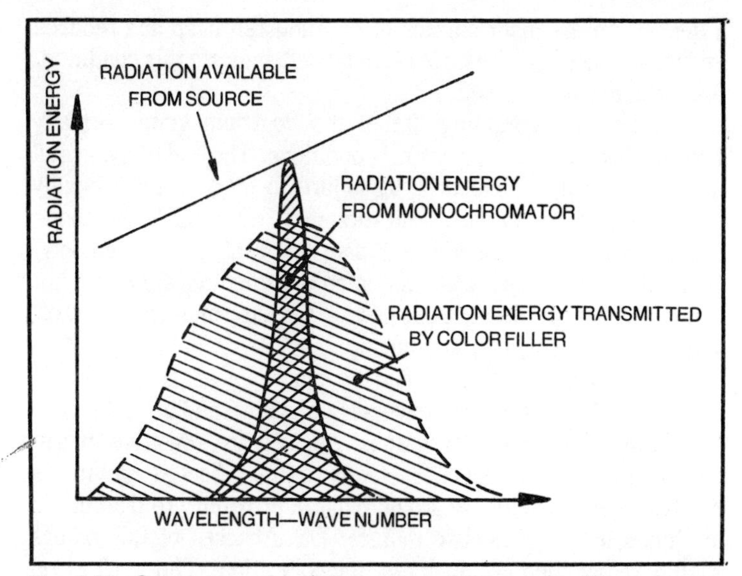

Fig. 2-21. Comparison of radiation energy from a color filter and monochromator.

calibrated from 0.1 A to 2.0 A or 10 to 200% T full scale. The recorder used with spectrophotometers has four wavelength scanning speeds (100, 50, 20 and 5 nanometers per minute) and seven chart speeds (10, 5, 2, 1, 0.5, 0.2 and 0.1 inch/minute). It has sensitivity of 100 millivolts per absorbance units or 100 millivolts per 100% T.

When scanning narrow wavelength range, it may be adequate to use a fixed slit width. This is usually kept at 0.8 mm. In adjustable slit width instruments, it should be so selected that the resultant spectral slit width is approximately one tenth of the observed bandwidth of the sample. If the absorption band is 25 nm wide at half of its height, the spectral slit width should be 2.5 nm. This means that the slit width set on the instrument should be 1.0 mm. This is calculated from the dispersion data as the actual dispersion in grating instruments is approximately 2.5 nm/mm slit width.

Spectrophotometers generally employ a 6 volt tungsten lamp which emits radiation in the wavelength region of visible light. Typically, it is 32 candlepower. These lamps should preferably be operated at a potential of 5.4 volts when its useful life is estimated at 1200 hours. The life is markedly decreased by an increase in the operating voltage. With time, the evaporation of tungsten produces a deposit on the inner surface of the tungsten lamp and reduces emission of energy. Dark areas on the bulb indicate this condition. It should then be replaced.

The useful operating life of the deutrium lamp normally exceeds 500 hours under normal conditions. The end of the useful life of this lamp is indicated by failure to start or by a rapidly decreasing energy output. Ionization may occur inside the anode rather than in a concentrated path in front of the window. Generally, this occurs when the lamp is turned on while still hot from previous operation. If this occurs, the lamp must be turned off and allowed to cool before restarting.

Spectrophotometers should be placed in an area which is reasonably free of dust and excessive moisture and not subject to significant temperature variations. As they are sensitive instruments, their performance is likely to be affected by strong electromagnetic fields, as would exist in proximity to diathermic machines or large electric motors. Disturbances of this nature should be avoided when determining location. The surface on which the instrument is to be placed must be stable and free from vibrations.

Single-Beam Null Type Spectrophotometers

A typical example of the single-beam optical null type spectrophotometer is that of Beckman Model DU. Although this is an old instrument, still a large number of these are in current use in various laboratories.

The schematic diagram of its optical system is shown in Fig. 2-22. Light from the source is focused on the condensing mirror. It is directed in a beam to the 45° slit-entrance mirror (C). The slit entrance mirror deflects the beam through the slit on to the collimating mirror. Light falling on the collimator mirror is rendered parallel and reflected to the prism where it undergoes refraction. The back surface of the prism is aluminized. The light reflected at the first surface is reflected back through the prism, undergoing further refraction as it emerges. This is called Littrow mounting and employs only one piece of quartz with the back surface of the prism metallized. Since the light passes back and forth through the same prism and lens, polarization effects are eliminated. Littrow mounting results in a compact instrument.

The desired wavelength of light is selected by rotating the prism mount. As the table on which the prism is mounted is slowly rotated by the wavelength drum, the prism provides a series of

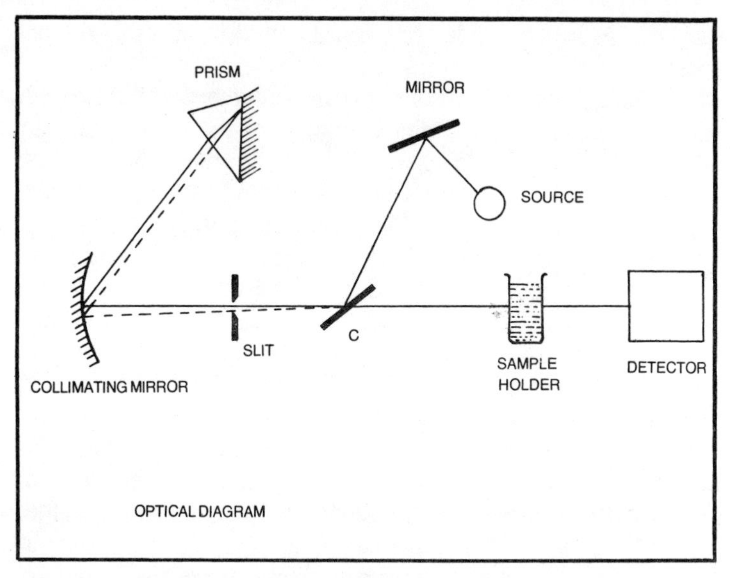

Fig. 2-22. Optical diagram of Beckman Model DU spectrophotometer.

images of the entrance slit at the exit slit as the spectrum is swept past. To achieve perfect correspondence of slit widths, the upper and lower parts of the same slits are generally used as entrance and exit slits.

The spectrum from the prism is directed back to the collimating mirror. The light passes through the sample and finally reaches the photodetector. The resulting current passing through the load resistor develops the voltage which is amplified in a dc amplifier using a high input impedance electronic circuit. The amplifier circuit is of a null type to provide absorbance or transmittance measurement. The electrometer plate current is indicated on a rugged 1 mA milliammeter.

Three potentiometric circuits are employed in order to counterbalance the photocurrent. Each of the potentiometer control circuit alters the grid potential of the electrometer tube to bring the needle of the milliammeter to zero. The first control is the dark current control which is used to offset the small phototube current when no radiation is falling on the detector. The second control is the sensitivity control which offsets the phototube current after passing through the reference or solution. The third control is the transmittance control which is used to adjust the position of the variable contact on the potentiometer slide wire to bring about null on the instrument when the sample is in the radiation path. The slide wire is calibrated in absorbance and transmittance units.

In such type of instruments, two phototubes are employed which are mounted on a sliding carriage. For the visible and near ultraviolet range (220-625 mμ), a blue sensitive tube is used. It has a cesium-antimony photosensitive surface and an insert of fused silica in the envelope. In the range 625-1000 mμ, a red sensitive phototube having cesium-oxide coated photocathode is utilized. When photomultipliers are used for increased sensitivity or for operation at smaller slit widths, they usually replace the blue sensitive tube. The slit width is often marked in millimeters to serve as a performance check and allow resetting a previous slit opening.

The load resistance in such type of circuits is very high (20 megohms to 200 megohms) so that the circuit is sensitive to extremely small currents of the order of 10^{-13} amps. Therefore, the load resistance and the detectors are housed in a compartment which is kept moisture free by putting silica gel crystals in a bag. In the presence of moisture, leakage current may develop and result

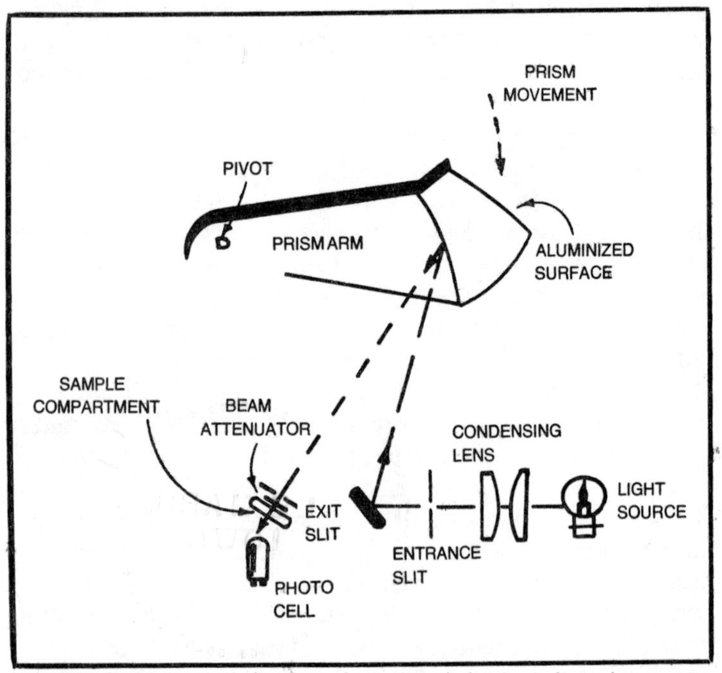

Fig. 2-23. Typical optical diagram of spectrocolorimeter using prism monochromator.

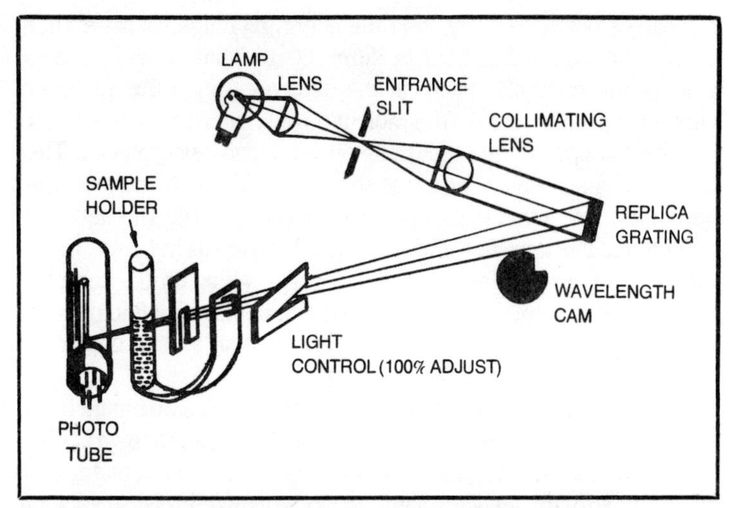

Fig. 2-24. Optical arrangement of Spectronic-20 spectrocolorimeter.

in drift of the null-balancing circuit. Silica gel crystals should be dried frequently and replaced to overcome this problem.

Direct Reading Spectrocolorimeters/Spectrophotometers

Direct reading instruments offer greater speed in operation and convenience. However, they have lower accuracy and precision than the null balance type instruments. Figure 2-23 shows the schematic of a direct reading type of prism absorptionmeter, Model SP 1400 of Unicam. The prism, in this instrument, is of the Fery type with an aluminized rear surface. Wavelength selection is made by rotating the prism about a pivot. The entrance and exit slits are each of 0.25 mm. The detector is a standard vacuum photocell. The output is developed across a 250 M ohms resistor and applied to a single cathode follower dc amplifier. The output from the photocell is applied to a sensitive spot galvanometer with taut suspension and mechanical zero adjustment. Linear and logarithmic scales are provided for direct readings in percentage transmission or optical density.

Bausch and Lomb Spectronic-20 is a direct-reading grating spectrophotometer/spectrocolorimeter. In this instrument, normal range of operation is 350 to 650 mμ and can be extended up to 900 mμ by the use of a red sensitive phototube. The monochromator comprises a reflection grating, lenses and a pair of fixed slits (Fig. 2-24). As the grating produces a dispersion that is independent of wavelength, a constant bandwidth of the order of 20 mμ can be obtained throughout the operating range. In the earlier versions of this instrument, a differential amplifier was used to amplify the photocurrent from the detector. Since the amplifier current is proportional to the radiant power, the scale of the meter can be calibrated to read transmittance and absorbance. The amplifier is so constructed that the current through the meter is zero under no signal (dark current) conditions. With the detector input fed to the amplifier, the unbalanced current as indicated in the meter is proportional to the radiant energy falling on the detector tube. The design of the amplifier is such that electrical fluctuations get canceled out. The wavelength scale is linear and is coupled to the grating with a sine-bar drive.

The *SPEKOL* is a single beam direct reading instrument from Carl-Zeiss Jena. The optical path of the instrument is shown in Fig. 2-25. The instrument makes use of a grating monochromator and can be used with tungsten lamp or arc stabilized mercury vapor lamp. A color filter is used to block the higher ordered wavelengths reflected from the grating.

Via the rigidly built-in condenser and the mirror, an image of the radiation source is produced on the entrance slit. The light converted into parallel rays by a collimator objective strikes the diffraction grating and is spectrally dispersed. By the achromat the diffracted light is focused in the plane of the exit slit. Monochromatic radiation separated from the spectrum through the slit passes the measurement sample and strikes the radiation detector. The photocurrent produced is amplified in the amplifier and fed to the indicating instrument. The wavelength is set by turning the grating by means of the drum provided with 1 mμ divisions in the range from 330 mμ to 850 mμ.

Signal Processing in Direct Reading Spectrophotometers

In single beam direct reading instruments, use is made of operational amplifiers for handling signals from the detectors which could be a vacuum phototube, photodiode, photomultiplier or photocell.

Figure 2-26 shows a typical circuit diagram for amplification and measurement of current from the phototube. The phototube anode is given 90 volts from a stabilized power supply and its cathode is connected to ground through a high resistance R_1. Operational amplifier A_1 is of very high input impedance, low drift and low noise. The op amp offers a very high input impedance as the input is connected to the non-inverting input. The inverting

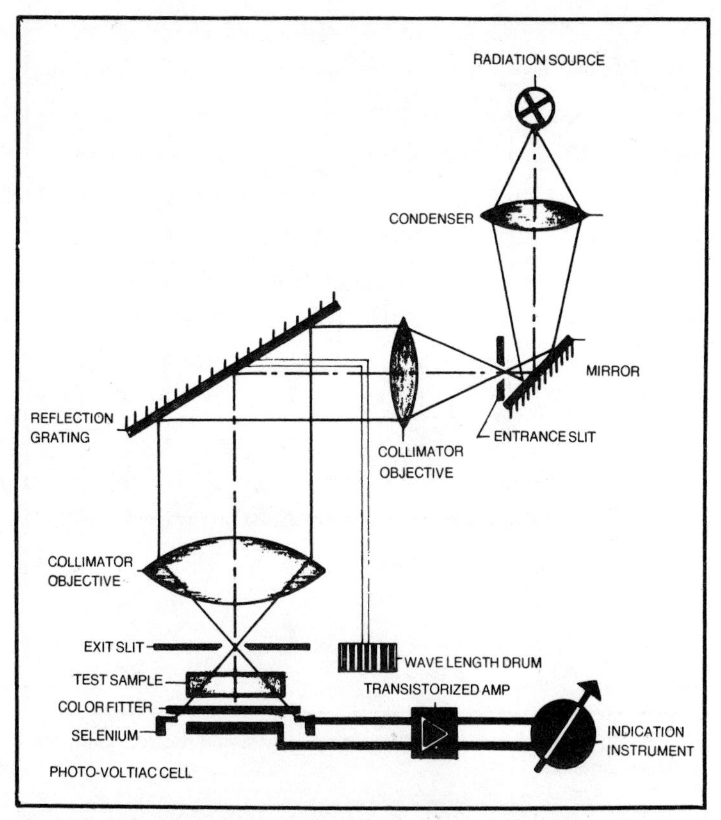

Fig. 2-25. Optical arrangement of SPEKOL (courtesy of Carl Zeiss Jena).

input terminal is connected to a variable supply voltage (± 15 V). This control is provided on the front panel to set zero initially when light is blocked from falling on the phototube. This control not only balances the dark current, but also nullifies offset voltage of the op amp.

The feedback resistance uses a T-resistor network, instead of a single resistor. This arrangement gives a very high effective feedback resistance and, at the same time, makes possible the use of resistors of lower values, which at the required precision of 0.1% are substantially cheaper and stable. The output of the amplifier gives linear readings on the transmittance scale. If absorbance is to be displayed, the output of the transmittance stage is given to a logarithmic converter whose output is equal to the minus log of transmittance.

$$A = -\log_{10} T$$

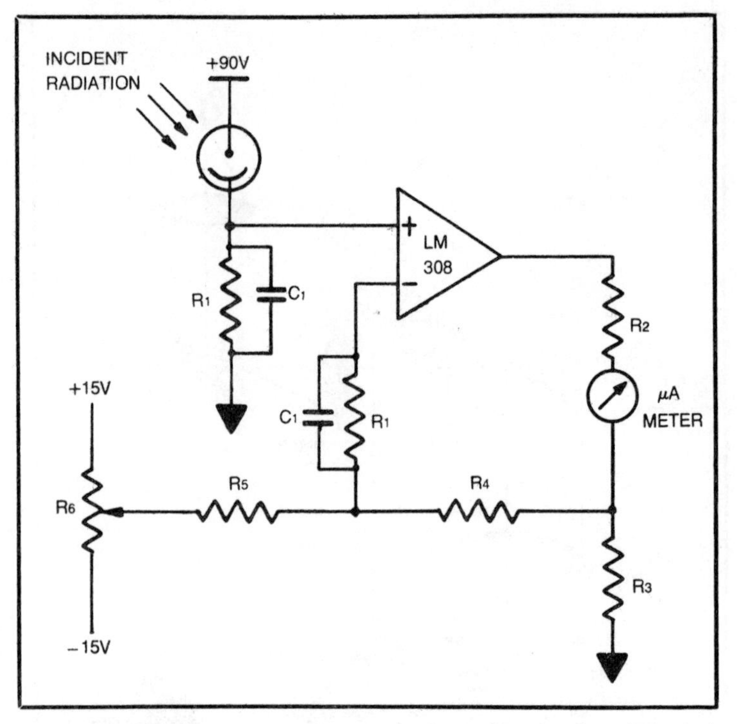

Fig. 2-26. Typical circuit used with direct reading spectrocolorimeter/ spectrophotometer.

The logarithmic modules are available in encapsulated form and can be directly used.

Currents from photomultipliers are usually very small. Therefore, considerable amplification is necessary before the current can be suitably recorded. This is done by using an operational amplifier having a field-effect transistor in the input stage and using a high value of the feedback resistance R_f (Fig. 2-27). As the input to the amplifier is connected by a shielded cable, there exists a stray or parasitic capacitance at the input terminals. It necessitates the use of capacitance C_f in the feedback to ensure amplifier stability. The capacitance also makes the bandwidth narrower, which significantly improves signal-to-noise ratio. The non-inverting amplifier input terminal is connected to a suitable voltage to compensate for the dark current.

Double-Beam Ratio-Recording Spectrophotometers

Single beam spectrophotometers have the main disadvantage that the instrument settings have to be adjusted to give a 100%

reading with the reference in the beam before the sample may be examined. This drawback is overcome by using a double beam instrument, wherein the arrangement is such that the radiation beam is shifted automatically to pass alternately through the sample and reference cuvettes. The cuvettes themselves are not shifted and remain in their fixed position. There are several ways by which this can be achieved. Most commonly, a single rotating sector mirror is used. Monochromators used in these instruments are similar to those used in single beam instruments.

Double beam instruments are generally of the recording type. This facility makes it easy to have rapid and accurate reproduction of spectrograms. The instrument compares, automatically, sample beam energy with reference beam energy. The ratio of the two would be the transmittance of the sample. This procedure is followed over a sequence of wavelengths. A graph is plotted with transmittance (absorbance) as ordinate and wavelength as abscissa, giving absorption spectrum of the sample under analysis.

In the prism type instruments, the Littrow monochromator disperses the radiation of the deutrium or tungsten lamp. The monochromatic light produced in this manner is modulated at 400

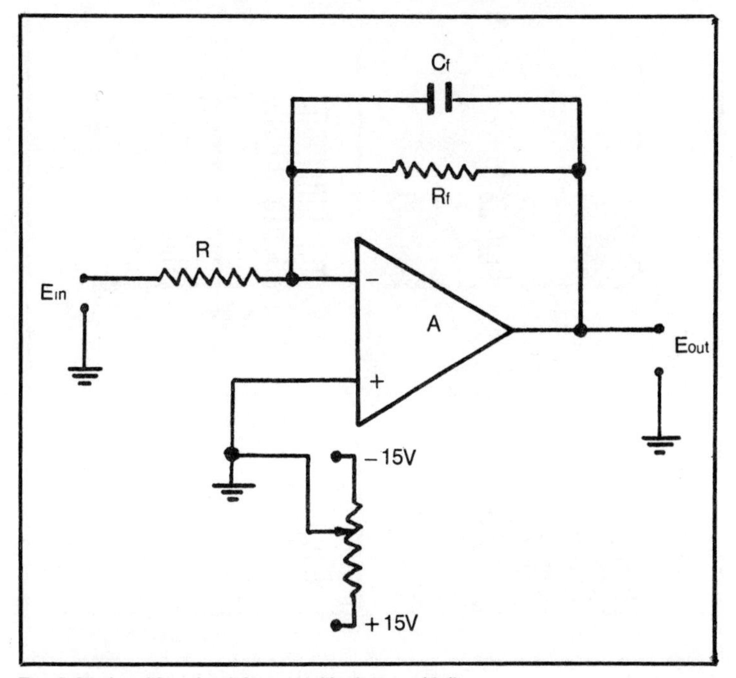

Fig. 2-27. Amplifier circuit for use with photomultipliers.

Hz and alternately led, via a rotary mirror through the measuring and reference cells, into a photomultiplier with multi-alkali cathode. The photomultiplier produces a 400 Hz ac voltage, the amplitude of which periodically oscillates between the measuring and reference value. After amplification, the signal is rectified in a phase-sensitive detector so that two dc voltages are produced that are proportional to the measuring and reference radiation. The quotient of these voltages is measured with an automatically balancing potentiometer and transferred to the recorder. The linear or logarithmical potentiometer permits recording of trans-

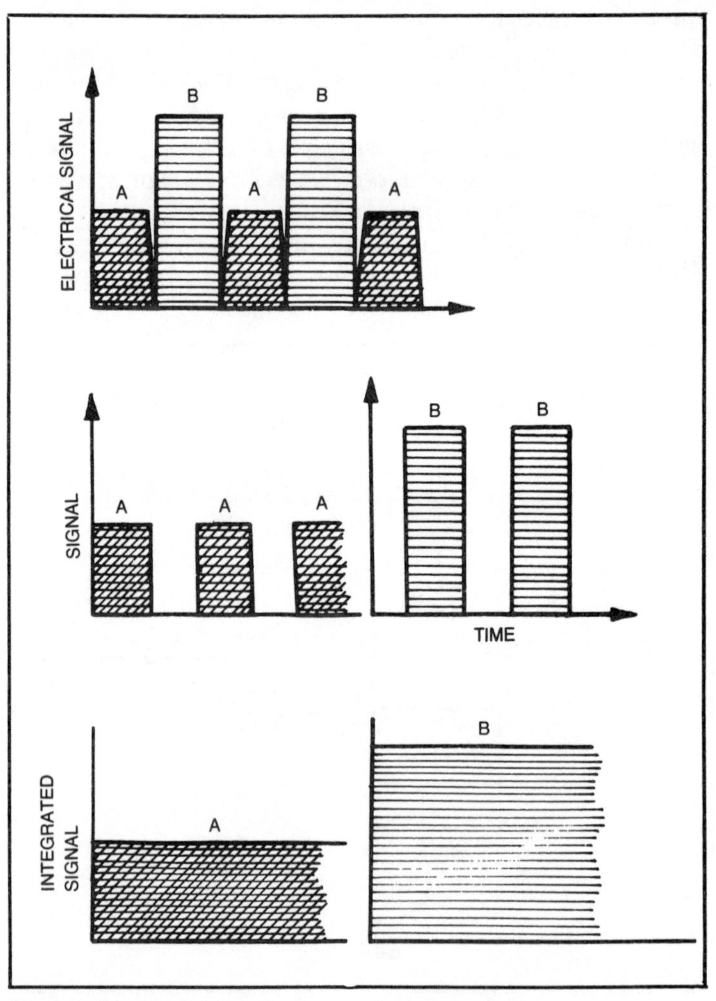

Fig. 2-28. Principle of ratio-recording spectrophotometric system.

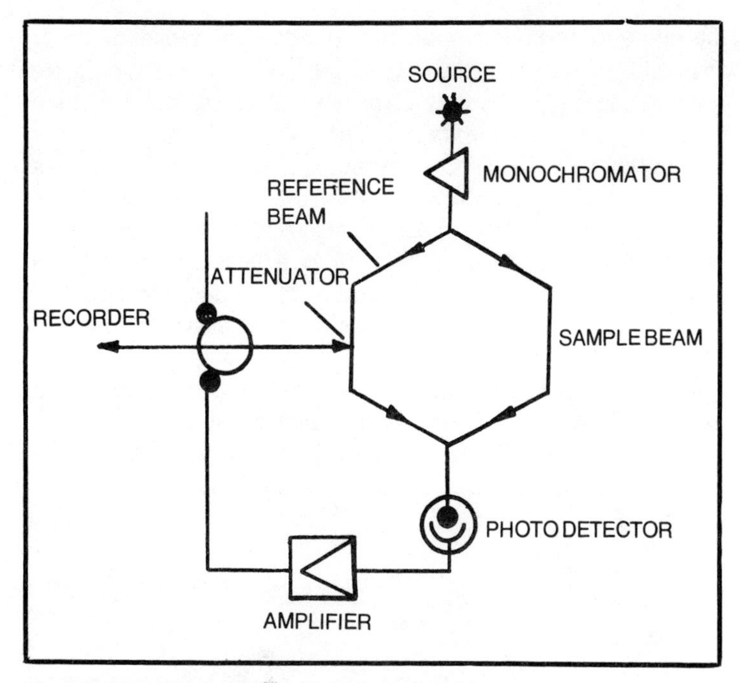

Fig. 2-29. Principle of optical-null spectrophotometric system.

mission or extinction. A slit program controls the monochromator slits so that there is approximately the same signal at the photomultiplier.

Figure 2-28 shows the signals in a ratio-recording double-beam spectrophotometer. The signals are in the form of pulses which are directly proportional to intensities of radiation passing through the sample and reference. These two signals are resolved electronically into two dc voltages corresponding to the sample and reference instruments and the latter is used as the standardizing potential on a potentiometric recorder.

Double beam instruments are also constructed on the optical null photometer system (Fig. 2-29). In these instruments, the alternating signal from the reference and sample is not separated into its components but is used directly to drive a servometer. The motor drives an optical attenuator or wedge into or out of the reference beam of the spectrophotometer, until it has the same intensity as the sample beam. The electrical signals produced at the detector from reference and sample sides are equal, and there is no signal to drive the servomotor and it stops. The optical attenuator arrangement is such that the distance it is driven into the reference beam is proportional to either the transmittance or absorbance. Therefore, the distance it has moved when the signals become equal and the motor stops is proportional to the relative intensities of the sample and reference beams. The absorbance or transmittance can be recorded by mechanically coupling the pen of a recorder to the attenuator.

Figure 2-30 shows optical arrangement of the double beam spectrophotometer due to Varian. Either a tungsten or a deutrium lamp can be switched in the system to cover an operating range from 200 nm to 700 nm, to cover the ultraviolet and visible regions. The monochromator consists of a diffraction grating and two narrow slits, one of which serves as an entrance and the other as an exit. It is possible to set the spectral bandwidth at 2, 4, 8 or 16 nm. The larger bandwidths are used for increased light output, the

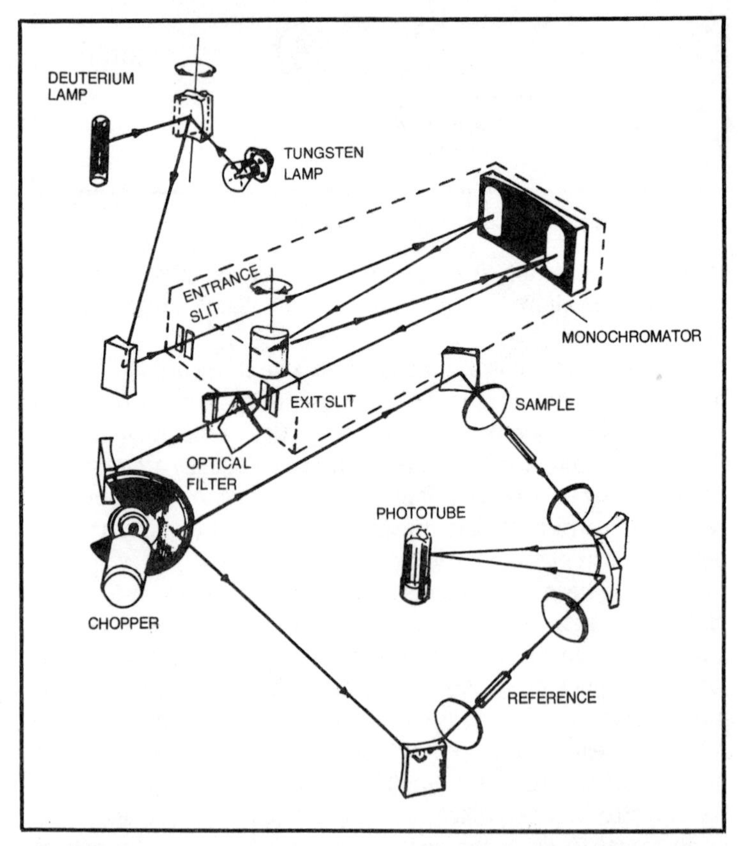

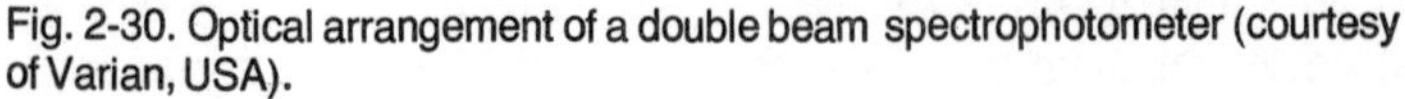
Fig. 2-30. Optical arrangement of a double beam spectrophotometer (courtesy of Varian, USA).

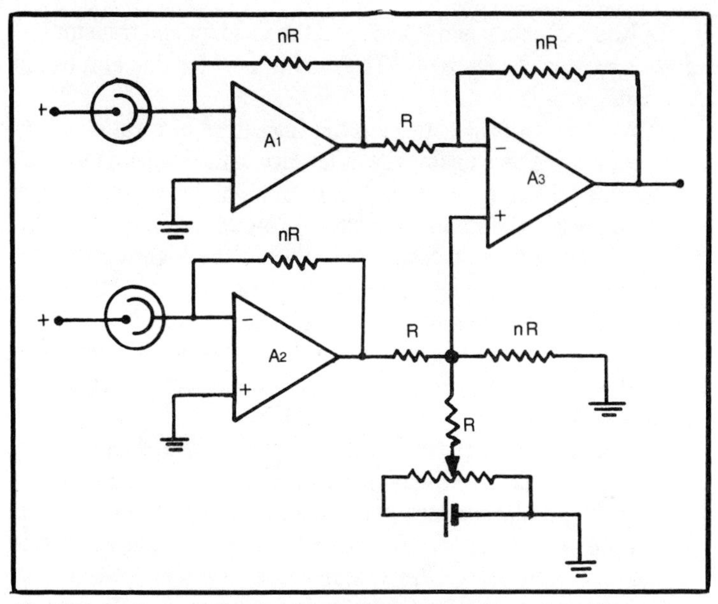

Fig. 2-31. Basic measuring circuit in double beam spectrophotometers.

smaller where spectral resolution is of importance. A four segment filter automatically performs order sorting and rejects stray light from the monochromator, holding it to 0.1% at 340 nm, giving low noise over the entire wavelength range.

The instrument chopper is so synchronized that light passes alternately through the sample and the reference cells. In this arrangement, the entire light passes through a given cell when the transmission of that cell is being measured. This offers a distinct advantage over beam-splitter arrangements where only 50% of the available light passes through each cell during measurement.

Figure 2-31 shows the arrangement employed in measurement of transmittance in a double beam instrument. The measuring process consists basically of comparing the photocurrents from two photocells or phototubes. The input circuits are current followers and their output signals are compared in a differential amplifier.

If it is required to measure absorbance A, the output signals from a current follower are given to logarithmic amplifiers. The logarithms of the signals are substracted in a differential amplifier. The output of the amplifier is given by

$$E_0 = K. \log \frac{I_0}{I} = KA$$

where K is a constant and I_0 and I are the incident and transmitted light intensities respectively. The circuit used for this purpose is shown in Fig. 2-32.

This circuit enables to take the logarithm of a ratio of two voltages or currents equaling the difference of the logarithms of the input signals. The outputs of the logarithmic amplifiers are brought to the inputs of a differential amplifier. This circuit is used mainly in spectrophotometric measurements when the logarithm of the ratio of the radiant power of the incident radiation to that of the radiation after passage through the absorbing layer is determined. In the circuits described above, both transistors should have identical properties and matched temperature dependent resistors should be employed.

Model 24-25 from Beckman Instruments are double beam instruments which have the optical system as shown in Fig. 2-33. Energy of the appropriate wavelengths is produced by the appropriate source lamp. This energy is converted to monochromatic light by using filter-grating optical system. The grating has 1200 lines per mm and is blazed at 250 nm. The filters are necessary to eliminate unwanted orders from the grating. Six different filters cover the wavelength regions from 300 nm to 900 nm.

The filter wheel is driven by a dc motor which is synchronized with the wavelength cam. As the wavelength cam moves, it causes

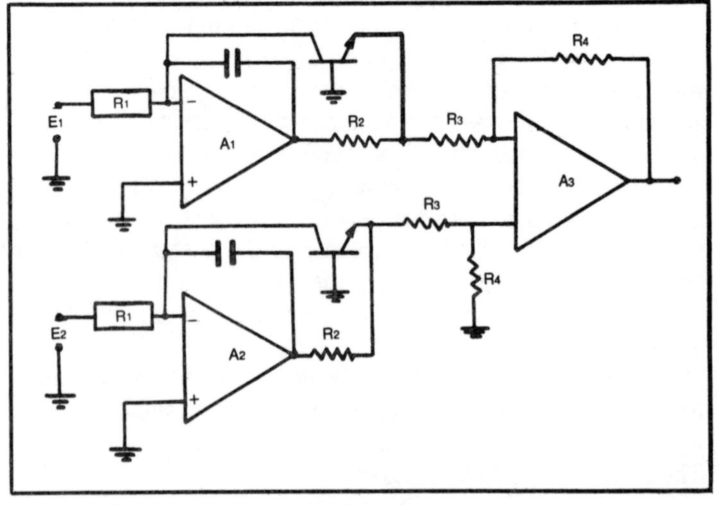

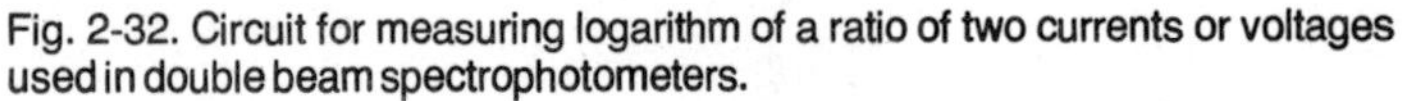
Fig. 2-32. Circuit for measuring logarithm of a ratio of two currents or voltages used in double beam spectrophotometers.

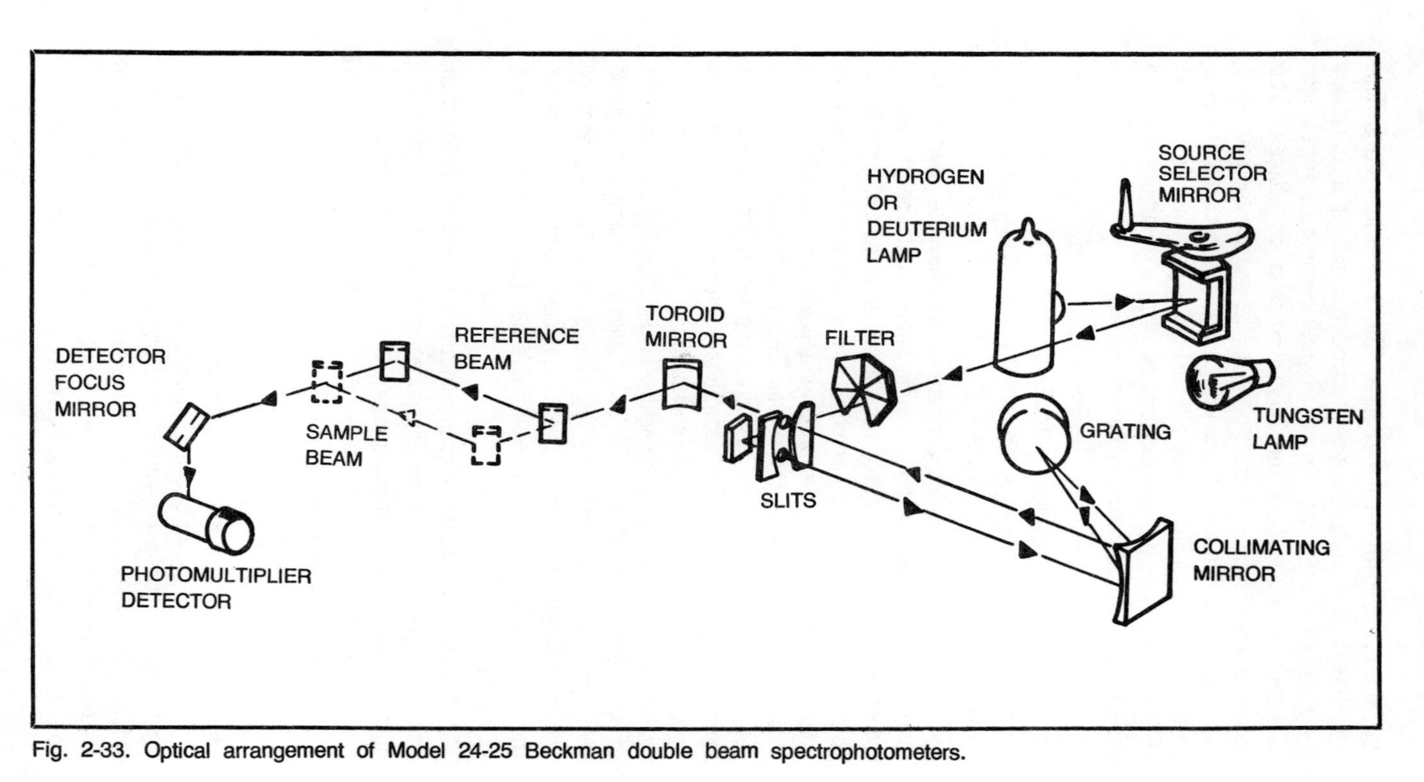

Fig. 2-33. Optical arrangement of Model 24-25 Beckman double beam spectrophotometers.

the filter motor to drive till the correct filter comes in position. The wavelength cam drives the wavelength arm which, in turn, causes the grating to pivot on its own axis, thereby causing the wavelength of light coming out of the monochromator to change. The monochromatic light is then directed to the sample and reference via a vibrating mirror bridge which vibrates horizontally at a certain frequency. This bridge allows light to pass into the sample and reference cell holders alternately, with a frequency equal to the displacement frequency of the bridge.

The vibrating bridge is controlled by the bridge drive circuitry. The reference and sample pulse train is then passed to the photomultiplier tube which converts the monochromatic light pulses to current pulses.

Figure 2-34 is a block diagram of the electronic part of the instrument. As the vibrating bridge chops light energy coming from the monochromator at 35 Hz, the output of the PMT will be an ac signal. This signal is passed to the preamplifier where it is amplified. The input of the preamplifier is a FET which offers a high input impedance to the signal. The amplified signal is then given to a demodulator which separates reference pulses from sample pulses and converts sample and reference pulses to a dc potential. However, the demodulation process requires synchronization so that when the bridge is directing light through the sample path, the electronics is demodulating the sample pulse and the same will hold true for the reference pulse. To achieve synchronization, the same signal that is used to drive the coil for the vibrating bridge is tapped off and used as the input to the demodulator.

The dc output of the reference side is also fed back to the high voltage power supply. If the reference signal should decrease, the high voltage would increase, thus restoring the reference channel to a constant potential. This would ensure constant energy through the reference channel during the scan.

For making absorbance measurements, the dc potential from the sample and hold amplifier is passed to the log converter. Differential absorbance can be measured between -0.3 A to $+0.7$ A by switching in bucking potential at the output stage. The analog dc potential is converted into binary coded decimal and this information is then displayed as an absorbance or concentration value on the nixie tubes. The output of the log converter is applied to a divider network, which in turn drives an external recorder.

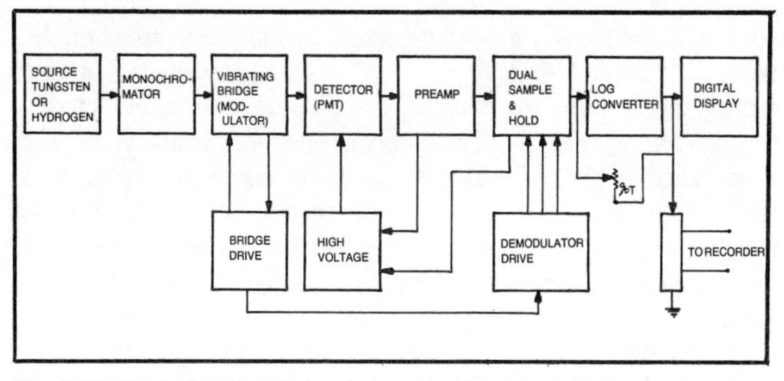

Fig. 2-34. Block diagram of electronic part of Model 24-25 Beckman double beam spectrophotometer.

Spectrophotometric technique has been used to a great extent for the investigation of biopolymer structures. Measurements within the ultraviolet spectral region are of special interest on account of the selective absorptions of the nucleic acids and proteins in this range. Wetzel et. al. (1969) describe a microspectrophotometer which permits the measurement of absorption spectra of objects of approximately 1 μm diameter, in the wavelength range of 240 to 360 nm.

The instrument is double beam recording type with automatic null balancing system with polarizers for attenuation. The source used is a xenon high pressure lamp having high radiance down to 230 nm. The monochromator is equipped with a Suprasil prism. The electronic circuit is similar to other double beam type spectrophotometers.

Dual Wavelength Spectrophotometer

The dual wavelength spectrophotometer permits the recording of absorbance changes in the same sample to be made at two different wavelengths alternately and virtually simultaneously. This function is performed by means of a control and reversible motor, which automatically adjusts the monochromator wavelength control to alternate between the two wavelengths selected. Following completion of the measuring cycle of the cuvette positioned at the first wavelength, the direction of the drive motor is reversed and the monochromator adjusted to the second wavelength. The two absorbance measurements are registered on the chart almost simultaneously.

There are three modes of operation for dual wavelength spectrophotometers which all depend upon chopping the light source in such a way as to time share two signal sources on the detector output. First is the dual wavelength mode. In this arrangement, two wavelengths of light are alternately passed through a single sample. The difference between transmittance or absorbance at the two wavelengths is measured as a function of time. This mode is used primarily to monitor the kinetics of reactions in the sample. In the second method, the two wavelengths of light are alternately passed through a single sample. In this case, one wavelength is scanned over some small range while the other wavelength is held fixed at some reference point. The third mode is called the split-beam mode. Here a single wavelength is alternately passed through two separate samples. The wavelength is scanned over any region and a transmittance or absorbance difference between spectra of the two samples is the desired output.

Figure 2-35 illustrates the optical system of a dual wavelength spectrophotometer. Light from the source is focused on the entrance slit of duochromator by a quartz condensing system L_1 and L_2. The lens L_3 causes an image of the mask in back of L_2 to be projected on the gratings G_1 and G_2. This mask has two adjacent rectangular windows. Light going through one window illuminates grating G_1 and light going through the other illuminates grating G_2. The rotating shutter blade mounted on the motor allows alternate opening and closing of the two windows in 60 Hz sequence, and the two monochromatic beams leaving the exit slit S_2 are time-shared and bear a fixed frequency and phase relationship with the driving line voltage.

In the split beam mode of operation, since both beams must be of the same wavelength, the gratings are driven angularly in synchronism by the scanning motor. This is achieved by setting the reference grating at any wavelength below the region to be scanned, for example 300 mμ. When this is done, the measure grating drives the reference grating at equal angle in a captive fashion. The wavelength during motor scan is continuously indicated by the measure side. The voltage developed across a potentiometer which is coupled to the lead screw that determines the grating position is used to drive the X-axis of the X-Y recorder. Upon leaving the duochromator, both beams are merged through reference and measure cuvettes by mirrors M_4 and M_5.

In the dual wavelength mode, each of the gratings are set by dialing the appropriate wavelengths by means of the two knobs on the panel of the instrument. Thus, radiant energy of these selected

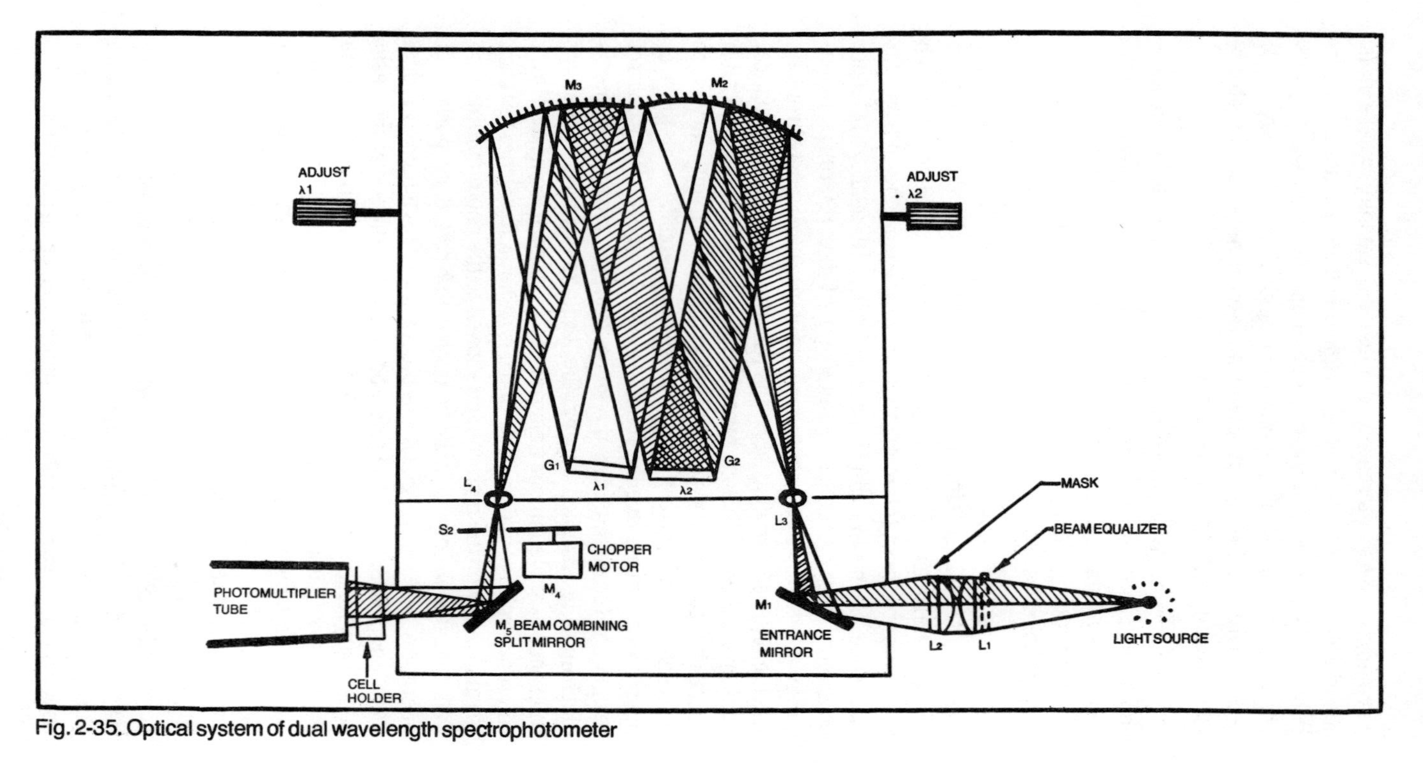

Fig. 2-35. Optical system of dual wavelength spectrophotometer

wavelengths pass alternately through the sample, where it is absorbed by or transmitted through the material under study. When a difference of transmittance along the two optical paths occurs, there is an alternating error signal which is amplified, demodulated and read out as the difference between the absorbance readings at the two wavelengths.

Monitoring at alternate wavelengths can result in considerable time savings on lengthy reactions. In additions, the comparison of absorbance measurements at two different wavelengths can result in more effective analysis of the effluent from chromatographic columns.

Figure 2-36 shows the schematics of the electronic system. It consists of two basic circuits, the *measure circuit* and the *reference circuit.* The measure circuit provides a signal to the recorder which is the difference of the intensities of the measure beam from the reference beam in optical density units. In dual wavelength operation, the difference in intensity of the reference monochromatic wavelength beam from the measure monochromatic wavelength beam is plotted on the recorder in terms of an optical density difference. In split beam operation, a difference or a ratio spectrum can be obtained. The reference circuit serves to internally standardize the instrument electronically at the frequency of the mains supply. It does this by dynode feedback control technique. In this method, the signal from the photomultiplier is electronically sampled during the time the reference cell or reference wavelength is being seen. It is then compared to a reference voltage and an error signal is generated which is fed back to raise or lower the high voltage applied to the photomultiplier so that the signal coming from it during the reference portion of operation is exactly equal to the reference voltage. In dual wavelength operation, the reference circuit assures that the basic sensitivity of the instrument does not change with time, since there is continuous re-standardization. This is important because changes in lamp intensity and photomultiplier instabilities would cause the instrument to drift. The signals are processed in a small analog computer consisting of operational amplifiers and given to the recorder.

Wade (1975) describes a detailed signal processing circuitry for use with dual wavelength spectrophotometers. Honkawa (1976) describes the use of a double beam dual wavelength spectrophotometer to analyze mutually interfering mixtures.

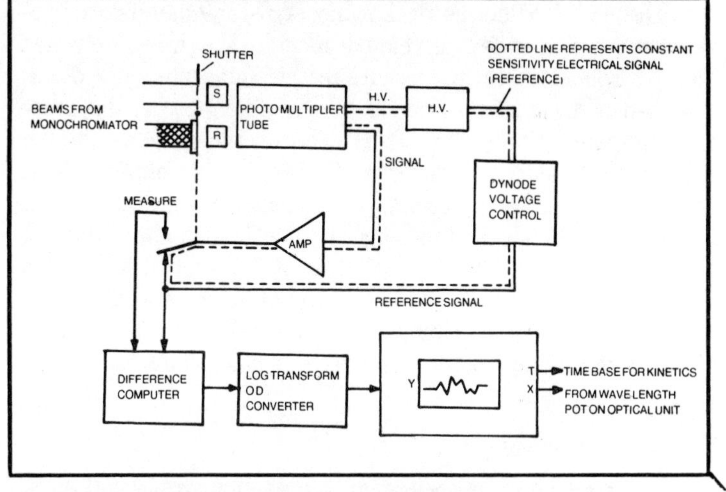

Fig. 2-36. Signal processor for dual wavelength spectrophotometer.

The Derivative Technique

The absorption spectrum of solids can consist of a large number of overlapping absorption bands. In order to isolate bands, it is necessary to use a system having high spectral resolution. However, in most of the cases, it is only possible to resolve the first few absorption bands and the higher quantum number bands appear, as points of inflection on the sloping recorder trace of transmitted light intensity due to overlap. In such cases, $dt/d\lambda$ can give a clearer indication of small spectral details.

The derivative technique as applied to grating spectrographs was described by Balslev (1966). Evans and Thompson (1969) describe a derivative attachment for a recording prism spectrophotometer which gives a direct measurement of the first or second wavelength derivative of the light intensity transmitted by a sample. The arrangement essentially employs a torsional scanner which consists of a small mirror mounted to one prong of a tuning fork which can be set into torsional oscillations at its resonant frequency (Fig. 2-37). The amplitude of oscillation can be varied by use of an electronic drive circuit which provides a pure sine wave reference signal. The oscillating-mirror moves a selected portion of the spectrum sinusoidally across the photomultiplier entrance slit. The original paper shows mathematically that the signal transferred to the recorder would be proportional to either the first or second wavelength derivative of transmittance. Calibration method is also illustrated.

The ACTA M-series Beckman spectrophotometers are provided with a plug-in derivative mode module. This module enables the spectrophotometer to measure the instantaneous slope of the absorbance curve at any wavelength. In other words, the spectrophotometer can measure $dA/d\lambda$. The module utilizes a passive resistor-capacitor electrical network to create the value $dA/d\lambda$ at each wavelength for the corresponding sample absorption curve. Due to a slight lag in the signal response with the electrical network, an absolute value of $dA/d\lambda$ is obtained only if scan speed is very slow. The derivative curve can be used as a qualitative analysis tool to ascertain impurities whose absorption peaks occur near or under the absorption peaks of the sample of interest.

Digital Spectroscopy

High resolution measurements, operating with small slits, typically operate with small flux. The method usually adopted is to chop the signal and gate two counters from the chopper signal. The first counter measures signal plus noise in the light period of the chopper and the second measures the noise during the dark period of the chopper. After preset interval, the difference of the two would constitute the signal. Based on this technique, Howard et. al. (1975) describe a method for scanning an absorption spectrum under digital control, particularly suited for very low flux levels. The scan is divided into small integral steps, digitally controlled, such that each step represents a constant number of input photons. High resolution spectra are obtained with a single beam at constant signal-to-noise ratio without any ratio devices. The necessity of measuring ratios is obviated because a constant reference signal is employed. A stepping motor is used to advance the chart paper in unison with the wavelength drive of the spectrometer. Hence, by abandoning the time base of the scan, one achieves spectra at constant incident intensity not requiring normalization. The output of the instrument is available in the digital form and can be directly transferred to a computer for analysis.

Calibration

Wavelength calibration of a spectrophotometer can be checked by using a *holmium oxide* filter as a wavelength standard. Holmium oxide glass has a number of sharp absorption bands which occur at precisely known wavelengths in the visible and ultraviolet

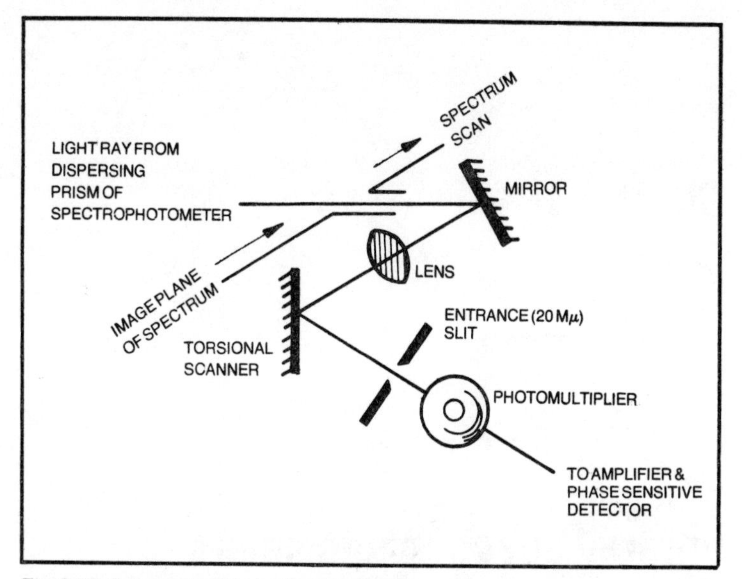

Fig. 2-37. Schematic of derivative attachment.

regions of the spectrum. Holmium oxide filter wavelength peaks are given below:

Ultraviolet range with deutrium lamp

279.3
287.6

Visible range with tungsten lamp

360.8
418.5
453.4
536.4
637.5

In double beam spectrophotometers, zero is adjusted with sample and reference beam. Then a holmium oxide filter is placed in a sample beam. The wavelength control is manually scanned through each wavelength until the absorption peak is found, always approaching each point from the longer to the shorter wavelength. The spectrum is then recorded. The wavelength calibration can also be checked in the visible region by plotting the absorption spectrum of a *didymium* glass which has been, in turn, calibrated at National Bureau of Standards, USA.

Chapter 3

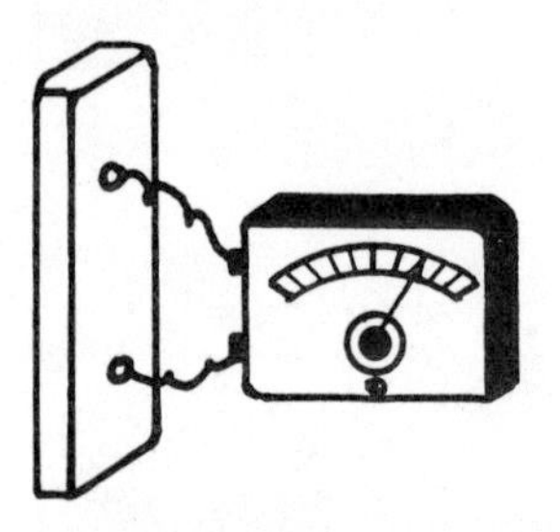

Infrared Spectrophotometers

The infrared spectrophotometer has become almost indispensable in the chemistry laboratory as it is ideally suited for carrying out qualitative and quantitative analysis, particularly of organic compounds. Its use in the applications to inorganic compounds is limited because of the strong absorption of infrared radiation by water. This constitutes a serious limitation in practical applications, since it necessitates the study of inorganic materials in the solid state.

Infrared region extends from 0.8 to 200 μ in the electromagnetic spectrum. However, most of the commercial instruments are available in the region from 0.8 to 50 μ. The position of absorption bands in the infrared spectrum is described both in wavelength as well as in wave numbers. The wavelength λ is generally measured in microns (μ). The wave number (ν) is the number of wavelengths per centimeter and is given by

$$\nu \text{ (in cm}^{-1}) = \frac{10^4}{\lambda \text{ (in } \mu)}$$

Infrared spectrophotometers using prisms produce spectra which are linear with wavelength, whereas instruments fitted with gratings generally deliver spectra linear with wave number. However, the results are preferably reported in wave numbers in either case, since these are proportional to molecular properties like frequency and energy, whereas the wavelength is a property of the radiation only. Sample transmittance is usually presented linearly on the vertical axis in an IR spectrum and absorption bands are generally presented pointing downwards (Fig. 3-1). Transmittance, in this case, is defined as the radiant power of the radiation which is incident on the sample, divided by the radiant power transmitted by the sample.

The energy acquired by a molecule can be utilized in three ways, namely *electronic excitation, vibrational change,* and *rotational change*; and the various molecular energy transformations that may take place are quantized. The amount of energy necessary to cause the various types of transitions generally correspond to definite regions of the electromagnetic spectrum. For a molecule, the average energy involved in electronic excitation is 5 eV. For molecules in a particular electronic state, the average energy involved in a vibrational excitation is 0.1 eV and a rotational excitation involves about 0.005 eV.

By definition

$$1\,\mathrm{eV} = 1.602 \times 10^{-19}\,\mathrm{J}$$
$$h = 6.626 \times 10^{-34}\,\mathrm{Js}$$
$$\text{and } E = h\nu$$

(1) 5eV corresponds to $\nu = 1 \cdot 2.10^{15}$ Hz
so $\nu = 40{,}000\ \mathrm{cm}^{-1}$
or $\lambda = 2500$ A°

(2) 0.1 eV corresponds to $\nu = 2 \cdot 4.10^{13}$ Hz
so $\nu = 833\ \mathrm{cm}^{-1}$
or $\lambda = 12\ \mu\mathrm{m}$

(3) 0.005 eV corresponds to $\nu = 1 \cdot 2.10^{12}$ Hz
so $\nu = 40\ \mathrm{cm}^{-1}$
or $\lambda = 250\ \mu\mathrm{m}$

This shows that light absorption at 2500 A° (ultravoilet range) produces electronic change.

Light absorption at 12 μm (infrared) produces vibrational change. Light absorption at 250 μm (far infrared) produces rotational change.

In infrared spectroscopy, we are interested mainly in the vibrations and rotations induced in a molecule by absorption of radiation. The infrared spectrum based on these absorption properties provides a powerful tool for the study of molecular structures and identification. Chemical identification is based on the empirical correlations of vibrating groups with specific absorption bands and the quantitative estimations are dependent on the intensity measurements.

Normally, the data obtained from an IR spectrum should be used in conjunction with all other available information like physical properties, elemental analysis, NMR, ultraviolet etc., as infrared data taken in isolation can at times be grossly misleading.

Infrared spectrophotometry is applicable to solids, liquids and gases. It is fast and requires small sample sizes. It can differentiate between subtle structural differences. On the other hand, interpretation of infrared spectra is highly empirical and requires huge libraries of reference spectra. Interpretation of spectra is a skilled art; nevertheless it is now part of the repertoire of many chemists. Infrared spectrophotometry provides means for monitoring many common atmospheric pollutants such as ozone, oxides of nitrogen, carbon monoxide, sulfur dioxide and others.

Modern infrared instruments employ microprocessors for control of spectrometer functions (Geary Stephen, 1975). This provides better control of the spectrometer, and in many cases additional features such as peak sensing or quantitative analytical capability. The internal representation of data in such systems is, of course, digital. Therefore, it would be quite feasible to produce data which can be accepted by more powerful computer systems.

BASIC COMPONENTS

Spectrophotometers for the infrared range are composed of the same basic elements as instruments in the visible and ultraviolet range, namely a source of radiation, a monochromator for dispersing the radiation, and a detector which registers the residual intensity after selective absorption by the sample. However, the materials used in the construction of the optical parts are quite different and they are suitably described in the sections to follow.

Radiation Sources

The source of radiation in infrared spectrophotometers is ideally a black body radiator. All practical sources fall short of this to a lesser or greater extent. The energy emitted by a black body radiator varies with wavelength and with temperature. In particular, increasing the temperature of the source raises the energy of emission enormously in the short wavelength region but has relatively small effect at long wavelengths.

The optimum infrared source is an inert solid heated electrically to temperatures between 1500°K and 2000°K. The maximum

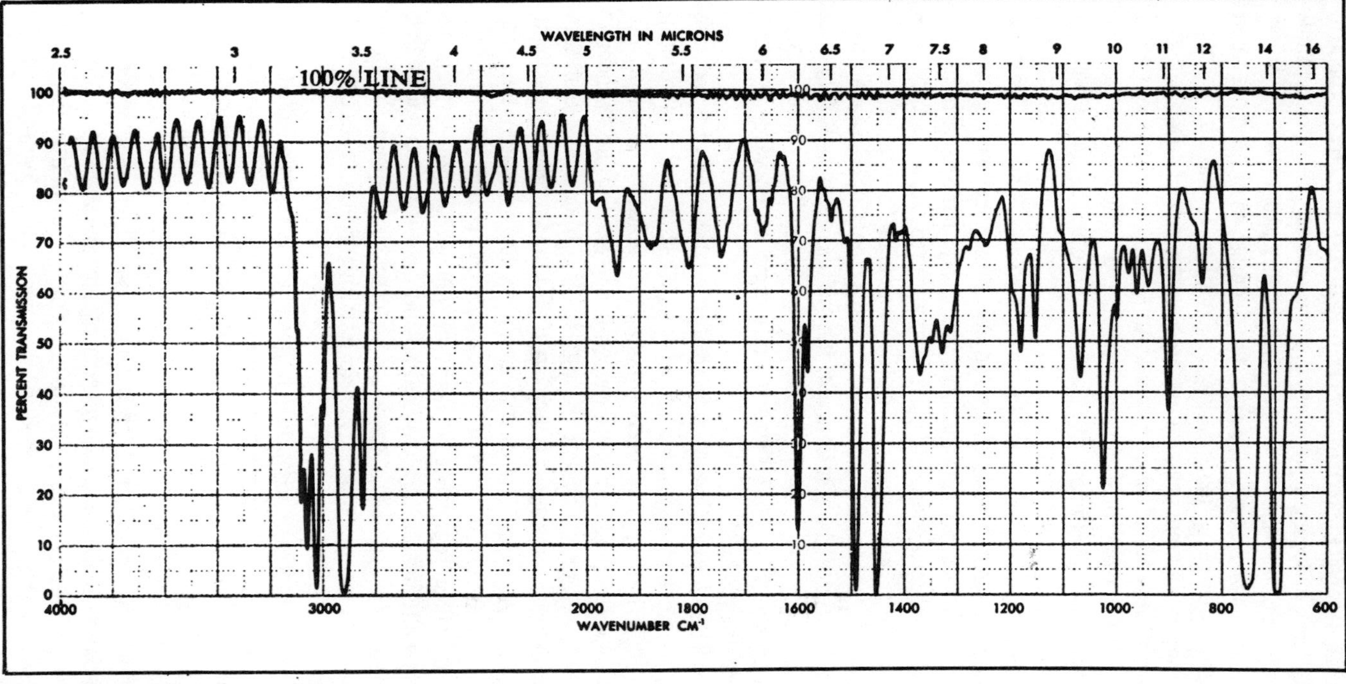

Fig. 3-1. Typical infrared spectrum.

radiant intensity at these temperatures occurs at 1.7 to 2 μ (5000 to 6000 cm^{-1}). At larger wavelengths, the intensity reduces continuously, until it is about 1% of the maximum at 15 μ (677 cm^{-1}). On the short wavelength side also, the decrease is much more rapid and a similar reduction is noticed at about 1 μ (10,000 cm^{-1}). There are three common practical infrared sources: the *Globar rod,* the *Nernst filament* and the *nichrome wire*.

The *Globar* is a silicon carbide rod which has a positive temperature coefficient of resistance. The rod is about 5 cm in length and 0.5 cm in diameter. It is electrically heated and run at a current to produce a temperature of about 1300°K. The heat dissipation at the ends of the rod is high as they are the coolest parts. This sometimes leads to arcing and burnout. Therefore, water cooling of electrical parts is required to prevent arcing. The Globar finds applications for work at wavelengths longer than 15 μ (650 cm^{-1}) because its radiant energy output decreases less rapidly. Since the resistance of the rod increases with length of time, provisions must be made for increasing the voltage across the element. It can be conveniently done with a variable transformer.

The *Nernst filament* is a small rod composed of fused rare earth oxides of *zirconium* and *yttrium*. The filament is of cylindrical shape having a diameter of about 1 to 2 mm and a length of 20 to 30 mm. Platinum leads are sealed to the ends of the cylinder to allow passage of current through it. The filament is heated to temperatures between 1500°C to 2000°C and produces maximum radiation at about 7100 cm^{-1}. The device has a negative temperature coefficient of electrical resistance and is operated in series with ballast resistance in constant voltage circuit. It must be externally heated to a dull red hot because it is non-conducting when cold and has a tendency to crack or separate from its connections on cooling. Therefore, it should be run continuously whenever possible.

The *nichrome strip*, though, gives less energy than the Globar or Nernst, is extremely simple and reliable in operation. In construction, it could be a tightly wound spiral of nichrome wire heated by passage of current. The temperature is about 800° - 900°C. SPECORD IR spectrophotometers of Carl Zeiss Jena employ a ceramic rod as a source in which is embedded a platinum-rhodium filament, producing a temperature up to 1200°C. This radiator consumes only 45 watts power. Light from any one of these sources is concentrated by a mirror and is focused onto the entrance slit of the monochromator.

Monochromators

Light from the entrance slit is rendered parallel after reflection from a collimating mirror and falls on the dispersing element. The dispersed light is subsequently focused onto the exit slit of the monochromator and passes into the detector section. The sample is usually placed at the focus of the beam, just before the entrance slit to the monochromator. Infrared monochromators generally employ several mirrors for reflecting and focusing the beam of radiation in preference over lenses to avoid problems with chromatic aberrations.

Both prisms and gratings are used for dispersing infrared radiation. However, the use of gratings is relatively a recent development. As a general rule, instruments operating below 25 to 40 μ have prism monochromators, whereas reflection gratings are utilized above 40 μ because transparent materials are not easily available in that range. Materials for prism construction which are found most suitable in the infrared region are listed in Table 3-1.

An ideal prism instrument would contain a large number of prisms made from different optical materials so that each could be used in sequence in its most effective region. Such an instrument would, of course, be extremely expensive. High resolution prism instruments contain combination of SiO_2, NaCl and KBr prisms. Low cost instruments use an NaCl prism over the full range. They give highest resolution in the vital fingerprint region. Some of these materials are very *hygroscopic*. Therefore, a good control of the humidity in the room in which an infrared monochromator is installed is essential. A prism monochromator is sensitive to temperature changes and must be thermostated to maintain constant wavelength calibration.

Prism monochromators employ the Littrow mount, which reflects the beam from a plane mirror behind the prism and returns

Table 3-1. Materials for Prism Construction.

Material	Optimum Range As Prisms
Glass (SiO_2)	300 mμ to 2 μ (5000 cm^{-1})
Quartz	800 mμ to 3 μ (12,500 to 3300 cm^{-})
Lithium Fluoride	600 mμ to 6.0μ (1670 cm^{-1})
Calcium Fluoride	200 mμ to 9.0μ (1100 cm^{-1})
Sodium Chloride	200 mμ to 14.5 μ (625 cm^{-1})
Potassium Bromide	10 to 25 μ (400 cm^{-1})
Cesium Iodide	10 to 38 μ (260 cm^{-1})

it through the prism a second time (Fig. 3-2), thereby doubling the dispersion produced. In a double pass system, the beam returns through the prism again, producing a total of four passes through the prism which results in improvement in resolution.

Figure 3-3 shows the optical arrangement of a SPECORD IR spectrophotometer. Emanating from the radiation source (1), the reference beam is reflected by spherical mirrors (2) and (3) through the attenuator diaphragm (4) and passes the reference cell (5). A plane (6) and a concave mirror (7) project it on to the rotating sector mirror (8). The sample beam takes an almost analogous course via mirrors (10) and (9), through the sample cell (11) and the 100% adjusting diaphragm (12), and via a concave mirror (13). The rotating sector mirror (8) alternatingly allows one of the two light beams to reach the entrance slit of the monochromator, via a toroidal mirror (14). The monochromator used is according to the *Ebert principle* and is equipped with a prism in Littrow arrangement. SPECORD 71 IR uses a 67° NaCl and SPECORD 72 IR a 67° KBr prism.

The instrument provides high spectral resolution achieved by long, curved entrance and exit slits. The scanning conditions within the entire wave number region are improved by flushing desiccated air through the path of rays. All hygroscopic elements are protected from the ambient air. In order to obtain wave number stability, the NaCl prism is temperature regulated to ± 1°C. An optical imaging system consisting of a field lens (22), two mirrors

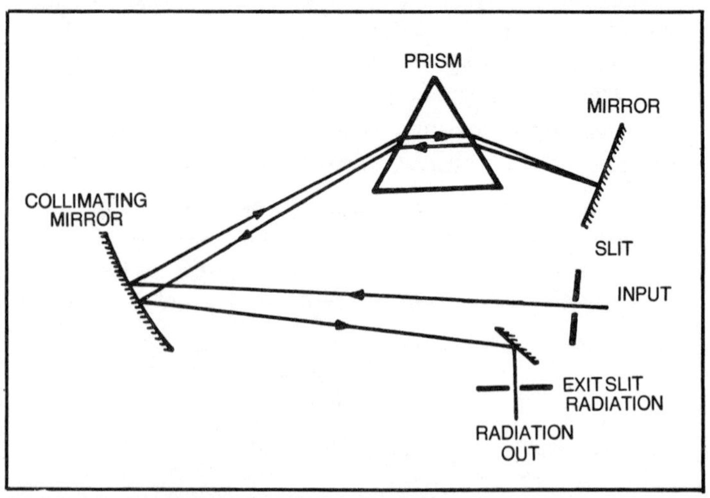

Fig. 3-2. Littrow mounting infrared monochromator.

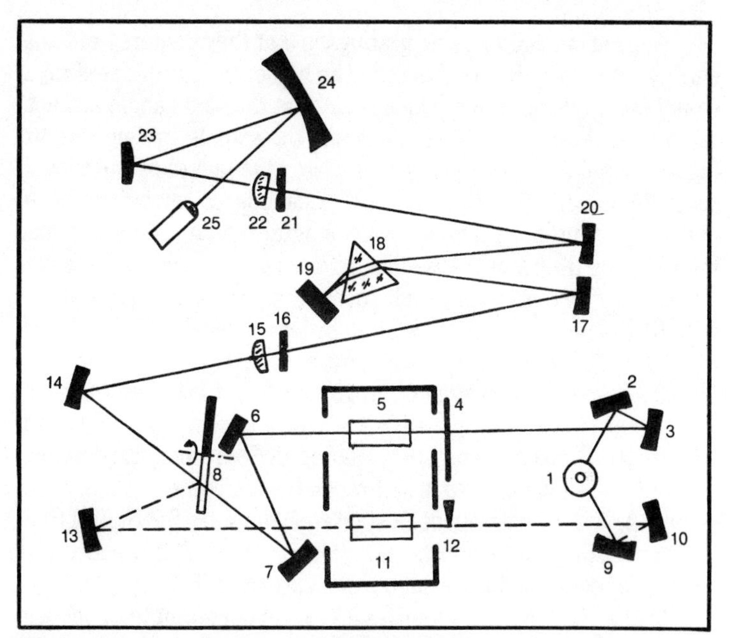

Fig. 3-3. Optical arrangement of SPECORD IR spectrophotometer (courtesy of Carl Zeiss, Jena).

(23 and 24) and the lens cemented on the detector (25) projects the beam leaving through the monochromator's exit slit on to the receiver surface. The detector is a thermocouple of high sensitivity and short time constant.

Efficient balancing of the intensities of the two beams within the entire wave number region is achieved by the 100% adjusting diaphragm (12). It is operated by means of a cam which allows to balance the 100% line at any part of the wave number scale.

Use of gratings as dispersing elements in the monochromators offer a number of advantages for the infrared region. Better resolution is possible because there is less loss of radiant energy than in a prism system. This facilitates employing of narrower slits. Also, gratings offer nearly linear dispersion. With these advantages gratings appear to be replacing prisms.

The standard grating used in infrared spectrophotometry is an echelon reflection grating. These gratings are usually constructed from glass or plastic that is coated with aluminum. Grating instruments incorporate a sine-bar mechanism to drive the grating mount when a wavelength readout is desired, and a cosecant bar drive when wave numbers are desired.

A disadvantage of using gratings is that they disperse radiant energy into more than one order. The higher order reflected rays would emerge from the monochromator at the same angle and act as unwanted radiation passing through the sample, giving rise to high transmission values. It necessitates the use of an additional separation method. There are two methods of removing these higher unwanted orders of wavelength, using either optical interference filters or a low-dispersion prism. Some instruments make use of two gratings, each covering a part of the range along with filters for removing unwanted orders (Unicam SP 200G and SP 1200). The two gratings have different construction, one having 60 lines/mm for the range 400 to 133 cm^{-1} and 180 lines/mm for 1200-1400 cm^{-1}. It is possible to maintain comparatively high resolution at a much lower cost by using an Ebert monochromator. It employs a single grating in its first and second orders. Figure 3-4 shows the optical diagram of IR spectrophotometer SP 1000 of Pye Unicam making use of an Ebert monochromator. The spectrum is scanned by rotating the grating about a vertical axis.

In the Beckman 4200 series IR spectrophotometers, monochromaticity is provided by a grating and set of circular variable filters (Fig. 3-5). Two annular filter segments are mounted on a rotating wheel to select the desired wavelengths from the continuous output of a nichrome wire energy source. This is designated as a wedge filter system and is coupled to a grating wavelength selector. The annular filter segments consist of IR transparent coated to transmit a continuously varying band of wavelengths when rotated in front of a V-shaped slit. The slit is positioned on the optical center line of the filter system.

A single element vacuum thermocouple receives energy transmitted by the wedge filter and generates voltage for amplification by the electronics system. In double beam operation, an amplified thermocouple output which is proportional to the difference of reference and sample beam intensities drives an optical attenuator into the reference beam until beam intensities become equal, i.e., until optical-null is achieved. Attenuator position in the reference beam is thus directly proportional to sample transmission at any given wavelength. Since the recorder pen is coupled mechanically to the attenuator, the pen records sample transmission as a function of wavelength. In single beam operation, with the reference beam physically blocked by the operator, sample beam transmission produces a thermocouple voltage that is proportional to sample beam intensity. The recorder

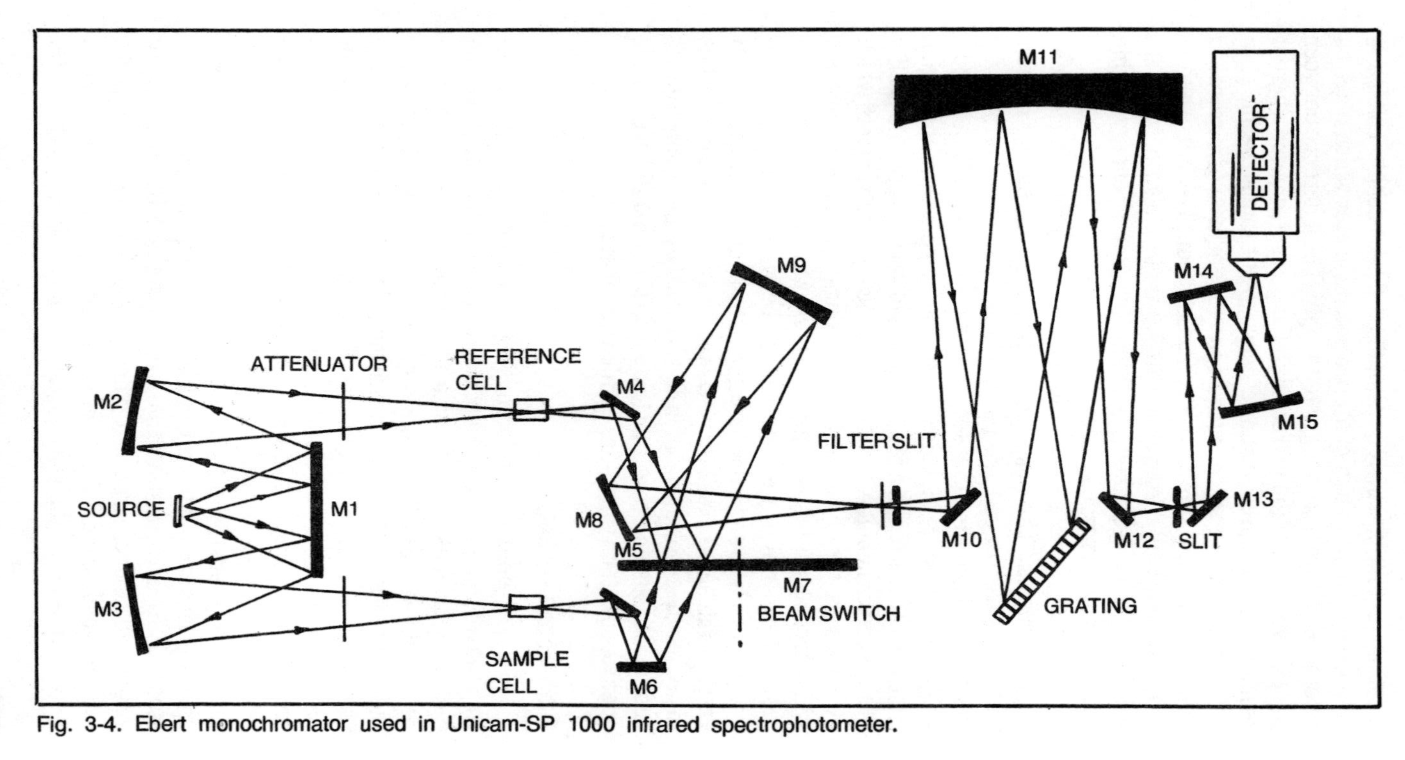

Fig. 3-4. Ebert monochromator used in Unicam-SP 1000 infrared spectrophotometer.

functions potentiometrically. Amplified thermocouple voltage is compared with the output of the pen potentiometer which receives reference voltage from a regulated supply. The difference between the two voltages is then directed to the pen servometer where it results in linear representation of energy in the sample beam.

The optical arrangement incorporates a number of mirrors and entrance and exit slits in the optical system. The slit width is adjustable manually from 0.005 to 7.0 mm. It also permits slit selection in programmed condition and keeps energy within dynamic range of servo system.

Several workers have reported the use of tunable lasers as energy sources. The major advantages are high energy concentrated in a narrow spectral band, a highly directional beam and a very short pulse time (5 n sec). Brosnan et. al. (1977) describe a single laser system in which the tuning range covers the region from 3.6-50 microns in a few minutes. However, so far the quality of the spectra is below that of the grating type instruments which still dominate the field. Diffraction gratings should not be touched, and unless absolutely essential, only blowing with air should be resorted to. In no case should these be washed with corrosive solvents. Prisms and lenses are generally hygroscopic and can get easily scratched. In case they become foggy, the only remedy lies in getting them polished, but in most cases they are beyond repairs and hence extreme care should be exercised if an instrument is opened, so that no vapors are exhaled near these.

Entrance and Exit Slits

In all continuous comparison techniques in spectrophotometry, part of the optical radiation emerging from the monochromator is directed onto the reference detector and the rest is allowed to fall into the sample. As the wavelength of the monochromator is scanned, it is arranged that the signal from the reference detector is maintained at an approximately constant level. It is possible to adjust the intensity of a tungsten lamp to obtain this condition, but when wide range of wavelengths in the infrared are to be covered, the only practical method is to control the monochromator slits. Most of the monochromators use a specially shaped cam to control the slit width as the wavelength is scanned. Besides this, a servo system can also be employed to achieve the same purpose. Collins (1975) describes a slit-servo system which maintains the output of an optical monochromator at a constant level as the wavelength is scanned. In his system after

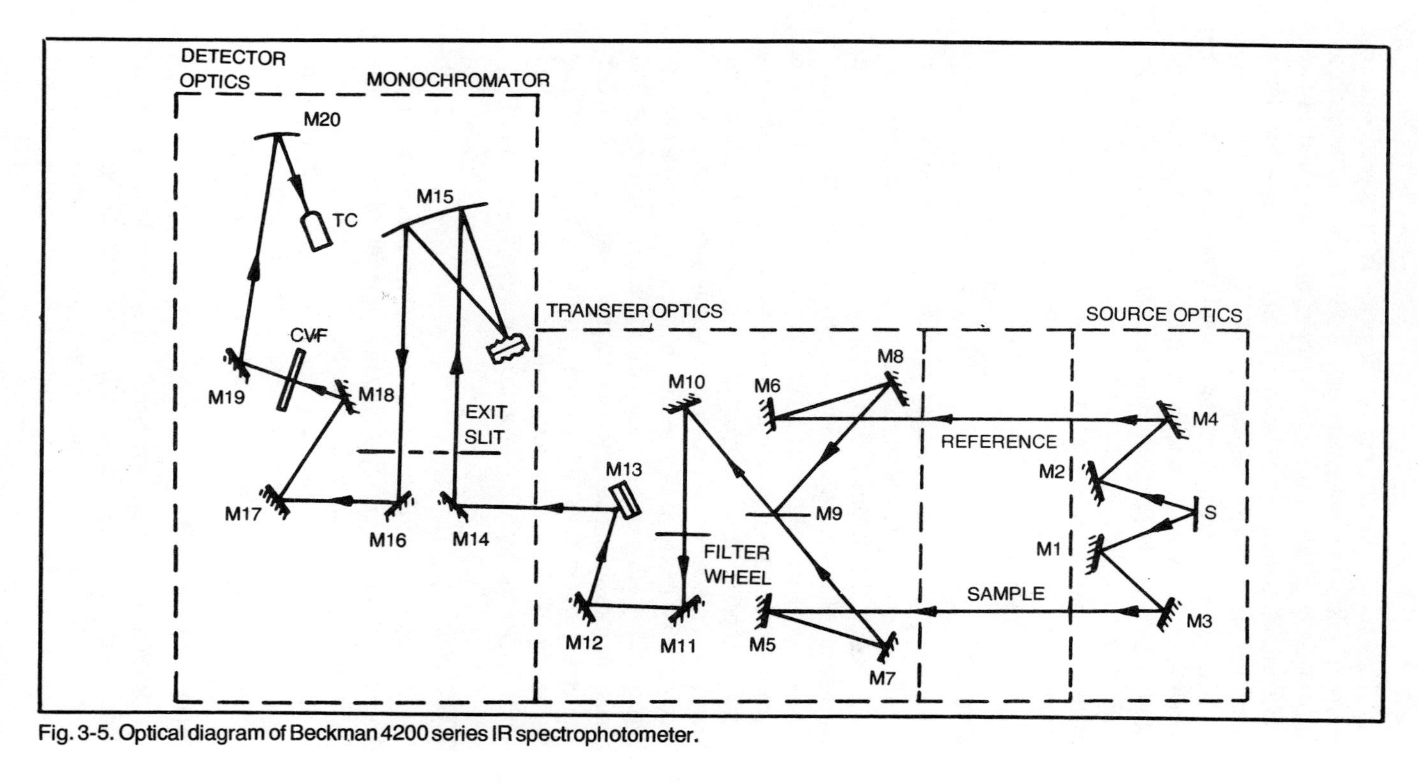

Fig. 3-5. Optical diagram of Beckman 4200 series IR spectrophotometer.

amplification, the signal from the detector is compared with a reference voltage and the error signal is converted to a frequency. The series of pulses drive a dc stepper motor coupled to the monochromator slits, and as balance is reached, the motor slows down, thus preventing overshoot.

The slit width control has the same importance as it has in ultraviolet and visible spectrophotometers, and a compromise of different factors is to be made while selecting the slit width to be used. Narrow slits produce smaller bandwidth, which consequently result in better spectral definition, whereas wider slits permit larger amount of radiant energy reaching the detector and consequently greater photometric accuracy.

Mirrors

As the materials that are used for lenses are not transparent to IR radiation over the entire wavelength range, lenses are generally not preferred. Front surfaced mirrors are usually used in the IR instruments of which plane, spherical, parabolic and toroidal types are most common. Although highly reflecting aluminum coated mirrors with a protective coating are usually employed, after long periods of operation, the mirrors are bound to become somewhat dusty. Under these conditions, the recording accuracy may not be affected very much, the reflective power will be reduced. As fingerprints or other contaminations may give their own absorption bands, mirrors should be cleaned, if required, only by blowing hot air over them. In case the mirrors are too dirty, these can be cleaned by washing with detergents and rinsing with distilled water. Under no conditions, should corrosive solvents be used.

DETECTORS

Detectors used in infrared spectrometers usually convert the thermal radiant energy into electrical energy which can subsequently be plotted on a chart recorder. The detectors used may be divided into thermal and photoconductive types. The photoconductive detectors are useful in the near infrared region, whereas thermal detectors can be used very much beyond this range.

Photoconductive Cells

Photoconductive cells are essentially electrical resistors which decrease in resistance in relation to the intensity of light striking their surface, and are characterized by greater sensitivity and rapidity of response. They are constructed from a thin layer of 0.1

μ of semiconductors like lead sulfide or lead telluride supported on a backing medium like glass and sealed into an evacuated glass envelope. These detectors are sensitive up to 3.5 and 6 μ for the lead sulfide and lead telluride cell respectively. Typical response time of these cells is 0.5 msec.

Thermal Types

Thermal detectors depend for their response on the heating affect of the radiation. They are useful for detection of all but the shorter infrared wavelengths. The infrared radiations are absorbed by a small black body and the resultant temperature increase is measured. Even under the best of circumstances, the temperature changes are extremely small and are confined to a few thousandths of a degree centigrade. *Thermocouples, bolometers* and *Golay cells* are the three commonly used thermal detectors.

Thermocouples. These are the most widely used detectors employed in infrared spectrophotometers. In these detectors the signal originates from a potential difference caused by heating a junction of unlike metals. They are made by welding together two wires of metals in such a manner that they form two junctions. One junction between the two metals is heated by the infrared beam and the other junction is kept at constant temperature. Due to difference in work functions of the metals with temperature, a small voltage develops across the thermocouple. The receiver element is generally blackened gold or platinum foil to which are welded the fine wires comprising the thermoelectric junction. The other junction is shielded from the incident radiation. Changes in temperature of the order of 10^{-6}C can be detected. It is possible to increase the output voltage by connecting several thermocouples in series. This arrangement is called the thermopile.

The average electrical output is about 1 microvolt. Amplification of such a low signal is difficult because of their low resistance (10-20 ohms) and the slow response of the average thermocouple. The thermocouple is enclosed in a magnetically shielded housing to reject spurious signals. Stray light can often be troublesome and should be avoided. Some of the combinations that have been used are Ag-Pd, Sb-Bi and Bi-Te.

Bolometers. These give an electrical signal as a result of the variation in resistance of a conductor with temperature. It consists of a thin platinum strip in an evacuated glass vessel with a window transparent in the infrared range. Irradiation by the infrared beam produces an increase in resistance of the metal strip which is measured with a *Wheatstone bridge*. Usually, two identical elements are used in the opposite arms of the bridge. One of the elements is placed in the path of the infrared beam and the other is used to compensate for the changes in the ambient temperature. Alternatively, the platinum strips may be replaced with thermistors which show a negative thermal-coefficient of electrical resistance.

Golay Cell. The *pneumatic detector* described by Golay (1947) essentially measures the intensity of infrared by following the expansion of a gas upon heating. The Golay cell (Fig. 3-6) comprises a chamber containing *xenon*, a gas of low thermal conductivity. It is sealed at the front end by a blackened receiver. The rear wall is a flexible membrane with a mirrored surface on its rear side. Infrared energy falling on the receiver warms up the gas in the chamber. A rise in temperature of the gas in the chamber produces a corresponding rise in pressure and, therefore, a distortion of the mirror diaphragm. Light from a lamp inside the detector housing can be focused on the diaphragm which reflects the light on to a photocell. Movements of the diaphragm corresponding to the amount of incident infrared energy changes the incident light energy on the photocell surface and causes a change in the photocell output. By periodically interrupting the incident radiation with a chopper, an ac signal is produced by the photocell, which can be amplified.

Pneumatic detectors with large receiver area are suitable for instruments in which wide slits are necessary. They function at all wavelengths throughout the infrared region (as far as 400 μ using a diamond window). They have a low response time. The whole receiver is sensitive to the radiations, thereby eliminating the need for optical alignment.

General Considerations Regarding Detectors

Due to the low intensity of available sources and the low energy of the infrared photon, the task of measurement of infrared radiation is particularly difficult. The electrical signal produced by the various detectors is quite small and requires a large amplification factor before it can be put to a recorder. To prevent the very small signals being lost in the noise signals, which might be picked up by the connecting wires, the preamplifier is located as close to the detector as possible.

In addition to this, the radiation beam is chopped with a low frequency light interrupter. By using narrow bandwidth electronic

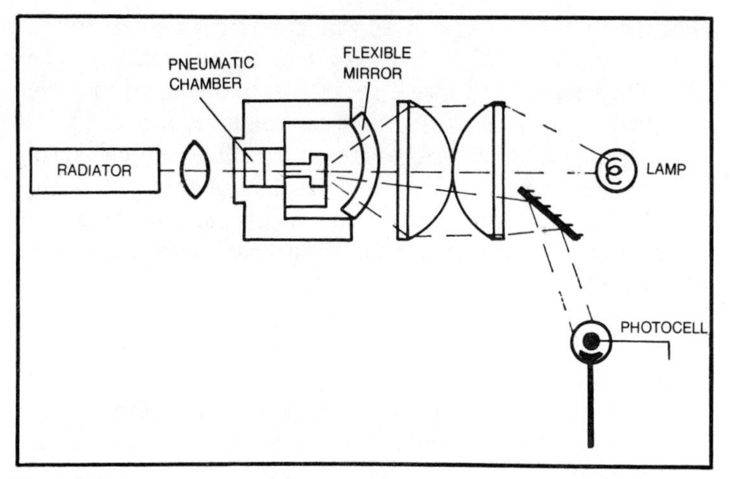

Fig. 3-6. Golay pneumatic cell.

amplifiers, the alternating signal corresponding only to the chopping frequency is measured. This arrangement minimizes stray light signals.

The response time of different types of detectors is typically as follows: photoconductive cells, 0.5 m sec; Golay cell, 4 m sec; bolometer, 4 m sec; and thermocouple, 15 to 60 m sec.

Sensitivity of photoconductive detectors is generally higher than thermal detectors. Nevertheless, there is no strong choice among thermal detectors on sensitivity considerations.

When using thermal detectors, it is essential to shield the detector to reduce its heating by nearby extraneous objects. Therefore, the absorbing element is placed in a vacuum and is carefully shielded from thermal radiation emitted by other bodies in the area.

Preamplifier For Use With Photoconductive Infrared Detectors

When photoconductive infrared detectors such as cadmium mercury telluride are used, it is necessary to employ a low noise preamplifier. Gore and Smith (1974) describe such an amplifier whose circuit diagram is shown in Fig. 3-7. Basically, the circuit consists of a FET input stage, which has sufficient gain to make the noise contribution of the next stage negligible. This is followed by a low noise integrated amplifier. The FET used is BF 818, which has a very low equivalent input noise voltage. A common source stage TR_1 is used with a bootstrapped drain load TR_2. To obtain the lowest noise, the FETs are used with drain currents near to I_{DSS} (drain current of FET with a zero source gate voltage). The feedback resistor R_5 is made as small as possible so that the minimum bias is applied to TR_1. The current in TR_2 is determined by the current in TR_1; therefore, to ensure that the gate junction of TR_2 is not forward biased, the FET with the higher I_{DSS} is used in this position. The amplifier was found to give, at room temperature, with a 51 ohm source impedance and a bandwidth from 8 Hz to 10 kHz, a noise level of 2 μV peak-to-peak.

The detector requires a 10 mA bias current supply, which must have a low noise level, if the system performance is not to be degraded. This current is obtained from TR_3, which is also a type BF 818 low noise FET. The noise voltage across the detector was found to be 19 mV rms, which is small compared to the theoretical amplifier noise.

ARRANGEMENTS FOR RECORDING THE INFRARED SPECTRA

Infrared spectrophotometers are produced by several instrument manufacturers. Almost all the commercial designs employ a double-beam system where the radiant energy passes alternately through the sample, and then a reference, to a single detector. They incorporate a low-frequency chopper (5 to 13 Hz) to modulate the output radiation from the source. Various sources and detectors described above are incorporated in one or more of the commercial instruments.

The double beam system offers the advantage that scans are not disturbed by absorption bands produced by atmospheric water vapor and carbon dioxide. Also, fluctuations in source output, detector sensitivity and gain have no influence on the record.

There are two types of arrangements for recording the infrared spectra. These are *ratio recording* and the *optical null* method.

Optical Null Method

In this method, the infrared radiation is passed simultaneously through two separate channels, one containing the sample and the other the reference. The two beams are recombined into a common axis and are alternately focused on the detector. If the intensities of the sample and reference beams are exactly equal, then no alternating intensity radiation goes through the slit. If the sample absorbs some radiation, an alternating intensity radiation is observed by the detector, which produces an ac signal. This ac

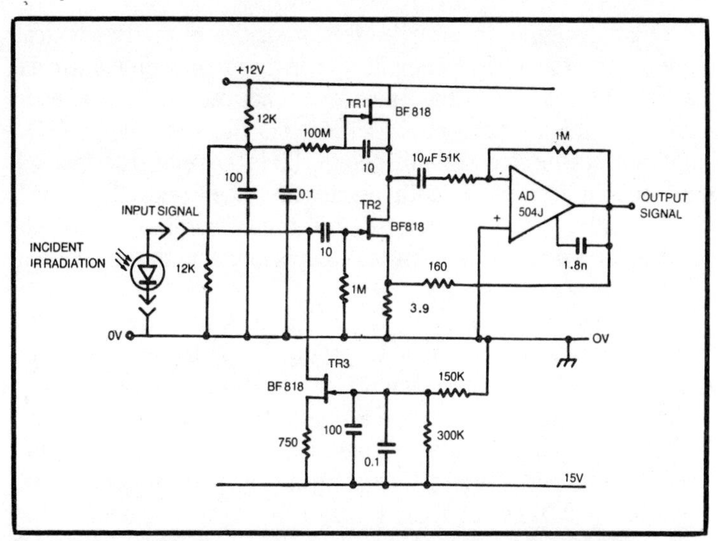

Fig. 3-7. Preamplifier for use with photoconductive infrared detectors.

signal can be selectively amplified in a tuned amplifier. As the amplifier is tuned to the chopping frequency of the light beam, signals of frequency different from the chopping frequency are not amplified. The alternating signal from the detector is used to drive a servomotor which is mechanically coupled to an optical wedge or a fine-toothed comb attenuator. Movement of the comb occurs when a difference in power of the two beams is sensed by the detector. The motor will stop when the reference and the sample beam intensities are exactly equal. The movement of the motor is synchronized with the recorder pen so that its position gives a measure of the relative power of the two beams and thus the transmittance of the sample. The record obtained, therefore, is of the sample absorption as a function of spectral frequency. The teeth of the attenuator comb are accurately cut so that a linear relationship exists between the lateral movement of the comb and the decrease in power of the beam.

Figure 3-8 shows the block diagram of a typical recording type double-beam optical-null principle infrared spectrophotometer. Emanating from the source (1), the two light beams pass through the sample and reference cells (3, 4). With a switching frequency of 12.5 Hz, the rotary sector mirror (5) alternatingly conducts the radiation of the sample and reference channels into the monochromator. Here, the light of the respective wave number is reflected and passed on to the thermocouple (7) which delivers a

12.5 Hz ac voltage signal, when the intensities of the two optical paths are not equal. After amplification in the preamplifier and main amplifier (8), and after phase-sensitive rectification in the demodulator (9), this ac voltage signal controls the servomotor (11), which adjusts the attenuator diaphragm (2) in the reference channel until the intensities of both beams are equal and the signal disappears. The servomotor also drives the recording pen, which records the sample's transmittance corresponding to the position of the attenuator diaphragm.

The Beckman Model IR 4200 series infrared spectrophotometers are automatic recording double beam optical-null instruments which cover a range of 4000 to 200 cm^{-1}. They employ a Nichrome/Nernst source and a thermocouple detector and are available with scanning speeds from 2 cm^{-1}/mm to 1000 cm^{-1}/mm. The monochromator is a filter/grating combination and provides a resolution of 0.5 cm^{-1} at 900 cm^{-1} and 1.4 cm^{-1} at 3000 cm^{-1} in the better models.

Ratio Recording Method

In the direct ratio recording system, radiation from the source is passed alternately through two separate channels, one passing through the sample cell and the other through the reference cell. The two signals from the detector are amplified, rectified and recorded separately. The sample signal is displayed as a proportion of the reference signal. In this method, attempts are not made to have a physical equalization of the sample and reference beam intensities. The reference signal is used to drive the slit width control so that the energy level reaching the detector is constant. This system is, however, not favored in infrared instruments as it requires a very stable amplifier system and the chopping system.

SAMPLE HANDLING

A wide variety of samples can be analyzed using IR spectroscopy. Solids, liquids or gases can all be handled. However, sample handling presents a number of problems, since no rugged window material for cells exists which is transparent over the entire IR range and is also inert. The most commonly used window material is NaCl which transmits down to about 650 cm^{-1}. KBr transmits down to about 400 cm^{-1}, CsBr to about 250 cm^{-1} and CsI to about 200 cm^{-1}. These materials being all water soluble, the surfaces of the windows made from them are easily fogged by exposure to

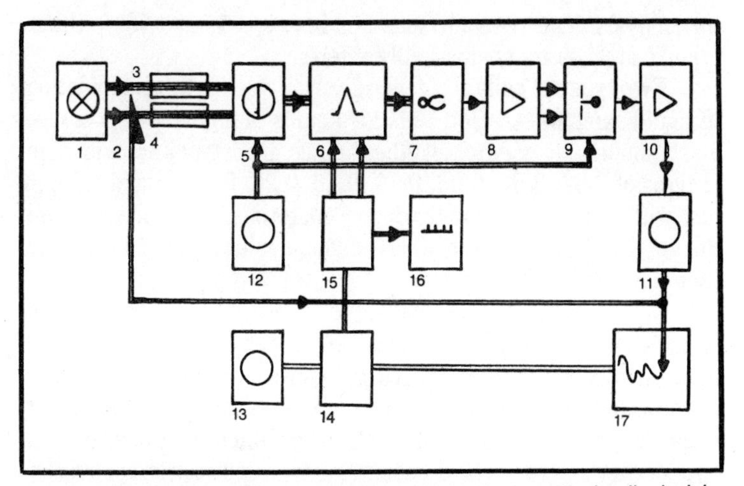

Fig. 3-8. Block diagram of a recording type double-beam optical-null principle infrared spectrophotometer.

atmospheric water vapor or moist samples. Therefore, they require frequent polishing when used under such conditions.

Gas Cells

Gas cells usually have a pathlength of 100 mm since this thickness gives a reasonable absorbance level for the majority of gases and vapors at the normally encountered partial pressures. The inner cell diameter is typically 40 mm and the volume is about 125 cm^3. The ends of the cells are ground square to which infrared transparent windows are glued with sealing gaskets. Two tubes with stop-cocks are attached to the cell for connecting the cell to the gas handling system. A typical glass cell may have a body of uniform cross-section, but it is preferable to use a tapered construction to conform closely to the section of the radiation beam and requires less sample volume for a given thickness. When weak bands are to be studied, it is necessary to use very long cell paths which are inconvenient to fit into the instrument. In such cases, multiple reflections using a combination of a mirror system with a beam condenser enables the spectra of minor components to be recorded.

Liquid Cells

The extinction coefficient of most liquid hydrocarbons in the infrared region are such that a pure sample of thickness between 0.01 mm and 0.05 mm gives an absorption spectrum quite suitable for analysis. The transmittance lies between 15 and 70%. Other liquid materials are not markedly different.

Two types of cells are generally available for the examination of liquid samples—sealed cells for liquids of high vapor pressure and demountable ones for all other liquids. Both types are with path lengths of 0.02, 0.04, 0.06, 0.10, 0.16, 0.25, 0.4, 0.6 and 1.0 mm and with the earlier stated window materials. The external cell diameter is 30 mm. About 100 μl sample volume per 0.1 mm path length is required.

Liquid cells of this thickness consist of IR transparent windows separated by thin gaskets of copper and lead which have been amalgamated with mercury before assembly is securely clamped together. As the mercury penetrates the metal, the gasket expands, providing a tight seal. The filling is done with a hypodermic syringe. The demountable type cells can be assembled for different path lengths, in which case circular spacing foils are placed between the windows.

Variable Path Length Cells

These are employed for determining the absorption coefficient and concentration of the test substance, as well as for establishing calibration curves. When using the cell in the reference beam, the solvent's absorption bands may be completely compensated by continuously changing the path length. The cell path length is usually adjustable (Fig. 3-4) in the range 0 to 5 mm via a differential thread with an accuracy of ± 1 micron. The path length can be read off the setting drum to an accuracy of 1 micron. In precisely fitted piston guideways, the cell windows are displaced against each other, thus achieving an extreme tightness of the cell. This permits long-time scans to be carried out with highly volatile solvents. This type of cell is normally supplied with KBr windows. CaF_2 or NaCl windows are also available. Length of the cell is typically 120 mm and diameter 75 mm.

To achieve repeatable cell thickness, it is important to be able to measure the path length. The path length can be conveniently measured by inserting the empty cell into the spectrometer and observing the interference pattern created by the reflection of part of the radiation beam from the internal faces of the cell. The pattern is a regular sine wave when recorded on a linear wave number scale and is of amplitude between 2 to 15%T. The path length of the cell is given by

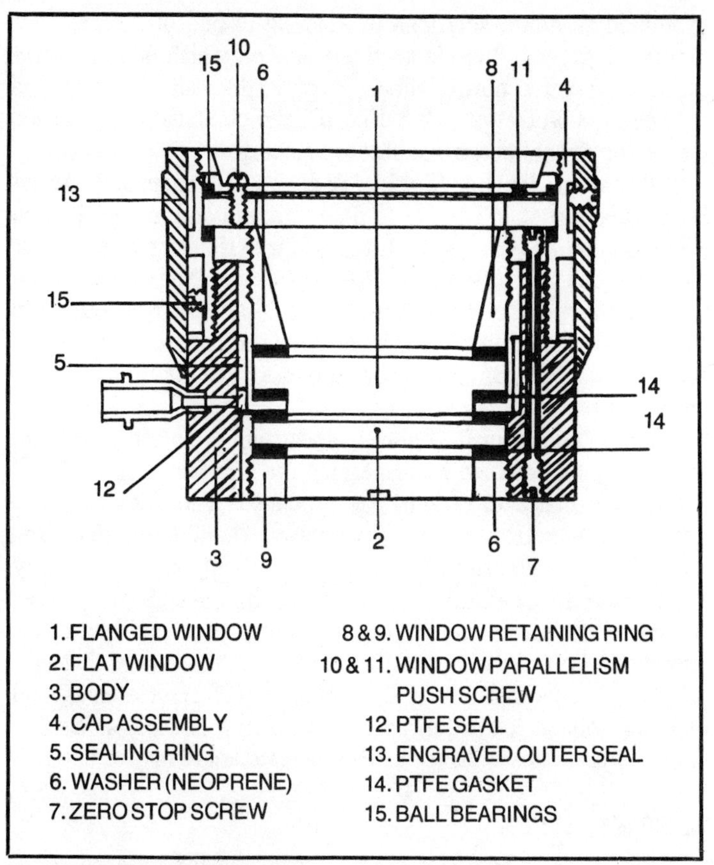

Fig. 3-9. Variable pathlength cell.

$$\text{path length (in centimeters)} = \frac{n}{2(\nu_1 - \nu_2)}$$

where n is the number of fringes between wave numbers ν_1 and ν_2. The method is not practical in case the cells are in a bad condition. Alternative methods like the use of a traveling microscope may be employed.

SAMPLING OF SOLIDS

Different methods are available for sampling solids for examination in the infrared spectrometer. They include the following.

Solids Dissolved In Solutions. In case it is possible to obtain suitable solutions of the solids, the solids are dissolved and examined as dilute solutions by running in one of the cells for liquids. However, there is no single solvent which is transparent through the entire infrared range. To cover the main spectral range between 4000-650 cm^{-1} (2.5-15.4 μ), the combination of carbon tetrachloride and carbon disulfide are employed.

Pressed Pellet Technique. In the pressed pellet technique, the solid sample is finely ground and mixed with an alkali halide like KBr and then pressed into the form of a disc for examination in the instrument. Pressing is done in an evacuable die under high pressure. The resulting disc or pellet is transparent to the infrared radiation and can be directly run. Other alkali halides like CsI and CsBr are used for measurements at longer wavelengths.

Mull Technique. Infrared spectrum of finely powdered solids may be obtained by dispersing the powder within a liquid medium so that a thick slurry is produced. The thick slurry is spread between IR transmitting windows. The liquid is so chosen that it has the same refractive index as the sample so that energy losses due to scattering of light are minimized. The most commonly used liquid mulling agent is a mixture of liquid paraffins, known as *nujol* (mineral oil). Although nujol is transparent throughout the IR spectrum, it cannot be used if C-H stretching and bending frequencies are to be observed. In this case, as second mull like *perfluorokerosene* or *hexachlorobutadiene* may be used.

MICRO SAMPLING

When the quantity of samples available for infrared spectrophotometric examination is very small, standard sampling techniques cannot generally be used. This is because the absorbance of a given sample is proportional to the number of grams of absorbing material and inversely proportional to the area of the sample. Thus, in order to increase the absorbance for a given weight of sample, it is necessary to decrease the area of the sample. The identification of gas chromatographic fractions, new compounds and materials isolated from natural products, frequently involve microgram quantities of samples and require the use of microsampling techniques for the measurement of their infrared spectra.

The simplest method of coping with small quantities of sample is to use cells, having reduced apertures, so that the spectrophotometer beam falls on all of the sample. These cells (for example the M-OX series of Beckman-RIIC microcells) are generally individually tailored to the beam dimensions in the sample compartment of a particular instrument so as to make the most efficient use of the limited amount of sample available.

In cases where the quantity of available sample is very small, or where the spectrophotometer design does not permit small sample areas to be used, it is necessary to employ a *beam condenser*. Basically, a beam condenser consists of an appropriate combination of lenses and mirrors which reduces the size of the spectrophotometer beam, focuses it on the sample, and then returns it to its normal size before entering the monochromator. Beam condensers are usually designed to fit directly into the sample compartment and to hold a variety of micro-liquid, -solid and -gas cells. Although most beam condensers waste some energy because of reflection losses and optical aberrations, most are highly efficient and conserve sufficient energy to permit extremely small samples to be examined in a nearly routine fashion. Typically, beam condensers have condensing factors of about four or six-to-one and are available either with all-reflecting optics or with refracting optics employing alkali halide lenses.

The normal sealed liquid cell has a nominal volume of about 0.1 ml with a 0.1 mm spacer. By using a cell with a specially shaped spacer for the size of the beam, this volume can typically be reduced to about 0.01 ml. Ultra-micro sealed cells for use in beam condensers have volumes of around 2 microliters with a 0.1 mm spacer. Powder sample techniques are probably the most widely used in the infrared analysis of micro solids. In micro sample applications, a small quantity of finely ground powder can be suspended in mineral oil and smeared on a demountable cell window; or it can be ground with KBr, pressed into a micro disk with a micro die, and mounted on a micro solid sample holder.

Of all these techniques, the micro KBr disk is by far the most popular because of its ability to deal with even the most minute samples. Samples as small as 1-100 μg can be examined routinely depending on the disk size.

SPECIAL TYPES OF CELLS

Buckland et. al. (1971) illustrate the construction of two types of infrared cells for studying the adsorption of gases on powdered solids. These cells do not involve metal components, greased taps or joints in contact with the adsorbing gas and are, therefore, free from the difficulties involved in the identification of infrared bands which may arise from contaminants in the cell. One of these cells allows high temperature pretreatment of the sample in a controlled flow of gas.

Poppelwell (1972) describes the construction of infrared absorption cells and gas handling apparatus for use with hydrogen fluoride.

Low temperature infrared spectroscopy is severely limited by the condensation of atmospheric moisture on cell windows and associated components. Interference from condensation is often prevented with a Dewar system with the sample situated between four halide windows. However, these Dewar systems have the disadvantage of scattering, enhanced by multiple halide windows and their sealing. Gooley et. al. (1969) describe the construction of an infrared cell which obviates the use of a Dewar system. In this cell, the cooling is effected by circulating a cold fluid while moisture condensation is avoided by a gas diffusion and heat gate system.

Sole and Walker (1970) developed a heatable infrared gas cell operable to at least 800°C. The problems concerning suitable window materials and the sealing of windows to the cell body were avoided by using inert gas curtains in place of solid windows. The cell can safely be used with wet or corrosive substances in standard commercial spectrophotometers.

Subramanian et. al. (1973) describe an assembly by the use of which the infrared spectra of radioactive samples such as plutonium could be measured without contaminating the spectrophotometer itself. Such a system would obviously be very useful in view of the difficulties involved in servicing a contaminated infrared spectrophotometer.

Passenheim et. al. (1974) describes the construction and calibration of a complete optical system, built entirely within a helium cryostat, for making absolute measurements of radiation-induced luminescence and optical absorption in the infrared range out to 25 μ. Helium cooling of the entire optical system reduced the thermal background radiation at the detector to approximately 1×10^9 photons/cm^{2}sec. ($\lambda < 25\,\mu$).

ATTENUATED TOTAL REFLECTANCE

Infrared methods can be used to study materials which are light scattering or opaque, or in the form of coatings on opaque materials. The technique usually employed is based on obtaining reflectance spectrum. However, a spectrum obtained by reflection of the radiation from the surface of a chemical material is generally very poor. To overcome this difficulty, a technique known as *attenuated total reflectance* was suggested by Fahrenfort in 1958. This enables one to obtain reflection spectra of satisfactory quality.

In this technique, energy from the source enters into a prism and is reflected almost entirely from face C when the beam strikes it at less than critical angle. The prism consists of a crystal of material of high refractive index and transparent to infrared radiation. Silver chloride and KRS-5 satisfy these conditions. If the sample is placed in contact with the face C, the internal reflection will be attenuated or reduced at the sample-crystal surface. The attenuation will take place at those wavelengths where the material absorbs infrared energy. Absorption spectra of coatings and liquids can be achieved even though they may be opaque. The information conveyed by the ATR spectra is essentially the same as that conveyed by transmission spectra. The optical arrangement of a simple ATR system is shown in Fig. 3-10.

Single reflection ATR is not always adequate because of the low sensitivity observed. Improvement in sensitivity can be obtained by the use of multiple internal reflection system in which internal reflections take place 25 to 50 times before emerging out of the crystal (Fig. 3-11). This strengthens the absorption pattern of a material placed on one surface of the optical flat. This is best achieved by reducing the width of the entrance face of the crystal and increasing its length. However, there is a practical limitation

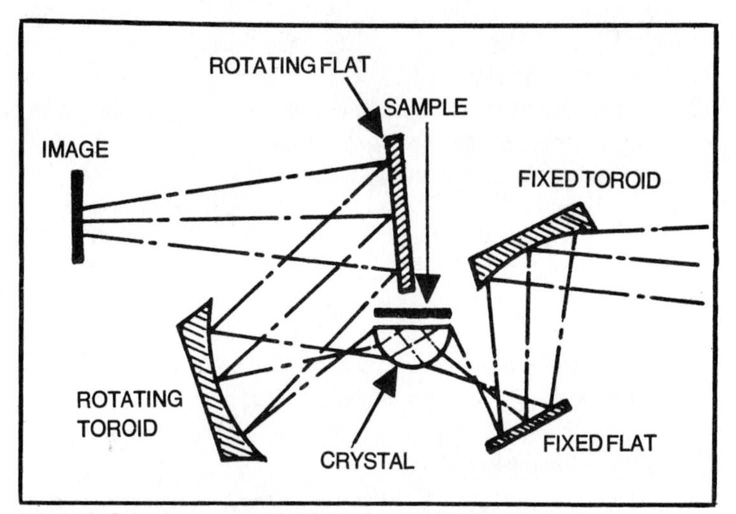

Fig. 3-10. Optical arrangement of a simple attenuated total reflectance system.

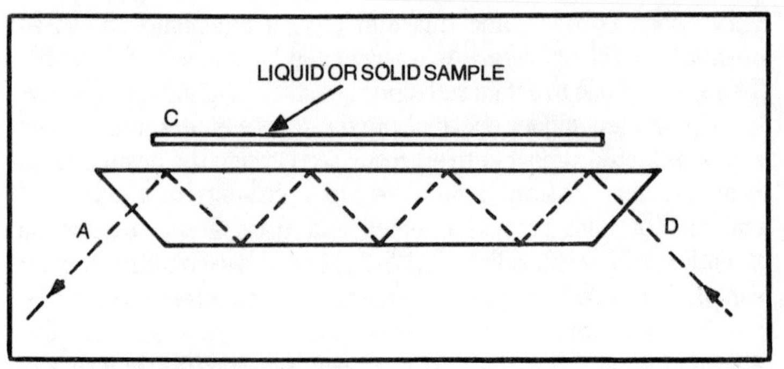

Fig. 3-11. Multiple internal reflections.

as the entrance face cannot be made smaller than the monochromator slit; otherwise it would result in a serious loss of energy. Carr-Brion and Gadsden (1969) describe an on-stream infrared analyzer using the multi-attenuated total reflectance technique.

One of the main advantages of the ATR technique is that the spectrum obtained is independent of the sample thickness. Typically, the reflected radiation penetrates the sample to a depth of only a few microns. Consequently, the method has been found to be particularly useful for establishing the surface characteristics of many materials. However, because of the small penetration depth, ATR spectra is inherently weak.

The prime objective in preparing samples for ATR analysis is to obtain intimate sample-prism contact. Since most of the commonly used ATR prism materials are fairly soft and easily deformed, only a limited amount of direct pressure can be applied to the sample in the attempt to obtain the necessary contact. Too much sample pressure can seriously distort the optical surface of the prism and result in lost energy and poor spectra. Films can, of course, be cast directly on the optically active surface of the prism, provided all traces of solvent can be removed without heat. Flexible, malleable samples, (e.q., rubber, plastics), can be simply pressed against the prism surface. Rigid, irregularly shaped samples are by far the most difficult to handle, since they frequently require grinding, sanding and/or polishing to achieve the flat surface necessary for this type of work.

INSTRUMENT CALIBRATION

In order to check on IR spectrophotometer, spectrum of a polystyrene film is run and the absorption spectra is compared with that provided by the company. In case any difference in the absorption spectra is detected, the instrument has to be thoroughly checked and recalibrated.

To facilitate speedy recalibration of low resolution far-infrared instruments, special broadband wavelength calibrators are often used. They usually employ polyethylene powder as the matrix. However, their fabrication is a little difficult procedure. Siddiqui and Stewart (1974) describe a broadband wavelength calibrator for use with low resolution far infrared monochromators which makes use of thin polyethylene sheets rather than polyethylene powder as the substrate. The method of construction consists in taking a clear polyethylene sheet and stretching it 12″ below a 250 W infrared lamp. Using an electric spray, a suspension of mercuric oxide powder in carbon tetrachloride is sprayed on to the polyethylene sheet in sprays of a few seconds each. It is ensured that carbon tetrachloride from one spray evaporates away before the next spray. When a uniform layer of mercuric oxide has been deposited, another polyethylene sheet is pressed on top and the sandwich is left under the infrared lamp for a further 15 mm. A filter of approximately 1.5 inch is cut from the uniform part of the sandwich and is mounted in a metal holder.

The transmittance of the filter obtained between 800 and 350 cm^{-1} on a Perkin Elmer 457 spectrophotometer shows that it provides three accurate calibration points: one due to the polyethylene matrix itself (at 722 ± 1 cm^{-1}) and the other two (at 590 ± 2 cm^{-1} and $500 \pm$ cm^{-1}) due to mercuric oxide.

Chapter 4

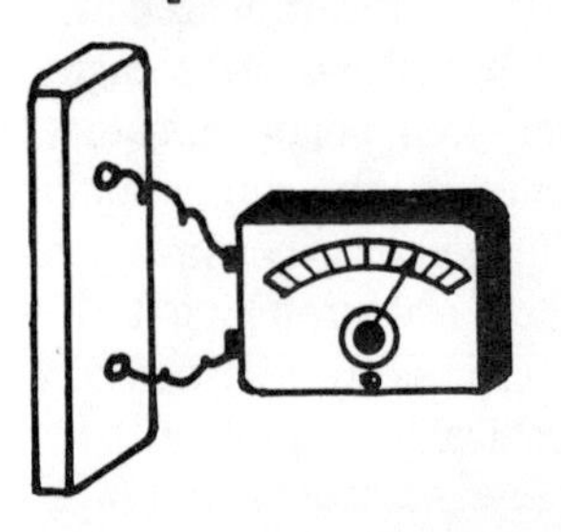

Flame Photometers

According to the quantum theory, the energy states of an atom or molecule are sharply defined and any change from one state to another, requires a sharply defined quantity of energy. When radiation falls on a material or the material is supplied with extra energy in some form, some part of the energy is taken up by the material and results in altering the state of the atoms or molecules of which it is composed. The atoms or molecules of the material are promoted to a higher energy state. However, the higher energy states are rather unstable. The particles at the higher energy levels tend to lose the extra energy and return to that original level or ground state either by undergoing a chemical reaction, dissipating the energy as heat or by emitting the energy as radiation. If it loses all or part of the energy as radiation, it will emit photons of an energy corresponding to difference between two energy levels. Since the levels are clearly defined for a given atom, the radiation will be emitted at clearly defined frequencies only. The frequencies are shown up as bright lines if the emitted light is dispersed as a spectrum. By measuring the wavelength of the emission, one can find out what atoms are present. Also, by measuring the intensity of the emission, one could compute the concentration of the element.

In short, the principle of flame photometry is based on the fact that if an atom is excited in a flame to a high energy level, it will emit light as it returns to its former energy level. By measuring the amount of light emitted, we can measure the number of atoms excited by the flame.

The method of flame photometric determinations is simple. A solution of the sample to be analyzed is prepared. A special sprayer operated by compressed air or oxygen is used to introduce this solution in the form of a fine spray (aerosol) into the flame of a burner operating on some fuel gas like acetylene or hydrogen. The radiation of the element produced in the flame is separated from the emission of other elements by means of light filters or a monochromator. The intensity of the isolated radiation is measured from the current it produces when it falls on a photocell. The measurement of current is done with the help of a galvanometer whose readings are proportional to the concentration of the element. After carefully calibrating the galvanometer with solutions of known composition and concentration, it is possible to correlate the intensity of a given spectral line of the unknown sample with the amount of the same element present in a standard solution.

Flame photometry is characterized by a high degree of constancy and reproducibility. The spectrum of an element as produced in a flame is relatively simple, consisting normally of only a few lines. Identification of the line is simple and spectral interference is less frequent. The most usual application of flame photometry is for the analysis of sodium and potassium in body fluids, and these analyses constitute the bulk of the determinations generally performed. However, the method is slowly replacing more troublesome methods for other elements also.

Flame photometry is concerned mostly with atoms. Molecules cannot normally survive the high temperatures employed in flame photometry.

The number of elements to which flame photometric methods can be applied depends mainly on the source temperature developed by the fuel mixture employed. This is because not all atoms are easily excited in the flame. The atoms of some elements have their energy levels so spaced that a large number of different lines are emitted, while for some others, almost all the light they emit is concentrated in one spectral line. The upper energy levels of atoms of some elements are so high above the ground state that they are very difficult to excite. Also, their emission lines could be of a wavelength which may not be directly usable. All these factors determine the lowest concentration of atoms which can be detected in the sample and thus varies very widely for different atoms.

Flame photometry offers the following advantages:

—The technique is very rapid. It does not require any chemical preparation except preparing a solution of suitable concentration.

—The method is most useful for analysis of some elements which are difficult to measure by other methods.

—The technique is most suited to analytical problems in which a large number of samples of similar types are to be measured.

—The method is quite cheap as it does not require any other expensive reagents.

EMISSION SYSTEM

A flame photometer has three essential parts (Fig. 4-1). An *emission system* consists of fuel gases and their regulation, consisting of the fuel reservoir, compressors, pressure regulators and pressure gauges; *atomizer*, consisting, in turn, of the *sprayer* and the atomization chamber where the aerosol is produced and fed into the flame; a *burner* which receives the mixture of the combustion gases; and a *flame*, the true source of emission. The *optical system* consists of the optical system for wavelength selection, filters or monochromators, lenses, diaphragms, slits, etc. The *recording system* includes detectors like photocells, phototubes, photomultipliers, etc., and the electronic means of amplification, measuring

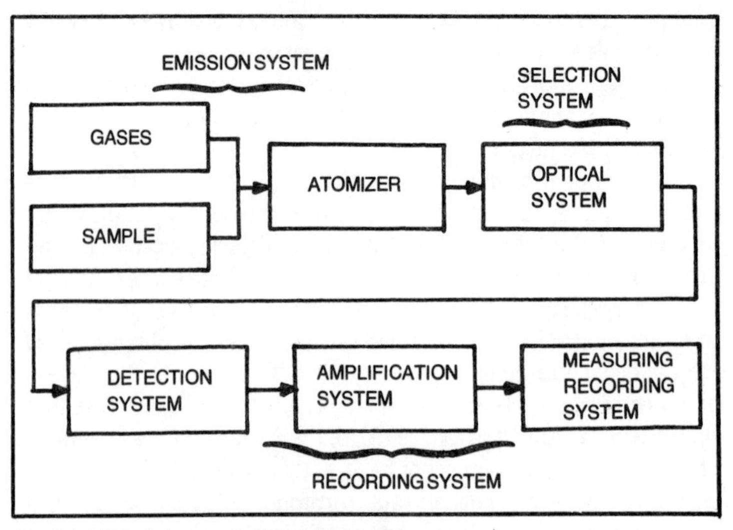

Fig. 4-1. Essential parts of a flame photometer.

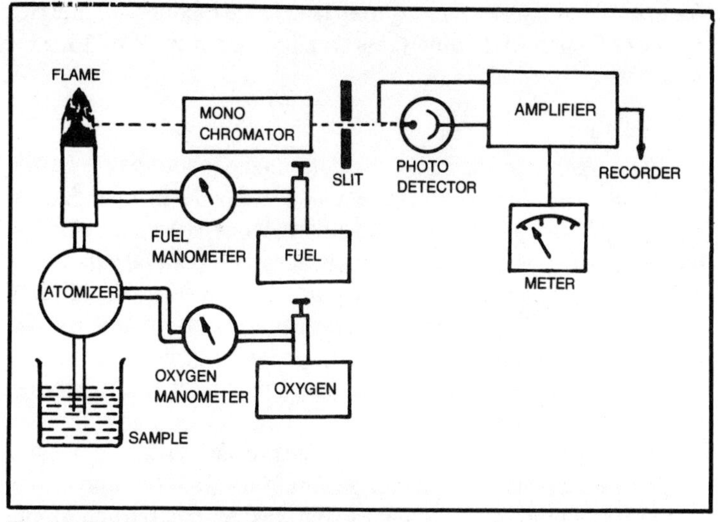

Fig. 4-2. Block diagram of a flame photometer.

and recording. Figure 4-2 shows a typical block diagram of a flame photometer. First, we will review the emission system.

Pressure Regulators

In order to obtain a steady emission reading, it is imperative to have a flame that is perfectly steady and free from flickers. To achieve this, the air or oxygen and fuel pressure is maintained constant during the operation of the instrument. Suitable pressure gauges are provided in the instrument to indicate the pressure actually present in the line. A 25-lb gauge for the oxygen or air supply and a 10-lb gauge for the fuel are generally used. Pressure regulators are usually followed by needle valves for control of flow. Gauges provided with the regulators are often not sensitive enough to detect small changes of pressure which have a profound effect on the flame photometer operation. Therefore, narrow range pressure regulators and manometers are installed in the line in order to observe small changes in pressure or gas flow.

Flow Meters

A *flow meter* may be inserted in the line from the gas reservoir to the atomizer in order to detect any clogging of the orifice. For the same flow rate, an appreciable change in the gauge pressure indicates a partially clogged orifice. By controlling the individual flow rates of the fuel and oxygen, the operator can choose various fuel-oxygen mixtures ranging from lean flame mixtures to fuel rich types of flames. The flow rates usually vary from 2 to 10 cubic feet/hour.

Fuel Supply

The fuel gas normally used in flame photometry is the acetylene gas, which is commercially available in cylinders of various sizes. Cylinder acetylene consists of acetylene gas dissolved in acetone which, in turn, is absorbed on a porous filling material. Consequently, after the flame is lit, it should be allowed to burn several minutes before adjustments are attempted in order to vent the excess acetone initially present in the vapor phase. Consumption of acetylene ranges from 1 to 5 cu. ft./hr. The other fuels used in flame photometry are propane, butane and hydrogen. When the available gas pressure is less or is variable, a booster pump is necessary. The pump generally used is a motor driven diaphragm pump which delivers the gas to the burner at the required volume and pressure.

Oxygen and Air Supplies

Oxygen from cylinders should be supplied to the burner through a regulator capable of delivering approximately 12 cu. ft./hour at a pressure of 12-15 lbs. per sq. inch. The air supply can be supplied from a cylinder of compressed air or from an air compressor through a tank held at about 10 lb. per sq. in. A pressure compensation valve is placed between the reservoir and the burner. When compressors are used, the air should be filtered through glass wool. Consumption of air is approximately 10 cu. ft. per hour for an acetylene air flame.

Atomizer

One of the most exacting problems in the flame photometer design is the manner in which the sample is fed to the flame at an uniform rate. The usual method is to prepare solutions of known concentrations and to spray these into the flame, using some form of aerosol production. Use of an aerosol permits the distribution of the same throughout the body of the flame rather than its introduction at a single point. In flame photometry, the name atomizer is given to a system which is used to form aerosol by breaking a large mass of liquid into small drops. This little device is responsible for introducing the liquid sample into the flame at a stable and reproducible rate. The atomizer must not be attacked by corrosive solutions.

There are two types of atomizers which are in common use. Some which introduce the spray into a condensing chamber and into the flame by the air of the combustible gas air mixture. Large droplets are removed in the condensing chamber. There are atomizers in which the sample is introduced directly into the flame, i.e., the atomizer and the burner are an integral unit.

Figure 4-3 shows the construction of the first type of atomizer. This is called the discharge-type atomizer and consists of two capillary tubes sealed into the walls of a glass chamber in such a manner that their bores are perpendicular to each other. The sample solution is poured into a funnel or drawn up from a container and is atomized by the blast of air from the tip of the other capillary. However, the atomized stream is composed of coarse spray with large droplets, which condense on the walls of the chamber and helical tube leading to the burner. The condensate flows down to the waste drain. The smaller droplets, in the form of a virtual fog, are carried by the air stream into the burner, where they are mixed with the burner gases and carried into the region of active

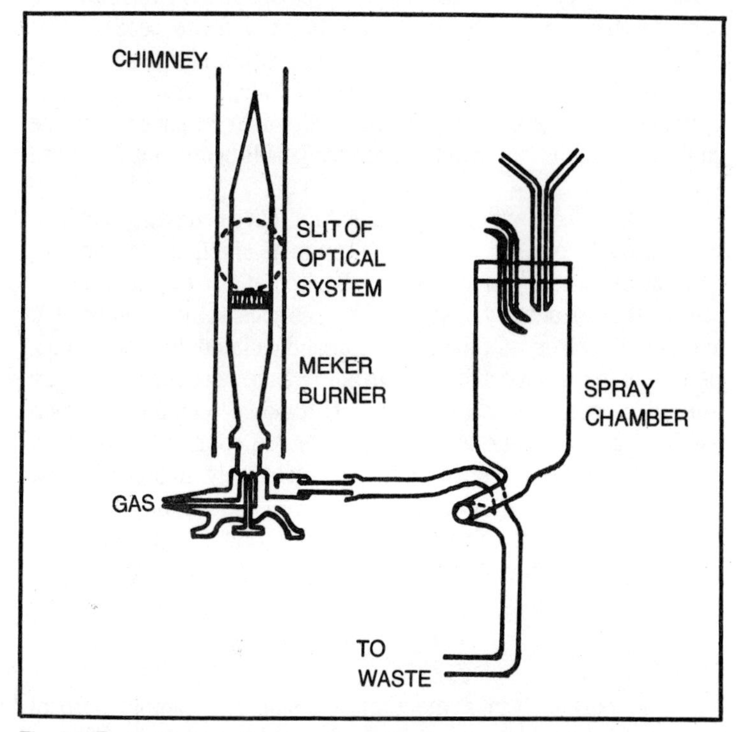

Fig. 4-3. Discharge type atomizer with condensing chamber and burner.

combination. Two removable hypodermic needles of stainless steel or glass are commonly used. With this type of atomizer, the consumption of the sample is comparatively high and ranges between 4 and 25 ml of solution per minute. Of this amount, only 5% actually reaches the flame. The sensitivity of this type of atomizer can be markedly increased by using a chamber which is heated by an electric heater placed around its walls. This hastens the process of vaporization of the solvent and produces an aerosol of very fine particles, all of which are swept into the burner. In this case, sample consumption is only 0.2 to 0.6 ml per minute and a substantial portion is carried directly into the burner to yield a much higher sensitivity.

The second type of atomizer is the integral burner-atomizer. The unit made of glass or metal is constructed of two concentric tubes. The sample solution is drawn through the innermost tube by the passage of oxygen through the orifice of the middle annulus. At the tip of the inner sample capillary, the liquid is dispersed into droplets. The outer annulus supplies the combustible gas to the flame. This type of atomizer is used in Beckman instruments. The body of the unit is machined from brass and the capillary for solution intake is of palladium. All droplets, both large and small, are introduced directly into the flame. Sample consumption is between 0.8 to 2 ml per min. Each atomizer requires separate calibration and is not strictly interchangeable with another of the same type.

Denton and Swartz (1974) maintain that pneumatic nebulization, though one of the most common methods of aerosol generation, produces a relatively wide droplet size distribution and limited aerosol density. Other techniques such as the spinning disk aerosol generator, atomizer impactor and isolated droplet generation techniques have also been used, but they produce only very limited concentrations of aerosol. Ultrasonic techniques, when used for aerosol generation, offer a unique combination of high solution to aerosol conversion rates relatively independent of carrier flow gas and have the ability to generate varying sized droplet populations. These workers describe the design of an ultrasonic nebulizer system for the generation of high density aerosol dispersions.

Burner

The burner brings the fuel, oxidant and sample aerosol together so that they may react safely and produce a good flame.

Burners used in flame photometry must fulfill the following conditions:

—When supplied with fuel and oxygen at constant pressure, the shape, size and temperature of the flame will remain constant.

—Assist in a perfect distribution of the mixture of gases and the aerosol which carries the atomized solution under analysis.

—Have a tip of suitable shape to produce a symmetrical flame and ensure a homogeneous flow and distribution of the gases, avoiding a strike back in the burner from accidental fluctuations in the feed system.

—Prevent condensation of the aerosol in the stem of the burner, which would reduce the effective quantity of the sample brought into the flame.

The most commonly used burner for low temperature flames is a Meker type. Here, the fuel gas issues from a small orifice and passes through a venturi throat where considerable air is entrained. The mixture of gas and the entrained air passes up the burner tube and burns at the top of the burner where the combustion is assisted by the surrounding air.

A deep grid of metal across the mouth of the burner prevents the flame from striking back down the tube. To screen the flame from air drafts, it is surrounded by a glass chimney which also protects the operator.

In the integral burner-atomizer unit, the sample solution is introduced directly into the flame. A cut section of this type of burner is shown in Fig. 4-4.

Flame

The flame is the most important part of the flame photometer since it forms the source in which the light radiations characteristic of the elements under analysis are produced. The flame performs the following functions. It converts the constituents of the sample to be analyzed from the liquid or solid state into a gaseous state, decomposes these constituents into atoms or simpler molecules and excites the resulting atomic or molecular species to emit light radiations.

In order to produce accurate results, the flame must be very stable. If its temperature or structure shows a change over a period of time, the emission produced by a given sample will also change.

The flame temperature must be high enough to excite the atoms to higher energy levels so that emission may take place. In

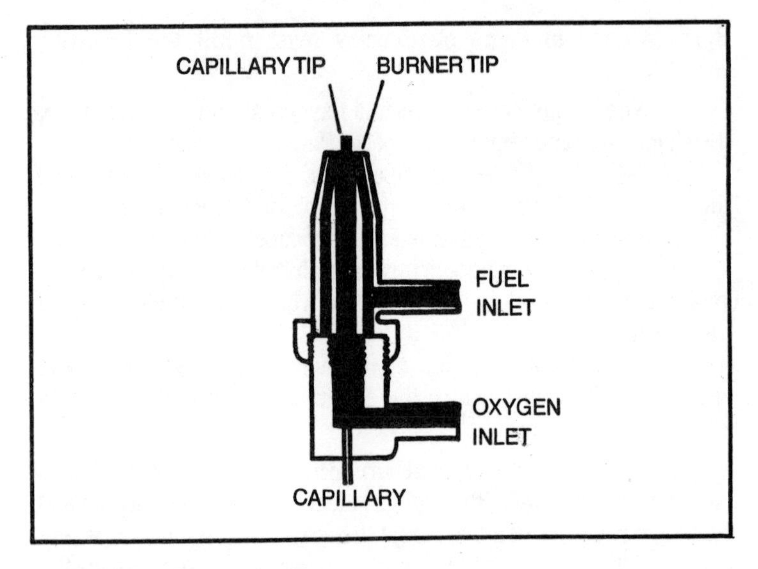

Fig. 4-4. Integral burner-atomizer.

general, the flames are produced by burning the gases given in Table 4-1.

All the fuels employed in flame emission spectroscopy produce both continuous background as well as certain band spectra. The nature and intensity of the spectrum for a given fuel is strongly dependent upon fuel-to-oxidant ratio and the flame temperature.

OPTICAL SYSTEM

The complex light emission produced in the flame by a solution containing one or more elements, makes it necessary to select the radiations of different wavelengths for each of the elements present in the solution. This job is accomplished by the optical system which collects the light from the steadiest part of the flame, renders it monochromatic and then focuses it onto the surface of the photodetector. A concave mirror is often placed behind the flame, with its center of curvature in the flame. In this way, the intensity of the emitted light is nearly doubled.

Filters

Wavelength selection may be done by optical filters or monochromators. Less expensive instruments make use of filters which may be absorption type or interference type. The usual glass or gelatin absorption filters have wide spectral bandwidths—35 to 45 mμ in width at one-half the maximum transmittance. Their overall transmittance is only 5-25%, decreasing with improved spectral isolation. Absorption filters do not give the degree of monochromaticity required for analyzing complex systems and, hence, the flame photometers so equipped are restricted in use to the determination of only sodium, potassium, calcium and lithium.

To improve the resolution, interference filters are often used. They are the most suitable for flame photometry as compared with absorption filters because they allow a much narrower band of wavelengths to pass. They are similar to monochromators in selectivity but are less expensive. Standard interference filters having spectral bandwidths from 10-17 mμ at one half the maximum transmittance are commercially available. The same characteristics are shown by continuous wedge filters which allow a continuous selection of different wavelengths. The continuity of an interference filter is achieved by using a spacer film of graded thickness between the two semi-transparent layers of silver. These filters usually have a working interval of 400 to 700 mμ. They provide a dispersion of 5.5 mμ per mm with a transmittance which is usually not higher than 35%.

Monochromators

Monochromators are incorporated in the instruments which are expected to provide better isolation of spectral energy. This requirement emanates when the spectrum lines are very close, very weak or both. By using narrow slit widths and sensitive detecting circuits, quite narrow bands of radiant energy can be isolated. There are two means of dispersing radiation, by refraction or by interference. The examples of the two types are prism

Table 4-1. Flame Temperatures for Various Gases.

Flame Temperature °C		
Fuel	**In Air**	**In Oxygen**
Propane	1925	2800
Butane	1900	2900
Hydrogen	2100	2780
Acetylene	2200	3050

monochromator and diffraction grating, respectively. In flame photometry, prism monochromators have been most widely accepted. They may be of glass or quartz. The glass prisms are only suitable for emissions essentially in the visible region whereas the quartz prisms enable many atomic lines to be studied that occur in the far ultraviolet region of the spectrum. In prism monochromators, emission bands or lines are chosen by local selection with movable exit slits or by prism rotation, where all the lines of the spectrum are passed through a fixed slit one after the other. The wavelength scale is non-linear as in prism spectrographs.

Dispersal of light may also be obtained by means of diffraction in a grating. A grating is produced by ruling grooves at extremely close intervals of the order of 15,000 grooves per inch on a highly polished surface. The dispersion is linear which permits a constant bandwidth to be used throughout the spectrum. The wavelength scale in this case is linear.

For the analysis of complex spectra, the resolving power should be very high. Resolving power is a measure of the ability of the monochromator to resolve two closely spaced lines of about equal intensity. Perhaps the most straight test of an instrument's resolving power concerns the separation of the manganese line at 403.3 mμ, the potassium line at 404.4 mμ and the lead line at 405.8 mμ from each other.

Other Optical Components

Reflectors and condensing lenses can be placed in the optical path to concentrate the light. A concave metal mirror with its focus in the flame itself is placed immediately behind the flame. Condensing lenses are used to render parallel beams of light before they fall on the interference filters. Some apparatus are fitted with an iris diaphragm in the optical path of the rays, which serves to reduce the sensitivity when using very concentrated solutions. Similarly, uniform metal grids with different size meshes may be used. The slit is of special importance in apparatus fitted with monochromators. Its width is more important than its length, because the sensitivity, the accuracy, the ratio of total emission/flame background and the resolving power depend on the width of the slit.

RECORDING SYSTEMS

After the isolation of the lines or bands emitted by the chemical elements under analysis in the optical system, the light intensities of these radiations have to be measured quantitatively. This is done by causing the radiation selected to fall on a photosensitive element. The electric current produced is then measured with a galvanometer, either directly or after amplification. The intensity of electric current produced is a function of the concentration of the solution under analysis. The photodetectors are either of photovoltaic or photoemissive type.

Photovoltaic cells have very little internal resistance and the current they produce cannot be amplified by dc amplifiers. The current is measured directly by connecting the terminals of the cell to a highly sensitive galvanometer. The spectral response of a selenium cell adequately covers various wavelengths encountered in the analysis of common alkali and alkaline earth metals. The sensitivity that can be achieved with these cells is considerable as they give currents of more than 5×10^{-10} amp, when they receive a light flux of more than 10^{-6} lumen.

Selenium cells are highly sensitive to temperature variations. The cells have a certain inertia and at least a quarter of an hour must be allowed for them to settle, with the burner lit before the readings are taken. A fall in efficiency, due to aging, is a well known characteristic of these cells, and they are also subject to fatigue upon exposure to prolonged illumination.

Phototubes of several types may also be used for detecting the radiations. They differ essentially from barrier layer cells in that they need an external electrical supply to provide a sufficient potential between the electrodes. Phototubes are less sensitive than barrier layer cells and the current produced usually needs amplification. Since they exhibit large internal resistance, amplification of the cathode current by an external circuit can be conveniently done.

The preferred detector for weak light intensities is the photomultiplier tube. This is many times more sensitive than the simple phototube which, in turn, permits greater spectral resolution without loss of signals. They are available with envelopes of ordinary glass or with quartz window inserts for use in the ultraviolet region below 350 mμ.

ELECTRICAL CIRCUITS

The final step in recording the light radiations produced in the source of emission consists in transforming the electrical signals produced by the selected radiation in the photosensitive element into a reading, which is related to the concentration of the solution under analysis.

Single Beam Instruments

For selenium photocells, a spotlight, multireflection galvanometer is quite suitable. The galvanometer may possess a maximum sensitivity ranging from 0.007 to 0.0004 μa per mm scale division and a coil resistance of approximately 400 ohms. As the cells are generally of low resistance when illuminated, rather low values of resistance must be used in the external circuit. Figure 4-5 shows a simple arrangement for measuring directly the output of a single barrier-layer cell. This type of circuit is suitable for a single-beam instrument which contains only one set of optics. The photoelectric current generated in the photo cell by the light is given to the galvanometer through the coarse and fine sensitivity controls at B and A respectively. Selenium cells exhibit no dark current which would require suppression, but a suppressor circuit might be useful in bucking out a reading resulting from the flame background.

Double Beam Instruments

In double beam instruments, provision is made for a second light path which is emitted by an internal standard element. The internal standard and unknown wavelengths are isolated by means of optical filters in a dual optical system and each beam is focused on a photoelectric cell (Fig. 4-6). The circuit is so arranged that the photoelectric currents produced by the emitted radiations of the unknown and internal standard elements oppose each other through a galvanometer. By means of an accurate potentiometer (P_1), the opposing photoelectric currents can be balanced. The settings of the potentiometer required to produce balance are

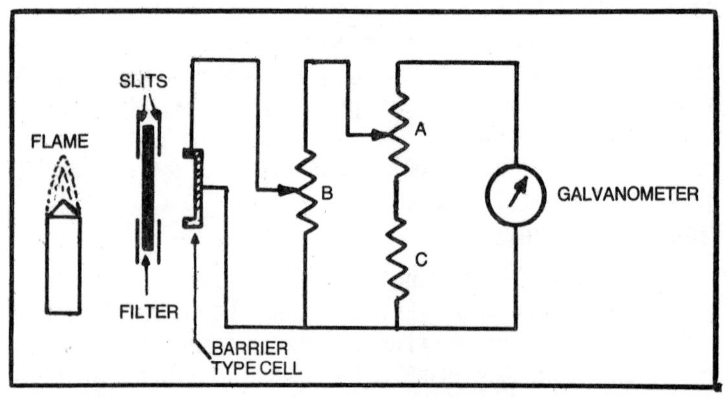

Fig. 4-5. Circuit diagram for single beam direct-reading flame photometer.

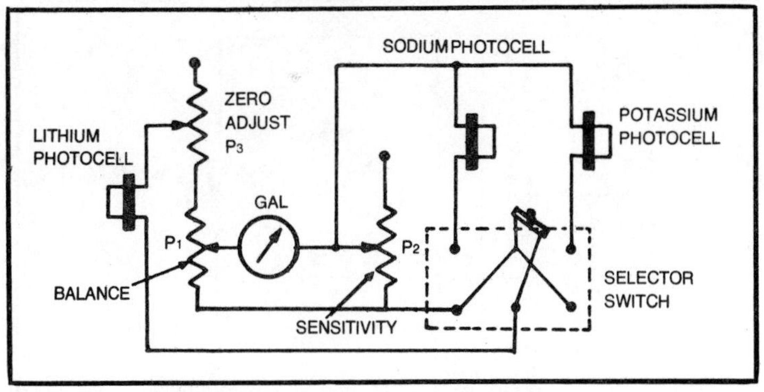

Fig. 4-6. Circuit diagram for double beam flame photometer.

calibrated in terms of known solutions. Sensitivity adjustments are made with another potentiometer (P_2) by shunting currents generated by the measuring cell. This makes it possible to read more concentrated solutions of the test element without changing the concentration of the internal standard used.

Circuits Used With Photomultiplier Tubes

The output current from the last amplifying stage in a photomultiplier tube for many solutions and concentration ranges normally employed often lie between 10 to 100 μa and can be directly measured by a spotlight galvanometer (Fig. 4-7) or simply a rugged pointer type micrometer. Photomultiplier tubes possess an appreciable output current even when the tube is dark and therefore, for direct measurements, it will be necessary to carry out a background correction before measuring. This is done by compensating with an auxiliary battery and adjusting R_2. The sensitivity of the photomultiplier tube can be adjusted by varying the voltage per dynode or preferably by variation of the voltage applied to the first dynode.

Amplifiers For Phototube Circuits

The current from the phototube is generally so small that it is necessary to amplify it, quite often by a factor of 10^5 to 10^6. The amplification of a direct current can be achieved with the help of vacuum tube amplifiers, with special electrometer tubes or other vacuum tubes operating under electrometer conditions used in the first stage. Vacuum tube amplifiers are voltage operated devices and therefore the changes of phototube current must be converted

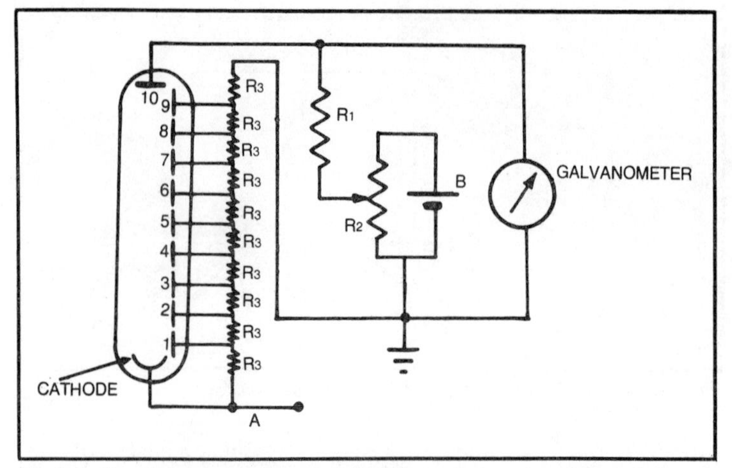

Fig. 4-7. Circuit diagram for direct reading flame photometer using a photomultiplier tube.

to voltage changes by having a resistance in series with the phototube. This resistance is usually kept very high and is of the order of 10,000 megohms. The amplifiers are throughout direct coupled. A serious drawback of dc amplifiers is the zero drift, which is a slow, continuous shift of the galvanometer pointer. Note that dc amplifiers are also sensitive to variations in supply voltage and temperature changes. Although amplifier circuits consisting of vacuum tubes are still found in the instruments, integrated circuits with high input impedance are now available which have greatly simplified the circuit construction.

Amplifiers For Photomultiplier Tube Circuits

Though the photomultiplier tube itself provides a large amplification, this can be appreciably increased by an amplifying circuit. The last dynode of the photomultiplier tube is usually connected to the grid of the amplifying valve and at the same time to a grid leak, thus causing a potential difference, the variations of which are amplified.

In order to reduce the rapid and random fluctuations, a long enough time constant is incorporated in the amplifier circuit to aid in the integrating process. Thus, the meter is rendered unresponsive to sudden flashes of light in the flame. Usually, the time constant is chosen to be approximately 1 second and is obtained by connecting a capacitor in parallel with the coupling resistor.

To overcome the problem of drift of dc amplifiers, the light from the flame can be interrupted, and the photoelectric current can be amplified by an ac amplifier. A synchronously rotating disc is placed before the flame and the photoelectric current is thus converted to a chopped dc signal which can then be amplified by an ac amplifier.

Figure 4-8 shows a circuit diagram using solid state components, which can be used to amplify current obtained in a photomultiplier. The output of the PMT is applied to the gate to FET Q_1. Q_1 inverts this negative going signal to positive and applies this to the non-inverting side of operational amplifier Z_1. Due to divider network R_9 and R_8, the gain of the operational amplifier Z_1 is held at 10. The output of the circuit can be directly connected to a moving coil meter. Digital display tube flame photometers are now available in which the analog voltage is converted to digital form and then given to nixie tubes or LEDs for numerical display.

Recording Type Flame Photometers

Recording flame photometers are useful when it is necessary to determine the background and consequently the emission intensity of the line or band more accurately than is possible on a direct reading instrument. In instruments of this type, the measured emission intensity of an element in a flame is recorded

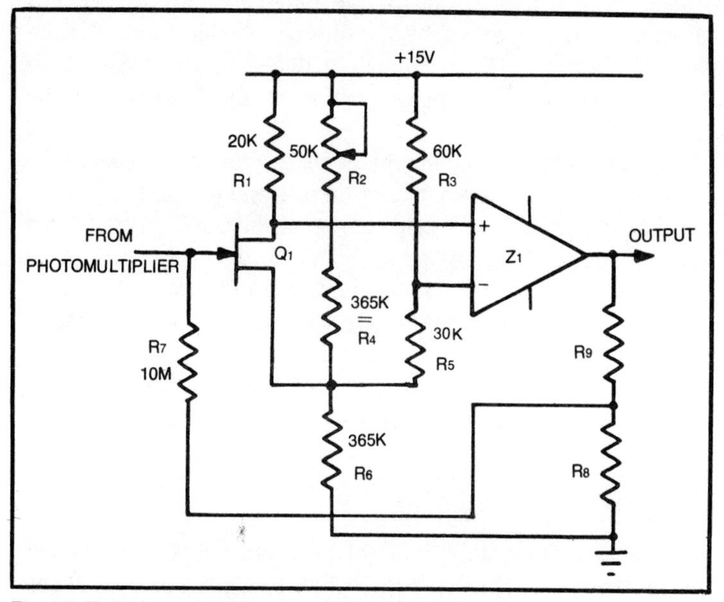

Fig. 4-8. Typical amplifier circuit used in flame photometers.

by means of an electronic recorder or a moving paper strip in the form of intercepts of various lengths. The wavelength selection is carried out by attaching a spectrum scanning mechanism. This mechanism consists of a synchronous motor, which rotates the shaft of the prism rotation drum of the monochromator through a reducing gear. The recorder normally used is a *potentiometric* type with a signal range of 0-50 mV.

COMMERCIAL DESIGNS

Fig. 4-9 shows the general arrangement of the EEL flame photometer. Air is introduced round the atomizer nozzle through the inlet (F) at a controllable pressure. The pressure is indicated by a pressure gauge provided on the front panel of the instrument. Gas enters the mixing chamber through the tube (M) and is likewise controllable. The air flow round the atomizer is applied to draw the sample (B) up the capillary tube (C) and through the atomizer, which injects it in a fine spray into the mixing chamber (D). Larger droplets falling from the air-stream pass through the drain tube (E). The baffles (G) produce an even mixture of gas, air and sample which passes to the burner (L) and is ignited to burn with a broad, flat flame within a well-ventilated chimney.

Light from the flame is collected by the reflector (H) and focused by the lens (I) on to the photocell (K), the critical wavelength being isolated by the interchangeable filter (J). The photocell output is then applied to deflect a taut suspension reflecting galvanometer in proportion to the intensity of the emitted light.

To maintain a steady supply of clean, compressed air to the instrument, it is necessary to use oil free and portable compressors. The compressors are generally required to give an output of 6 liters/minute at 0.7 Kg/cm^2 (10 lb/in^2) and make use of an asynchronous motor 1/20 hp and running at 1400 rpm.

The instrument provides full scale deflection with 2 ppm sodium, 3 ppm potassium, 15 ppm lithium and 40 ppm calcium. It needs only 2 ml sample and is provided with a sampling attachment, which enables the operator to make a series of tests with rapidity.

CLINICAL FLAME PHOTOMETERS

The flame photometer is one of the most useful instruments in clinical analyses. This is due to the suitability of the flame photometer for determining sodium, potassium and calcium which

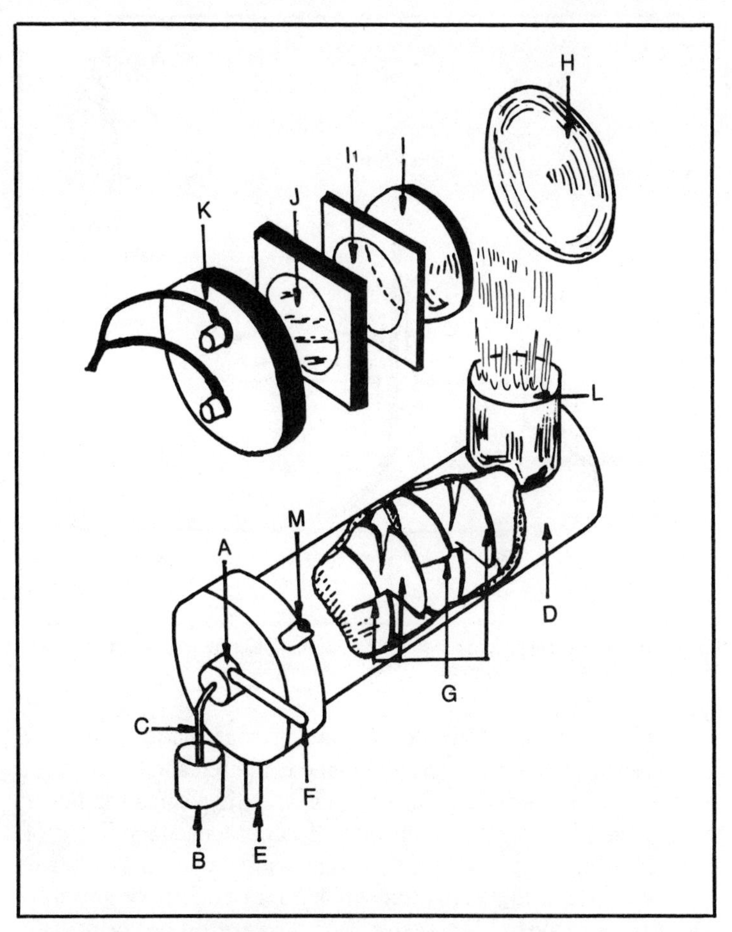

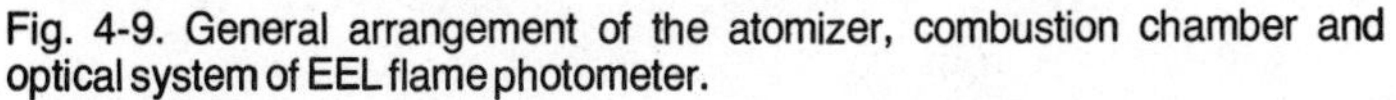
Fig. 4-9. General arrangement of the atomizer, combustion chamber and optical system of EEL flame photometer.

are of immense importance in the development of the living being and indispensable to its physiological functions. In the clinical analysis of sodium and potassium, the flame photometer gives, rapidly and accurately, numerous differential data for normal and pathological values.

KLiNa flame system of Beckman is a dedicated instrument for simultaneous analysis of sodium and potassium. In this instrument, sample handling is automatic, as the system has a turntable which will hold up to 20 samples in cups and an automatic positive piston displacement diluter that dilutes the sample prior to entering the spray chamber (Fig. 4-10).

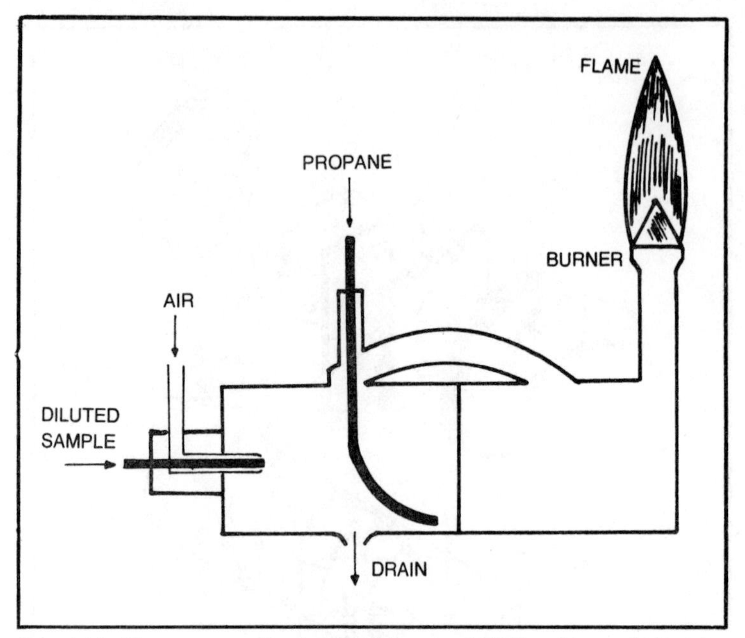

Fig. 4-10. Flame spray chamber and burner of KLiNa flame photometer (courtesy of Beckman Instruments, USA).

Ignition and shutdown of the flame are automatic. When the calibrate button is depressed, the flame is ignited and the circuits are energized. Pressing a standby button extinguishes the flame but maintains thermal equilibrium. Twenty-four microliters of sample are aspirated for simultaneous sodiums and potassiums, providing microsample analysis suitable for pediatric or geriatric work. After anaylsis, the results are displayed directly in millimoles per liter or milliequivalents per liter.

For the precise and accurate determination of Na and K concentrations, use is made of the fact that lithium normally not present in significant concentrations in serum exhibits about the same flame emission characteristics as Na and K. Lithium ions are added to the diluent used for samples, standards and controls. The lithium in the diluent is referred to as the internal standard.

The diluted sample containing the fixed known amount of lithium, in the form of a dissolved salt, is nebulized and carried by the air supply into the first of two compartments in the spray chamber (Fig. 4-11). Heavier droplets fall out of the stream onto the chamber walls or separate from the stream upon striking a partition in the chamber and flow to a drain from the compartment.

Propane enters the first chamber to mix with the air and sample stream and carry it through a tubular glass bridge into a second compartment. The aerosol and propane mixture travels up from the chamber to the burner head, where the mixture is burned. Exhaust gases are vented to room air from a cover located on top of the instrument.

To provide internal standardization, the response of the sodium and potassium detector is a ratio function of the response by the lithium detector. Thus, any change in air flow rate or fuel pressure that may affect the sample would proportionately affect the lithium detector.

The KLiNa flame photometer determines and provides digital display of sodium and potassium or lithium concentrations in a sample by responding optically and electronically to the intensity of the principal emission lines that characterize each ion as it is excited in the propane and air flame.

The flame is monitored continuously by three photomultiplier detectors. Each detector views the flame through an optical filter that passes only the wavelength band of interest to the particular detector. The sodium detector, therefore, responds only to wavelengths in a narrow band centered at 591 millimicrons; the potassium detector responds only to wavelengths in a narrow band centered at 768 millimicrons; the lithium detector responds only to wavelengths in a narrow band centered at 671 millimicrons.

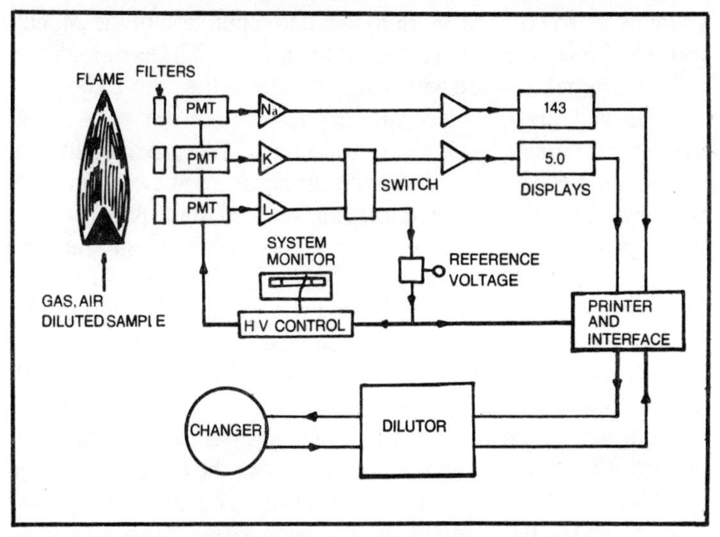

Fig. 4-11. Schematic diagram of Beckman KLiNa flame photometer.

The precision specifications for the KLiNa flame system are ± 0.2 mmol/liter for potassium and lithium, and ± 2.0 mmol/liter for sodium. Potassium and lithium both show linearity to 20 mmol/liter, while sodium is linear to 200. In addition to the 0-20 scale, potassium may be rescaled to readout to 200 mmol/liter for convenient analysis of urine samples.

By observing simple precautions and maintenance procedures, flame photometers have been found to give a long working life. The following procedures may be followed to correct any malfunctioning of the instrument. If the instrument shows low sensitivity, it may be due to a blocked atomizer, low gas pressure or faulty photocell. Necessary corrective steps should be taken to clean the blockage of the atomizer, restoring the gas pressure or changing the photocell. The reasons for varying or intermittent readings are usually a contaminated burner, faulty photocell, low gas pressure, blocked atomizer, improper air supply and excessive vibration. The instrument must be placed on a shock proof base. Appropriate checks must be made for the above faults.

The Corning EEL Model 450 is a low temperature internal standard flame photometer (Fig. 4-12) designed specifically for the determination of sodium, potassium and lithium in clinical samples. Results are read directly from two Digital Panel Meter (DPM) displays. Lithium determinations are performed separately and the result is presented on the potassium DPM display, the sodium DPM display being blanked off. The Model 449 Diluter is available as an optional extra to facilitate operation of the Model 450. The Model 449 is a peristaltic pump, which dilutes the sample and the internal standard with distilled water in the ratio 1:100.

Air and fuel are fed into the instrument via rear panel connectors. The fuel is filtered, reduced in pressure and fed to a ratio regulator which maintains the air pressure at twice the fuel pressure. The fuel is also fed to a fuel solenoid which is only open, and kept open; when the air pressure is above 9 psi, the fuel pressure is above 22 psi, the power supply is connected, the power push-button is on and the flame is alight (except for the 20 seconds ignition sequence when the fuel solenoid is held open although the flame is not lit). The fuel is fed from the fuel solenoid to the fuel needle valve and then to the mixing chamber. The air supply is filtered and fed to the ratio regulator, where the pressure is regulated to 14 psi. From the ratio regulator the air is fed to both the air sheath needle valve and the nebulizer. The air sheath needle valve provides fine control of the flame shape. The nebulizer

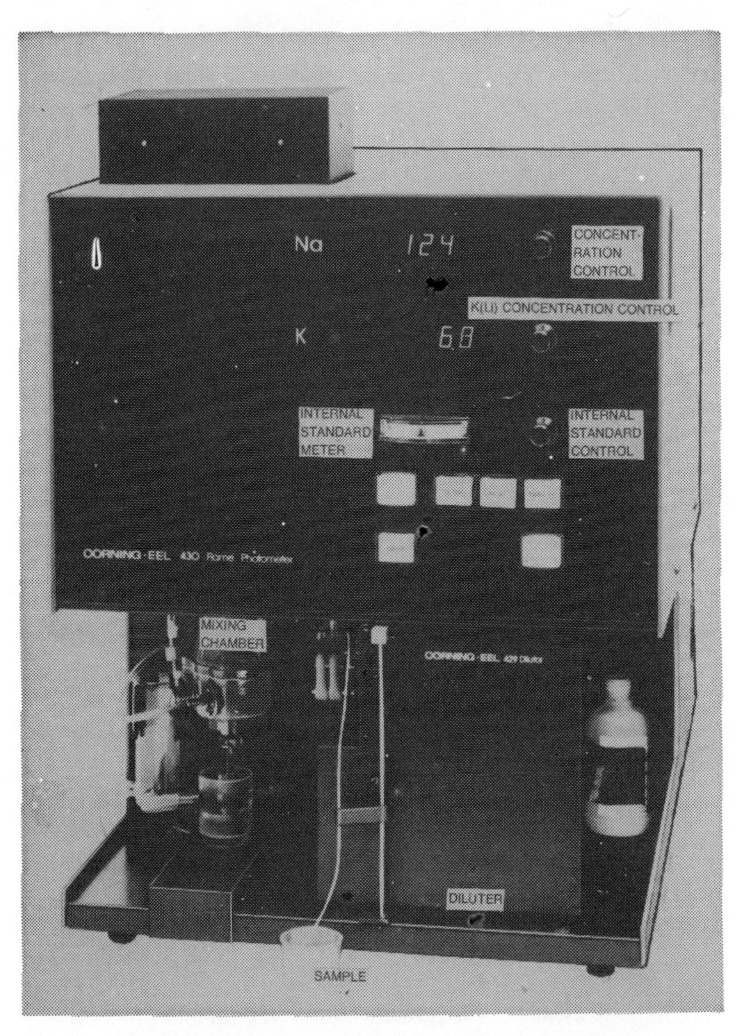

Fig. 4-12. Digital display type flame photometer (courtesy of Corning EEL, UK).

aspirates the sample, with the air supply at 14 psi, and a fine mist is formed which is mixed with the fuel in the mixing chamber. The combined sample/fuel mixture is burned in the burner assembly within an air sheath that provides extra oxidant and prevents the ingress of impurities into the flame.

Lithium is added to all samples, as a reference element, to overcome the effects of random fluctuations due to variations in flame temperature, aspiration rate and various chemical interferences. The unknown sodium and potassium signals are ratioed against the lithium signal to ensure that a steady readout is obtained. When lithium is to be measured, internal standard principles are used, but potassium is added as the reference element. The concentration of potassium as the internal standard is far in excess of that present in sera. To avoid incorrect readings, due to the potassium already present in serum samples, the standards contain potassium (and other elements) at the serum level.

Light radiated by the flame is filtered by three separate optical filters. There is one filter for each photomultiplier tube and the filters will only pass light at a specific wavelength. This ensures that light radiation emitted at the K wavelength (766 nm) only impinges on the potassium photomultiplier, light radiation at the Na wavelength (589 nm) only impinges on the sodium photomultipliers, and light radiation at the Li wavelength (671 nm) only impinges on the lithium photomultiplier. The photomultipliers convert the light inputs into electrical signals. These three signals are amplified and linearized. The internal standard signal (lithium when sodium and potassium determinations are being performed) is ratioed against the unknown sample. Thus, an accurate result is displayed on DPM because any increase or decrease in the sample signal, due to flame variations, is also present in the internal standard signal.

A moving coil meter, mounted on the front panel, gives a constant indication of the level of the internal standard contained in the sample. If the reading is outside the prescribed limits, the DPM displays are blanked off.

The technical specifications of this instrument are as follows.

Sample Requirement:	150 μl with diluter, 6 μl manual.
Response Time:	22 seconds with diluter, 10 seconds manual.
Aspiration Rate (50 Hz):	Nominally 0.58 ml/minute with diluter, 3.5ml/minute manual (diluted sample).
Air Supply:	Flow rate approximately 12 liters/minute at 1.4 kg/cm^2 (20 psi).
Fuel:	Commercial propane, at a flow rate of approximately 400 cc/minute, operating pressure 2.1 kg/cm^2 (30 psi).
Linearity:	Nominally 1% on all ranges.
Dilution Ratio:	1 in 100 for all determinations and for the internal standard.

ACCESSORIES FOR A FLAME PHOTOMETER

Modern flame photometers come with many useful accessories. For example, the KLiNa flame photometer of Beckman comes with a diluter sample changer and paper tape printer.

The diluter is a motor driven cam-programmed system that functions through a cycle of operations. These operations involve sample pickup and transport to an internal mixing cup, the addition of a measured volume of diluent, mixing to ensure a properly prepared sample *aliquot*, coupling of the mixing cup to the spray chamber, so that sample aspiration can occur, and finally the washing and draining of the mixing cup. The diluter uses positive displacement pumps to assure exact sample dilutions in the operator selectable ratios of 50:1, 100:1 or 200:1.

The automatic changer enables automatic presentation to the diluter sample probe of up to 20 successive samples. It is a turntable which rotates stepwise to locate each sample cup under the extended and down position of the diluter sample probe. The probe is extended from the diluter once for each sample determination. The probe tip enters the sample and a measured volume is taken for transport to the diluter mixing cup. Individual sample trays are placed on the changer turntable. Each tray can hold 20 sample cups. The sample cup could be 0.25 ml, 0.5 ml or 2.0 ml size. The 2.0 ml size is generally recommended.

The printer provides a paper tape printout to record the result of each sample determination. Each printout is a data line that, from left to right, reads the sample position number in the changer sample tray, the type of determination performed and the results expressed in milli-equivalents per liter.

EXPRESSION OF CONCENTRATION

The concentration of solutions is usually expressed in parts per million (ppm) in flame photometry. This type of expression enables one to make easy calculations on dilute solutions and the concentrations can be expressed in weight/weight, weight/volume and volume/volume ratios. The ratios can be expressed in volume if the density of the liquid is near unity and hence 1 ml = 1g. This is true especially in dilute solutions whose density is approximately that of distilled water. The equivalents are as follows:

$$1\text{ ppm} = 1\text{ mg/l} = 1\text{ mg/kg}$$
$$1\text{ ppm} = 1 \times 10^{-3}\text{ g/l} = 1 \times 10^{-3}\text{ mg/ml}$$
$$1\text{ ppm} = 1 \times 10^{-6}\text{ g/ml}$$

INTERFERENCES

Flame photometry is subject to several types of interferences which arise during the course of analysis through variations in any of the parts of the instrumental system or in the composition of the samples. These interferences give rise to experimental errors.

Flame Background Emission

A flame is an extremely complex mixture of reacting gases. Since these gases are hot, they emit a certain amount of light even when no sample atoms have been injected into the flame. This light is the flame background. The lines produced by a sample are superimposed on this background. Failure to correct properly for the background reading can be a source of serious error. It is desirable to eliminate, as far as possible, the flame background and other emission lines from the measurements. The narrower the bandwidth of the monochromator used, the more efficient is this elimination.

Direct Spectral Interference

This interference occurs when the emission lines are so close together that the monochromator cannot separate them. For example, the emission of the orange band of CaOH interferes with the sodium line at 589 $m\mu$. Interference is more serious when absorption filters are used in place of monochromators. In this case, the two lines overlap partially or completely and will be read together in proportion to the degree of overlap. If the interference cannot be obviated by increased resolution, the difficulty must be overcome by prior removal of one element, perhaps by selective solvent extraction.

Self Absorption

Interference due to self absorption occurs due to the absorption of radiant energy through collision with atoms of its own kind present in the ground energy level. If some of the radiant energy is self-absorbed, the strength of the spectral line is weakened. Self absorption primarily depends upon the number of atoms present in the ground state, the concentration of atomized solution and by the probability that these atoms will be excited by the incident radiation from excited atoms of its own kind. Self absorption is insignificant at very low concentrations of a test element.

Effect of Anions

Anionic disturbances result from a radiation in the number ot free metallic ions by stable combination in the flame with the anions. There is usually no interference with concentrations less than 0.1 M. Above this concentration, sulfuric, nitric and phosphoric acids, in particular, show a marked effect in lowering metallic emission. For example, the strong depression of the calcium emission in the presence of phosphate, aluminate and other similar anions has been fairly well known. In practice, over limited intervals of concentration, the depression is linear. This forms the basis for indirect determination of the depressant in the presence of a standard amount of calcium.

Effect of Ionization

The easily excitable alkali metals such as potassium show ionization through excessive dissociation and consequently result in the depletion of the number of available neutral atoms that can be excited. This, in turn, weakens the intensity of the atomic spectrum, whereas any ionic spectrum is strengthened. When small quantities of easily ionized metals are added to a flame, the number of neutral atoms tend to increase more rapidly in comparison to the ionized atoms than proportionally to the concentration of metal sprayed into the flame. As a result, the curve of intensity versus concentration may initially be concave upward.

Solution Characteristics

The physical properties of the solutions strongly influence the atomization rate, which in turn influences the luminescence of the flame. The main factors responsible for this are the differences in viscosity, surface tension, density, volatility and temperature between standards and samples. For example, an increase in viscosity lowers the atomization rate, with a consequent weakening of the emission, causing errors and loss of sensitivity. Similarly, variations in surface tension are considered as another source of error. They affect the emission in a similar manner to viscosity through their effect on atomizing conditions. Added salts and acids hinder the evaporation of solvent. Large droplets result in a diminished quantity of aerosol reaching the burner because the larger droplets are more likely to settle upon the walls of the condensing chamber.

There are several standard ways of eliminating or avoiding the influence of the previously mentioned factors. The techniques of overcoming interferences demand high skill and experience. With the development of atomic absorption spectrophotometry, in which the problems of interference and flame stability are not as severe, flame emission analysis has been largely limited to the determination of sodium, potassium and, less frequently, lithium, calcium, manganese and copper in relatively simple mixtures.

PROCEDURE FOR DETERMINATIONS

In order to determine the concentration of an unknown element X in a given sample or a series of samples, the instrument is first standardized by making a calibration curve. For this, it is desirable to know the probable range of concentration of X in the sample so that a series of calibration standards can be made up, having a suitable spread over the range expected. The instrument is first adjusted to give a zero meter reading when pure distilled water is aspirated into the flame. The instrument settings are then adjusted to give the full scale reading with the most concentrated standard sample. Without altering this adjustment, emission intensities for the remaining standards and the unknown are measured. Calibration curves are thus plotted and the concentrations read off against this calibration.

Calibration Curve Method

A background correction to the calibration data can be made with a spectrophotometer by measuring the radiation intensity on either side of the peak at points sufficiently removed, so that the line spectrum contributes nothing to the radiation. Otherwise, the background is estimated from a blank containing all components of the sample except the one being determined. Instrument response is proportional to the concentration and a straight line is obtained. The concentration of the unknown is obtained from the calibrated curve.

Since it is difficult to obtain a flame of high long-term stability, it is usual to check the calibration at frequent intervals by spraying a standard. This ensures that drift can be measured and taken into account. In fact, calibration curves are difficult to reproduce on the same day—virtually impossible on different days.

The concentrations to which the samples are diluted for measurements can often be selected if the concentration in the original solution is much greater than the detection limit of the element concerned. Careful choice of the concentration is very important as it has a considerable effect on the accuracy of analysis for the following reasons. Calibration curves are not necessarily linear. It is found that the curves for a given element are linear over a limited range. At stronger concentrations, the same element would give a curved calibration. In case of a high total concentration of solids in the flame, the flame background is greatly increased and the burner may be fouled. If the viscosity of the sample is too high, it may not spray efficiently. If the solution is diluted excessively, errors of dilution may occur and the signal to noise ratio may decrease.

Standard Addition Method

A number of radiation interferences can be avoided by using the method of standard addition. In this method, net emission readings are obtained on two solutions, solution A containing an aliquot of the unknown solution and solution B containing the same quantity of unknown solution plus a measured amount of a standard solution of the element. The amount of test element in each of the two solutions is then determined from their measured emission intensities and the standard calibration curve. Subtracting the quantity of unknown found in solution A from that found in solution B gives the amount of test element equal to that added, when there is no depression or enhancement. When one of these effects is present, the corrections are applied which correct for the interference. It is necessary that the calibration curve be linear over the range of concentrations employed.

Internal Standard Method

The flame system is a critical part of the flame photometer and its long term stability is dependent upon fuel flow, air pressure, nebulization rate and burner/atomizing chamber temperature. While the first two factors can be controlled effectively to within acceptable limits by suitable instrument design, the other two are less easily controlled in this way. To overcome variation from these sources, an internal standard procedure has been developed.

In this method a fixed quantity of the internal standard element is added to the blank sample and standard solution. The radiation of the element are measured together with those of the elements under analysis by dual detectors or by scanning successively the two emission lines. The ratio of the emission intensity of the analysis line to that of the internal standard line is plotted against the concentration of the analysis element on log-log paper to prepare the calibration curve for a number of standards. Corrections for background radiations are made for each line. The plot of log (emission ratio) versus log (concentration of test element) will produce a straight line whose slope ideally is 45° over limited concentration intervals. On most double beam instruments, this ratio is given by the reading of the balancing potentiometer. Calibration curves may be drawn on a linear coordinate paper. As a ratio, rather than an absolute light intensity that is being measured, fluctuations due to variable operation of the nebulizer—burner system are greatly minimized. This procedure also reduces errors due to differences in the composition of the test and standard solutions.

The choice of a suitable element and spectral lien depends upon the following analytical conditions.

- The element should give spectral lines that are close to those of the analysis element so that it is not necessary to use a different type of photodetector. Thus, lithium 670.5 mμ line is a suitable reference for the analysis of potassium at 760.5 mμ and sodium at 589 mμ, but it is less satisfactory for the determination of calcium at 422.7 mμ.
- The element must be one that is not present in the samples and must be available in a high state of purity.
- It must be added in concentrations very near to those of the analysis element so that radiation of approximately equal intensities may be obtained. For this reason, the concentration of lithium is normally adjusted to 100 $\mu g/cm^3$ (15 meq) in solutions where the sodium and potassium concentration lies in the 0 to 100 $\mu g/cm^3$ range.
- The reference element must produce radiations that are perfectly selectable.
- When added, the element must not alter the physical properties of the solution.

For analysis of alkali metals and even of other elements, lithium has been preferred as an internal standard, as it satisfies most of the previously mentioned requirements.

Chapter 5

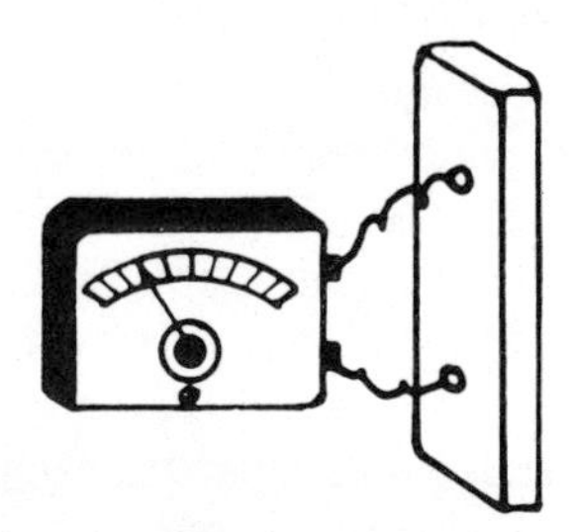

Atomic Absorption Spectrophotometers

Atomic absorption spectroscopy is an analytical technique based on the absorption of radiant energy by atoms. It was explained in the chapter on flame photometry that when a dispersion of the atoms of a sample is produced in a flame, some of these atoms get thermally excited and emit characteristic radiation as they return to the ground level. Most of them, however, remain in the ground state. When a beam of light is made to pass through the flame, a portion of it will be absorbed by dispersed atoms in the same manner that a beam of light passing through a solution will be absorbed by the dispersed molecules of a solute. It is possible to find a series of absorption bands corresponding to the energy levels of the atoms sprayed into the flame. The wavelength of the bands is characteristic of the atoms of the element concerned, and the absorbance of the band is proportional to the concentration of the atoms in the flame.

The potential of atomic absorption spectroscopy for the determination of metallic elements in chemical analysis was first realized by Alan Walsh who, during the mid 1950s developed it into its modern form. Standard commercial equipment became available in about 1960. Since that time, the use of this technique in routine analysis has become widespread where it replaces many traditional wet methods for the estimation of metals in solution.

The versatility of atomic absorption spectroscopy is demonstrated by the fact that it permits the estimation of between

60 and 70 elements at concentrations that range from trace to macro quantities. It is applicable to the estimations of metals in organic and mixed organic-aqueous solvents as well as to those in aqueous solution.

In atomic absorption spectrophotometry, the absorption lines are very narrow, approximately 0.02°A wide. This is because, being atomic bands, they do not have broadening due to rotational or vibrational structure, as is the case in molecular absorption bands. However, the lines are very weak because there would be relatively few atoms in the light path. Also, a narrow absorption band of the atoms in the flame necessitates the use of a very narrow source line and, therefore, a continuous radiation source like a tungsten or hydrogen lamp cannot be used in atomic absorption spectrophotometers. Since a different hollow cathode lamp is required for each element to be determined, atomic absorption is practically useless for qualitative analysis. Quantitative analysis is easily performed by a measurement of the radiation absorbed by the sample. As little as 0.01 to 0.1 ppm of many elements can be determined.

In technique, atomic absorption spectrophotometry is quite similar to flame photometry. In principle, the two methods are complementary and in combination provide a powerful analytical tool. Whereas the flame emission technique measures energy that is emitted by atoms, atomic absorption measures the energy that is absorbed by the atoms. Atomic absorption is able to make use of a much larger number of metal atoms in the flame that do not acquire sufficient energy to emit light. These are the atoms in the *ground state*.

Each element has its own unique absorption wavelengths. A *source* is chosen for the element to be tested which emits the characteristic line spectrum of that element. The sample, normally in liquid solution or fine suspension, is sprayed into a *flame*. The flame vaporizes the sample and puts the atoms in a condition where they can absorb energy. The atoms must be chemically uncombined and in their minimum energy or ground state. A *monochromator* selects a characteristic sample line and eliminates adjacent lines. A *photodetector* measures the light passing through the flame both before and after introduction of the sample into the flame.

The amount of light absorbed by the atoms in the flame is proportional to the concentration of the element in the sample solution. By comparing the absorbance reading on a meter with a calibrated chart, showing the relationship of absorbance versus concentration made from data of known concentrations of the element, the unknown concentration of the element in the sample may be determined.

CONSTRUCTION DETAILS

Figure 5-1 shows a block diagram of the atomic absorption spectrophotometer. The source of radiation is a hollow cathode lamp whose cathode is made of the element to be investigated. This lamp emits the line spectrum of that element (Fig. 5-2A).

The sample is sprayed into the flame and the atoms of the element are dispersed in the gaseous phase. The atoms absorb the radiation only at a particular line called the resonance line (Fig. 5-2B) giving the result as shown in Fig. 5-2C. The resonance line gets diminished after passing through the flame containing the sample vapor, while all other lines remain unaffected. The other lines are removed by a monochromator which selects a band of wavelengths around the resonance line and rejects all other lines (Fig. 5-2D). The photodetector measures the diminished resonance line (Fig. 5-2E) and displays it on a suitably calibrated scale. Atomic absorption obeys Beer's Law and calculations for concentration can be performed accordingly.

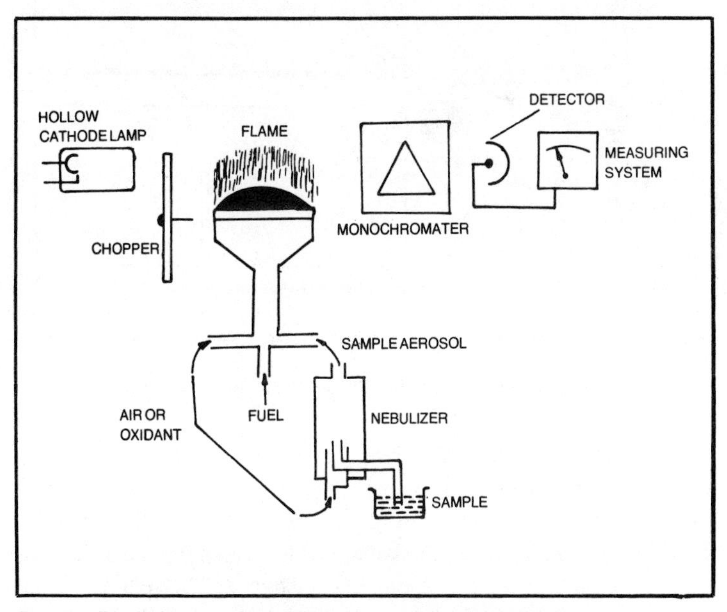

Fig. 5-1. Block diagram of an atomic absorption spectrophotometer.

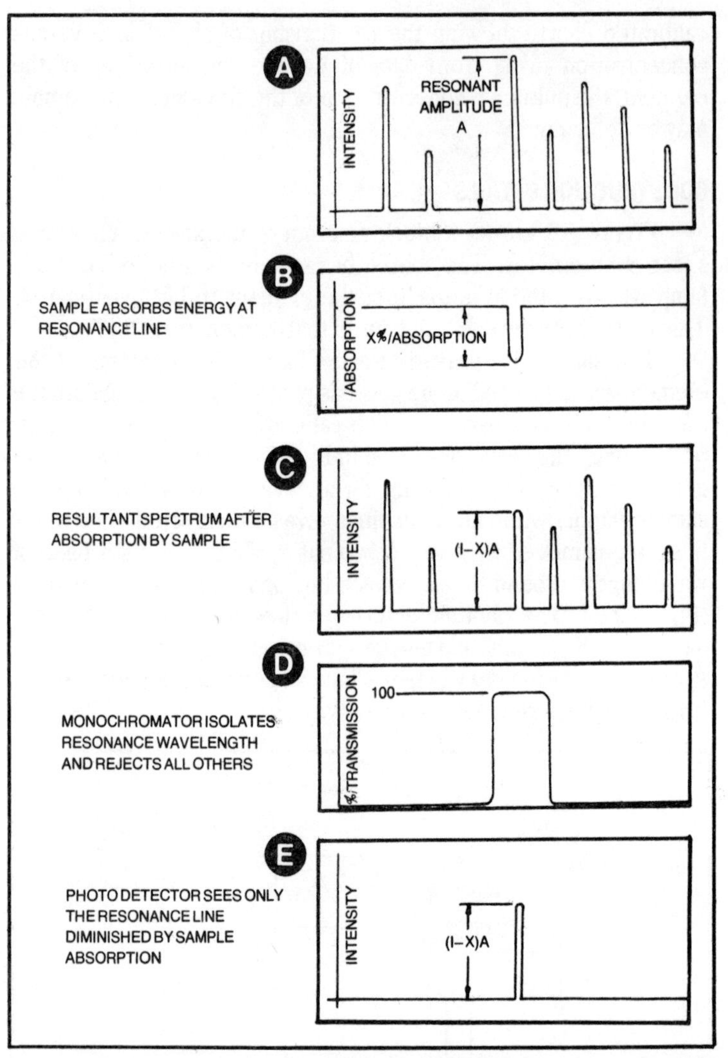

Fig. 5-2. Principle of an atomic absorption spectrophotometer (A) Lamp emits the line spectrum of the element. (B) Resonance line. (C) Resultant spectrum. (D) Monochromator isolates resonance wavelength. (E) Photodetector displays the diminished resonance line.

Many elements emit strongly in the flame at the same wavelength at which they absorb. This creates a serious error in measurement of the absorption. This error can be eliminated by modulating the light from the lamp, either by using an ac discharge or by using a mechanical or electronic chopper between the lamp and the flame. The detector circuit is so designed that only the modulated light is recorded and the unmodulated flame emission gets eliminated.

HOLLOW CATHODE LAMPS

The *hollow cathode lamp* is the most commonly used source of radiation for atomic absorption spectroscopy. It is a discharge lamp which emits the characteristic light of the element to be analyzed.

The lamp consists of a cylindrical thick walled glass envelope which has a transparent window of glass or silica affixed to one end (Fig. 5-3). It contains one anode and a cup shaped cathode which are both connected to tungsten wires for taking out the electrical connections. The cathode is a hollow metal cylinder 10 to 20 mm in diameter and is constructed of the metal whose spectrum is desired, or it may simply serve as a support for depositing a layer of the element. The tube is filled with a highly pure inert gas at low pressures of 1 or 2 mm. The gases generally used are helium, neon, or argon. Mica sheets are placed inside the lamp to limit the radiation to within the cathode.

Choice of the window material depends upon the wavelengths of the resonance lines of the element concerned. For wavelength shorter than 250 nm, a silica window is chosen whereas for wavelengths longer than 250 nm, either silica or ultraviolet, glass is employed.

Hollow cathode lamps have been constructed for almost every naturally available element. However, for atomic absorption spectroscopy, only lamps of the metallic elements are used. They usually have a long operating life, of the order of 1000 hours. The operating current requirements of a particular lamp would depend

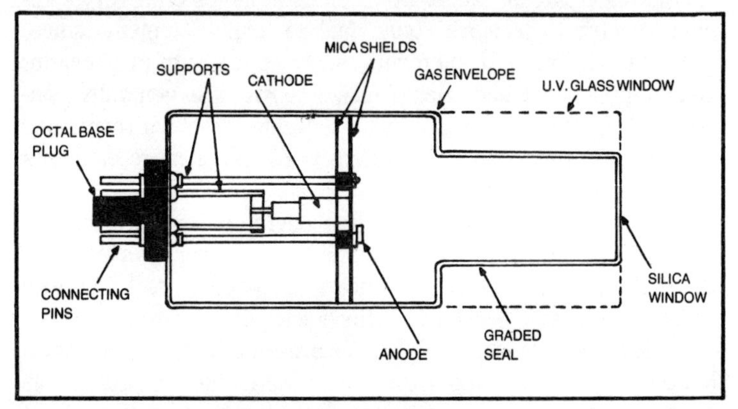

Fig. 5-3. Construction details of a hollow cathode lamp.

on the element concerned and it is usually between 5 to 25 mA. The warmup period of such type of lamps is normally between 5 to 30 minutes. A lamp which requires longer than 30 minutes may be regarded as unsatisfactory.

When a potential is applied between the electrodes, a discharge is struck and ionization of the gas takes place. Current would flow as a result of movement of the ions to the electrodes. If the potential is sufficiently large, the gaseous cations acquire sufficient kinetic energy to dislodge some of the metal atoms from the cathode surface and produce an atomic cloud. This process is known as *sputtering*. Some of the sputtered metal elements in the excited state give out light characteristic of the excited atoms. If we excite an atom of element X by an electrical discharge into a higher energy state, and if it returns to the ground state and emits a photon, this photon would have exactly the correct energy to be absorbed by another atom of X and to raise it to the same excited state.

In atomic absorption spectrophotometers, delay is encountered while changing from one element to another due to the warmup time for the hollow cathode lamps. It may be possible to overcome this lag by warming up several lamps at once, but lamp life gets wasted. It is possible to combine certain elements in the same hollow cathode in such a way so as to provide very nearly equal performance to single elements lamps. Sintered cathodes are used for some of these lamps. Highly pure powders of the elements are mixed in the correct proportions and press formed. These lamps are called multi-element lamps. The instruments are provided with a switching system which puts each lamp into its approximate current range. Multi-element lamps emit very complex spectra. Therefore, only dual or triple element lamps, specially for related elements such as calcium-magnesium, sodium-potassium and copper-nickel-cobalt, are generally constructed. However, they have the disadvantage that with continued operation, the resonance line intensity of one or more elements may fall off more rapidly than others.

The hollow cathode lamps are operated from a regulated power supply which provides chopped or ac power to ensure that the emission from the flame which is unchopped, and therefore results in dc signal, is ignored by the electronic system.

Slevin and Harrison (1975) have presented a useful review on hollow cathode discharges oriented toward their application as spectrochemical emission sources. Kabanova and Sautina (1977) observed less noise in atomic absorption measurements performed with a pulsed hollow cathode lamp. Lamp pulsing is most useful for weak emission lines and when high temperature atom cells are employed.

BURNERS AND FLAMES

The essential requirements of a burner for atomic absorption are unique. The flame must extend over a reasonable length to improve sensitivity. This is directly analogous to the use of longer path cells in normal spectrophotometry. This follows from the fact that if the flame can be spread out into a long, narrow band, more of the atoms it contains may be brought into the light beam. In this way, absorption bands become strong and improve the sensitivity. Also, in order to avoid problems due to scattering of light by unburned droplets and large sample droplets, which cannot be burned, they must be removed before reaching the flame. In addition to this, the burner must be able to take up solutions with a high concentration of dissolved solids, limit clogging problems and reduce requirements for sample dilution.

The burners employed for the flame emission instruments are of the total consumption type. In these burners, the sample solution, the fuel and the oxidizing gas are carried through separate passages and meet at an opening at the base of the flame. However, for atomic absorption studies, a premix type of burner (Fig. 5-4) is mostly preferred. In this type of burner the sample is aspirated

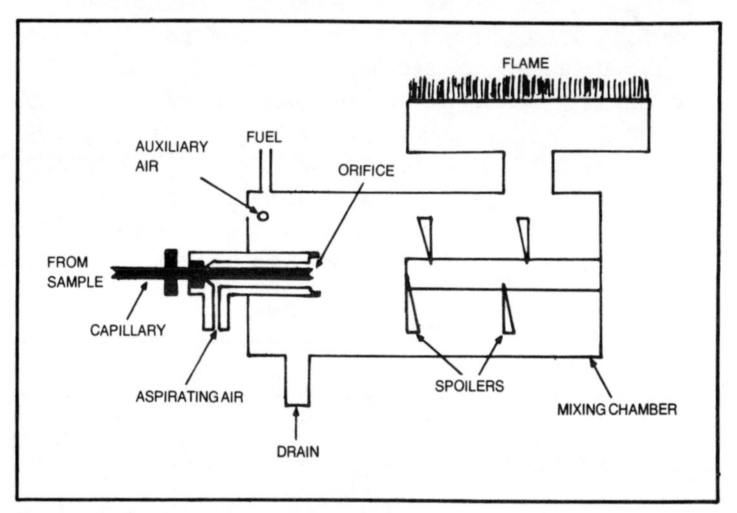

Fig. 5-4. Construction details of a premix burner.

through a thin capillary tube by the air flowing into the atomizer section. The air sample mixture emerges from the atomizer as a fine spray of droplets which is then mixed with the fuel, usually acetylene. The mixture is rendered turbulent by the flow spoilers and is then forced up into the burner head. Larger drops of the sample collect at the bottom of the chamber and are drained off.

Nicholas (1971) explains the design and construction of a spray chamber for use in atomic absorption spectroscopy and flame photometry. In this arrangement, gas mixing and large drop removal is achieved by passing the aerosol through a constriction which functions as a turbulent jet. The geometry suggested can perhaps lead to improved detection limits in atomic absorption spectroscopy. Because of long path available, a slot burner is used in atomic absorption spectrometry. The slot dimensions are dependent upon the fuel and oxidant gases. Usually, a 10 cm slot is used for air-acetylene flame and a 5 cm slot for a nitrous oxide-acetylene flame. The port is usually 0.4 -0.6 mm in width. Very thin flames are used so as to obtain high analyte concentrations in the flame for longer slot. The burner head is fabricated from aluminium, stainless steel or titanium.

Hell and Ramiraz-Munoz (1970) described a new design for a burner system, denoted as *Autolam-burner*. As with other premix burners, it consists of a nebulizer, spray chamber and burner head, but it differs in that the nebulizer and burner head are 180° out of the alignment; the analyte droplets must complete a semicircle from the inlet to the flame. The design enabled uniform mist to reach the flame and no clogging of the slot was noticed.

The desirability of avoiding flash back has led to the development of capillary burners, Aldons, et. al. (1970). They are more stable and for a number of elements the emission signal is greater than that from a conventional burner. With a suitable adapter, they can be made to fit any premix-burner assembly.

The mechanism of atom production is quite complex. Apparently, it is not the same for all solvents and flames. One of the

Table 5-1. Flame Temperatures.

Flame	Temp °C
air/propane	1925
air/acetylene	2300
N_2O/acetylene	2955

Table 5-2. Elements Which May Be Determined by Atomic Absorption Spectroscopy With Different Flames.

Air/propane or Air/acetylene	Air/Acetylene	Air/Acetylene or N_2O/acetylene	N_2O/Acetylene
arsenic	antimony	barium	aluminium
bismuth	chromium	calcium	beryllium
cadmium	cobalt	molybdenum	germanium
copper	iron	strontium	silicon
gold	lithium	tin	tantalum
lead	magnesium		
manganese	nickel		titanium
mercury	platinum		tungsten
potassium			
silver			
sodium			
zinc			zirconium

important factors that needs consideration is the possible formation of oxides. Many metal oxides are very stable and once formed are very difficult to reduce, even in a high temperature flame. Therefore, the reducing nature of the flame (amount of fuel to oxidant) assumes critical importance.

A new technique for nebulizing sample solutions has been described by Dresner (1975). It employs a device shown as Babington nebulizer. In this device, the nebulizing gas is passed into a hollow sphere in whose surface exists a small slot. The liquid to be nebulized is merely allowed to flow over the outer surface of the sphere, past the slot and is thereby disrupted into a fine aerosol. Required air pressure is in the range of 5 to 20 psi; common to flame spectrometry, and a droplet size range between 2 and 50 μm is produced.

Fuel gases used in atomic absorption studies are acetylene, propane, butane, hydrogen and natural gas. The common oxidants are air, oxygen enriched air, oxygen and nitrous oxide. The nitrous oxide-acetylene mixture is used whenever hot flame is required. See Tables 5-1 and 5-2.

If a high standard of analytical performance is to be expected from the instrument, the burner must be kept in a clean condition. The burner head can usually be disassembled for cleaning purposes. Inorganic deposits on the surfaces of the burners can be removed by soaking them in water. Layers of organic material may be removed with the help of a fine emery paper. Soaking in acid solutions is generally not recommended. The burner should be cleaned after every day of operation.

PLASMA TORCH

A major occurrence in the area of atomic absorption spectroscopy during the recent years has been the introduction of commercial inductively coupled plasma torches in place of chemical flames. The ICP torch provides extremely high sensitivity, low matrix interferences and multi-element analysis capabiltiy for samples in solution. However, the ICP sytems are extremely expensive and less convenient to operate than typical flame spectrometers. Nevertheless, their preference cannot be overlooked in laboratories where high sample throughput is imperative and where determination of at least five elements in each sample is required. In most other cases, the chemical flame will remain the choice for the majority of determinations (Hieftge and Copeland, 1978).

There has also been an increasing interest in the use of non-flame techniques like electrothermal and cold vapor atom formation devices in the recent years. They primarily include carbon filament and furnaces, metal filament atomizers, cold vapor devices, etc. Sturgeon (1977) has reviewed carbon furnace atomization devices and has also highlighted problems associated with commercially available devices.

OPTICAL SYSTEM

The optical system of an atomic absorption spectrophotometer is shown in Fig. 5-5. The system employs a photometer which is of the single beam type. The source lamp is a hollow cathode lamp which is ac operated and produces a modulated light beam. The light beam is passed through a limiting stop and focused at the midpoint of a slot burner. After traveling through the flame, the beam is refocused on the entrance slit of the Littrow grating monochromator. The slit width is kept continuously variable and may be controlled by a cam and cam follower mechanism.

The Littrow mounting uses only one piece of quartz or silica with the back surface of the prism metallized. Since the light passes back and forth through the same prism and lens, polarization effects are eliminated. This type of monochromator provides adequate resolution to separate the resonance line in the lamp from the other lines. Also with its high dispersion, it is possible to open the slits wide and admit high energy. Conventional prism systems

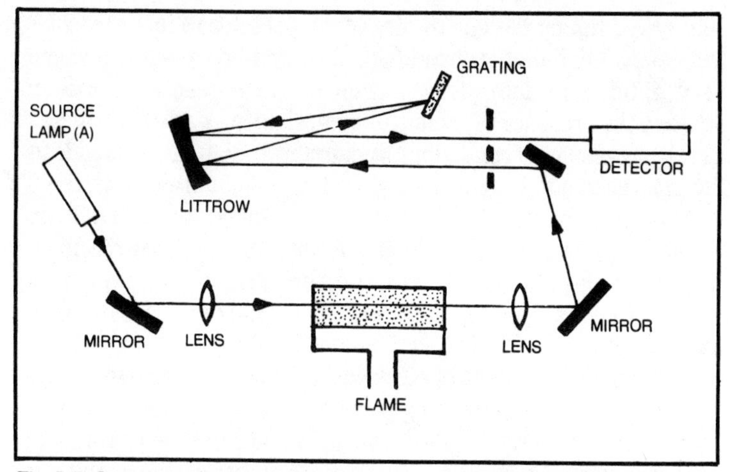

Fig. 5-5. Optical system of an atomic absorption spectrophotometer.

do not meet these specifications. Figure 5-6 shows the optical diagram using a silica prism as a monochromator.

A Littrow grating monochromator with a dispersion of 16 Angstroms/mm provides a bandpass as low as 2 millimicrons at the standard slit width of 1 mm. Instruments using grating monochromators not only provide good uniform dispersion over the whole spectrum but also high resolving power. They have a superior dispersion to a prism instrument at high wavelengths and are at least as good at low wavelengths.

For atomic absorption determination of most of the common metals, a very moderate monochromator is quite suitable. For determinations of metals for which the lamps possess poor output and complex spectra, a better monochromator is necessary.

The radiation emerging from the exit slit falls on the cathode of a photomultiplier, sensitive over the range 210 - 800 mμ. The lenses used in the system are of fused silica whereas the mirrors are coated with MgF_2 to preserve reflectivity in the UV region.

ELECTRONIC SYSTEM

The emission system not only produces light of a particular wavelength, originating from the lamp falling upon the photodetector, but also light of the same wavelength arising from the flame. It is necessary to distinguish between these two sources of radiation, since it is the measurement of that from the lamp only which is required. This requirement is attained by modulating the light from the lamp, with either a mechanical chopper or by using an ac power supply and tuning the electronics of the detector to this particular frequency. On the other hand, with a dc system, it would become very difficult to determine calcium in the presence of sodium because the presence of sodium or potassium in a flame contributes a variable amount of continuous background radiation which adds to the measured intensity of all spectral lines. Sodium also interferes with magnesium when the dc system is employed, because there is a sodium emission line at 2852.1°A, only 0.7°A from the magnesium line at 2852.8°A. On the other hand, the ac system is entirely free from spectral interference and the dc signals from the flame and other extraneous sources are eliminated.

The main advantage of electronic modulation over a mechanical chopper is that no moving parts are involved. However, in a double beam system in which the light from the measuring and reference beams is passed onto the same photomultiplier, the use of a mechanical chopper becomes essential.

The random background noise that originates from the flame is generally noticeable at low frequencies. Also, the high frequency photomultiplier noise increases above about 1000 Hz. The optimum modulation frequency of the chopper is therefore selected around 350 to 400 Hz.

The most innovative method for background correction developed recently is the one involving the *Zeeman effect*. In this method, a magnetic field is applied to either source or generated atoms to split the Zeeman components of each spectral line. Since these components are polarized, it becomes possible to isolate them using stationary or rotating optical polarizers. This method, known as Zeeman modulation, can thus be employed for high precision background correction. Koizumi and Yasada (1976) explain this technique in detail. Accurate instrumental subtraction of flame background can also be accomplished by directly measuring the background in a reference flame and subtracting it from the signal produced by an analytical flame. This dual flame approach has been employed by Haraguchi, et. al. (1976).

Figure 5-7 shows a block diagram of the electronic system of an atomic absorption spectrophotometer.

The low level signals developed across a PMT anode resistor are either chopped by mechanical interruption of the light beam directed onto the PMT or are electronically chopped by a transistor switch. In most of the spectrophotometers where a transistor switch is used, it is operated by the waveform supplied from the synchronizing circuits to ensure that the chopped amplified signals

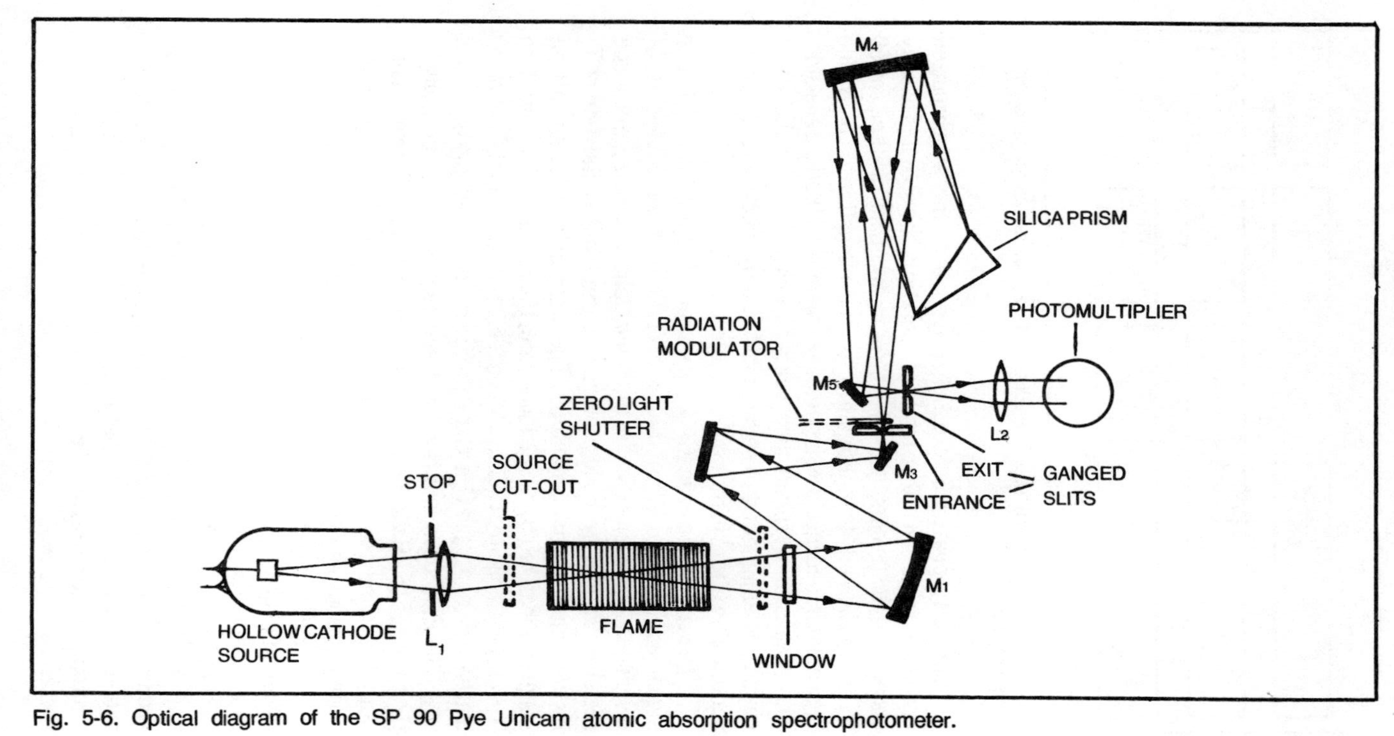

Fig. 5-6. Optical diagram of the SP 90 Pye Unicam atomic absorption spectrophotometer.

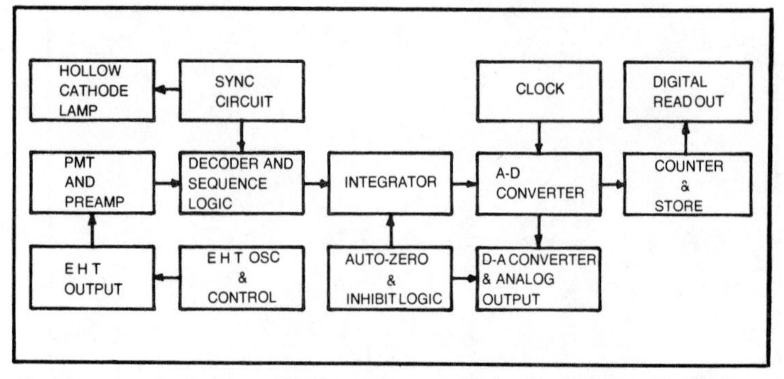

Fig. 5-7. Block diagram of the electronic circuit of an atomic absorption spectrophotometer.

are correctly phased before being applied to the synchronized decoder circuits. In either mode of operation, the signal is usually chopped at the frequency used in the synchronizing system, say 50 Hz or 60 Hz, and is amplified in a low noise amplifier. To provide high input impedance, the input stage contains a bipolar transistor. The preamplifier circuit also contains signal limiting diodes, a circuit for eliminating of spikes produced by electronic chopping and suitable filters in the power supply line.

The circuit shown in Fig 5-8 has a gain of approximately 4. The signal applied via C_1 to the gate of field effect transistor T_1 originates at the anode of the photomultiplier, R_1 being the anode resistor. The zener diode protects the gate emitter junction of T_1 by limiting the signal amplitude. The stage gain of the amplifying stage is linearized by dc feedback from the junction of R_3 and R_4 to T_1 source. Capacitor C_2 provides ac feedback between the collector and base of T_2. R_5 and C_3 act as a filter in the power supply line.

The EHT oscillator is a high frequency oscillator operating in the range of 15 kHz to 20 kHz. It is an astable multivibrator whose output is connected to the primary winding of the EHT transformer through an output stage consisting of a Darlington stage.

The EHT requirements for PMT is of the order of 2 KV at a maximum current of 200 μA. The transformer primary windings may be bifilar wound to reduce losses and are connected in series to permit a push-pull signal. The secondary winding is connected to four EHT rectifying diodes (Fig. 5-9) arranged in a quadrupler rectifier circuit which is followed by smoothing filters. In most of the spectrophotometers, the positive EHT line is earthed at the photomultiplier preamplifier and the polarity of the output is, therefore, negative.

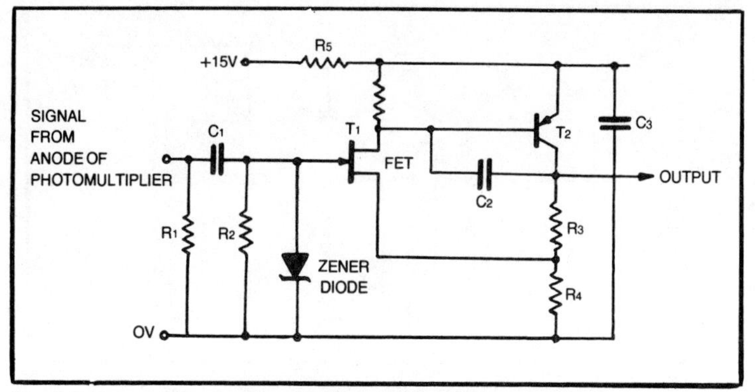

Fig. 5-8. Preamplifier circuit used with a photomultiplier detector.

The main synchronizing circuit derives its input from the mains supply sine wave signal of 50 Hz frequency. The signal is squared in Schmidt trigger circuit to get a lamp modulation waveform at 50 Hz pulse frequency. The same circuit also, provides for the sample decoder drive at 50 Hz and reference decoder voltage which are connected to a PMT signal processing circuit in decoder and a sequence logic circuit.

The amplified signal from the PMT preamplifier is given to the decoder circuit where it is amplified, encoded and given to the integrator. The sequence logic circuit controls various timings of the cycle. The operation can be manual or it could be remote controlled.

The integrator provides dc voltage corresponding to signal and reference signals. The integrator could be a conventional op amplifier connected in integrator configuration having a capacitor in the feedback circuit. Auto zero potentiometer is connected in feedback in the amplifier following the integrator.

Basically, the analog to digital converter employs a single ramp principle. It consists of a comparator circuit which evaluates the difference between the reference and sample signals. It produces a pulse of width proportional to the signal difference. The conversion is carried out according to linear or logarithmic law; the latter is required in the absorbance mode.

The comparator is fed with a ramp generator and a switch circuit. The ramp waveform is determined by the voltage charge on the ramp capacitor. When a linear ramp is required, the ramp voltage decreases at a constant rate, the capacitor getting discharged at a constant current. A logarithmic ramp is produced

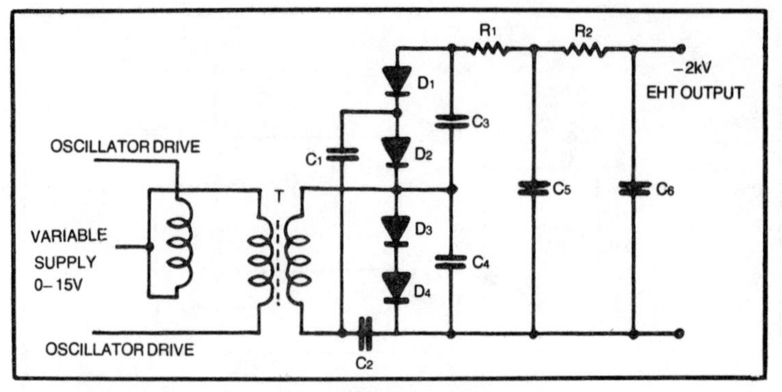

Fig. 5-9. EHT supply circuit for a photomultiplier.

when the capacitor discharges towards a preset voltage, and the discharge current is passed via a resistor.

The output of the comparator is given to a sign circuit of logic gates which produces two output signals in antiphase. The output signals indicate the polarity of the signal differences evaluated in the comparator.

An additional circuit provides for timing pulses like 10 Hz reset pulse output, a 5 Hz set zero pulse output and a 5 Hz transfer data pulse output. These pulses are obtained by dividing the mains frequency in a divider.

The basic clock frequency for the ADC is obtained from an astable multivibrator. The clock pulses are responsible for causing operation in digital display. The clock frequency can be varied for scale expansion. The logic circuit provides gating of the clock output with the comparator output pulse. The resulting bursts of oscillator pulses provide the digital equivalent to the voltage difference of the dc analog signals.

The bursts of clock pulses are counted in decade counters and given to BCD to seven segment decoder/drivers which contain the decoding logic and drive transistors for each of the seven segments of each numeral of the display.

SIGNAL INTEGRATION

In atomic absorption spectroscopy, the signal-to-noise ratio limits the work at low concentrations. This can be greatly improved by signal integration. If the detector signal is integrated for a certain time, the output voltage of the integrator would be the sum of the signal and the noise. Since the noise is mostly random, its total component will be zero in the ideal case (Nishita et. al., 1972).

Another type of instrument for atomic absorption spectroscopy, in which the integrals of the photo multiplier signals are divided and displayed digitally in % absorption or directly in % concentration, is described by Boling (1965).

Integrated readings are obtained by feeding the amplified current from the photomultiplier to a capacitor, for a period of time precisely controlled by an electronic timer. At the end of this period the voltage attained by the capacitor is applied to the readout meter. The reading obtained remains steady until a reset control is operated.

The integration time can be varied from 2 seconds to 40 seconds. The accuracy and repeatability of measurement increases with integration time. The random variations in absorbance value of aspirated sample vary from instant to instant due to variations in the number of excited neutral atoms in the flame. These instantaneous variations from the average level are summed up and presented at the DVM or recorder. The integration starts as soon as the start button is actuated. Immediately, the rising signal can be observed on the recorder or DVM. At the end of the integration time, the memory holds the reading for a short duration and then returns the pen to zero.

A signal recorded on the integration mode will be about four to five times greater than that recorded on the conventional mode. The advantage of integration mode is clearly shown in Fig. 5-10.

In the memory mode, the integrated value of absorbance is held on the digital display until the beginning of next analysis. But the potentiometric recorder registers zero until the measuring time or integrating time elapses. At the end of the integrating time, the pen is given the signal to record the value held by the memory for the same period as memory time. The advantage of memory is that it is noise free (Fig. 5-10) and manual control settings such as scale expansion can be adjusted to bring the readings on scale. Such changes do not effect the measured value.

SAMPLING SYSTEM

A sampling system is used to supply a constant feed rate of sample into the burner. The most common method employed makes use of a solution which is introduced into the flame by an atomizer. The design and construction of the atomizer are similar to that used in emission spectroscopy. A complete sampling system consists of a nebulizer and a spray chamber (Fig. 5-11). The nebulizer reduces the sample solution to a spray of droplets of various diameters by a pneumatic action. The spray is directed into the spray chamber where the larger droplets are precipitated and the air stream carries the remaining spray into the burner and thence to the flame. Normally, the sample uptake rate of the system is 3-4 ml/min. It has been experimentally established that an improvement in analytical sensitivity, when using non-aqueous solvents, occurs by employing a more efficient nebulization of the sample solution. Organic solvents are often used in place of water when the sample may be an organic liquid or a solid soluble only in organic solvents. The choice of a suitable organic solvent is very important because the solvent can affect the physical properties of the flame and may influence the precision and sensitivity of the analysis.

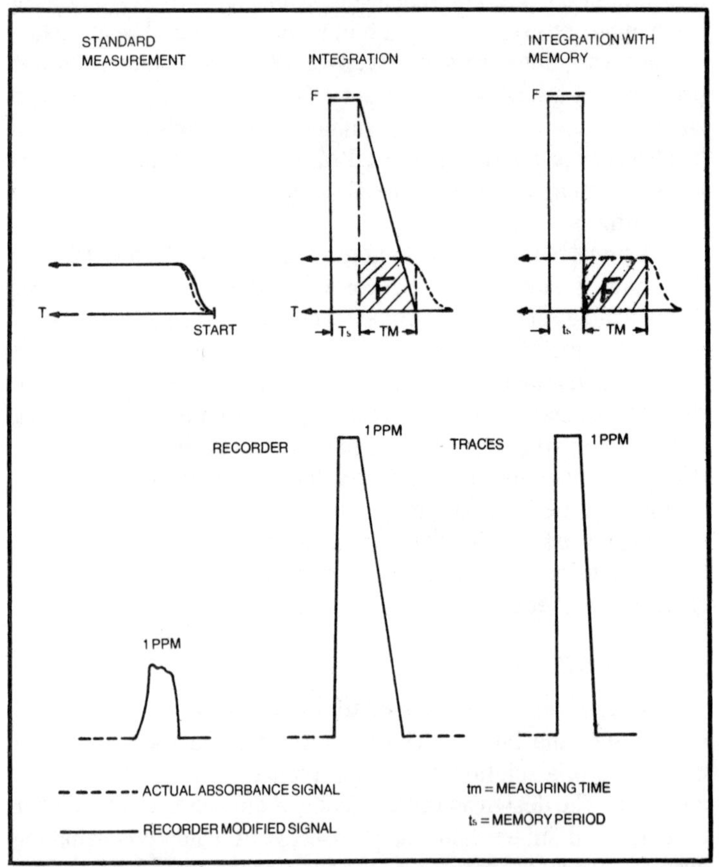

Fig. 5-10. Principle of operation of different modes in atomic absorption spectrophotometers.

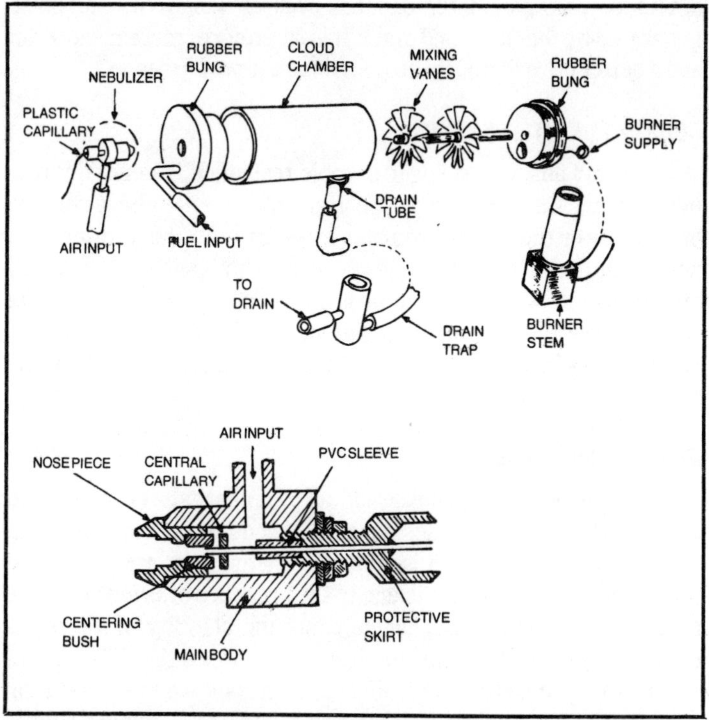

Fig. 5-11. Nebulizer and spray chamber used in atomic absorption spectrophotometers.

The sampling system may be coupled to an automatic sample changer which may have a number of sample holding polyethylene cups. Each cup passes the sampling point in turn and contents are sampled by an automatic head.

PERFORMANCE ASPECTS

The performance of an atomic absorption spectrophotometer is usually described by sensitivity and detection limit. By definition, *sensitivity* is the concentration of an element in solution which absorbs 1% of the incident light. Its value depends upon atomizer efficiency and the design of the burner and optical system. It is independent of amplifier gain and noise from flame or lamp.

Detection limit is defined as that concentration of an element, in ppm, which gives a reading equal to twice the standard deviation of the background signal, the standard deviation being computed from at least 10 readings. In practice, the detection limit is the minimum concentration of the element in solution that can just be detected. This parameter depends both on sensitivity and noise generated by the flame and lamp. It is, therefore, more meaningful and a better criterion of spectrophotometric performance.

SOURCES OF INTERFERENCE

Atomic absorption spectroscopic technique is generally recognized to be free from spectral interference because the bandwidth of the monochromator is essentially the width of the source emission line. Thus, closely spaced lines such as Mg 285.21 and Na 285.28 would present no problem in absorption work. The wavelengths at which individual elements absorb are well defined and the possibility of two elements absorbing at exactly the same wavelength is quite remote.

Anionic Interference

However, the response of an element at its resonance wavelength may sometimes depend on another component in the same solution. This effect is known as chemical interference. The major source of this effect is due to the presence of *anions* with the metal ions in the sample, which affect the stability of the metal compounds formed during atomization and hence the efficiency of metal atom production. The effect is important when determining calcium and magnesium in biological fluids and aluminum alloys, etc. Sulfate, phosphate, silicate, aluminate and a number of other radicals all interfere with these metals by tending to bind the calcium into a compound that is not associated in the flame. Generally, all these interferences may be completely removed by the addition of an excess of *lanthanum*, which competes with the calcium for the interfering anions. Lanthanum is also useful in removing similar interferences from strontium and barium.

Viscosity Interference

Viscosity interference is caused by the fact that solutions of widely differing viscosities enter the burner at different rates. Fortunately, this effect is negligible in most of the analyzes. However, if it is desirable to overcome this interference, it may be necessary to match the standards to the samples. The method usually adopted is that of additions. In this method, a crude determination of the concentration of the desired element is made, after which the sample is divided into three aliquots. Different additions are made to two of the aliquots, while the third is left blank. All three are then determined by atomic absorption, and a graph is plotted for absorbance versus ppm added. The intercept of the curve on the ppm axis gives the concentration in the unknown. With the method of additions, atomic absorption measurements are independent of sample characteristics, and the making up of standards is not necessary.

Ionization Interference

Ionization interference is usually observed at high flame temperature and manifests itself as an enhancement in the response of the element under determination. At any temperature in the flame, there is an equilibrium between neutral atoms and ions. For an element having relatively low ionization potentials, the proportion of ions will be high. If another easily ionizable element is added into the flame, the equilibrium for both elements is shifted towards the atomic state, producing an enhancement in analytical sensitivity. The interference is generally overcome, by addition of the same quantity of the interferer to the standard solutions, as is present in the samples or by addition of an easily ionizable metal, like sodium or potassium, to both standard and sample solutions.

METER SCALE

The linear relationship between concentration and absorbance (Beer's Law) is followed only approximately in atomic absorption spectroscopy. In general, calibration curves tend to approach an *asymptote* drawn parallel to the concentration axis. The calibration curve correlating concentration and percent transmission can also be drawn, but it shows more pronounced curvature.

The meter scale is usually calibrated in absorbance (0-1.0A) which approximates to 0-100%T. Scale expansion facility is also added in order to improve readability of the scale. Scale expansion readings are marked in different ways by manufacturers. In some instruments, scale expansion is marked in absorbance units such as 2.5, 1, 0.5, 0.25, etc., which means the full scale reading of the digital readout is selected to read these values of absorbance at its maximum. Some manufacturers mark the scale expansion as X1, X5, X10, etc., which means a reading presented at scale X1 will be five times greater than selected at X5 range. That is to say the first one-fifth range is extended to read as full scale.

Shepherd and Hedgpeth (1973) describe a circuit which enables one to continuously record the output from a Perkin-Elmer

303 atomic absorption spectrophotometer directly as optical density (Fig. 5-12). The circuit has a pair of operational amplifiers (A_1, A_2) as its first stage which operate as unity gain buffers. These amplifiers receive the signal from sample and reference detectors and are followed by linear-to-log converter. This is a circuit consisting of op amps having transistors in their feedback circuits. There is a provision for offset adjustment and impedance matching output stage before the circuit is connected to the recorder.

CURVE CORRECTION FOR LINEARIZATION

When analyzing solutions with higher concentrations, or if the element being analyzed is a strong absorber, a linear plot of concentration or absorbance is not normally obtained. This is because, at higher concentration, the indicated absorbance is less than the expected value. The lower concentrations of the same element, however, show a linear behavior.

The earlier instruments did not incorporate the facility for curve correction. Due to the advent of modern advance in electronics using integrated circuits, and more recently computer circuits (microprocessors), curve correction facility is introduced

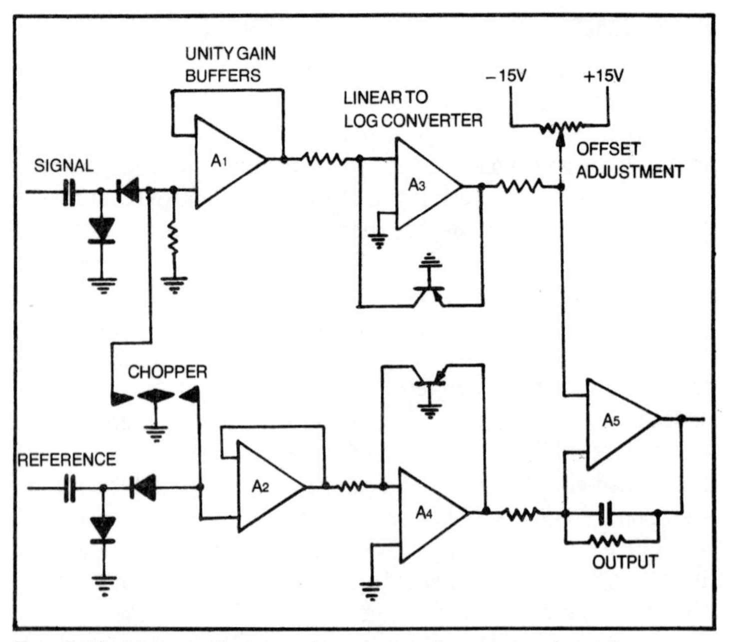

Fig. 5-12. Linear-to-log recorder adapter for atomic absorption spectrophotometric measurements.

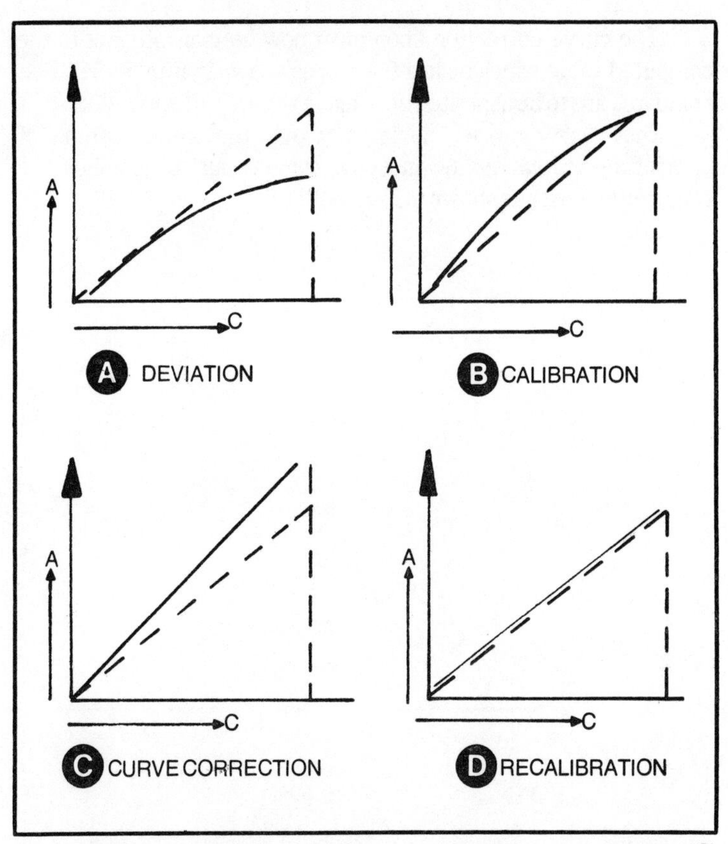

Fig. 5-13. Curve correction for linearization. (A) Deviation. (B) Calibration. (C) Curve correction. (D) Recalibration.

as a manual setting after computation or automatic in programming of the microprocessor.

Figure 5-13A shows a deviation of plot from linearity. To linearize the plot, after zeroing with blank, aspirate a standard solution of highest concentration of interest and set the digital display to the required reading by Auto-Cal control such that the value is registered with memory for internal calibration as shown in Fig. 5-3B. Now the mid-scale standards will show a higher reading than the dotted values. The required correction can be calculated

$$\text{Percent deviation} = \frac{\text{Readout concentration deviation of mid-range standards}}{\text{Actual concentration of mid-range standards}} \times 100$$

The curve correction knob must now be manually set to the computed value which makes the plot as shown in Fig. 5-13C. The standards are to be aspirated once again to verify linearization.

Should any further deviation exist, the same method of computing and correcting must be carried out to establish the required linearity as shown in Fig. 5-13D.

Chapter 6

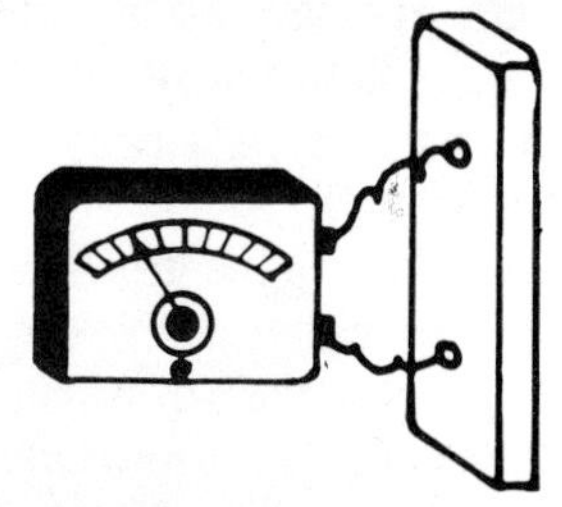

Fluorimeters And Phosphorimeters

The analytical technique based on the absorption of infrared, visible and ultraviolet light has found extensive applications in chemistry. However, a limitation of this technique is the difficulty of differentiating between different substances with similar absorption spectra. Many solutions when irradiated with visible or ultraviolet light not only absorb this light, but re-emit light of a different wavelength. This effect is known as *fluorescence* and its exploitation opens up possibilities in the discrimination and accurate determination of many substances in very dilute solutions. The difference in wavelength between incident and re-emitted light enables efficient discrimination to be made between substances with similar absorption characteristics.

Fluorescent analysis makes valuable contributions to many branches of research and industry. However, it has found widest application in biomedical laboratories. Clinical laboratories employ it for such studies as adrenalin and noradrenalin in urine and blood, and in screening tests for tumors of adrenal glands. Tests for transminase, lactic dehydrogenase, DPN (diphosphonucleotide), steroids, DNA (deoxyribonucleic acid), histamine and many others are routinely made with fluorimeters. The value of the fluorimetric technique for such studies as air pollution and analysis of alkaloids has tremendously increased. Industrial health laboratories use fluorimeters for routine determination of beryllium. Agriculture and food chemists use this technique for vitamin and insecticide residue studies. Water, air pollution and sanitation engineers find it effective and convenient for studying flow and diffusion of air and water.

PRINCIPLE OF FLUORESCENCE

Normally, in the ground state of molecules, molecular energies are constant and at minimum values. When a quantum of light impinges on a molecule, it is absorbed and an electronic transition to a higher electronic state takes place. This absorption of radiation is highly specific, and radiation of a particular wavelength and energy is absorbed only by a characteristic structure. In the ground state of most molecules, each electron in the lower energy levels is paired with another electron whose spin is opposite to its own spin. This state is called a *singlet* state. When the molecules absorb energy, they are raised to an energy level of an upper excited singlet state S_1, S_2 or S_3 (Fig. 6-1) represented by the transition G-S_1 or G-S_2. These singlet transitions are unstable and the molecular energy tends to revert almost immediately to a lower level. In going to a lower energy level, the absorbed energy is lost by steps from the upper singlet state to the lowest singlet state through the transition S_3-S_1 or S_2-S_1. These singlet transitions are responsible for the visible and ultraviolet absorption spectra.

The excited singlet state persists for a time which is of the order of 10^{-8} to 10^{-4} sec. During this time interval, some energy in excess of the lowest vibrational energy level is rapidly dissipated. In case all the excess energy is not further dissipated by collisions with other molecules, the electron would return to the ground state with the emission of energy. This phenomenon is called fluorescence. Since this transition involves less energy than in the original absorption process, the fluorescence is emitted at wavelengths longer than those of the exciting source. Many vibrational levels are actually involved, so molecular fluorescence spectra are generally observed as continuous bands with one or more maxima. Filters are used to filter out the exciting wavelengths, so that only fluorescent energy reaches the detector.

Any fluorescent molecule has two characteristic spectra, the *excitation spectrum* and the *emission spectrum*. The excitation spectrum shows the relative efficiency of different wavelengths of exciting radiation to cause fluorescence whereas the emission spectrum would indicate the relative intensity of radiation emitted at various wavelengths. Normally, the shape of the excitation

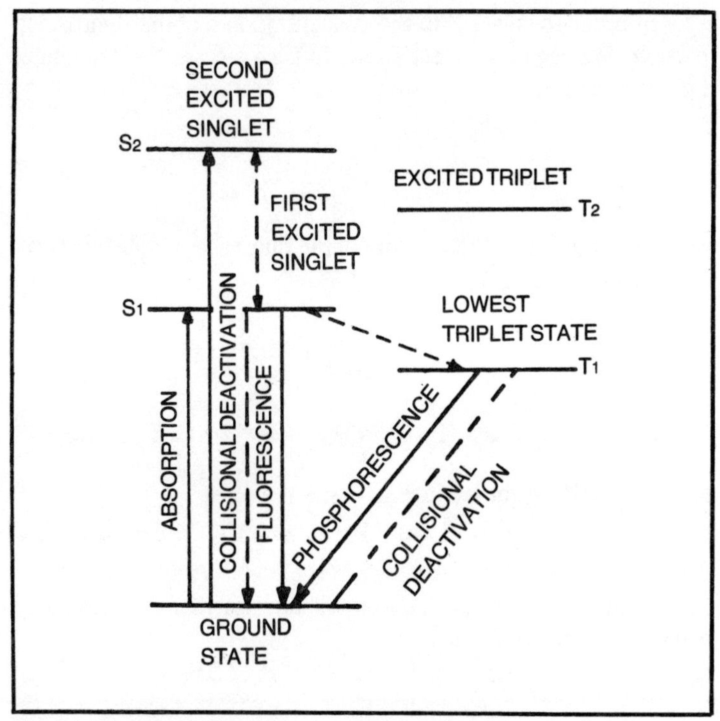

Fig. 6-1. Principle of fluorescence and phosphorescence.

spectrum should be identical to that of the absorption spectrum of the molecule and independent of the wavelength at which fluorescence is measured. However, due to instrumental errors, this is seldom the case. Also, the shape of the emission spectrum is independent of the wavelength of exciting radiation. If the exciting radiation is at a wavelength different from the wavelength of the absorption peak, there will be less absorption of radiant energy and consequently less energy will be emitted.

RELATIONSHIP BETWEEN CONCENTRATION AND FLUORESCENCE INTENSITY

The total fluorescence intensity is equal to the rate of light absorption multiplied by the quantum efficiency of fluorescence (η), that is

$$F = \eta (P_0 - P) \qquad \textbf{Equation 6-1}$$

where P_0 is the power of the beam incident upon the solution and P is its power after traversing a length b of the medium.

In order to relate F to the concentration c of the fluorescing particle, we can apply Beer's law. In such a case, the fraction of light transmitted is

$$\frac{P}{P_0} = e^{-\tau bc} \qquad \textbf{Equation 6-2}$$

where τ is the molar absorptivity of the fluorescent molecules and τbc is its absorbance A.

$$1 - \frac{P}{P_0} = 1 - e^{-\tau bc}$$

$$\therefore (P_0 - P) = P_0 (1 - e^{-\tau bc}) \qquad \textbf{Equation 6-3}$$

substituting Equation 6-3 in Equation 6-1, we get

$$F = \eta \times P_0 (1 - e^{-\tau bc}) \qquad \textbf{Equation 6-4}$$

For very dilute solutions in which A (τbc) < 0.05 and only a small fraction of light is absorbed, the exponential term can be expanded, and Equation 6-4 reduces to

$$F = K' \times \eta \times P_0 (2.3 \tau bc)$$

$$\text{or} \quad F = K.C.$$

This equation shows that for very dilute solutions, a plot of the fluorescent power of a solution versus concentration of the emitting particles would be linear. However, at higher concentrations (Fig. 6-2), linearity is lost and the curve bends downwards as the concentration increases.

While making fluorescent measurements, it is always advantageous to make the exciting energy as large as possible. However, photodecomposition should be avoided.

FACTORS AFFECTING FLUORESCENT YIELD

Fluorescence intensity important for quantitative work depends upon numerous factors, some of which are characteristic of the fluorescent substance and some of which are dependent upon the measuring technique employed. Failure to realize the significance of these factors can seriously impair quantitative work.

Some substances are strongly fluorescent while others have weak or undetectable fluorescence. The efficiencies are defined by the fluorescent yield, that is, the ratio of radiation emitted to that absorbed. Some substances, such as certain dyes, are highly

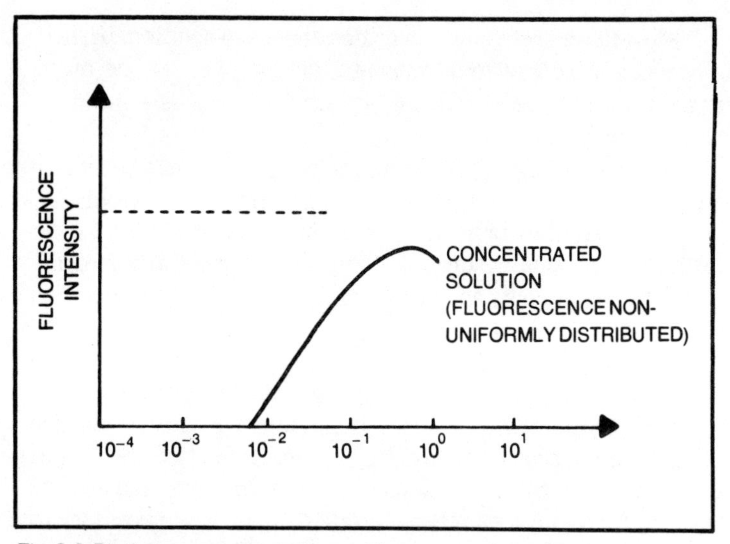

Fig. 6-2. Plot between concentration and fluorescence intensity.

efficient and provide yields approaching 1, whereas other substances have fluorescent yields less than 0.0001. The efficiency for a given substance is not constant but depends upon environmental effects, some of which will be described.

Fluorescence intensity depends upon both the intensity and the spectral distribution of the exciting radiation. Therefore, for reliable analyses, it is essential to employ stable sources which provide intense radiation in the proper wavelength regions. Since intensities vary between sources and with time, reference materials are commonly utilized to maintain uniformity between readings.

The temperature of the sample has a considerable effect upon the fluorescence. As a result of the reduced probability of collisions and reactions, the fluorescent yield is increased by cooling the sample. Reduced temperatures also enhance the fine structure observed in many spectra.

Solvent effects similar to those noted in absorption work are common in fluorescence work, as changes in both the wavelength position and intensity of fluorescence peaks are observed. Quenching effects are greatly influenced by the solvent employed.

The pH of the sample solution is critical for many substances. In some instances, highly fluorescent materials become completely non-fluorescent when the pH is altered by only a few pH units.

Most fluorescent substances decompose in solution, particularly when subjected to ultraviolet radiation. This is one of the major disadvantages of utilizing powerful sources as a means of increasing sensitivity.

In some instances, the fluorescence spectrum of a substance overlaps the region in which it shows absorption. Consequently, part of the emitted radiation is reabsorbed before it leaves the solution. Since the absorption occurs at selected wavelengths, both the total intensity and the shape of the fluorescence spectrum are affected.

Various foreign substances present in the sample can quench the fluorescence radiation. Thus, oxygen diminishes the fluorescence of polyaromatics, and transition elements such as iron and manganese quench the uranyl fluorescence. Dilution techniques are commonly employed to detect impurity quenching and in some cases to correct for its effect. In other cases, it is necessary to remove the interfering substance before reliable measurements can be obtained.

The observed spectral fluorescence spectrum is dependent upon the dispersion, stray radiation and transmitting properties of the optical system. For a given instrument these factors are essentially constant. So simple calibration curves can be utilized for quantitative work. However, significant differences may exist between instruments, so caution should be exercised in comparing interlaboratory results.

MEASUREMENT OF FLUORESCENCE

Fluorescence measurements are carried out in instruments called fluorometers or fluorimeters. The various components of these instruments are similar to those in photometers. These are analogous to photometers as they, too, make use of filters to restrict the wavelengths of the excitation and emission beams. Fluorescence measuring instruments making use of monochromators like spectrophotometers are called spectrofluorimeters. These are of two types. In one type, a filter is employed to limit the excitation radiation, and a grating or prism monochromator isolates a peak of the fluorescent emission spectrum. The other type of instruments are equipped with two monochromators. One of the monochromators restricts the excitation radiation to a narrow band whereas the other permits the isolation of a particular fluorescent wavelength. The latter type are the true spectrofluorimeters.

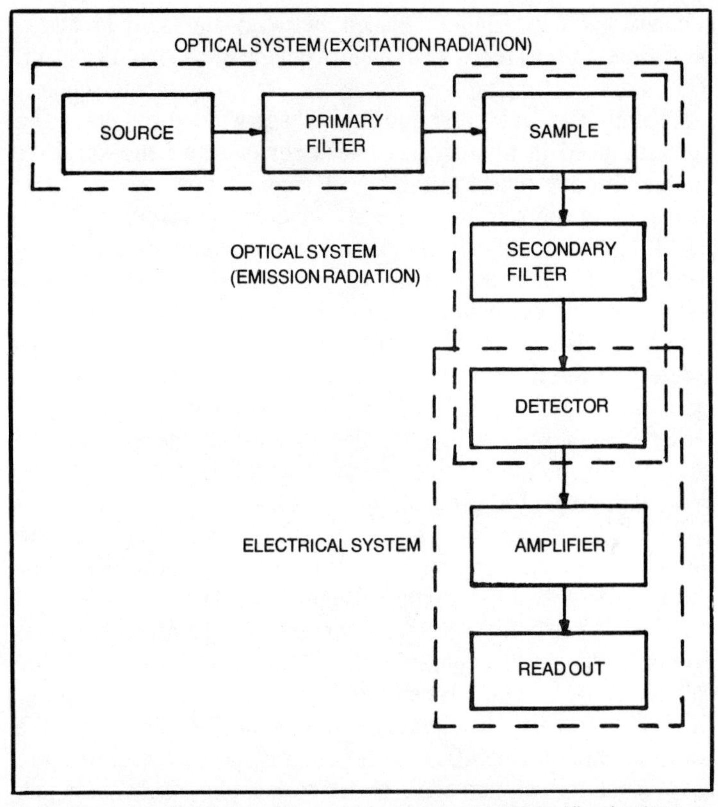

Fig. 6-3. Schematic diagram of the optical components of typical fluorimeter.

Single Beam Filter Fluorimeter

Figure 6-3 shows a schematic diagram of a single beam filter fluorimeter. The source of light is a high pressure mercury discharge lamp with glass or fused silica envelope. Selection of the exciting wavelength is made by inserting a primary filter in the incident beam. It is possible to isolate one of the principal lines at 365, 405, 436, 546 or 579 mμ and to have intense line spectrum with good spectral purity. Low pressure lamps, equipped with a silica envelope, can produce intense radiation at 254 mμ.

Both cylindrical and rectangular cells constructed of glass or silica are employed in fluorescence measurements. Adequate care is taken in the design of the cell compartment to reduce the amount of scattered radiation reaching the detector. The fluorescence emitted by the sample is measured by selecting the wavelength of fluorescence radiant energy by a second optical filter, called the secondary filter. This is placed between the sample and a photodetector located at a 90° angle from the incident optical path. The fluorescent signal is generally of low intensity and requires large amplification for carrying out necessary measurements. The detector used in fluorescence measurements are the sensitive photomultiplier tubes. After signal amplification, the signal is displayed on a duly calibrated microammeter or recorder.

It is always difficult to measure absolute fluorescence intensity. Therefore, measurements are usually made with reference to some standard substance which may be chosen arbitrarily. The circuit is balanced at any chosen setting by placing the standard solution in the instrument. Without making any adjustment, the standard solution is replaced by additional standard solutions of lesser concentrations. Lastly, fluorescence of the solvent and cuvette alone is measured to establish a true zero concentration. With the help of these settings, a curve is plotted between concentration against scale readings which gives the fluorescence-concentration curve. Unknown concentrations can be estimated from this curve if the scale reading is known.

It is known that the dark current of a photomultiplier is primarily due to the emission of thermionic electrons from the cathode and it is generally reduced by cooling the photomultiplier. In the case of very weak light measurements, the reduction of the dark current is an important factor in improving the signal-to-noise ratio. A number of methods are used to cool photomultipliers to a temperature where the thermal component of the dark current is virtually eliminated.

Aoshima and Sugita (1974) describe a photomultiplier cooler with a high stability of ± 0.005°C and the photomultiplier temperature is adjustable between room temperature and −20°C. The electronic circuit basically employs a temperature sensor, an amplifier and a triac which is triggered by a unijunction transistor. The triac controls the thermal element.

The detection system used in the commercially available spectrofluorimeters is almost exclusively based on the use of the photomultiplier tube in a single channel configuration. Yet the parallel multichannel optoelectronic image detector should be clearly superior because of its ability to record the entire spectrum simultaneously. Parallel detection can result in either a significant increase in signal-to-noise ratio or a corresponding reduction in analysis time. The detector is a silicon intensified target vidicon whose control is achieved with a computer. Talmi et. al. (1978)

give an overview of experimental setups and data interpretation regarding use of optoelectronic image detectors in fluorescence spectrometry.

In order to study the dynamic properties of excited molecules, one must know not only the decay time of the luminescence but also its quantum yield, that is, the number of emitted photons divided by the number of absorbed photons. The number of absorbed photons can be determined easily by any absorption measurements, but the number of emitted quanta cannot be found in a straightforward manner. One has to correct the recorded emission spectra for the spectral response of the detection system. The spectral response of the detection system is not always known and it is considered too much trouble to redraw the spectra point by point. The latter can be conveniently done by interfacing the detection system to a computer. Gruneis et. al. (1975) showed that a simple and cheap solution to the problem is possible by the use of programmable read only memories (PROMs). The original article describes schematically how the detection system of a conventional fluorimeter can be supplemented to produce automatically the corrected emission spectra.

For studying the decay of fluorescence on the nanosecond time scale, impulse technique is generally preferred. In this method, the sample is excited by a short pulse of light of suitable wavelength and the emitted fluorescence is detected and recorded as a function of time. The excitation and detection of the fluorescence are repeated many times and the results are averaged in order to obtain acceptable signal to noise ratios. However, if any drift occurs in the excitation light source or in the detection electronics, the measured profile of the light source will not truly represent the correct profile and may introduce serious errors. Hazen et. al. (1974) suggest a solution to minimize this problem by alternate collection of the data for the fluorescence and for the excitation lamp.

Boutilier et. al. (1977) have presented a comparison of pulsed source excitation with continuous wave source excitation in molecular luminescence spectrometry. They report that in general pulsed source excitation does not offer significant signal to noise advantages over continuous wave excitation and detection. There is a considerable activity in adopting lasers as excitation sources (Bradley et. al. 1976).

White (1976) claims a thirty-fold increase in sensitivity for a fluorescence spectrometer obtained by rotating the monochromator slit images by 90° to permit viewing the sample along the length of the excitation slit image, and then adding concave retromirrors behind the sample in both the excitation and emission beams.

Automated luminescence spectrometers using a microprocessor (Itoh et. al. 1976) for correcting and recording absorption, excitation and emission spectra are greatly extending the qualitative and quantitative analytical capabilities of the technique. Computer-assisted methods for qualitative analysis and chemical characterization based on fluorescence spectra have been described by several workers.

Parker (1976) describes a calibrated tungsten lamp in a silica envelope for obtaining spectral sensitivity curves for the calibration of fluorimeters.

Double Beam Filter Fluorimeter

Double beam instruments are the direct reading type and are generally preferred because of this obvious advantage. In these instruments, the circuit is so arranged that the current from the reference detector opposes that generated by the measuring photocell. The reading on the calibrated scale at which the two currents neutralize each other directly gives the fluorescence reading.

Figure 6-4 shows a schematic diagram of a double beam fluorimeter. A mercury lamp is used as a source of radiation. A collimated beam is passed through the primary filter which then falls on a cuvette containing the sample. The measuring photocell is mounted at right angles to the excitation beam. Fluorescent radiation passes through secondary filters which remove scattered ultraviolet radiation but pass visible light.

A part of the excitation beam is passed through a reduction plate for reading the intensity of the reference beam in order that its power be of the same order of magnitude as the weak fluorescent beam. The attenuated reference beam is reflected onto the surface of a reference photocell, mounted on a rotating table. By adjusting the angle of the photocell, the amount of radiation falling on the photocell, and thus the current output, can be adjusted.

The outputs of the reference and the measuring photocells is compared and a sensitive galvanometer is used to check the null point. The circuit is arranged so that the current from the reference cell opposes that generated by the measuring photocell, which

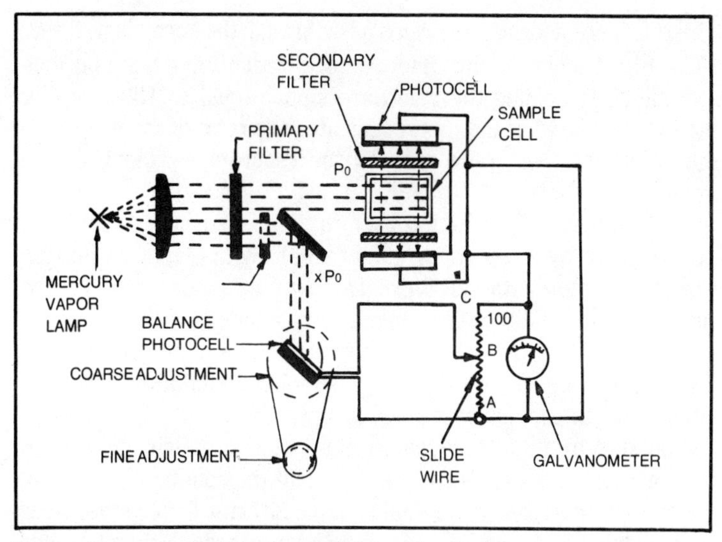

Fig. 6-4. Schematic diagram of a double beam fluorimeter.

respond to fluorescent radiation. In this arrangement, the concentration C is given by

$$C = K \frac{R_{AB}}{R_{AC}}$$

where AC is the total resistance. The constant K is independent of P_0.

Quite often, two photocells, connected in parallel, are used to measure the weak fluorescent radiation. This arrangement increases the sensitivity of the instrument.

Ratio Fluorimeters

A *ratio fluorimeter* which enables direct readings for rapid analysis of multiple samples, is due to Beckman. In this instrument (Fig. 6-5) the exciting energy is matched to the sample by the two primary filters located on both sides of a specially designed source lamp. The lamp cycles in phase with the ac line voltage and has a dark time as the lamp anodes alternately reach zero potential during the change in phase of the line voltage.

During one phase, light reaches the reference test tubes and produces fluorescence which strikes the photomultiplier and produces the reference signal. During the other phase, light falls on the sample and causes it to fluoresce. This fluorescent energy when it strikes the photomultiplier produces the sample signal.

The reference and sample beams pass through the secondary filter. This filter isolates the fluorescence radiation from spurious radiation. By setting the reference signal equal to 100% on the meter, the signal at the meter indicates the ratio of the sample to the reference. The ratio can also be recorded by using a high impedance recorder.

The source of excitation used in the Beckman instrument is divided into two sections by a light shield with electrons flowing from a common cathode alternately to two anodes, one on the reference side and the other on the sample side as the lamp cycles in phase with ac line voltage. When one beam is on, the other is completely off. Thus, the lamp supplies excitation radiation alternately to the reference and sample photocells. The lamp is surrounded by phosphor coated glass sleeve which has a clear portion plus coatings with three different phosphors. By rotating the sleeve, mercury wavelengths above 237 mμ through the clear portion, plus peaks at 310, 360 and 450 mμ through the phosphor coated portions are achieved. This dual source lamp provides double beam operation without light choppers, beam splitters or other similar devices. The sample beam and the reference beam consist of fluorescent energy of the same wavelength, thus making readings independent of lamp temperature variations. Two types of filters are employed: two primary (excitation) filters and one secondary (fluorescence) filter. Both types of filters can be easily changed for analysis of different samples. Glass test tubes can be used as sample containers for most of fluorescence that the photomultiplier receives. Since fluorescence is greater at lower temperatures, provision is made for water cooling of the sample so that sample fluorescence is as strong as possible. The sample tube area may be cooled by a continuously flowing supply of tap water.

Reference bars are supplied with fluorimeters for establishing upscale meter values. These are glass rods fused with varying concentration of uranium salts. Each bar emits about three times more fluorescence intensity successively from numbers 1 through 6. They cover an approximate range of 5-1000 ppb of quinine sulfate equivalent. In some cases they may be more useful than standard solutions, because their fluorescence values are stable indefinitely, if not scratched or chipped.

Double beam fluorimeters using two photomultipliers to compensate for fluctuations in light level from the exciting source are generally used. Nevertheless, the stability of these instruments is not particularly good and is generally not better than ± 1%

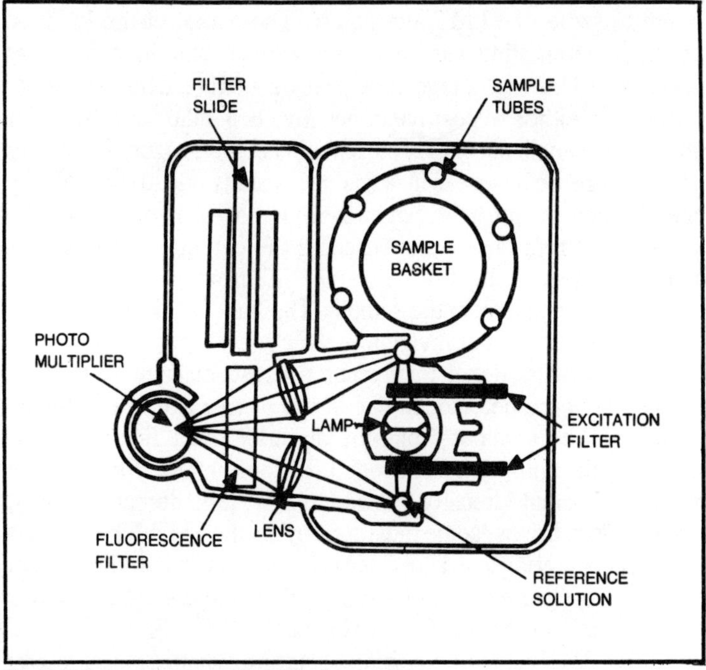

Fig. 6-5. Optical diagram of the ratio fluorimeter.

over a 10 minute period. This poor performance has been found to be due to gain changes and drifts in dark current of the two photomultipliers, caused by variations in high voltage, ambient temperature and fatigue effects. A simple solution to this problem is to use a single photomultiplier for both reference and sample beams. Melhuish (1975) describes a double beam, single photomultiplier spectrofluorimeter for the precise measurement of fluorescence emission or excitation spectra, with a stability better than ± 0.2% over a period of 10 minutes, when measuring photomultiplier anode currents between 5×10^{-8} and 5×10^{-6}A.

Another version of the double beam type fluorimeter makes use of an optical bridge that is analogous to the Wheatstone bridge, used in measuring electrical resistance. The bridge detects and measures the difference between the light emitted by a sample and that of the calibrated rear light path. This is the principle of the Turner Model 110 fluorimeter (Fig. 6-6).

The instrument employs a single photomultiplier tube as a detector. A rotating mechanical light interrupter surrounds the photomultiplier and causes the reference beam and the fluorescent beam to strike the PMT alternately. The output of the PMT is, thus, an alternating current which can be amplified in an ac amplifier. The next stage is a phase-sensitive detector whose output will either be positive or negative depending upon whether the fluorescent beam or the reference beam is stronger. The power of the reference beam can be adjusted manually or automatically by the rotation of the light cam which increases or decreases the intensity of reference beam reaching the detector. This is done until the amount of light reaching the PMT from the rear light path equals the amount from the sample. The light cam is attached to a linear dial, each small division of which corresponds to an equal fraction of light. For a totally non-fluorescent sample, no light would reach the detector in one of the phases and the null-point could then be obtained from one direction only. For accurately adjusting the null-point from both directions, a third (forward light path) of constant intensity is made to fall on the detector so that, under all conditions, some radiation strikes the PMT. This permits a correct operation both above and below zero with non-fluorescent blanks. Two lucite light pipes are employed in the optical system for the rear light path and forward light path.

The PMT is mounted inside a rotating cylindrical interrupter with two sets of light slots. The upper slots allow light to pass alternately to the PMT from the rear calibrated light path and the combination of the sample and forward light path. The lower slots interrupt the light path from the reference beam, which in turn generates the reference signal. This circuit arrangement cancels out variations in line voltage, light source and photomultiplier sensitivity.

The PMT produces an ac signal at about 400 Hz that is proportional to the light unbalance. This signal is amplified and fed as signal to the phase-sensitive detector.

The reference signal is generated by a lamp with a special glass filter around it, which passes only infrared light and directs it on to a red-sensitive photo tube. The reference lamp and phototube generates an ac signal. This signal is amplified and is given as reference to the phase detector, its phase depending upon the position of the light interrupter. The phase detector generates a dc signal proportional to the signal received, which is applied to the chopper, where the dc signal is converted to ac. The ac signal is amplified. In automatic instruments, this ac signal is used to drive the servomoter, which in turn drives the light cam so that a null-balance is obtained.

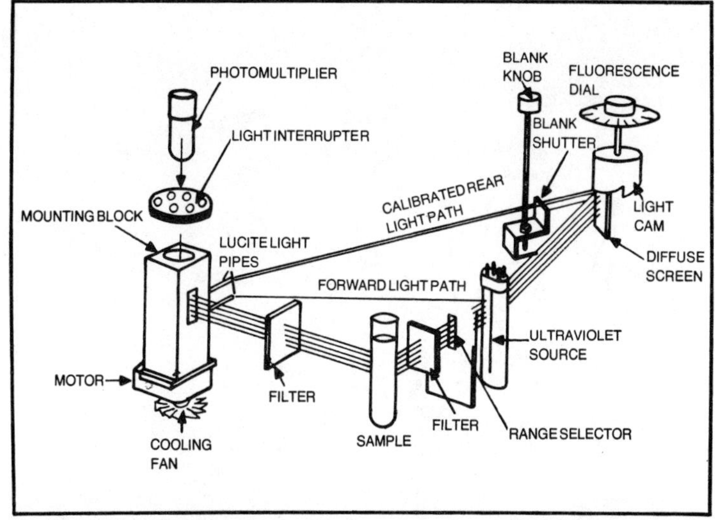

Fig. 6-6. Double beam fluorimeter based on the optical bridge principle (principle of Turner fluorimeter).

Spectrofluorimeters

Spectrofluorimeters are true fluorescence spectrometers which normally make use of two monochromators, one to supply selectively the excitation radiation and the other to isolate and analyze the fluorescence emission. The dual monochromator system thus permits one to determine and selectively utilize the peaks of excitation and fluorescence wavelength. The spectrum of the fluorescence emitted by the sample can be studied at any frequency of exciting light. Also, the variation of the intensity of fluorescence with the frequency of the exciting light used can be determined.

Figure 6-7 shows a schematic of the optical system used in a spectrofluorimeter. The source of light is a 150 watt high pressure xenon arc lamp. The ellipsoidal mirror collects radiation from this lamp and focuses an enlarged image of the arc upon the entrance slit of the excitation monochromator, completely filling the f/3.5 mirrors.

Monochromatic radiation of specific wavelength is selected by angular positioning of the grating. A desired band-pass is achieved by the selection of appropriate interchangeable slits for the entrance and exit positions of the monochromator. The radiant energy emerging from the exit slit of the excitation monochromator activates specific molecules within the sample, result-

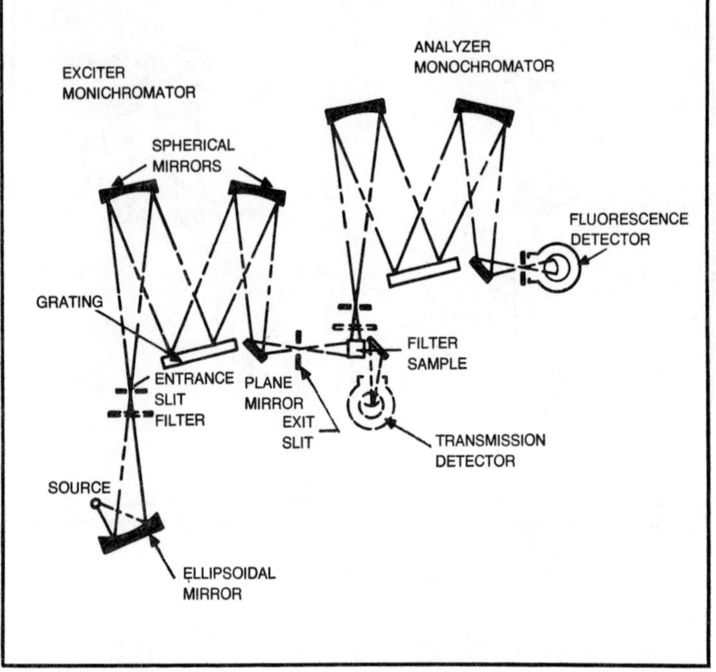

Fig. 6-7. Optical system of a typical spectrofluorimeter.

ing in fluorescent emission. The emitted radiation is viewed by the analyzing monochromator which permits only a characteristic sample emission to reach the detector, blocking all undesired spectral regions.

When fluorescent energy falls on the photomultiplier, the registered photocurrent can be indicated on the microammeter, recorder or oscilloscope. The spectrum can be manually or automatically scanned.

Many compounds exhibit far greater fluorescence at reduced temperatures than at normal ambient or higher temperatures. Therefore, spectrofluorimeters are usually accompanied by cryogenic accessories to cool the sample by means of an immersion probe which is cooled by easily replenished reservoir of refrigerant. Using liquid nitrogen, for example, provides a sample temperature very close to 77°K. A spectrofluorimeter operating on the above principle is available from Farrand, USA.

Langelaar et. al. (1969) report the construction of a sensitive spectrophosphofluorimeter. The sensitivity of the detection system was increased by the use of a cooled, selected photomultiplier and a lock-in amplifier and time averaging computer. To minimize fluctuations in the output, the xenon arc was stabilized.

A Baried Atomic Model SF 100E spectrofluorimeter (Fig. 6-8) employs a unique optical system comprised of four grating monochromators. Two are used in tandem for dispersing excitation energy, and two are used for observation of fluorescent spectra. This double grating system in conjunction with folded optics offers a significant advantage in the reduction of physical dimensions. Each monochromator incorporates a 600 line per mm plane reflection grating. Wavelength control is provided by means of a cam which may be operated manually or automatically. The calibrated operating range of each pair of monochromators is 220 mμ to 700 mμ. Wavelength settings are indicated by a scale which is calibrated in 2 mμ increments. The spectral widths of the entrance and exit slits to each pair of monochromators are each independently variable in steps of 2, 8 and 32 mμ.

The excitation source is a specially selected 150 watt dc xenon arc lamp which gives a very stable, high intensity light output in the ultraviolet, visible and infrared regions. An off-axis ellipsoidal mirror forms an enlarged image of the source arc at the entrance slit of the excitation monochromator. The lamp can be very simply focused and laterally adjusted.

The detector used is a photomultiplier which is supplied with a stabilized solid state power supply. The detector signal is amplified in an operational amplifier.

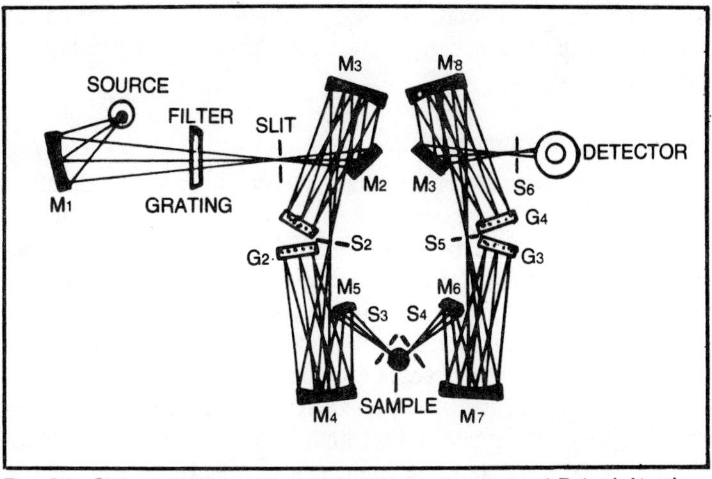

Fig. 6-8. Optical diagram of spectrofluorimeter (courtesy of Baired Atomic, USA).

MEASUREMENT OF PHOSPHORESCENCE

In the phosphorescence process, two transients are involved between the lowest excited singlet state and the ground state. This involves an intersystem crossing or transition from the singlet to the triplet state. A triplet state results when the spin of one electron changes so that the spins are the same or unpaired. The first is a nonradiative transition as the molecular energy level lowers to the vibrational level of a metastable triplet state (S to T_1). From the triplet state, the molecular energy level drops to the ground state; the radiation emitted during the transition T_1-G is called phosphorescence (Fig. 6-1). The time involved in the transition is of the order of 10^{-3} seconds which gives a relatively long period of phosphorescence emission. A characteristic feature of phosphorescence is an after glow, i.e., emission which continues after the exciting source is removed. Since the energy level of the triplet state lies below the excited singlet levels, the phosphorescence is emitted at wavelengths longer than those of fluorescence. Because of the relatively long lifetime of the triplet state, molecules in this state are much more susceptible to deactivation processes, and only when the substance is dissolved in a rigid medium can phosphorescence emission usually be observed.

Phosphorescence Emission

Phosphorescence emission is characterized by its frequency, lifetime, quantum yield and vibrational pattern. These properties form the basis for qualitative analysis, whereas the correlation of intensity with concentration gives the basis for quantitative measurements.

Phosphorescence Spectrometer

The measurement of phosphorescence is made in a similar manner as that of fluorescence except for employing a mechanical method of distinquishing between them by their time delay as shown in Fig. 6-9. Light from the xenon lamp is passed through an excitation monochromator. It falls on the sample via a fixed slit system and through a slotted-cam type shutter which surrounds the sample cell. The slotted cam is driven by a variable speed motor having the sample cuvette at the axis of rotation. The slots are so arranged that the sample is first illuminated and then darkened. While it is dark, its phosphorescence is allowed to pass and go to the phosphorescence monochromator placed at right angles to the

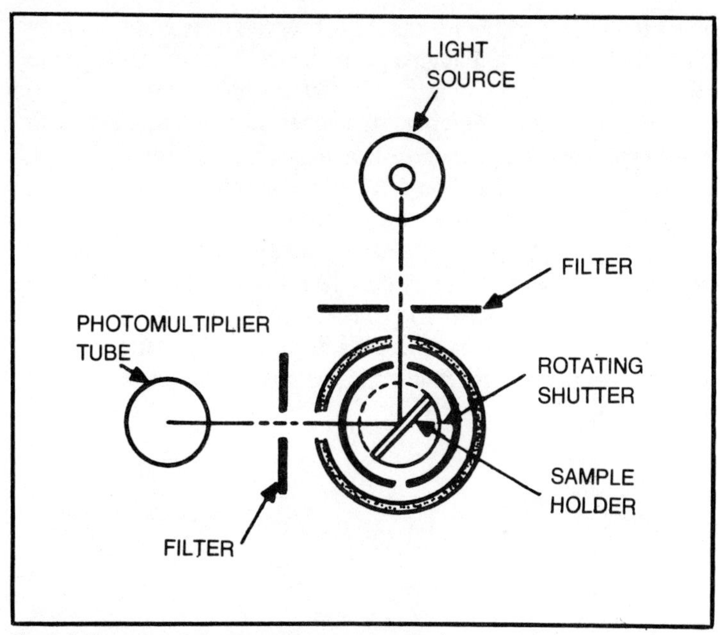

Fig. 6-9. Phosphorescence measurement method.

excitation radiation. The mechanical device accomplishing the modulation of the radiation from the light source incident on the sample, and simultaneously modulating the luminescence radiation from the sample which is incident on the photodetector, is known as a *phosphoriscope.* The modulation is periodic and out of phase, so that no incident exciting or luminescent radiation reaches the photodetector during one phase, whereas only long decaying phosphorescence radiation reaches the photodetector during the other phase. Hence the phosphoriscope allows measurement of phosphorescence in the presence of fluorescence and other scattered radiation. The phosphorescence spectrum is obtained by setting the excitation monochromator at the wavelength corresponding to a phosphorescence maximum and allowing the emission monochromator to sweep throughout its wavelength range.

Phosphorescence is not observed at room temperature as the energy of the triplet state is readily lost by a collisional deactivation process involving the solvent. Therefore, phosphorescence is normally observed at reduced temperatures in solidified samples. The sample is generally placed in a small quartz tube which is then placed in liquid nitrogen (77°K) and held in a quartz *Dewar flask.* The incident radiation passes through the unsilvered part of the

Dewar flask and phosphorescence is observed through the sample part of the flask at right angles to the incident beam. The sample cell is immersed directly into the coolant which is usually liquid nitrogen. Several commercial fluorimeters are provided with phosphorescence attachments. For example, the Farrand MK1 spectrofluorimeter, the Baird Atomic Fluoroscope and Aminco Boroman SPF have the provision for these attachments.

Pailthorpe (1975) describes the construction of a spectrofluoriphosphorimeter which is fully compensated for both excitation intensity variations and the spectral sensitivity of the detecting monochromator and photomultiplier. He successfully detected the phosphorescent emission from tryptophan in wool keratin at room temperature.

Chapter 7

Gas Chromatographs

Chromatography is a physical method of separation of the components of a mixture by distribution between two phases—of which one is a stationary bed of large surface area and the other a fluid phase that percolates through or along the stationary phase. Chromatography was first reported by a Russian botanist Tswett, who separated leaf extract into colored bands using a column of *inulin* and a solvent of *ligroin*. The technique remained largely ignored until the 1930s, when chromatographic separations of carotenes and xanthyl further demonstrated the possibilities of the technique and accelerated its development. The high separating power of gas chromatography has made possible the analysis of samples which were hitherto regarded as difficult or impossible. This is evident by the wide range of applications found in routine and research work in medical and industrial fields. However, the applicability of the technique is limited to those substances which may be vaporized without decomposition or which may be thermally decomposed in a reproducible manner. Several techniques are used in chromatography.

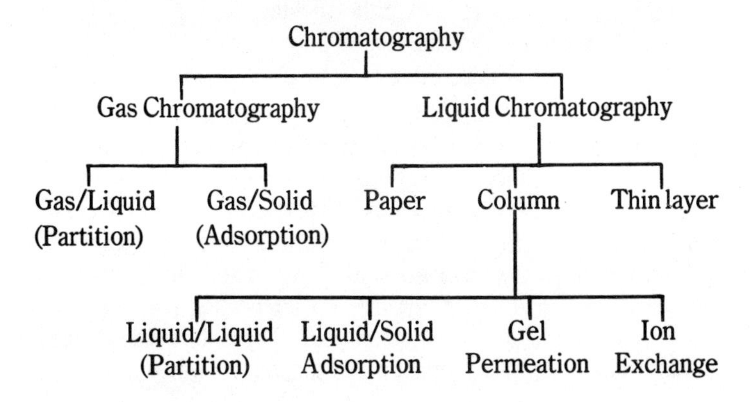

The process of chromatographic separation involves transport of a sample of the mixture through a column. For this purpose, the mixture may be in the liquid or gaseous state. The stationary phase may be a solid adsorbent or liquid partitioning agent. The mobile phase is usually a gas or liquid and it transports the constituents of the mixture through the column. During such transport, the material in the column (stationary phase) exercises selective retardation on the various components of the sample. This retardation may be due to adsorption, solubility, chemical bonding, polarity or molecular filtration of the sample. Therefore, the components of the mixture tend to move through at different effective rates and result in tending to segregate into separate zones or bands. In general, all chromatographic procedures isolate, detect and characterize these bands at some point, usually the column exit. Upon emerging from the column, the gaseous phase immediately enters a detector attached to the column. At this place, the individual components register a series of signals, which appear as successive peaks above a baseline on the recorded curve called *chromatogram*. A typical chromatogram is shown in Fig. 7-1. The area under the peak gives a quantitative indication of the particular component and the time delay between injection and emergence of the peak serves to identify it.

CONSTRUCTION DETAILS

The basic parts of a gas chromatograph are shown in Fig. 7-2. It consists of the following parts:

—Carrier gas supply along with pressure regulator and flow monitor.

—Sample injection system.

—Chromatographic column.

—Thermal compartment or thermostat.

—The detection system.

—The strip chart recorder.

The carrier gas, normally N_2, Ar or He is usually available in a compressed form in a cylinder fitted with a suitable pressure regulator. The gas is conducted from the cylinder through a flow regulator to a sample-injection port maintained at a certain temperature T_1, which is such that it ensures rapid vaporization but not thermal degradation of the solute. Gas and liquid samples are almost always injected by syringe through a self-sealing silicon rubber diaphragm in the injection port. The solute vapor mixes almost instantaneously with the flowing carrier gas and is swept

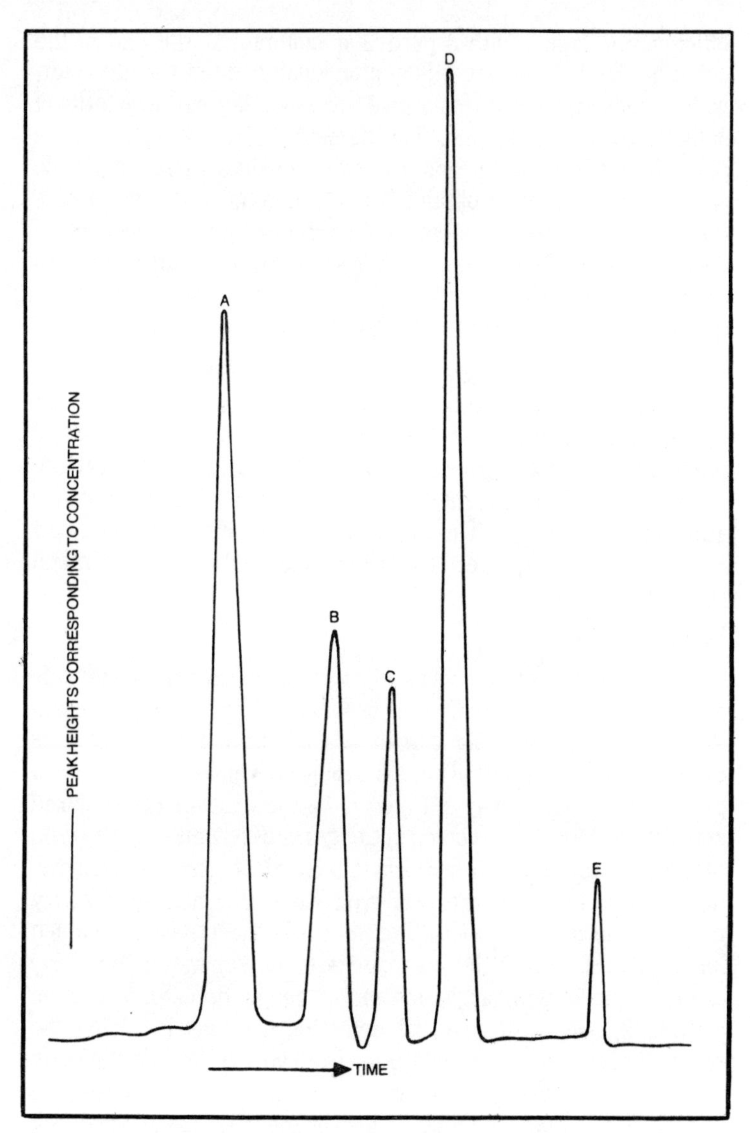

Fig. 7-1. Typical gas chromatogram.

into the chromatographic column, which is the heart of the chromatograph. It is here that the different solutes in the vaporized sample are separated from each other by virtue of their different interaction with the column packing. The column is maintained at another temperature T_2. This temperature determines the time for the passage of the solutes and, to some extent, the resolution and efficiency obtained with a particular column. At the end of the column, the solutes emerging individually enter the detector, which produces an electrical signal corresponding to the quantity of solute leaving the column. The detector signal is supplied to a potentiometric recorder and a plot of the time-signal amplitude called chromatogram is obtained. This record is used to determine the identity of the components in the mixture and their respective concentrations. The various parts of a gas chromatographic system are described below.

CARRIER GAS SUPPLY OR THE MOBILE PHASE

In a gas chromatograph, the mobile phase is formed by a continuous supply of a carrier gas. This supply is taken from commercially available cylinders in which they are stored at pressures up to 2500 lb/sq in. They pass through the column at low rates of flow (20-50 ml/min), at pressures not much greater than atmospheric pressure. The carrier gas supply system comprises a needle valve, a flowmeter, a pressure gauge and a few feet of metal capillary restrictors.

Types of Gases

Several gases like helium, nitrogen, argon and carbon dioxide have been tried as carriers—hydrogen, helium, nitrogen, argon and carbon dioxide. The carrier gas affects column as well as detector performance. The carrier gas, which is best for a particular detector, may not always be the best for the required separation. Mostly the choice of the gas is determined by the type of the detector and the ready availability of the gas. For example, helium and hydrogen are preferred when thermal conductivity detection is employed, since their thermal conductivities are much higher than those of the compounds to be separated. Similarly argon is used with argon-ionization gauge detectors. Carbon dioxide is used with integral detection systems involving the removal of carrier gas by absorption in alkali solution. On the basis of separation power, nitrogen, argon and carbon dioxide are slightly better than the lighter gases, as the latter have a tendency to enhance axial diffusion of the solutes, a factor that could seriously affect the efficiency of the column. Nitrogen is particularly used where separating power is more important than high detector response.

Purity of Gases

The presence of contaminants in the carrier gas may affect column performance and detector response, particularly when

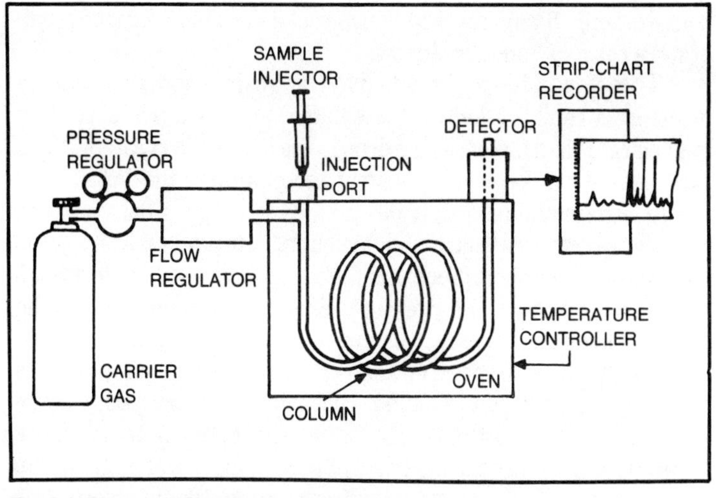

Fig. 7-2. Block diagram of a gas chromatograph.

ionization detectors are used. Carrier gases may be purified by inclusion of a trap containing a molecular sieve 5A. This is usually adequate for removal of hydrocarbons and water vapor. Ultra pure N_2 for use with flame ionization detector can be generated by commercially available apparatus. A low oxygen content of the carrier gas is essential for obtaining repeatable retention times and peak widths, because a small amount of oxygen is enough to change the liquid phase and affect the overall partition ratio. In programmed temperature gas chromatography, a high purity carrier gas is even more important than in isothermal operation. The column is saturated at every temperature with the carrier gas impurities. During the heating up cycle, the saturation level changes and the impurities are given off to a certain extent. These impurities, which are liberated, are eluted and will temporarily increase the zero signal and noise resulting in a broad peak.

The Gas Flow Rate

The rate of gas flow to be used in a particular analysis, among other factors, depends upon column diameter. The flow is generally in the range 10-400 ml/min—very low and very high flow rates may affect the efficiency adversely. Flow rate should be controlled within 1% in order to reduce analytical errors. It should also be constant in order to give reproducible retention times. The flow rate of the carrier gas also affects the detector signals because fluctuations would produce variable heat removal from the katharometer filaments, and hence cause variable filament temperature response and sensitivity.

The flow of the carrier may be maintained constant by inserting a capillary before the column, so that a pressure drop much larger than the pressure drop in the column is created. When this capillary is kept at a constant temperature, the flow in the whole system will be mainly determined by this capillary, and not by varying flow resistance of the column. Alternatively, the gas is passed through a short piece of capillary tubing, resulting in a small pressure drop. A regular pressure regulator connected across the capillary keeps the pressure drop constant, resulting in a constant flow. With a view to speed up analysis, flow programming in which the flow through the column continuously increases can be adopted. This would have the same effect, as in temperature programming, of producing closely spaced peaks at the end of the chromatogram. For columns of different sizes, the volumetric gas flow rate should be varied in proportion to the square of the diameters, so as to maintain the average linear rate at approximately the same value.

SAMPLE INJECTION SYSTEM AND THE SIZE OF THE SAMPLE

The purpose of the sample injection system is to introduce a reproducible quantity of the sample to be analyzed into the carrier gas stream. The transfer of the sample should be made rapidly to ensure that the sample occupies the smallest column volume and thus prevents excessive peak broadening, which affects the overall resolution of the system. Samples can be introduced in their gaseous, liquid or solid states and many methods have been suggested for the purpose. However, the choice of the method of sample injection depends upon the pressure in the column at the point of introduction, the type of detector to be used and the source of the sample. Theoretically, for optimum separation efficiency, the sample must be introduced into the column as rapidly and in as concentrated a band as possible. This is called ***plug insertion***, something that chromatographers consider ideal. The purpose of the sample injection system is to insert, volatilize and move the resulting gaseous sample into the column. The design of this system is a critical gas chromatograph performance factor, because a column capable of efficient component resolution may appear ineffective due only to the inadequacy of the sample insertion system.

Theoretical considerations enjoin that the efficiency of separation in a GC column improves as the size of the sample is reduced. With a normal analytical column of 4-8 mm (internal diameter), a liquid sample of 2-20 μl and a gas sample of 0.5-5 ml at atmospheric pressure is generally satisfactory. Exception to this general practice is in the case of trace analysis, when liquid samples up to 200 μl are necessary to obtain sufficient response.

Liquid Samples

The usual method is to inject liquid samples with a microsyringe through a silicon rubber septum. Syringes of various capacities are commercially available and are generally employed for injection of samples between 0.1-10 μl. A sample is injected into the hot zone of the column, so that the liquid gets rapidly transferred into the gaseous phase.

A typical arrangement for injecting liquid samples is shown in Fig. 7-3. The metal block containing capillary is heated by a controlled resistance heater. Here the sample is vaporized and carried into the column by the carrier gas. Care should be taken to insert, inject and remove the needle quickly.

Gas Samples

Gas samples are injected by a gas tight syringe suitable for delivering 0.1-10ml of sample. They are usually difficult to handle and often cause inaccuracies.

The other method of injecting gaseous samples is the bypass system, also called a stream splitter. This system has been found most valuable and gas sampling valves using this technique are used extensively. The principle of the system is to fill a loop of known volume with the sample. By operating a valve, the loop is placed directly in the carrier gas line. The valves were earlier made of glass, but they have now been replaced by polytetrafluoroethylene designs.

Figure 7-4 shows a schematic of this type of arrangement. Basically, it is an arrangement of three stopcocks, between two of which there is a standard volume in which gas sample is enclosed. Gas from the bypass capillary loop is introduced into the column by a rotating or sliding valve, so that the loop is connected with the stream of the carrier gas.

German and Hevnen (1972) describe a sampling valve for gases or liquids, which is capable of operating at high sample pressure (up to 75Kgf cm^{-2}) or at temperatures (up to 300°C). For the high pressure operation, a Kel F-Teflon-Molykote composite sealing material was developed and used. The valve is particularly

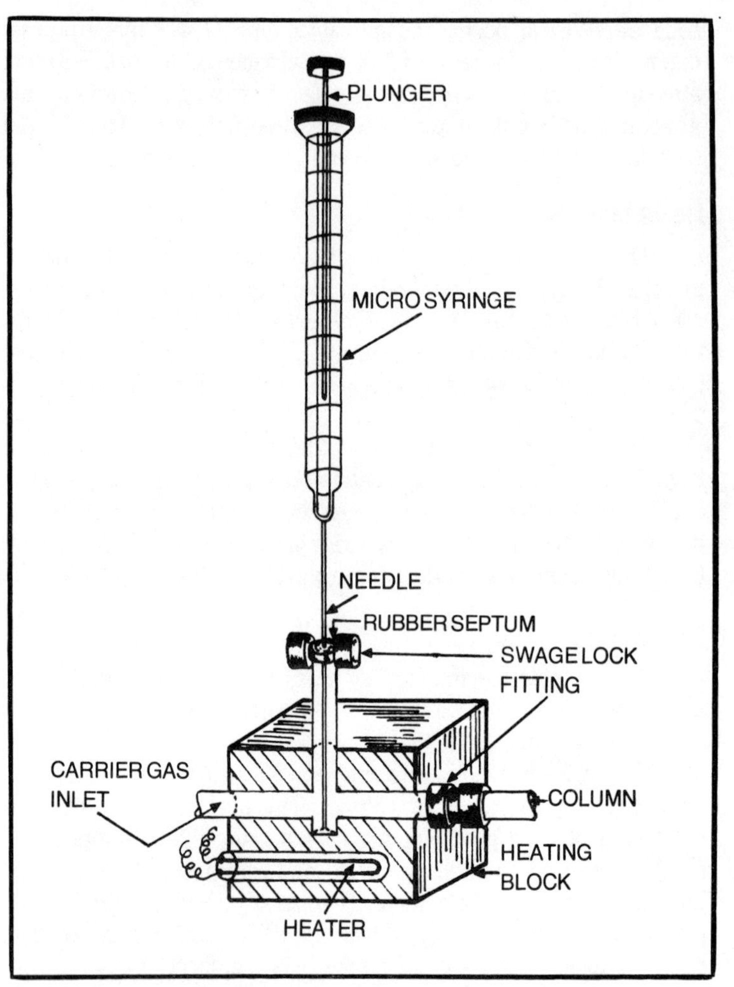

Fig. 7-3. Arrangement for injecting liquid samples.

suitable for studies in gas kinetics, since very small vapor or liquid samples can be injected directly from a reaction system into the gas chromatographic column.

Solid Samples

Solid samples may also be injected by using solid injection syringes where the sample is deposited on the end of the plunger and withdrawn inside the needle. After piercing the injection septum, the plunger is extended to place the solid in the hot zone of the column where it is vaporized. Alternatively, the solid is deposited in a glass tube or gauze from a solution. After the evaporation of the solvent, the sample holder is dropped into the column, thus making the injection. Another method is to dissolve the solids in volatile liquids or temporarily liquefy them by exposure to infrared heat.

Pyrolysis

Pyrolysis offers a technique for injection of certain types of materials which are low or non-volatile, but may be thermally decomposed in an inert atmosphere to offer a qualitatively and quantitatively reproducible mixture of volatile fragments. The pyrolysis products are transferred to a chromatographic column and separated in the usual manner. Pyrolysis has been accepted as a valuable technique for sample injection in rubber, plastic, polymers and adhesives industries.

CHROMATOGRAPHIC COLUMN

The column is the heart of a gas chromatograph where the fundamental process of separation takes place. Its action is based on the fact that when a sample of gas or vapor is introduced into a column, it spreads by molecular diffusion to yield a concentration profile. As the sample moves through the column, additional spreading takes place, but the band maintains its general shape, which is detected and recorded as the familiar chromatographic peak. The degree of peak broadening with respect to time and column length is an indication of column efficiency. Column performance is usually measured by the number of theoretical plates which may be determined from the dimensions of peaks.

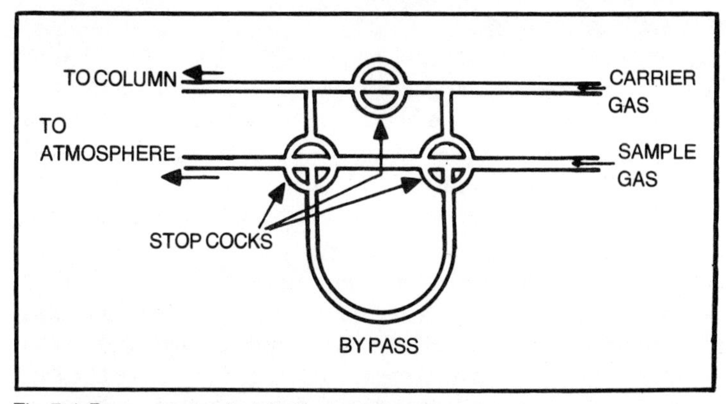

Fig. 7-4. Bypass system for injecting samples.

A theoretical plate is defined as a layer, at right angles to the column of such a thickness that the solute at its mean concentration in the stationary phase in this layer, is in equilibrium with its vapor in the mobile phase leaving the layer. Laboratory columns of 20-100 plates are widely in use and normally have a height equivalent to a theoretical plate (HETP) of about 1 cm. Longer practical columns of up to 100,000 plates have been reported in literature.

Packed Column

There are two types of columns which are commonly used. They are packed columns, capillary columns or open tubular columns.

The packed column is a tube packed with a suitable material which performs the separations. Columns may be made from any suitable tubing. Glass, stainless steel or copper are the materials most frequently used for making columns. For moderate temperatures, polyvinyl-chloride tubing is satisfactory. Internal diameter of the column is usually between 4 and 8 mm. The length of the column may be between 1 to 50 meters. However, for most of the applications a length of about 2 meters is adequate. Columns longer than 3 meters are difficult to pack uniformly. Therefore, very long columns are best constructed by coupling short (less than 3 meters) sections to obtain any desired length. Sample sizes for packed analytical columns vary between 0.1-10 μl. By increasing the column diameter, greater sample sizes, of the order of 100 μl to 3 ml, can be used for preparative purposes. For this type of chromatography, the column has internal diameter from 6 to 25 mm and the packing material specifications are also changed to cope with increased loading.

For convenience, the column is made in the form of a U or helix or it can even be straight. Straight and U-shaped columns can more easily be repacked. A helical column is normally difficult to fill. For this reason, copper tubing column is preferred because it can be filled while straight and then bent to helical shape afterwards. A definite advantage of coiled or helical column is that they are more compact for a given length of column and therefore easier to heat to an even temperature. The helix type of column is usually of 50-250 mm in diameter and 2 meters in length. As long as the column is packed sufficiently tight before coiling, there is no significant difference in performance compared to a straight tube. Columns are quite difficult to pack after they are coiled.

Instruments designed for glass columns have tall ovens to accommodate U shapes that can be easily packed for maximum performance. Columns larger than 12.5 mm in diameter are made in straight sections connected by small diameter tubing.

Capillary Column

Capillary columns are open tube columns of tubing approximately 0.25 mm in diameter. Their lengths may run anywhere from 30-300 meters. Very high efficiencies have been achieved with capillary columns, since the cross diffusion of sample molecules is minimized by the narrow diameter. Capillary columns contain no packing and the stationary phase is coated directly on the inside of tubing. Capillary columns cannot handle samples more than 0.1 μl. Larger samples are handled by the use of inlet splitters. With capillary columns, better separations can be achieved at lower temperatures and in a shorter time.

Capillary column gas chromatography is rapidly developing to meet many of the analytical needs such as environmental separations, analysis at trace levels for clinical and bio-medical diagnostics, the separation of natural products and other complex mixtures. On the basis of glass capillary columns, Desty (1975) suggested that a sub-picogram high speed analyzer will soon be possible. The theoretical limits of resolution for such columns can be calculated although this has to be considered in light of column properties such as adsorption, retention polarity, thermal stability, acidity and separation efficiency (Grob, 1974). Schomburg and Husmann (1975) have made an excellent case for the use of capillary columns in preference to packed columns and have illustrated the power and versatility of many of the capillary techniques.

Support Material

A *support material* is used in partition chromatography to provide a thin liquid film with as large an interface as possible between the gas and liquid phases, so as to facilitate a partition between them. A primary requirement of solid support material is that it should not possess adsorptive properties towards sample components. Besides being inert, the support should have the structure and surface characteristics to hold the liquid phase uniformly over a large area.

A number of supports have been tried including glass balls, sodium chloride and pumice powder. However, only diatomaceous earth has been found satisfactory. It has been in use under the trade name CELITE with an average particle diameter of 40 μ. Trade names also include GC-22 Super Support, Sil-O-Cell C-22 firebrick and chromosorb P, W or G. Chromosorb is prepared by calcining diatomaceous earth obtained from a marine deposit in Lompoc, California. CELITE is the most widely used support material.

Glass beads are also available in a variety of mesh ranges. Because of their low surface area and light liquid phase loadings required, they will elute high boiling point compounds at a lower temperature than is normally required. Teflon 6 is another material which is one of the most inert and non-adsorptive. It is best suited for the analysis of highly reactive materials which are difficult to chromatograph on other supports. However, they offer lower efficiency and a temperature limit of 250°C. Particle sizes may range from 10 to 100 mesh.

The most commonly encountered problem due to support participation with the column is that of peak tailing, which appears as a long tailing edge of a peak due to adsorption. In extreme cases, if the peaks tail badly, separation is impaired and the determinations based on peak area measurements become difficult to make, thus giving unreliable results. In these situations samples may be totally adsorbed. It is possible to minimize tailing effects by modification of the support. The CELITE surface, which is prone to cause sample adsorption problems, resulting peak tailing, may be covered in hydroxyl groups. The adsorption effects can also be minimized if the CELITE is treated with a silating agent such as hexamethyldisilazine or dimethyldichlorosilane.

The column must be so loaded that it has an even packing and the gas flow does not vary either across the column or irregularly along its length. Experience is needed to achieve an even packing of the column to obtain a high efficiency.

The Stationary Phase

The separation of the sample into its components is achieved by a partition process involving the stationary phase and the moving carrier gas phase. The stationary phase is either liquid or solid. Therefore, there are two possible methods with the gas as the mobile phase, gas-liquid chromatography and gas-solid chromatography.

Gas-Liquid Chromatography

Here the stationary phase is liquid which is distributed on a solid support material. The stationary phase must be involatile at all temperatures at which the column will be operated for the analysis and should be coated as a thin even film onto the support. It is chosen for the selective retention characteristics of components in the sample that it will be used to separate. In general, highly polar stationary phases are employed to selectively retard polar compounds. On the other hand, non-polar stationary phases offer little selectivity and components tend to be eluted due to differences in boiling points of the sample components.

At temperatures above 150°C, special difficulties arise, as normal solvents become highly volatile or even unstable. In such situations, substances like silicon polymers may be used, especially for temperatures above 250°C.

An important requirement of the stationary phase liquid is a certain compatibility with the components of the sample under analysis. Generally it is found that a polar substance is most satisfactorily analyzed on a polar stationary phase. Similarly, a non-polar compound will give the best results on a non-polar phase. For example, for separations of polar components like alcohols, amines etc., it is preferable to choose a polar liquid like polyethylene glycol. Nevertheless, departure from the above rule of similarity is quite often necessary.

Normally, for analytical packed columns, 1 to 10% w/w of stationary phase on the support is employed. The analysis time is approximately proportional to the quantity of stationary phase in the column. Higher rates induce diffusion phenomena that would impair the separation. On the other hand, at low ratios, the inert support might manifest considerable residual absorptivity to cause tailing of elution peaks.

For applications in the biomedical field for the analysis of sugars, bile acids and steroids, a low percentage of stationary phase may be used to produce a very fast column, which will pass high boiling point samples at moderate temperatures to avoid thermal decomposition. The choice of stationary phase is extremely important for the successful analysis of each sample mixture. Sometimes, in practice, brief experiments are necessary with a number of trial columns in order to make a suitable section. It is here when the experience of the operator would prove useful. Ridgeon (1971) gives a detailed table, which lists a number of the commonly used stationary phases with details of their maximum operating temperatures and suitable solvents from which they can be coated on to the support.

Gas-Solid Chromatography

The stationary phase in this type is solid material with surface active properties. The separating principle is based on the variation in the extent to which constituents of a mixture are adsorbed on the adsorbant packed in the column. Therefore, the separation is obtained because of the different adsorption affinities, which the column packing has towards the sample components. This type of chromatography is used in the analysis of inorganic gases and low molecular weight hydrocarbon gases. Among the most commonly used adsorbants in gas-solid chromatography are silica gel (SiO_2), alumine (Al_2O_3), charcoal and molecular sieves (sodium or calcium aluminum silicates).

THERMAL COMPARTMENT

The column is not normally operated at room temperature because it would then be suitable only for the analysis of gases or extremely volatile liquids. Therefore, it must be heated in some form of thermostat. Moreover, it is desirable to keep the column at a precisely constant temperature. This is essential because the quantitative response of the detector is often affected by column temperature. For this purpose, the column is housed in an oven, whose temperature is controlled to an accuracy of 0.1°C.

Isothermal Operation

Various methods have been tried; namely vapor jackets, electrically heated air baths, liquid baths or metal blocks. Usually, an air bath chamber surrounds the column and air is circulated by a blower through the thermal compartment. The temperature of the oven may be controlled accurately using a proportional temperature controller with a platinum resistance thermometer as a sensing element. The oven is thermally insulated so that heat loss to the atmosphere is minimized. However, this factor is balanced against the thermal capacity of the insulating material, which if too high would affect the rate of cooling of the oven.

Normally, the temperature is so chosen that it gives a satisfactory time for analysis. Approximately, a temperature in the vicinity of the average boiling point of the components in the sample will be convenient, so as to effect an elution period of 10 to 30 minutes.

Figure 7-5 shows a schematic diagram of an oven temperature controller. The temperature sensing is done by the platinum resistance R_1 which is placed in the oven. The temperature setting

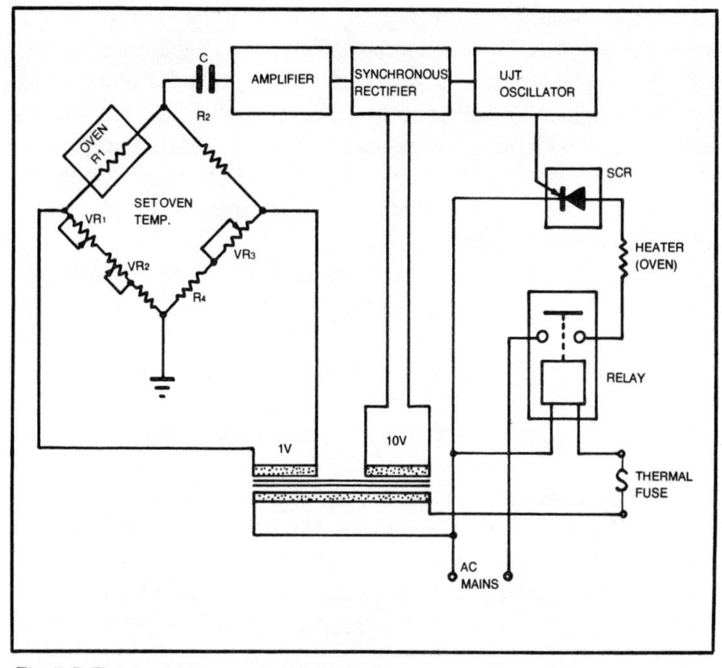

Fig. 7-5. Temperature control circuit for ovens.

is done by adjusting the potentiometer (VR_1) calibrated in terms of temperature. This control is provided on the front panel of the instrument. When a setting is made, the bridge gets unbalanced and the amplifier, the synchronous rectifier and the UJT oscillator are actuated to open the gate of SCR. Thus, the current is supplied to the oven heater and the oven temperature begins to rise.

As the oven temperature approaches the preset value, the sensor resistance becomes higher. The bridge approaches nearer to the balanced state and the heater current decreases. When the oven temperature reaches the preset value, the heater current would not flow and the bridge would be balanced. The oven temperature is thus kept constant. A thermal fuse placed in the circuit prevents the oven from overheating.

Temperature Programming

When the column temperature is kept constant, it is difficult to analyze samples having components of a wide boiling range. This difficulty can be overcome by using programmed heating of the column, so that its temperature is not kept at a constant temperature, but is subjected to an exactly controlled temperature rise while a separation is in progress. The technique combines in it the advantage of a low temperature for better separation of low boiling components with that of high temperature for more rapid elution of high boiling components, thereby shortening the time of analysis and sharpening the resultant chromatographic peaks. Temperature increase may be programmed to be carried out either continuously or in steps or abruptly to a predetermined higher level between two peaks.

Programs are available which give linear and non-linear temperature programming of ovens. The temperature can be raised at various rates. Generally, linear rates of temperature programming in the range of 1 to 20°C/minute are used. The rates of 5-7°C/minute are most typical. Some applications require non-linear temperature programming in an exponential or ballistic manner.

For temperature programming, the program according to which the temperature is to be varied is taped or ink recorded on special mylar format sheets. Curved rates of temperature rise, linear sections and isothermal operation can be plotted as desired. The recorded sheet is fixed on a rotating drum and the program line is followed by an optical scanner. The scanner is linked to a servo system which continuously controls the wattage supplied to a proportional heating system. At the completion of a run, column temperature must be dropped from about 300°C to less than 100°C in a few minutes so that the column may become ready for the next run.

Use of Two Columns

When using the programmed temperature technique, the behavior of the column itself is influenced with the change in temperature. There is an increasing tendency for the stationary phase to bleed from the column as the temperature rises and it is reflected as a baseline drift, with a chromatogram superimposed on it. When working at high sensitivity, it is possible that this drift in baseline may severely limit the use of temperature programming. This problem can be partially offset by using two matched columns (Fig. 7-6) and operating two detectors in a differential mode. One column is called the sample column and the other is a reference column, to which no sample is added. The signals from the detectors are combined to balance the bleed effects and give a straight baseline. However, a careful setting up procedure is necessary to balance the two columns.

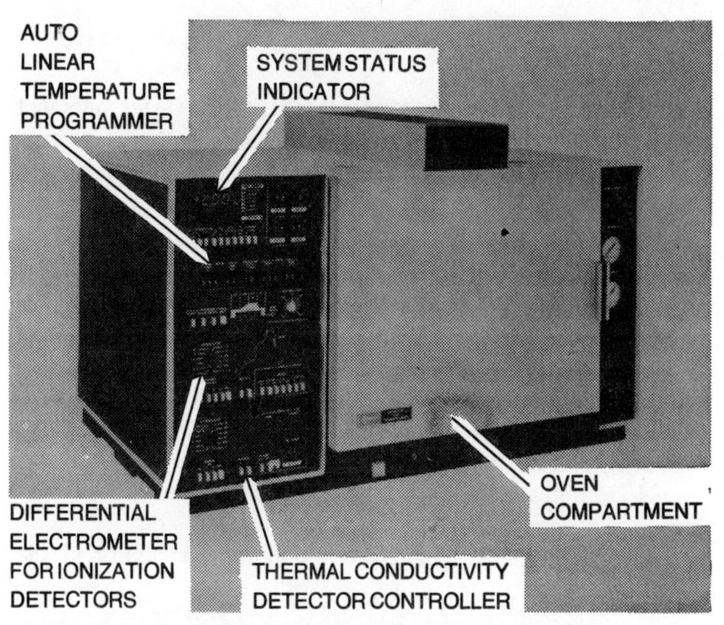

Fig. 7-6. Dual column gas chromatograph (courtesy of Varian, USA).

The following are important considerations, which are kept in view when designing column ovens:

—The oven must have minimum thermal gradients so that the temperature is uniform over the whole column.

—The oven must have a fast rate of heating. For this, it must be constructed from low mass materials. This requirement is particularly significant for changing the column temperatures rapidly, as in operations involving temperature programming.

—Temperature controlling facility up to 400°C is necessary.

—Power consumption should be kept low. For that, heat loss by all means must be minimized.

—The door of the oven should be large enough to facilitate installation and removal of the column and its accessories.

DETECTION SYSTEMS

The detector is placed at the exit of the column. It is employed to detect and provide a quantity measurement of the various constituents of the sample as they emerge from the column in combination with the carrier gas. The detector, in fact, acts as a transducer and converts the changes in some physical property to changes in an electrical signal which can be conveniently recorded.

The choice of a particular type of detector is governed by the following factors:

■ The detector should have a sensitivity to be sufficient enough to provide an adequate signal for all components with a small sample. It should also permit the use of lower column temperatures.

■ It is desirable to be able to measure components from the fractional ppm to almost 100% in one sample. The response of the detector should be linear over the whole range.

■ A small internal volume ensures that the resolution of components which are separated by the column is not lost, and that the shape of peaks is not distorted by the detector.

■ Detector temperature should be such that an appreciable amount of the eluted vapors does not condense in it.

■ The detector should be insensitive to changes in the rate of flow of the carrier gas.

■ The detector should give good reproducibility of the baseline.

There are several detection systems which are used in gas chromatography. Quite often, the fields of application of these detectors overlap to a certain extent, but one of the detectors will usually have characteristics, making it most suitable for a particular analysis.

Signals from various types of gas chromatographic detectors can be conveniently amplified by employing operational amplifiers. With flame ionization detectors, where the signal levels are of the order of 10^{-11} A and even lower, it is necessary to utilize electrometric input operational amplfiers in the input stage. These amplifiers have input bias currents of the order of 10^{-13} to 10^{-14}A. A wide range of signal amplitudes can be handled by a logarithmic electrometric amplifiers. Linearity of the overall response is then restored by an exponential converter following the logarithmic amplifier.

The Katharometer or Thermal Conductivity Detector

The *thermal conductivity detector* is a simple and most widely used type of detector. It is based on the principle that all gases have the ability to conduct heat, but in varying degrees. This difference in heat conduction can be used to determine quantitatively the composition of a mixture of gases. By definition, the thermal conductivity of a gas is the quantity of heat (in calories), transferred in unit time (seconds) in a gas between two surfaces 1 cm^2 in area, and 1 cm apart, when the temperature difference between the surfaces is 1°C.

In its simplest form, the detector may consist of a hollow tube with an electrically heated coil mounted axially in its center. When only the carrier gas flows over it, a thermal balance can be attained at a certain temperature. However, when a gas or vapor differing in thermal conductivity from the carrier gas flows past the heated coil, the temperature of the coil gets altered and a proportionate change in the electrical resistance of the wire takes place. Such changes in resistance arising from the components of the sample are used for detection and estimation of the unknown sample components.

In actual practice, the detector consists of two temperature sensing elements arranged in a Wheatstone bridge circuit, one in the reference and the other in the measuring arm. The heat sensitive elements are either thermistors or resistance wires like platinum or tungsten. Figure 7-7 shows a typical circuit arrangement for measuring the changes in the resistance produced in the katharometer cell elements. Resistances R_1 and R_2 are the katharometer wires and resistances R_3 and R_4 are the ratio arms of the bridge. Resistances R_7 and R_8 are used for making baseline adjustment and are made of manganin wire. The output of the bridge is fed to the recorder through an attenuator so that if signal is greater than the span rating of the recorder, full scale reading may be adjusted.

For the balanced bridge conditions, when the carrier gas flows through the two cells, no current would be flowing between A and C and

$$\frac{R_1}{R_2} = \frac{R_3}{R_4}$$

However, when the resistance R_1 changes due to the components of the sample gas, it causes an unbalanced current to flow from A to C. The magnitude of the current serves to detect and measure the magnitude of the gas component vapor passing over the measuring cell. If the Wheatstone bridge is excited with ac it can be made many times more sensitive because the ac signal can be conveniently amplified before it is given to the recorder.

The sensitivity of a thermal conductivity detector depends on the nature of the carrier gas. When helium gas is used, 10^{-7}g of inorganic gases can be detected. A katharometer is fairly satisfac-

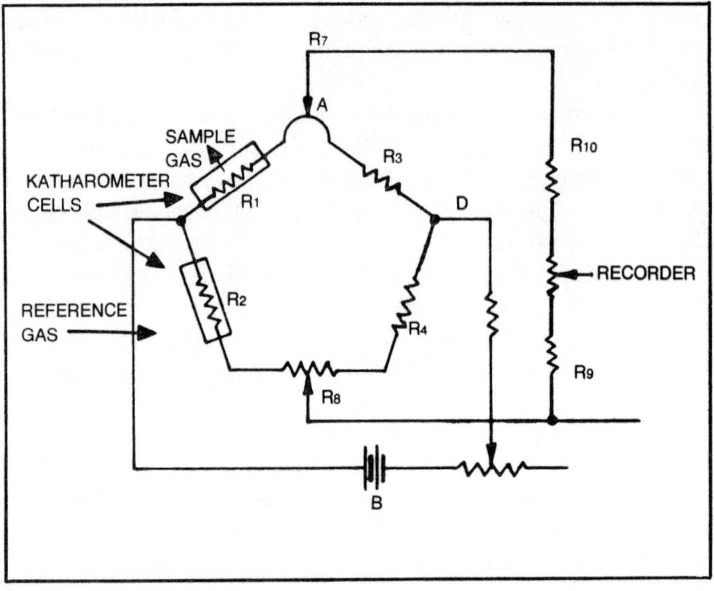

Fig. 7-7. Katharometer cell.

tory for a wide variety of analytical applications. The technique is non-destructive and the components of the sample can be further trapped for other forms of analysis.

Flame Ionization Detector (FID)

A *flame ionization detector* is by far the most widely used detector in gas chromatography. It responds with high sensitivity to almost all organic compounds. Its linear dynamic range is approximately 10^7, which is much wider than that obtained from other detectors.

In this detector, the effluent from the column is led into an oxy-hydrogen flame (Fig. 7-8). An electrical potential is applied across two electrodes placed in a stainless steel housing. The hydrogen flame burns at the tip of a capillary, which also functions as the cathode and is insulated from the body by a ceramic seal. The collector electrode consists of a loop of platinum and is located at about 6 mm above the burner tip.

The current across the electrodes remains constant when only the inert carrier gas passes the flame. However, when the vapor of a compound emerging from the column passes the flame, the vapor molecules are broken into ions by the hot flame. These ions result in the ionization current, and there would be a consequent change in the current flowing across the electrodes. The magnitude of the variation in current would be directly proportional to the number of ions or electrons formed in the flame gases which in turn would be proportional to the carbon content of the organic molecules in the vapor.

The flame-ionization detector has a high output impedance similar to that obtained with glass electrodes when making pH

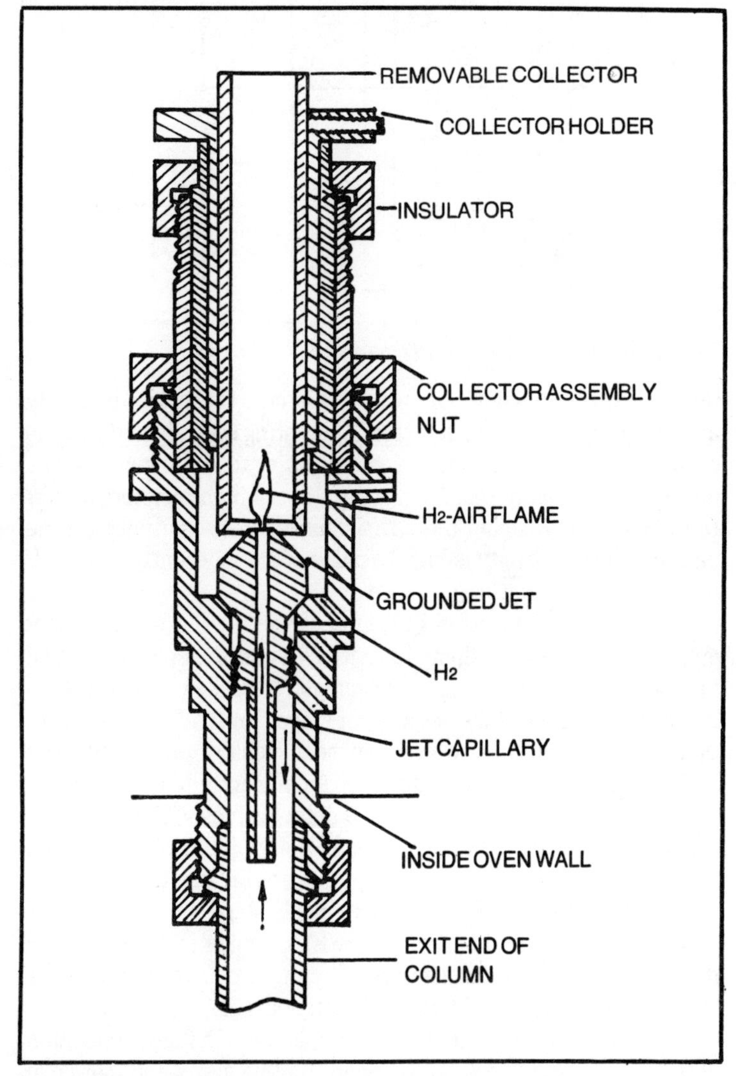

Fig. 7-8. Flame ionization detector.

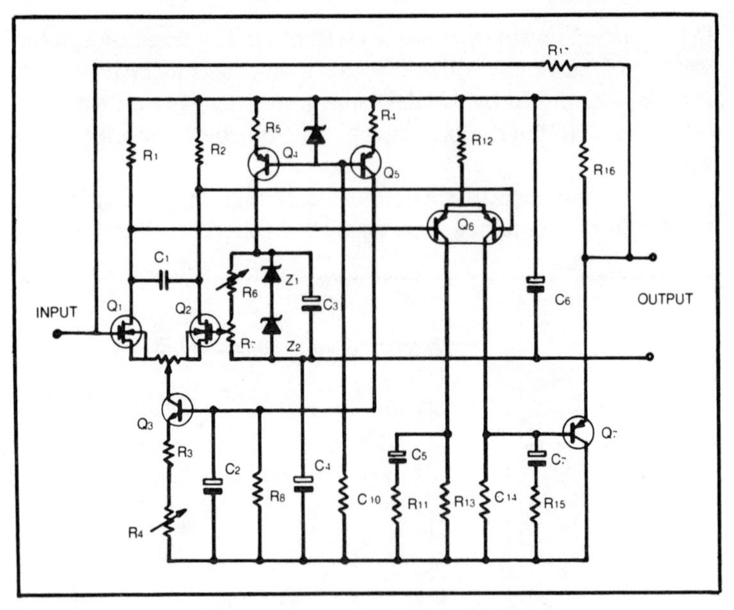

Fig. 7-9. Amplifier circuit used with flame ionization detectors.

measurements. Commercial pH meters can, therefore, be easily adapted for use with this detector. A vibrating reed electrometer is often used in the input stage of the amplifier to attain sensitivities up to 5×10^{-13} amperes. By placing a set of high resistors across the flame and changing their resistances, the sensitivity can be varied. The sensitivity is high because of the inherently low noise level of this detector.

Meigh and Oetzmann (1971) describe a solid state current amplifier with which a flame ionization detector can be used at all practical sensitivity levels. The amplifier makes use of metal-oxide-silicon transistors in the input stage (Fig. 7-9) in place of an electrometer pentode, which had been usually employed. The amplifier gives a degree of stability significantly better than that achieved with thermionic electrometer valves. The amplifier constructed by these men has a sensitivity of 100 mV pA^{-1} and a noise output equivalent to 2×10^{-15}A rms at the input; the thermal drift is typically 1 mV $°C^{-1}$. The input stage of the amplifier consists of a matched pair of MOSFETs, which is followed by a matched pair of p-n-p transistors. The output stage is an emitter follower, while the first stage is a long tailed pair.

There are certain limitations to the use of a flame ionization detector. The FID does not respond to inert gases and inorganic compounds. The emerging components gets destroyed in the flame. The response to sample weight has to be separately determined for each component.

In a flame ionization detector, it is not only the sample components which are ionized in the hydrogen flame, but the liquid phase escaping from the column also participates in producing ions. The ionic current due to the liquid phase is recorded on the chromatogram as a background signal. In case of the programmed temperature analysis, the background signal changes with time because the amount of liquid phase escaping from the column increases gradually in response to the raised column temperature. This causes great inconvenience to the programmed temperature analysis of trace substances. To eliminate this difficulty, a differential hydrogen flame ionization detector has been devised. It consists of two FIDs based on the same working principle. The schematic diagram of the DFID is shown in Fig. 7-10.

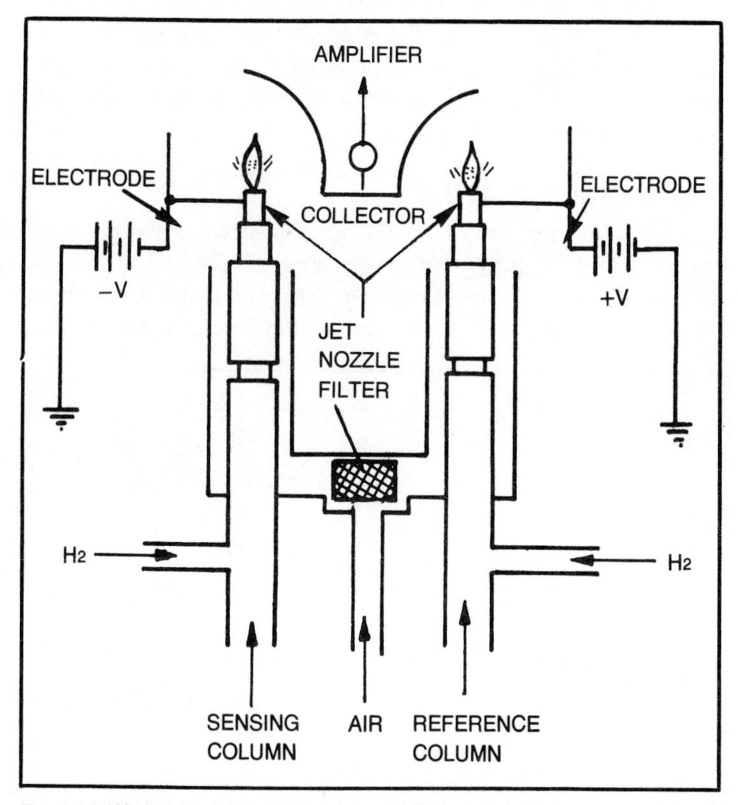

Fig. 7-10. Working principle of differential flame ionization detector.

Two columns are used in this arrangement. If the same packed column as the sample-side one is employed on the reference side, the carrier gas is fed through the reference-side column at the same speed as on the sample side. If the reference-side column is mounted very near the sample side one, under these conditions, the amount of liquid phase flowing from the reference-side column can be considered as the same as that of the liquid phase flowing from the sample side. On applying the voltages as shown to the electrodes, the signals produced are shown in Fig. 7-11, indicating that the sample side and reference side signals offset each other, producing the baseline as straight. However, a certain degree of baseline drift cannot be avoided even with DFID, when the sensitivity of FID or the amount of liquid phase differs between the sample side and the reference side.

The Varian model 3700 G.C. system employs a dual flame FID. In this design two hydrogen-air flames are used to separate the region of sample decomposition from the region of emission. One of the flames provides the energy to break down the large molecules in the effluent while the other flame can use all of its energy for emission. Consequently, there is no quenching and much less compound dependency.

Electron Capture Detector (ECD)

This detector works on the principle that the ionization current set up by certain radioactive sources like Ni^{63} or H^3 gets reduced when an electron capturing compound is introduced into the cell. In effect, the ECD measures the loss of signal due to

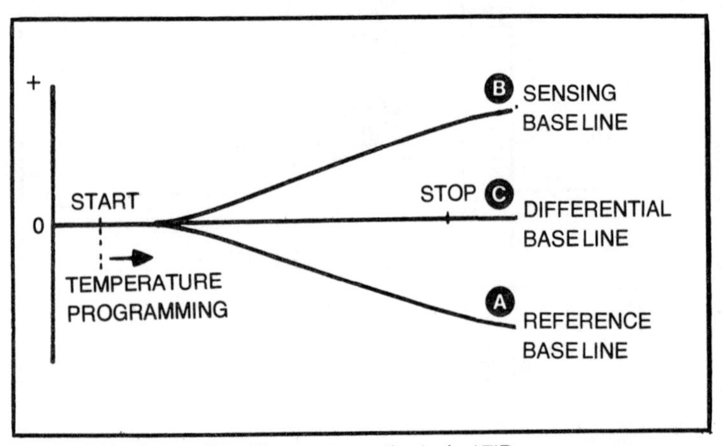

Fig. 7-11. Ideal case of baseline compensation in dual FID arrangement.

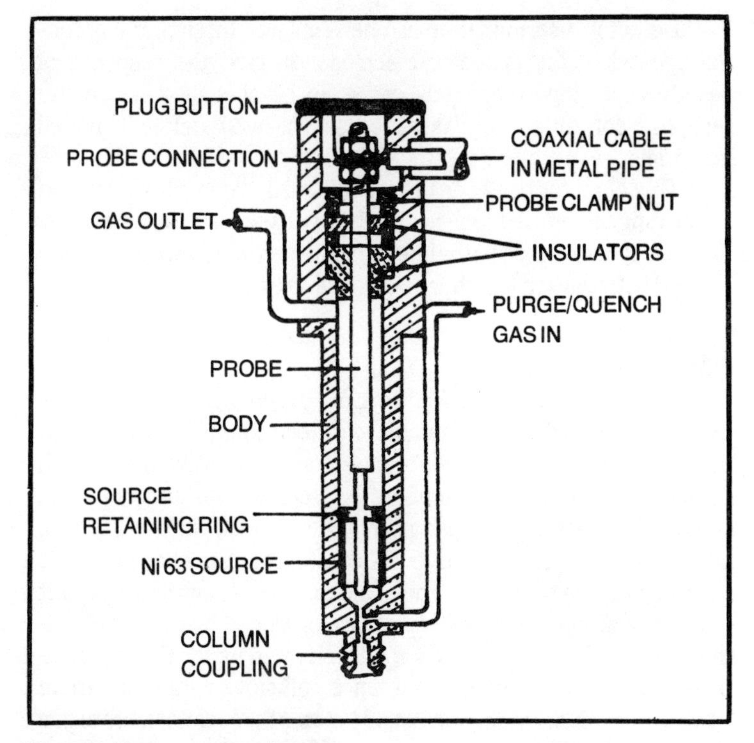

Fig. 7-12. Electron capture detector.

recombination phenomenon rather than measuring a positively produced electrical current.

The detector consists of two electrodes (Fig. 7-12) across which a potential difference of 10 to 100 volts can be applied. A radiation source of β-rays (tritium) is mounted on a tantalum wire saturated with the radioactive isotope of hydrogen, so that the emitted β-rays encounter the effluent from the G. C. column.

As the carrier gas (nitrogen) flows through the detector, β-particles from the tritium source ionize the nitrogen molecules and form *slow electrons*. These slow electrons migrate to the anode under a fixed voltage. When these electrons are collected at the collector electrode, they produce a steady current, which provides a base line on the recorder.

The organic compounds containing halogen, nitrogen and phosphorous have the property of capturing electrons, resulting in a variation in the number of electrons reaching the collector electrode, thereby producing proportionate signals in the detection device.

Detector discrimination can be regulated through the potential applied to the collector electrode. In fact, the response of weakly capturing compounds can even be abolished, since the response for different classes ceases at well defined applied potentials.

This detector has a sensitivity of 3×10^{-14}amp. However, the linear range is limited to less than 10^3. Nitrogen and hydrogen are the best carrier gases with this type of detector. Hydrogen should be used with caution lest there should be an explosion.

Argon Ionization Detector

For the *argon ionization detector*, argon gas is used as the carrier gas. The detector contains two electrodes placed parallel to each other and a potential difference is applied across them. With the carrier gas emerging out of the G.C. column, no current passes across the electrodes under normal conditions, as the gas is a non-conductor. A radioactive source (tritium) is placed in the approach region to the electrodes, so that the β-rays emitted by it excite the argon atoms. Electrons are produced by this bombardment action. These electrons are accelerated under the influence of potential of about 1000 V and, upon collisions with other argon atoms, raise them to the metastable state. Such metastable argon atoms collide with organic molecules of the samples emerging from the G.C. column, resulting in these molecules becoming ionized and consequently conducting. This results in the flow of proportionate current across the electrodes and produces signals, the magnitude of which would depend on the quantity of organic samples passing through the detector.

The argon detector responds to most of the organic and inorganic compounds although it is inert to water vapor, oxygen, methane, carbon dioxide and oxygen. The sensitivity of the detector is 0.08 μg/ml and the linear dynamic range is 10^5. The detector is suitable for measurement of organic molecules present over a wide range of concentrations.

Cross-Section Ionization Detector

The *ionization cross section detector* is one of the most useful detectors for gas chromatography, which has proved to be a very reliable method for the separation of gases in mine air. It is precise, reliable, robust, insensitive to changes in the carrier gas flow rate and characterized by a response which is linearly dependent on the concentrations of the components under investigation over a wide range of changes in concentration. The vapor concentration for any given molecule can be calculated from the known properties of its constituent atoms.

The ionization current in the chamber is small when it is filled with light gas such as hydrogen, but increases with the addition of any other gas. This increase in current is because of the increased total ionization cross-sections of the gas mixture inside the chamber. The ionization cross section of a gas is a quantity determined by the size of a gas molecule and the number of electrons in the atoms forming the molecule. Therefore, the action of the cross section ionization detector is based on differences in the cross section affecting the ionization of the analyzed components.

Otvos and Stevenson (1956) and Lovelock et. al. (1963) describe the physical basis of the ionization cross section method. In this type of detector, a radiation source (Sr^{90}) mounted in the approach region of the two electrodes separated by the carrier gas generates an ion pair from the organic molecules passing with the carrier gas. The application of a potential of 300-1000 V ensures collection of the electrons. A variation in the current flowing across the electrodes leads to proportionate signals, and the magnitude of the signal depends on the concentration of the component emerging from the G.C. column.

In this detector, hydrogen or helium is usually the carrier gas. The detection is non-destructive, though its sensitivity is low (about 10^{-7}gm/sec). Response to any substance can be calculated from the values of the atomic cross sections of its constituent atoms. Pursall and Dugar (1972) describe a dual chamber ionization cross section detector with tritium source for detection of the permanent gases in mine air. This detector is capable of detecting a minute quantity of the test gas in the parts per million range or less.

To eliminate the signal baseline drift in the ionization detector, electrometer amplifier circuit, both in isothermal measurements and with temperature programming, a device has been proposed (Riggs, 1971) consisting of a voltage follower, from the output of which is derived a signal for a comparator yielding a correction voltage, which is fed back to the follower.

Some Recent Detectors

A number of new detectors have been introduced recently. There has been a resurgence in the use of electrochemical detectors for gas chromatography. The majority of these include electrolytic conductivity detectors, micro-coulometry and ion-selective electrodes. Some workers have used radiochemical radio-ionization detectors for gas chromatography. Brawn et. al. (1976) described a system in which the column effluent is split between a mass spectrometer and a gas proportional counter. This system can determine the solute containing the radioactive tracer, which is important for identifying metabolites from metabolism. Another novel detector, which has recently been developed by Nakajima (1976), is the *catalytic reaction detector*. The basis for this detector is to measure continuously the temperature in and above a catalytic bed as the column effluent passes through it. When a solute reacts on this bed, an exothermic or endothermic response is observed. The detector has been found to give sensitivities comparable to those of thermal conductivity analyzers in the monitoring of unsaturated hydrocarbons. Karasek (1976) considered the design and temperature related characteristics of the piezoelectric detector. Bubowski (1976) developed a solid state metal-oxide semiconductor detector. This was found to give a linear response for acetone and various alcohols. Pellizzari (1976) analyzed nitrosamines using mass spectrometry. Novak and Janak (1977) discussed the properties and the results obtained with the flow impedance bridge detector, which is one of the few detectors that can be operated at temperatures in excess of 1000°C. A large variety of spectrometric detectors have also been used as selective detectors for gas chromatography. Infrared, fluorescence and atomic spectrometry have been employed with varying degree of success. Other techniques employed include colorimetry, nuclear magnetic resonance (Dankelman 1976), ultraviolet absorption spectrometry and chemiluminescence.

Calibration of the Detector

Before the analysis of the unknown sample is carried out, calibrate the detector and calculate response factors for components to be determined. This is done by preparing mixtures of known composition by accurate weighing or mixing and then analyzing them under the same conditions, which would be used for the unknown samples. From these results, calibration curves can be drawn and the response factors determined, which are applied to the samples to be analyzed.

RECORDING INSTRUMENTS

Chromatogram recording is usually done on the self-balancing type potentiometric single pen graphic recorders. The span of these recorders may be 0.5 or 1 mV. These recorders require a low impedance input and, therefore, impedance converters are used with the high impedance detectors. The chart widths are 25 to 30 cm and response time of about 1 second. For multispeed operation, a gear box for changing the speed is necessary.

All ionization detectors generate some background signal (with the carrier gas only) ranging from 10^{-8} to 10^{-11} amp. The maximum signal in the presence of vapor is in the range of 10^{-6} to 10^{-8} amp. Therefore, the apparatus for measuring current must respond to all current in the range of 10^{-6} to 10^{-13} amp. It should also have means of offsetting the background current of the detector in use. The response over a large current range is achieved by putting a series of high stability resistors across the input of the potentiometric recorder.

Qualitative Analysis

The elution of a component from the chromatographic column appears as a peak on the graphic recorder. Under specified column conditions, a component has a characteristic retention time (the time a component is retained in the column) or retention volume (the volume of carrier gas passed during the retention time). This forms the basis of qualitative analysis.

Quantitative Analysis

Quantitative analysis by gas chromatography depends on the measurement of areas under the component peaks. The area contained by a peak is proportional to the quantity of the component present in the sample. Every peak is measured and the areas calculated. These are summed and components are expressed as a percentage. The following conditions are necessary for this. The flow rate must be constant so that the time abscissa may be converted to volume of carrier gas. The output of the detector system must be linear with concentration.

Retention Measurement

The identity of an unknown component in a mixture when analyzed with GLC can be established by knowing its time of elution or retention volume. It is known that under given operating conditions of column temperature and flow, the time of elution or retention volume for a component is constant. In practice, retention times of known and unknown materials are compared and a tentative identification of the component is established. However, when more than one component has the same retention time, it is possible that wrong results may be deduced.

Relative Retention Measurement

In this technique, relative retention data is made use of instead of the absolute values of the retention parameters. The method is more reliable, does not depend upon analysis conditions to be set very precisely and is suitable for comparisons. Unknown components are compared with the standard to give a relative retention volume. In this way, an operator may build up a library of retention data for future use.

Methods of Measurement of Peak Areas

For accurate determination of the concentration of the sample, the peaks should be completely resolved. This is slightly difficult as one would have to depend upon extrapolation. Peak areas may be determined by various methods.

■ **Width At 50% Height × Height.** In this method (Fig. 7-13) the peak height (h) is measured and multiplied by the width at half-height (W),

$$\text{Area} = W \times h$$

■ **Triangulation.** The method consists in drawing tangents to each side of the peak and to extend them so as to form a triangle with the baseline. The area of this triangle is approximately proportional to the area under the peak

$$\text{Area of the triangle} = \tfrac{1}{2} \times W \times h$$

where W = peak width
h = height

■ **Weighing Of The Chart Paper.** In this method, the peak area is determined by cutting out the peak from the chart and weighing the paper. The method assumes a uniform weight distribution in paper so that

$$\text{Area} : \text{Weight}$$

Plain paper is preferable to a lined chart, because the latter varies in weight. Some check about the uniformity of the paper must be made.

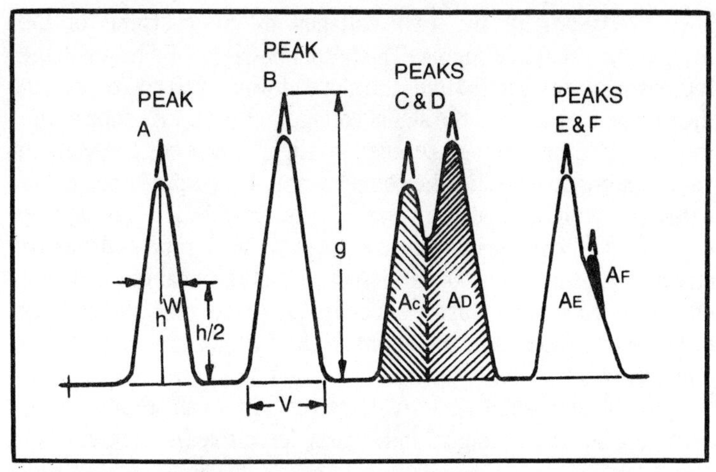

Fig. 7-13. Peak area measurement by width at 50% height × height method.

■ **Use Of Planimeter.** This method is tedious and requires practice for obtaining satisfactory results.

■ **Electronic Integration.** Electronic computing circuits may be used for integrating the area under the peaks. This is described in detail in Chapter 20.

The area under the chromatographic peaks can be measured by analog or digital techniques. In the analog technique, the detector current is amplified in an operational amplifier and then integrated. It gives a curve with waves, the wave heights corresponding to the peak areas. In the digital method, a circuit shown in Fig. 7-14 can be employed.

The output from the electrometer amplifier after the detector is brought to resistance R_1 of the integrator. To the E_s point is brought a voltage used to eliminate zero shift. As soon as the

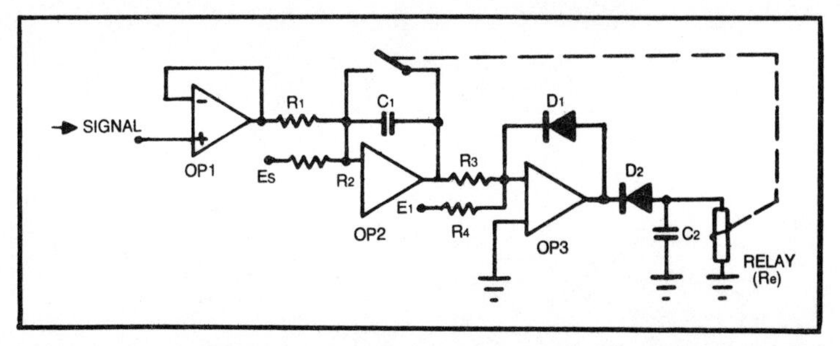

Fig. 7-14. Circuit diagram for measurement of area under the chromatographic peak.

absolute values of the two voltages differ, charging of the integrating capacitor begins. The output of OP_2 would be a voltage increasing linearly with time, the slope of which will correspond to the difference between the signal voltage and the E_s. As soon as the integrator output voltage reaches or slightly exceeds the voltage value applied to input E_1, the comparator (OP_3) output jumps to the saturation voltage and the relay (R_e) is closed. The integrating capacitor is thus discharged. This process can be repeated and the circuit can be used for conversion of an analog signal into a digital signal. As each relay closure is recorded by a counter, the number of pulses registered in the counter gives a measure of the area under the peak.

Chromatographic peaks distorted by tailing can be modified by summing the original signal and its first derivative (Ashley, 1965).

Chapter 8

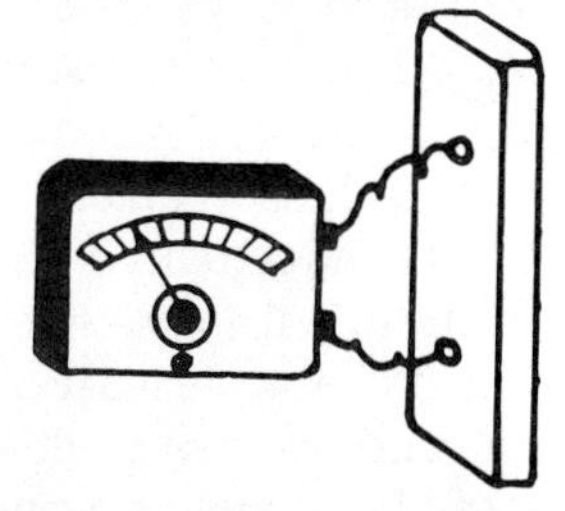

Liquid Chromatographs

The early separations by chromatographic technique were performed using a bed of solid, powder absorbent such as alumina or charcoal through which the sample was passed in a stream of solvent. Since these techniques used liquid as the percolating agent, liquid chromatography can be considered as the oldest of all chromatographic processes. Later developments included liquid/liquid partition chromatography, paper chromatography and the ion-exchange chromatography. Gel permeation chromatography and thin layer chromatography were developed in the 1950s.

Until a few years ago, liquid chromatography was not quite commonly used because of the non-availability of high sensitivity detection systems. With the introduction of such detectors, the analytical potential of liquid chromatography is greatly enhanced and sophisticated liquid chromatographs are now commercially available.

High pressure liquid chromatography (HPLC) is similar to gas chromatography in that the chemical components of a mixture are separated as the mixture is forced through a column, packed with fine particles. In gas chromatography, the substance is carried through the column in vaporized form by an inert gas, whereas in HPLC it is carried through in liquid form by a solvent. Because the substance need not be vaporized, HPLC can be used on a broad range of substances that are not analyzable by gas chromatography.

High pressure liquid chromatography presents some unique problems such as the maintenance of the solvent flow accurately for obtaining repeatable results. Problems in controlling solvent flow arise because the solvents differ in viscosity, compressibility and other characteristics. In addition to this, the volume of the solvent mixture is not necessarily equal to the sum of the volumes of the individual solvents. Also, as the column must be tightly packed with small, uniform particles to obtain adequate separation of the component substances, high pressure of the order of 3000 psi or more is needed to force the substance through the column in a reasonably short time.

Liquid chromatography has been performed in a column and on an open bed (paper chromatography and thin layer chromatography). Note that HPLC has been performed almost totally in columns. However, thin film chromatography was introduced recently as a high speed method for thin layer chromatography.

TYPES OF LIQUID CHROMATOGRAPHY

Liquid chromatography can be classified as under:

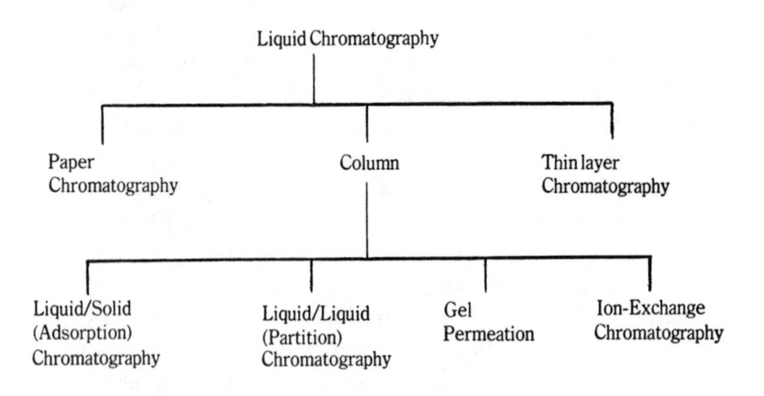

Adsorption Chromatography (Liquid/Solid)

In *adsorption chromatography*, a solid adsorbent, usually in powder form, is the stationary phase through which a mobile liquid phase carrying the mixture to be analyzed is allowed to percolate. Adsorption chromatography is carried out in columns with the adsorbent supported by a plug of glass or cottom wool or by a sintered glass filter.

Partition Chromatography: (Liquid/Liquid)

In this technique, the mobile liquid phase is made to percolate through a column containing the stationary liquid phase which is deposited on a solid surface as a thin film. Solid supports usually used are silica gel, porous glass and cellulose.

Ion-Exchange Chromatography

Ion-exchange chromatography involves the exchange of ions between a solution and a solid insoluble material in contact with the solution. Many naturally occurring solid materials have the ability to exchange ions. Also, many artificial ion exchange materials have been developed. The ion exchange process is reversible and this fact is made use of in ion-exchange chromotography. When a sample is introduced at the top of an ion-exchange column, the ions exchange rapidly with the ions in the resin. If a mobile phase is used, the sample ions are displaced into the solution again and then re-exchange on to the resin. This process continues until the sample ions emerge from the end of the column. If the various sample ions are held on to the resin to different extents, then the time taken for them to pass through the column will be different and a separation will be achieved.

This type of chromatography depends upon molecular ion-exchange instead of liquid-surface adsorption. Separation of acidic and basic organic substances from mixtures is achieved by using synthetic resins, which have highly selective action for certain substances, particularly amino acids and allied compounds. Essentially, the process involves interchange of anions and cations between the components of certain resins.

Gel-Permeation Chromatography

Gel permeation chromatography is a recently developed separation technique in liquid chromatography. The separation is based on molecular size and shape. The gel permeation column is packed with a stationary phase in the form of a gel which contains pores of a specific size. As the sample is carried through the column bed by the carrier liquid, the sample molecules penetrate the pores in the packing gel depending upon the size and shape of the molecules. Large molecules do not penetrate the gel and are consequently quickly eluted. Elution takes place in inverse order of their degree of gel permeation and consequently of decreasing molecular size.

Thin Layer Chromatography

In *thin layer chromatography* (TLC), the stationary adsorbents are applied to a planar glass or plastic surface and the solvent flows over them. All of the basic types of action like adsorption, partition, ion-exchange and gel filtration can be used on TLC plates, while solvents are applied in a chamber similar to that used in paper chromatography.

Paper Partition Chromatography

Paper partition chromatography is a simplified version of column chromatography which makes use of strips or hollow cylinders of filter paper to hold both the solid and liquid phases. Drops of the solutions containing unknown mixtures are applied to a number of parallel strips, a few inches from the end of each test paper, and allowed to dry. The strips of paper are placed in a chromatography chamber with a saturating and equlibrating vapor and hung from a solvent reservoir, so that the movement of the solvent downward can be timed and the relative partition of the different substances measured.

PARTS OF A LIQUID CHROMATOGRAPH

A liquid chromatograph consists of the following parts as shown in Fig. 8-1.

- A high pressure pump system to force the liquid mobile phase through the column. A gradient elution or solvent programmer.
- The sample injection system.
- The column.
- The detection system including display or recording devices.

As in other chromatographic techniques, the sample is introduced into the column with the help of a sample injection system. Various components of the sample are fractionated during their passage through the column. The detection system senses these components as they elute from the column and produces a signal proportional to the amount of solutes passing through the detection system.

HIGH PRESSURE PUMP SYSTEM

Liquid chromatographs of the early type made use of wide diameter columns packed with coarse mesh packing material. They

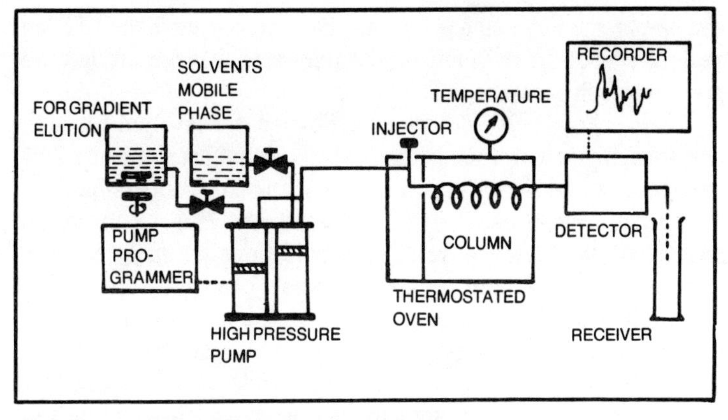

Fig. 8-1. Block diagram of a liquid chromatograph.

required very little pressure to obtain an adequate flow rate of the mobile phase liquid. Modern instruments, which employ smaller diameter columns filled with fine mesh particles, have necessitated the use of high pressure solvent delivery systems.

The most commonly used methods for solvent delivery are gravity feed system and various types of pump systems incorporating piston pumps, peristaltic pumps, diaphragm pumps and syringe pumps, etc. The gravity feed systems, though simple, are not able to deliver solvent at high pressure. They are, therefore, not used with narrow bore columns packed with fine mesh particles, which need high inlet pressures to yield the required flow rate.

High pressures of several hundred atmosphere are required in high resolution high speed liquid chromatography. Varian Aerograph Model 8500 liquid chromatograph (LC) employs a pump system which gives pressure up to 600 atmospheres. Generally used pumps are the piston type, which provide very high solvent pressures. The flow rate can be set to the desired rate by adjusting the pump stroke length and the motor speed. The pressure is observed as a dependent variable. The flow of solvent from a piston pump is usually in the form of a series of pulses. This type of ripple in the flow is likely to affect column resolution and detector stability. To smooth out the ripple, a long nylon tube of about 1.5 mm diameter may be used between the pump and the chromatographic column. Ripple can also be reduced by using bellows, restrictors or multipiston pumps, where the action of the individual pistons is arranged at regular intervals of a complete stroke cycle.

A most commonly encountered piston type pump is the syringe pump. In these pumps, a constant and reproducible flow can be obtained by using a gear mechanism. Spring loaded Teflon seals are used in the plungers to minimize leakage around the pistons at high pressure.

Schrenker (1975) describes a flow control system for high pressure liquid chromatography which maintains constant flow irrespective of differing solvent viscosities and compressibilities. The system utilizes a hydraulic capacitor (Fig. 8-2) which smoothens the high pressure pump pulsations normally encountered in piston operated pumps. It consists of a rigid vessel filled with fluid of known compressibility. A small fraction of the space is separated from the compressible fluid by an impermeable membrane. The solvent mixture passes through this separated space.

A restriction in the solvent flow path, which could be the chromatographic column itself, is the hydraulic analog of a resistor, resistor, so the unit can function analogously to an RC filter. With a sufficiently large fluid volume (large C), adequate smoothing of the pump pulsations, can be obtained with a relatively low value of F on the output side. The flow measurement and control system used by H.P. in their model 1010B HPLC is shown in Fig. 8-3. A pressure transducer is installed in the hydraulic capacitor. Measurement of the pressure is synchronized with the pump, so that measurement is made only during the discharge phase. The time integral of the ac component of the transducer output is proportional to changes in pressure and is thus proportional to flow. The ac component is fed

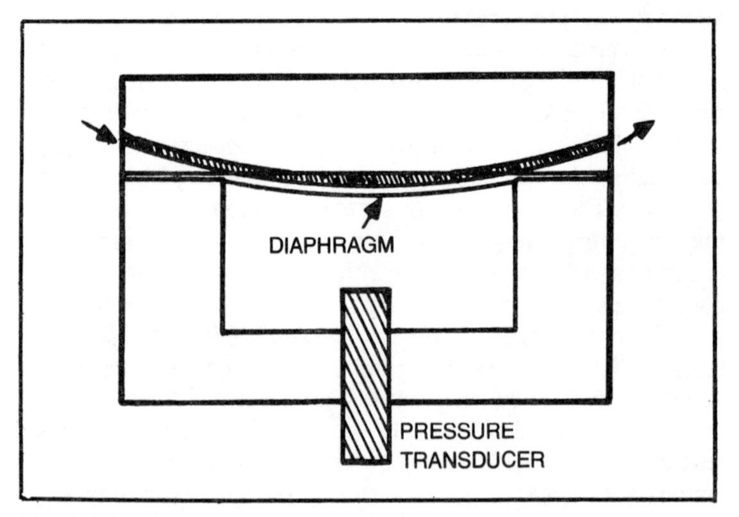

Fig. 8-2. Hydraulic capacitor for smoothening of flow pulsations (after Schrenker, 1975).

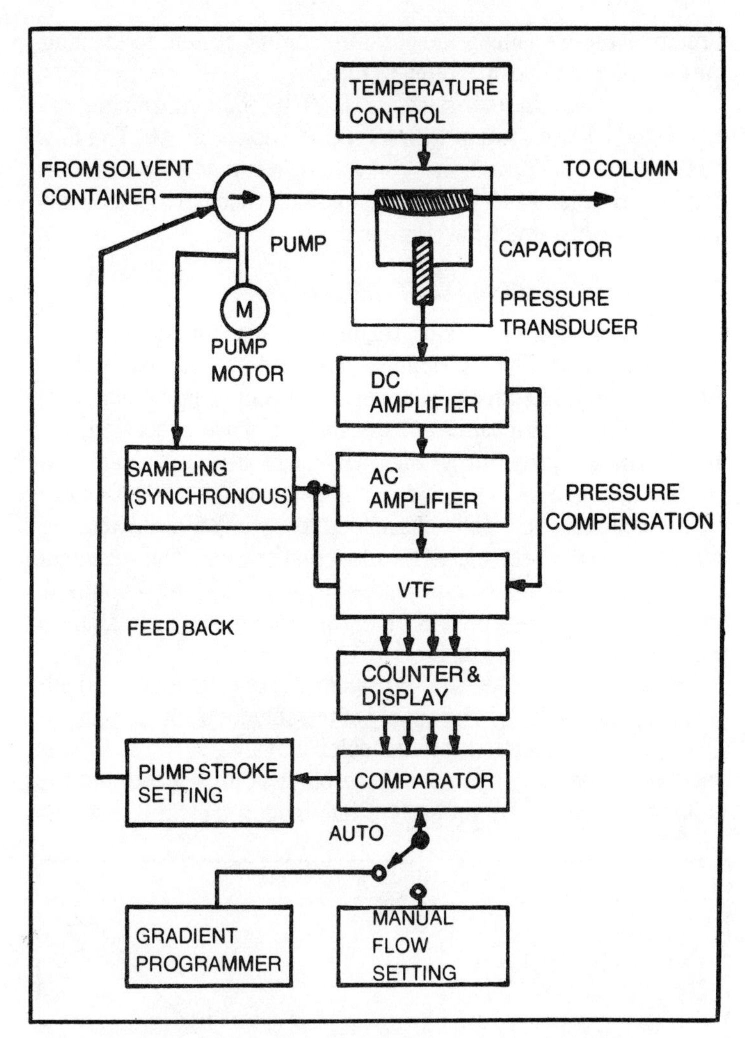

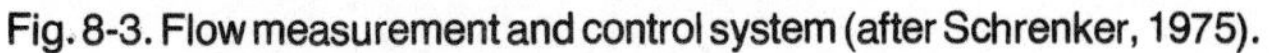
Fig. 8-3. Flow measurement and control system (after Schrenker, 1975).

to a voltage-to-frequency converter. A counter totals the output of v-to-f converter, effectively integrating the transducer output. At the end of each 12 second integration period, the counters contents are compared digitally to the set point. Any error is then used to adjust the pump stroke setting to bring the flow rate to the set point value. The averaged dc output of the pressure transducer represents absolute pressure, and is used as a constant to compensate for the influence of pressure on the compressibility constant of capacitor C. The effects of temperature on fluid compressibility are circumvented by using a temperature control system to maintain the capacitor at a constant temperature.

Flow rates in the range of 0.10 to 9.99 ml/minute are controlled by this system with a repeatability of ± 1%. The flow rates of the two pumps are controlled independently and then outputs are mixed in a low volume mixing chamber immediately upstream of the sample injection part.

GRADIENT ELUTION OR SOLVENT PROGRAMMING

Gradient elution is often required to resolve complex mixtures, especially those containing components with significantly different chromatographic behavior. A solvent programmer helps to control the composition of the mobile phase according to a predetermined program as the analysis proceeds. Solvent programming is generally carried out by continuously adding a more polar solvent to the mobile phase feed reservoir, thereby increasing the polarity of the eluant as a function of time. The technique involves the use of separate pumps feeding different solvents or solvent mixtures concurrently into the column and programming the output of each pump.

Figure 8-4 shows the arrangement usually employed for solvent programming. The supply of the solvent from the pump is given to a T connection through solenoid valves. One solvent is used as a feed to the pump and the other is introduced into the bellows assembly. By properly programming the time intervals

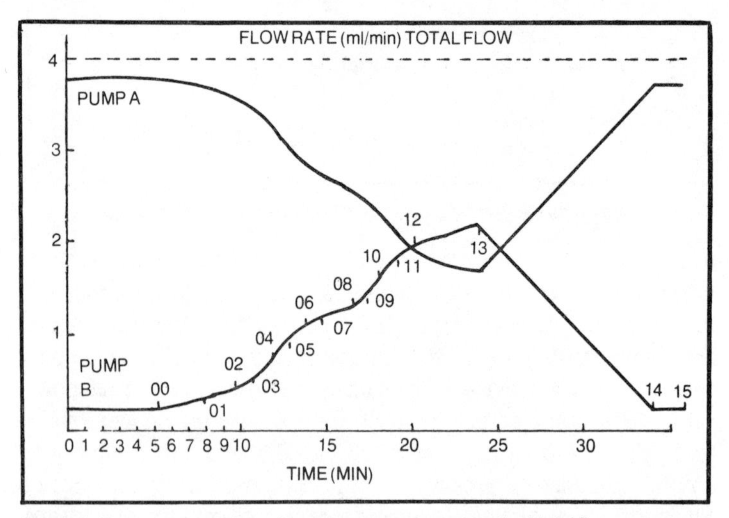

Fig. 8-4. Flow programming of two pumps to give a constant total flow output.

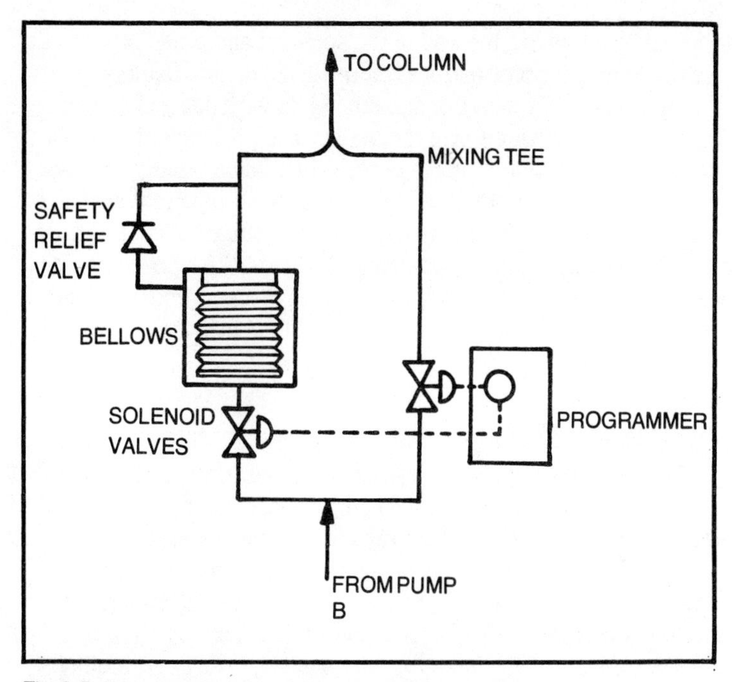

Fig. 8-5. Arrangement employed for solvent programming.

between opening and closing of the solenoid valves, the desired gradient to the column can be obtained.

Standard linear, convex and concave gradients are usually sufficient. However, some manufacturers also provide facilities for non-linear gradients. With the help of this system, it becomes possible to generate gradients of virtually all complexities that one may need to resolve difficult mixtures. In the Varian Aerograph Model 8500 L. C. system, the rate of change of solvent composition is selected very easily with the help of proper combination of pins in the program board. The solvent strength can be increased, decreased or held constant. It is held constant at any point or points in the program simply by omitting a rate pin in a given step. This ability to hold constant solvent strength at any point in the program is highly useful for maximizing resolution in difficult parts of an analysis, for example, in separating isometric compounds.

In the recent instruments, a gradient program is entered by way of a calculator-like keyboard on the gradient programmer. The program is entered as a series of linear segments (Fig. 8-5) that approximate the desired program curve. Each segment is specified by three program entries: flow rate at the beginning of the segment, flow rate at the end of the segment and time duration of the segment. This information is stored in the digital memory.

Because the flow rate programming for each pump is independent of the other, three program modes are possible: change the mixing ratio from the two pumps, while maintaining constant column flow by programming equal but inverse flow rate changes for the two pumps (gradient programming); change the column flow rate while maintaining a constant mixing ratio by programming equal percentage changes of flow in A and B (flow programming); and combine the first two modes, resulting in a flow program superimposed on a gradient program.

SAMPLE INJECTION SYSTEMS

There could be several methods for the introduction of the sample on the top of a liquid chromatographic column. One method is to disconnect the solvent supply to add the sample in solution and reconnect the solvent supply to the column. The mechanism is simple, but the method is tedious to operate. More recently developed methods fall into two categories, namely the *syringe injection method* and the *injection valve method*. Both these methods enable the sample to be introduced directly into the column packing without interrupting the solvent flow.

The *syringe injection method* involves the insertion of the syringe needle through a rubber septum at the top of the column (Fig. 8-6). This method, however, cannot be used for the injection of large sample volumes into high pressure solvent systems. At pressures greater than about 3 atmospheres, the pressure has to be reduced by turning off the solvent supply before the injection of the sample can be carried out.

In the injection valve method, the injection valve containing sample loops are connected in the solvent supply pipe work at the top of the column. The sample loop can be introduced into the solvent stream when desired, without turning off the solvent flow. After sufficient flushing of the loop with the solvent has taken place, the sample gets completely carried to the column. The loop can then be removed from the solvent stream for refilling with the next sample. Injection valves can be used for sample introduction into very high solvent pressure systems. By changing the volume of the sample loop, the sample size can be easily varied. Also, this method can be conveniently automated for automatic injection of the samples.

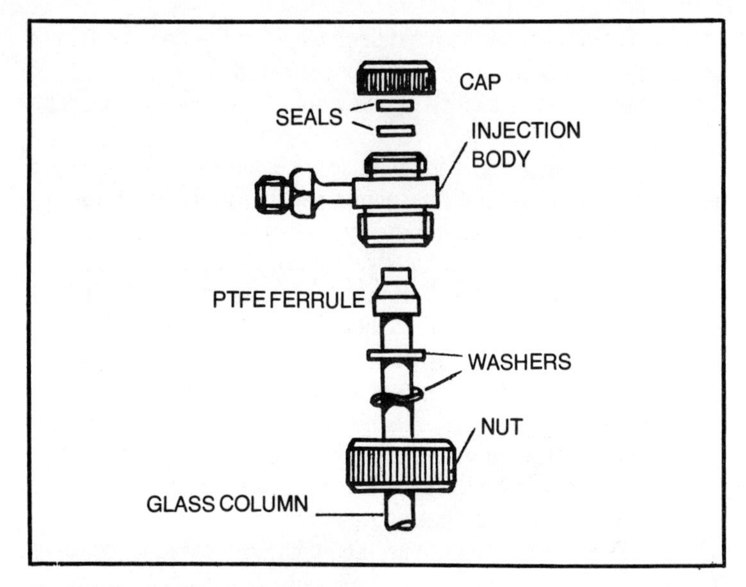

Fig. 8-6. Syringe injection method.

THE COLUMN

The *column* is, by far, the most important part of any chromatographic system since the ultimate performance of the chromatograph is determined largely by the column. Most of the early analysis work, which was carried out by liquid chromatography, made use of large columns with an internal diameter of 1 cm or more. With the development of highly sensitive detection systems, it has become possible to analyze minute quantities of sample and to reduce the column diameter. Reduction of sample size and the column diameter results in an improvement in separating efficiency. There is another factor which necessitates the use of small diameter columns. A large contribution to the band broadening in the chromatographic peaks is known to be due to large scale unevenness of flow, which becomes worse as the column diameter increases. The effect of uneven flow may be reduced by decreasing the diameter of the column. The columns in current use are generally in the 0.1-2.0 cm internal diameter range. However, there are practical problems when column diameter becomes less than 0.05 cm because very small particle sizes and very high pressure drops will have to be used. As regards the column length, they may be 1-4 meters long, but most of the applications can be performed on columns of less than 100 cm length. Various materials have been used for the construction of columns. Glass columns are usually preferred on account of their inert nature and the facility of being able to observe the packing visually.

Theoretical considerations have revealed that much smaller particles should be used for column packing in liquid chromatography than in gas chromatography. Generally, the particle size ranges from 20-50μ. Occasionally, even smaller particle sizes are used. For example, in the Varian 8500 L.C. system the columns are packed with particles of diameter less than 10 μF. Columns packed with small sized particles are more efficient since the solute mass transfer takes place at a much rapid rate. Small packing particles used in liquid chromatography, however, present great problems in getting homogeneous packing of the columns. Also, the method of column packing is dependent on the type, regularity and the particle size of the packing used.

In most of the separations, liquid chromatographic columns are operated under ambient temperature conditions. However, some researchers have shown that improvements in column efficiency can be achieved in certain cases by working at elevated temperatures. Columns are therefore placed inside ovens capable of operation up to 250°C. The temperature of these ovens are controllable to a high degree of constancy.

DETECTION SYSTEMS

High sensitivity detection systems are necessary for achieving optimum column separating performance by the use of small sample volumes. The absence of a versatile and economical detector has been one of the main reasons for the limited development of the liquid chromatography in the past. The earlier methods of detection used in liquid chromatography were of non-continuous nature. However, continuously monitoring detection systems are more convenient in operation and, therefore, all modern liquid chromatographic detectors incorporate this feature.

Several detection systems have been developed, which are mostly dependent upon the measurement of a physical property of the column elute. These physical properties could be changes in ultraviolet absorption, infrared absorption, heat of adsorption, refractive index or electrical conductivity.

UV-Visible Spectrophotometric Detectors

With this type of detector system, it is possible to detect and analyze compounds that absorb at any wavelength in the UV-

Visible range from about 200 nm to 800 nm with a bandwidth of 5 nm. Almost any spectrophotometer suitable in this range can be modified to work as a detector. The cell in this case would be of a flow-through type, with a path length of 1 cm and cell volume as low as 8-20 μl. Cells may be made of quartz, Teflon and KELF. They are usable up to a pressure of 500 psi. Stainless steel cuvette assemblies are used for higher pressures. When using a recording spectrophotometer, it is possible to stop flow and scan the spectrum of individual peaks in the chromatogram. Just like conventional spectrophotometers, a choice of light sources is usually available depending on the wavelength desired. Deutrium for the wavelength range of 200 to 400 nm and tungsten for 350 to 800 nm can be selected when required. These instruments are calibrated in absorbance units over these ranges: 0.01, 0.02, 0.04, 0.1, 0.2, 0.4, 1.0 and 2.0 absorbance units. Three transmission ranges may also be provided to measure 100%, 10% and 1% full scale on any suitable recorder having a full scale sensitivity of 10 mV. For resolving small events in peaks, the instruments are provided with a 10-turn fine adjustment to achieve zero suppression over the full range of 2 absorption units or 120% transmission range.

Spectrophotometric instruments used in liquid chromatography are often called *spectroflow monitors*. Figure 8-7 shows the block diagram of spectroflow monitor Model SF 770 of Schoeffel Inst. Corp., USA, used for UV-visible liquid chromatography. The system makes use of a double beam principle and a chopping system to measure the transmission through the sample and reference cells alternately. A reflection mode attachment is also available with this instrument. Conventional log function generating amplifiers are employed to get direct readings of optical density. This instrument gives a noise figure which is better than 5×10^{-4} O.D. (at $\lambda = 280$ nm) and stability better than 5×10^{-4} O.D. per hour (at $\lambda = 280$ nm).

Detection systems are also available for the UV range along. Most of these systems operate at fixed wavelengths of 254 nm or 280 nm. These wavelengths offer excellent sensitivity for many compounds, yet permit the use of a wide range of solvents without interference, even with gradient elution. A UV detector employed by Varian aerograph Model 8500 L.C. has a special thermal isolation design which gives a low noise level of only $\pm 5 \times 10^{-5}$ absorbance units, equivalent to low nanogram sensitivity. The output of the detector is linear in absorbance and thus linear in concentration for solutes obeying Beer's Law. The wide linear dynamic range (10^4) enables one to measure both trace and major components in the same chromatogram.

Fluorescence Detector

Fluorescence measurements of minute quantities, as encountered in liquid chromatography, differ greatly in technique and behavior if compared to standard flow through absorption monitors or spectrophotometers. In a fluorimeter, the presence of a fluorescence emitting substance is measured. The emitting substance is only present occasionally in liquid chromatography and its emission of energy is detected by a highly sensitive photomultiplier.

The Schoeffel Model 970 spectrofluorimeter is an example of an instrument designed specifically for L.C. applications. It offers continuously selectable monochromatic excitation energy over the entire UV-visible spectrum, utilizing a highly stabilized deutrium or tungsten-halogen lamp. The monochromator makes use of a grating system. Its continuously variable wavelength extends analytical application far beyond the limited areas normally offered by line spectra of Hg lamps. Excitation energy from the monochromator enters the cuvette and emission from a 5 μl cavity is collected and directed towards the photomultiplier. A set of easily interchangeable filters is provided to select the emission spectra of interest.

The filter set contains filters of wavelengths 370 mm, 389 mm, 418 mm, 470 mm, 550 mm and 580 mm. Transmittance is greater than 0.9 in these areas compared to less than 10^{-5} below the cutoff point and, because of this high efficiency, virtually no emission energy gets absorbed before reaching the photomultiplier. Figure 8-8 shows a fluorescence detector for L.C. applications.

In normal fluorimeters which utilize a standard cuvette, volumes of several ccs are illuminated by larger excitation light beams of at least 1 cm^2. The emitting material acting as a secondary light source to be detected originates from a much larger area or volume than what is available in the microliter type cuvette of fluorimeters used in high pressure liquid chromatography. Since emission occurs in all directions from an excited sample, surrounding this sample with an efficient optical collector is a must. Schoeffel has introduced a two Steradians cuvette (Fig. 8-9) which has provided a solution to this problem. This cuvette is of stainless steel and is a flow-through type.

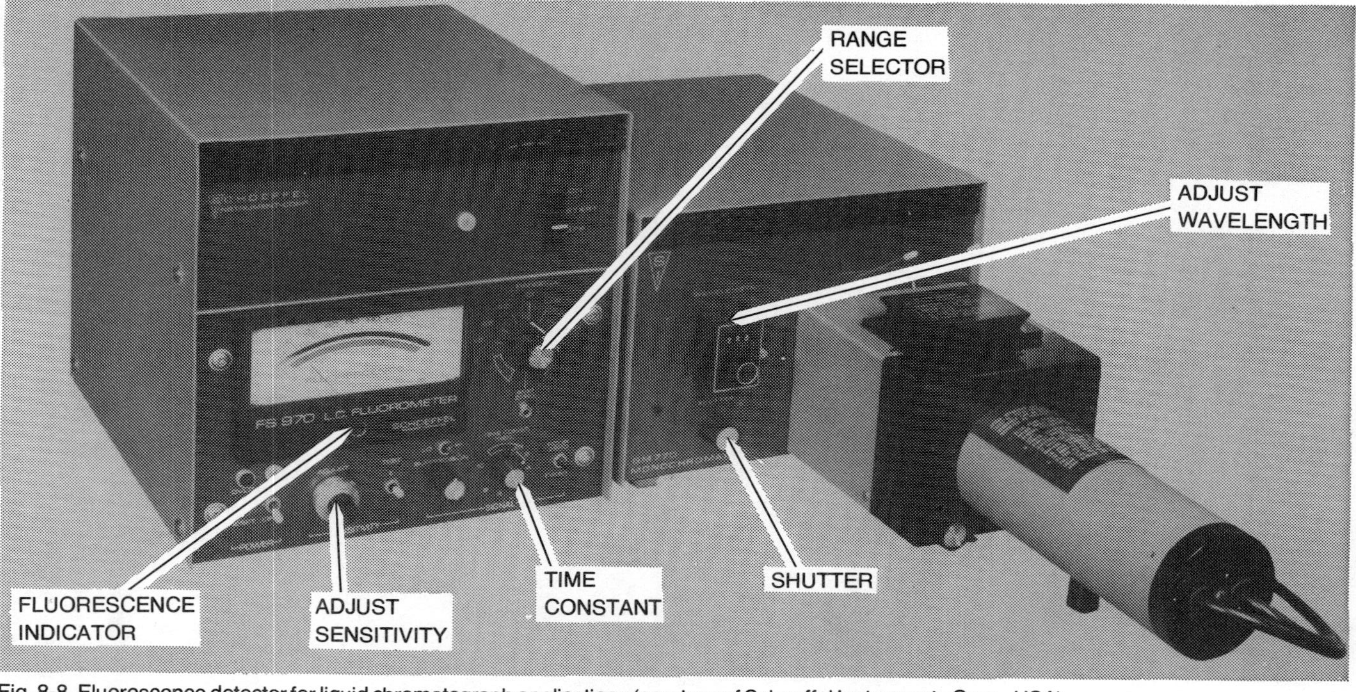

Fig. 8-8. Fluorescence detector for liquid chromatograph applications (courtesy of Schoeffel Instruments Corp., USA).

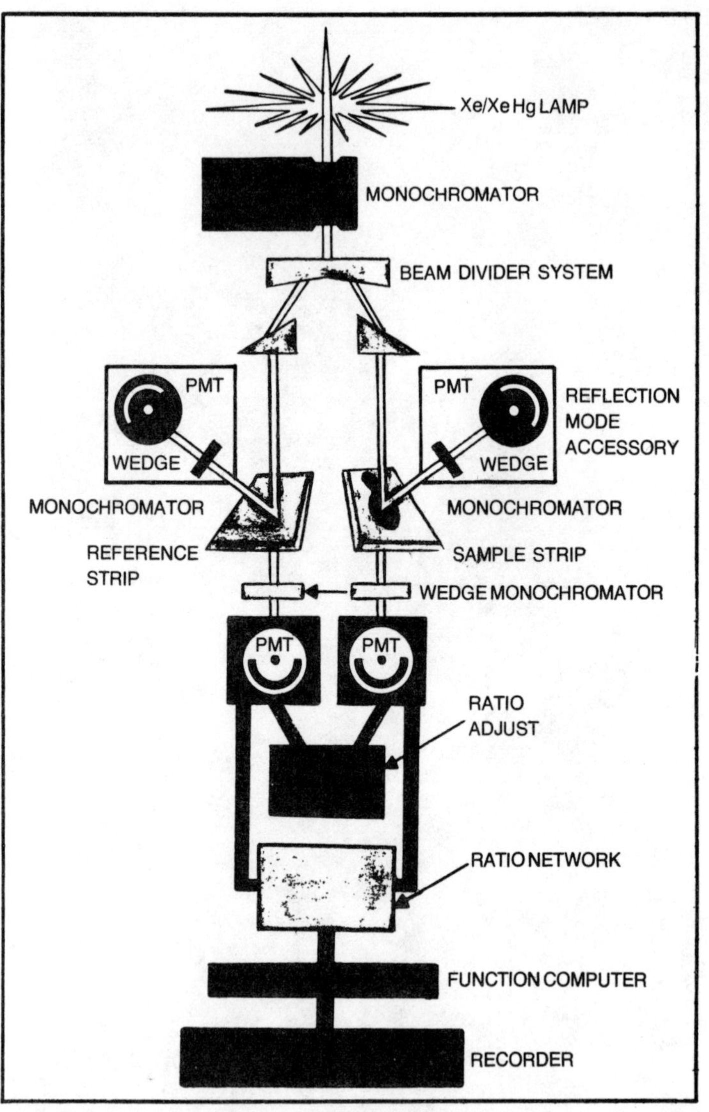

Fig. 8-7. Optical diagram of Spectroflow Model SF770 (courtesy of Schoeffel Instruments Corp., USA).

Refractive Index Detectors

Compounds without significant UV or visible absorbance can be monitored with a *refractive index detector*. This detector may be a dual beam refractometer of the Fresnel type. These instruments

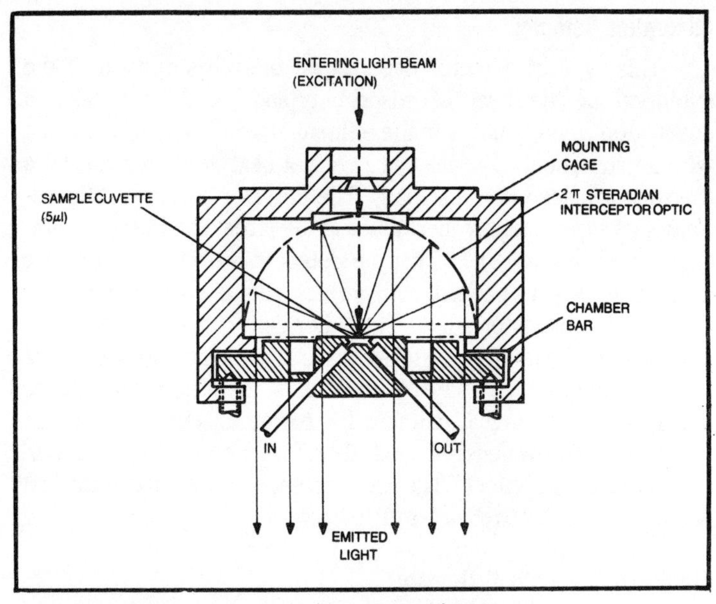

Fig. 8-9. Steradian cuvette used in fluorescence detector.

measure the intensity of reflectance, which in turn, is inversely proportional to the refractive index. Careful control of the temperature is very necessary in these instruments. Small cell volumes of the order of 5 μl are possible.

Differential refractometers are based on the measurement of the difference in refractive index between the solution and of the pure solvent. This is usually carried out by measuring the bending of monochromatic light beam as it passes through a sample cuvette. Temperature control to within 0.01°C is necessary for measuring absolute refractive indices to within 10^{-5} units. Cell volumes of the order of 75 μl and higher are necessary. The sensitivity of these refractometers alters with variations in flow. The range of refractive index covered is 1.30 to 1.60.

Hazebrock (1972) explains the construction of a laser interferometric differential refractometer for detection of liquid chromatographic effluents in small diameter packed capillary columns. The instrument has a sensing volume of 4 μl and a lower detection limit of refractive index of less than 2×10^{-7}, and makes use of the narrow beam of a small helium-neon gas laser. The instrument incorporates a special and more versatile phase-to-voltage converter for recording the optical phase shift in the interferometer.

Adsorption Detector

This type of detector depends on the measurement of the evolution of the heat of adsorption and the heat uptake at desorption as the solutes in the effluent stream come into contact with an adsorbent. The measurement of heat is carried out by a thermistor placed in the stream. The detector operates differentially by having another thermistor in the stream and temperature changes as low as 10^{-4}°C can be detected by this system. One detector cavity is packed with an adsorbent, such as alumina, porous glass beads or silica. The other cell, called the reference cell, is packed with inactive glass beads. The thermistors placed in these detectors form a part of the Wheatstone bridge which gives out an electrical signal suitable for recording purposes. These detectors are flow sensitive and, therefore, center discs are used to decrease this effect. The dual detection system cancels out baseline shifts due to changes in flow rate.

Electrical Conductivity Detector

Electrical conductivity is claimed to be one of the most important and promising detectors in the future of liquid chromatography. Areas in which this mode of monitoring are particularly attractive are aqueous and non-aqueous gel filtration, ion exchange chromatography and many applications of liquid-liquid chromatography. The conductivity cell used in the Nester/Faust liquid chromatograph model 1200 has three electrodes with an internal diameter of under 2 mm and length of 10 cm. The electronics consists of a stabilized and isolated ac bridge and phase detector system. The unit is sensitive to a change of approximately 1 part in 10,000 in conductivity and a measuring range of 10 to 100,000 micromhos per cm.

Figure 8-10 shows a typical arrangement of a conductivity detector for L.C. applications. It is a three electrode design. Dimensions of the cell are 10 cm length and 2 mm id. The cell constant of the cell used by Nester/Faust Manufacturing Corporation is 10 cm^{-1}. The total effluent stream passes through the flow-through cell, although the stream can be split to pass only the required amount through it.

Thermal Detectors

Thermal detectors are also known as micro adsorption detectors. The principle of operation of these detectors depends on

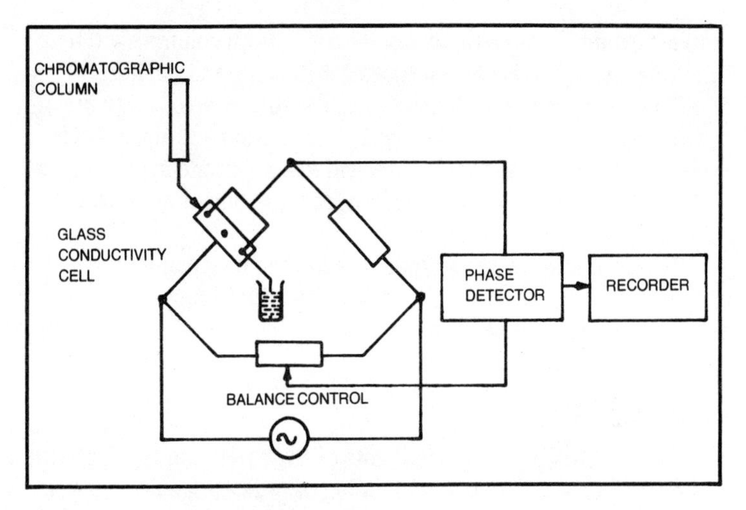

Fig. 8-10. Conductivity detector.

temperature changes taking place due to the heat of adsorption on an active solid surface. These detectors were first reported by Claxton (1959) and Hupe and Bayer (1967) and are now manufactured by a number of companies. Generally, a portion of the adsorbent column packing is contained in a small chamber at the column outlet. A second chamber filled with an inert material such as glass microbeads is used to achieve a reference signal. This assembly is carefully thermostated. Both chambers contain matched thermistors which constitute the measuring arms of a Wheatstone bridge. As the eluted sample is adsorbed on the solid, a local temperature change takes place which initiates a signal.

These detectors can be used in applications involving liquid-solid, liquid-liquid, ion-exchange and gel-permeation chromatography. They are non-destructive. However, they are subject to other thermal effects such as thermal conductivity and heat capacity of the solvent. They, therefore, require accurate calibration before use.

Programmers and Readouts

Liquid chromatographs are finding on-line applications for monitoring chemical processes and providing information for desired adjustments. Thus, it must operate automatically. Unless it is computer controlled, it requires a programming unit. The programmer controls the analysis time, injects the sample, selects and measures the peaks and presents the results for display or control. In addition, the programmer controls all column switching functions, including values used for back flushing and washing.

The programmer is usually placed in the control room along with a strip chart recorder. The computerized system also includes a teletype and CRT terminal. A dedicated computer can preferably replace a programmer, which can normally control several L.C. instruments.

The computer or programmer calibrates the readout for each component of interest. Furthermore, the programming unit controls all functions related to calibration including re-zeroing the baseline at selected times.

AMINO ACID ANALYZERS

Amino acids are the components of all proteins and as such are essential to the growth and well being of living organisms. Some are metabolized from the diet by hydrolysis resulting from enzyme activity. Others are synthesized within the organism. Therefore, the presence, distribution or absence of any amino acid in living tissues is significant of normality or abnormality in the organism. Over 40 of these complex molecules are known to exist naturally, and many more have been synthesized.

Amino acids are distinguishable from each other only by one or two atoms in their structure. However, they may be found bonded together in manifold diversity in the organization of proteins. Although the concentration of any free amino acid can be accurately measured colorimetrically, all except two must be measured at the same wavelength. Therefore, they cannot be distinguished from each other unless they have first been separated in predictable sequence. The microtechnique of paper *electrophoresis* when applied in this application lacks resolution and reproducibility. Gas chromatography, though potentially fast and simple in operation, loses accuracy in the conversion of amino acids to the gaseous phase and cannot be used for samples containing proteins or polypeptides.

Resin Column Chromatography

The most widely accepted technique employed till 1955 was ion-exchange chromatography on sulfonated polystyrene resin, followed by colorimetric analysis of the separated fractions. The method gave high accuracy, but the inconvenience of handling many individual samples and the long and laborious process of cleaning glassware limited the deployment of the technique in routine analysis. Spackman, Stein and Moore (1958) developed an apparatus in which ion-exchange resin chromatography was followed by the automatic introduction of color reagent into the separated fractions, as they emerged from the column. So, a continuous flow of solution, colored at intervals according to the concentration of each fraction as it occurred, could be passed through a colorimeter to produce a graphical record on a multipoint recorder.

Resin column chromatography depends upon the characteristic strength of the ionic bond formed with the resin by each acid as it is applied to the top of the column. This bond breaks, remakes and breaks again successively, causing a characteristic delay as the acid is eluted through the column by a buffer of known pH and ionic strength. Thus, the column effluent consists of a series of discrete amino acids, separated by a buffer in which no amino acids are present. Ninhydrin color reagent is metered into the column effluent. The color reaction develops during a known period at a controlled temperature. The flow energizes a multipoint recorder. The record is a continuous chromatogram, interpreted quantitatively against the chromatogram of a standard amino acid mixture.

Proline and *hydroxyproline* are measured at 440 nm and all other amino acids at 570 nm. Therefore, two photometric systems are employed. One photometer serves the high precision section, measuring at both wavelengths mentioned, through a path length of typically 2 mm. A secondary 570 nm channel has a path length of only 0.7 mm to accommodate more highly-concentrated fractions. Two other photometers measure at the same wavelengths through 20 mm. They are used to obtain the same sensitivity from the high speed section, as the high precision section provides, from samples 10 times more concentrated. Alternatively, they may be used to improve tenfold the detection limit of the high precision section.

During measurement two columns are used, with a short one for the slow moving basic amino acids, which elute more quickly. At the end of a separation, basic amino acids remain in the long column and must be removed before the next hydroxide, followed by automatic changeover to 3.25 pH buffer to restore the equilibrium of the column.

Apart from the amino acids, which are commonly known to be the constituents of proteins, the substances that can be analyzed on analyzers based on the ion-exchange chromatography on resin are amines, imino compounds, peptides and any other chemical compounds which are ninhydrin positive.

Automatic Amino Acid Analyzer

For automatic separation, identification and quantitation of amino acids and related compounds, automatic instrumentation has been introduced. These instruments permit complete protein or peptide hydrolysate analyses with reduced sample amounts and increased sensitivity of 0.05 micromole or less.

Figure 8-11 shows the schematic diagram of the components and the flow system of a typical amino acid analyzer. The system may employ seven chromatographic ion exchange columns having a precision bore 0.9 cm inner diameter. In the bottom of each of these tubes is a porous Teflon disc, upon which the resin is supported. The disc can be easily removed for cleaning or replacement and facilitates draining of the resin from the column.

The sample containing 0.05 to 2.0 micromoles of each amino compound is introduced at the top of an ion-exchange column. A buffer is supplied to the column from a reservoir by an accurate non-corrosive metering pump at a constant rate of 80.0 ml per hour.

The flow pumps are the most important of all the functional elements in the liquid chromatographic system. Unless pumps are provided that are capable of producing a constant liquid flow, any attempt to perform a quantitative or qualitative substance is impossible. The highest mechanical perfection is very essential in order to attenuate all adverse effects due to changes in the kind of liquid pumped, its viscosity and back pressure, etc.

The pump used is a double action, piston-driven one in which each stroke performs both intake and discharge functions simultaneously, and in which a valve is switched in synchronization with the piston movement so as to achieve a constant liquid flow. The flow rate is varied by the replacement of a set of gears. Before the buffers enter the pumps, trapped air is eliminated with a bubble trap type deaerator to ensure constant volume delivery by the pump and to prevent release of air in the resin bed of the ion-exchange column. All columns are enclosed in a thermostated water jacket.

The purpose of the circulating bath is to keep the column temperature at a constant level. For this purpose, the water inside the bath, which is controlled to a fixed temperature, is circulated around the column by a circulating pump. The bath is equipped with a 400 watt heater, which switches on and off under the control and action of a temperature sensing electronic circuit. Another heater of 500 W is provided to raise the temperature faster than usual.

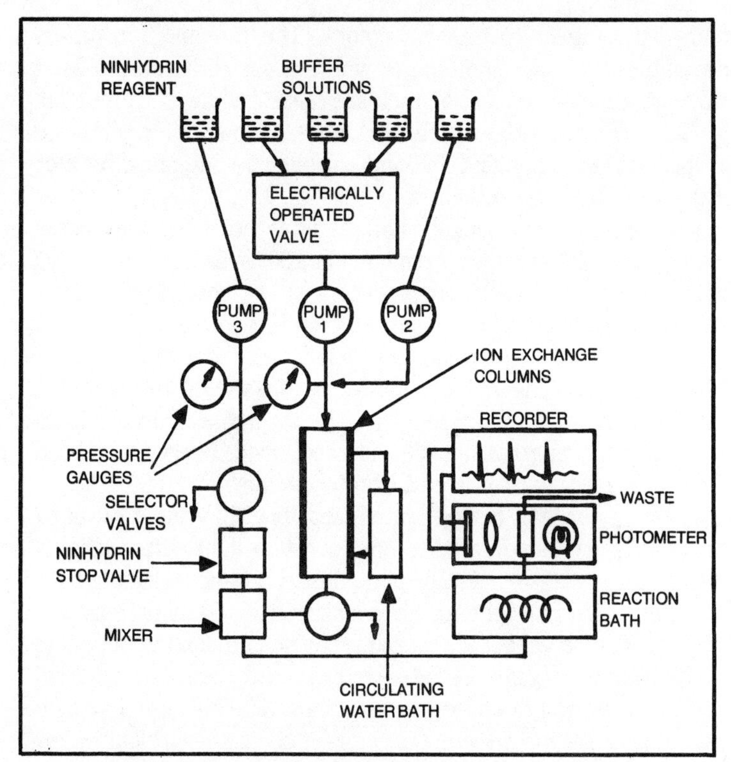

Fig. 8-11. Block diagram of amino acid analyzer.

The circulating water is pumped at a flow rate of about 2 liters/minute. The bath is equipped with a cooling coil which is used in case the water temperature lower than room temperature is required. The complete analysis proceeds unattended after the sample is applied and initial adjustments are made.

As the buffer and sample are pumped through a column, the amino compounds in the mixture separate according to the difference in the affinity that they have for the adsorbent. The resolved amino compounds emerge from the bottom of the column and flow through capillary tubing to a column selector manifold. Here the column effluent is mixed with ninhydrin reagent supplied by a pump at a constant rate (typically 40.0 ml per hour). The mixture then flows through a reaction coil of Teflon capillary tubing contained in a reaction bath maintained at a temperature of 100°C. The reaction coil is volumetrically calibrated so that the time required for a given portion of solution to pass through the reaction coil is optimum for complete reaction between the amino acids and ninhydrin reagent to produce colors. The stream then passes through a three unit photometer where the absorbance values at wavelengths 440 mu and 570 mu are measured for each reaction product. Two of the three photometer units have different absorption path lengths at 570 mu, enabling a wide range of color concentrations to be measured.

Alternatively, a single photometer is equipped with three interference filters which are mounted, 120° apart, on a revolving wheel. They are switched in cycles by synchronizing signals generated by the recorder. A special wheel is used to turn the filter revolver 120°, as a servomotor which drives the cam makes one complete turn. The time required for the motor to make one complete turn is about 0.5 seconds. The absorbance flow cell has a standard thickness of 2 mm. This is a continuous flow type cell and the liquid flows upwards from the bottom to the top.

A tungsten lamp having a rated capacity of 10 V and 3 A is used as a source and is operated to emit constant light output using a current regulating device. The photoelectric element is a selenium cell, whose output goes to the recorder after due amplification. The absorbance values are measured by sensitive photodetectors which transmit signals to the strip chart recorder. The different peaks correspond to different amplitudes. The baseline is established at zero absorbance on the recorder chart when a blank buffer ninhydrin solution is passed through the photometer. The recorder used is a self-balancing type potentiometric recorder having a chart width of 12 inches, which graphically records the electrical outputs of the photometer for each wavelength. The recording sequence associated with the wavelength is generally printed out in a different color. In addition, the recorder has a marker pen which automatically marks the chart whenever a new set is advanced to the filling position. The entire analytical process is automated with the aid of a programming device and a complete array of electrical controls, which automatically safeguard the operation day and night.

Chapter 9

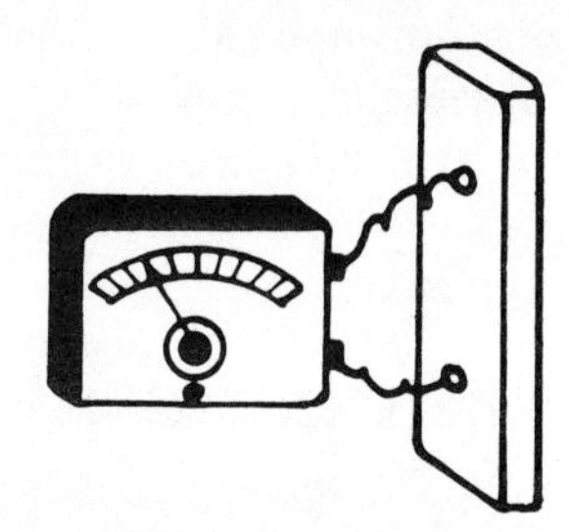

Electrophoresis Apparatus And Densitometers

Electrophoresis is an old established method of analytical chemistry. It is based on the principle that the individual components of the colloidal solution migrate in a liquid at different speeds when subjected to an electric field. When the current is passed for a certain time through such a solution, various components present in the solution would move through different distances in their effort to migrate towards the electrodes. Therefore, a substance which may be a mixture is thus separated into its components along the migration distance according to a definite law. Measurement of the concentration along this migration distance, therefore, would provide the quantitative result of the analysis. Historically, some of the earliest reports described characteristic electrophoretic mobilities of bio-colloids such as proteins or enzymes. However, the technique received little note until 1937, when Tiselius published a paper introducing the moving boundary concept. The moving boundary method utilizes the migration of particles in free solution and observation of the various molecular boundaries through sensitive refractometric techniques. With this, the value of electrophoresis in obtaining distinct and measurable fractions of a variety of substances got well established particularly in clinical laboratories.

Normally, with a moving boundary method, only two components of a mixture, one with the highest mobility and the other with the lowest mobility, can be separated in pure form. If it is desired to recover components other than those of highest and lowest mobilities in pure form, multiple separations have to be carried out.

The apparatus designed by Tiselius was quite cumbersome. It was later modified by Moore and White (1948), who described an instrument which is simplified but retains the essential aspects of Tiselius apparatus. The use of the classical U-tube was abandoned in favor of cells of rectangular cross-section, which promote good heat exchange and do not distort the image formed by the optical system.

PAPER ELECTROPHORESIS

The moving boundary method has the disadvantage that complete separation of the components of a mixture is never achieved. There is always a dead volume of the original solution in the bottom of the U-tube and only the fastest and slowest moving components may be obtained in one operation. This limitation is due to the very manner in which boundaries are stabilized, namely by gravity. The density of the solution must increase downward at every boundary or else there will be some convection tending to destroy the boundaries.

If electrophoresis is carried out in columns packed with powders, gels, filter paper or on a piece of paper, the undesirable effects of gravitational convention may be eliminated as the supporting medium itself would stabilize the various boundaries. Complete separation of the various components leads to the formation of separate and distinct zones on the supporting medium. All that is required is to apply a potential difference to the ends of a strip of paper or a column of gel impregnated with some suitable buffer solution. If a drop of the solution to be analyzed is put on it, the differently charged particles will move with different velocities. The migration process can be followed and the velocities measured by suitable means.

Of all the media used at present for supporting electrophoresis, the most practical and widely used is the filter paper. The basic principle of the method using paper as the supporting medium is the same as that of the moving boundary method. Under the action of the electric field, the charged molecules migrate through the capillaries of the paper, just as they might through an unbounded solvent. However, the advantage of the method is that it is possible to obtain a complete separation into zones of different migration and not as boundary separations of overlapping zones in the liquid phase. The separation zones are located by applying various active reagents.

Paper electrophoresis has developed into a generally applied and valuable routine clinical method. Its main advantage is the small amount of substance required, which may be 1 mg of albumen per analysis and on the smaller outlay on apparatus and technology to carry out a larger number of electrophoretic analyses simultaneously.

In the past few years, several new approaches have been tried in the field of electrophoresis. Among these are the use of hydrolyzed starch and polyacrylamide as *molecular sieving* type gels which yield numerous fractions from serum and other heterogeneous proteins. A great deal of interest has also been evinced in the matter of supports such as cellulose acetate, but they are also subject to many of the same deficiencies as the filter paper methods.

Methods of Zone Localization

The procedure for localizing the zones after an electrophoretic separation on paper has been achieved plays an extremely important role in detection and quantitative estimation of the components. The simplest and the most convenient method of detection and estimation is a color reaction, and the main problem then remains to find a suitable reagent. A large number of reagents for color reactions have been proposed. For example, amino acids are detected by the use of ninhydrin either in an alcohol or acetone solution. For visualizing the amino acids, the paper at the end of electrophoresis is first dried and dipped into the reagent. The coloration appears after drying and develops slowly. The color may then be preserved by fixing the paper in copper nitrate in alcohol solution which has been acidified by addition of nitric acid. The most commonly used reagents for proteins are dyes. Some of the commonly used dyes are bromopheno blue, naphthalene blue black and azocarmine G.

Quantitative Considerations

After staining or dyeing, the next step is to measure the quantity of the dye bound. This may be done by elution or by photometric scanning of the dyed stip.

In the elution method, the paper is cut into thin sections, say 5 mm strips, and the substances eluted from them. The technique is accurate but the major difficulty is to decide into how many sections the strip should be cut to obtain good results. After the sections are cut, they are eluted for about 15 to 30 minutes with 0.01 M caustic soda in dilute alcohol solution. The amount of dye present in each section is then measured in a spectrophotometer in terms of optical density.

Direct photometer recording in transmitted light through the strip is based on the Beer-Lambert Law. This law applies only to a clear transparent, homogeneous and isotropic medium. In practice, these conditions are not generally fulfilled by colored filter paper and, therefore, the measurements are confined to limited range of the calibration curve, which approximates closely a straight line.

Since optical density is a function of the wavelength of the incident light, measurement is made approximately at the absorption maximum of the dye stuff by using color filters. The recording can be more conveniently made with reflected incident light. In that case, the strips do not need to be transparent. This method is based on the relation between the color intensity and the intensity of the light reflected from the illuminated paper. Measurement by reflected incident light is very easy and produces curves free from distortion.

After measuring the optical density photometrically, the readings are plotted against the distance of paper section from the origin. The curve obtained expresses the variation in color density along the paper strip and, hence, the variation in protein concentration.

Evaluation of the Curves

The area under the peaks of an experimental extinction curve obtained may be easily measured on a planimeter or the curve is extrapolated according to Gaussian distribution curves. Vertical lines up to the integral curve are drawn (Fig. 9-1) through the points of intersection of the curves and from these lines parallel to the zero lines are drawn. A ruler of 100 mm length should be placed with one end at the highest point (A) of the integral curve and with its other end on the zero line (B). Now it is possible to read from the ruler the percentage proportions of the individual components of the mixture as differences between the points of intersection with the parallel lines.

There is always an error present when we try to determine the concentration of the components from the area under each of the peaks. This is because all the proteins may not bind the dye in

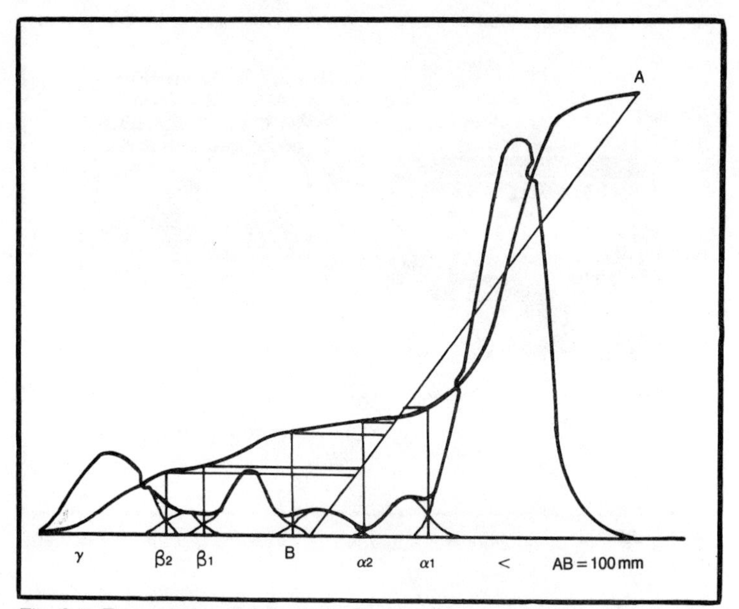

Fig. 9-1. Extrapolation of the optical density curve according to the Gaussian distribution technique.

the same proportion. For exact determinations of protein concentration, it is necessary to multiply the photometric value for a particular protein by an affinity factor of the dye used. This factor is usually obtained experimentally by measurements of the color density on highly purified protein solutions of known concentration.

ELECTROPHORESIS APPARATUS

A complete electrophoresis apparatus (Fig. 9-2) consists of an electrophoresis cabinet, a power supply and a densitometer or scanner.

Electrophoresis Cabinet

An *electrophoresis cabinet* consists of a methacrylate plastic cabinet and gable cover, and a carrier rack of phenolic plastic. The gable cover prevents condensation droplets from falling on paper strips. The carrier supplied for use with paper or cellulose strips up to a total width of 7¼ inches is suitable for supporting paper strips. End surfaces of the carrier are roughened to grip wet paper.

Some electrophoresis baths are designed to take up to 6 strips, each 34 cm long and 5 cm wide. The strips are supported in

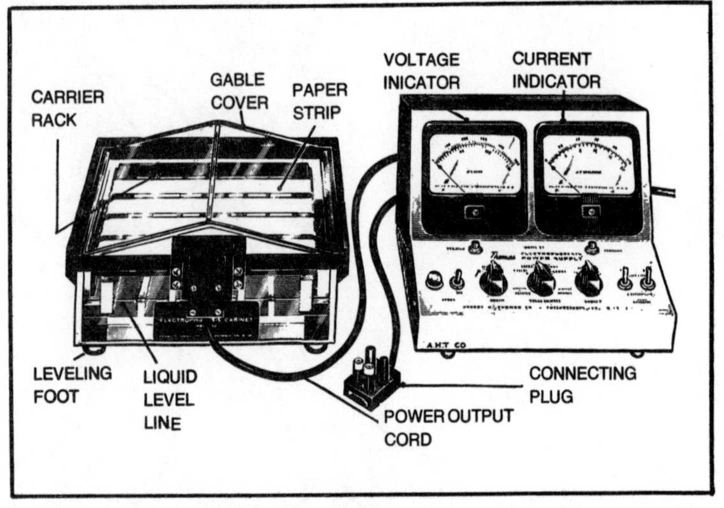

Fig. 9-2. Paper electrophoresis apparatus (courtesy of Arthur H. Thomas, USA).

pairs on three removable plastic bridges which are provided with a number of supports to carry the paper and to prevent undue sagging. At each end of the bath there are two compartments to contain the buffer solution. The ends of the paper strips dip into the solution in the outer of these two compartments, while three platinum electrodes are mounted in the inner compartment. The electrodes are wired to a lead and polarized plug for connecting to the power unit.

The tank should not leak electrically as well as mechanically. To check electrical leakage, the buffer solution is poured into the tank without placing any filter strip. The electrodes are connected to a power supply at 500 volts and a microammeter is connected in series with the supply. The current should not exceed 1 microampere in a good cell. A tank with a higher leakage should be discarded or cemented at its inter-surface joint.

Regulated Power Supply

A *regulated power supply* allows you to select either constant voltage (0-250 volts or 0-1000 volts) or constant current (0-20 ma or 0-100 ma range). The supplies are provided with two meters, one for indicating current and the other for voltage. The polarity reversal facility and overload protection are provided in the power supplies. The ripple content in the power supply should be less than 0.1%.

Constant Voltage Power Supply

A constant voltage power supply maintains a constant voltage across the load, irrespective of the current drawn by the load. Therefore, it should have output impedance as close to zero as possible.

Figure 9-3 shows a voltage regulated power supply. It consists basically of a conventional rectifier and filter circuit to convert ac from the mains to dc with low ripple content. The rectified dc is fed through a series regulator to the output terminals of the load. A reference circuit applies a voltage to one terminal of a comparison amplifier equal to the desired output voltage. If the output voltage does not equal the reference voltage, the input voltage to the comparison amplifier is not zero, and the amplified output of the comparison amplifier changes the conduction of the series regulator. This results in a change of current through the load resistor until the load voltage is equal to the desired output value.

The voltage and current requirements of power supplies required for electrophoresis are usually quite high. Due to the non-availability of transistors suitable for operation at high voltages and high currents, vacuum tube circuits are still preferred in these power supplies. High load currents are provided by using a number of high current valves like E 286/E 284 in parallel.

Constant Current Power Supplies

An ideal constant current source is the one for which the current remains constant regardless of the value of output voltage

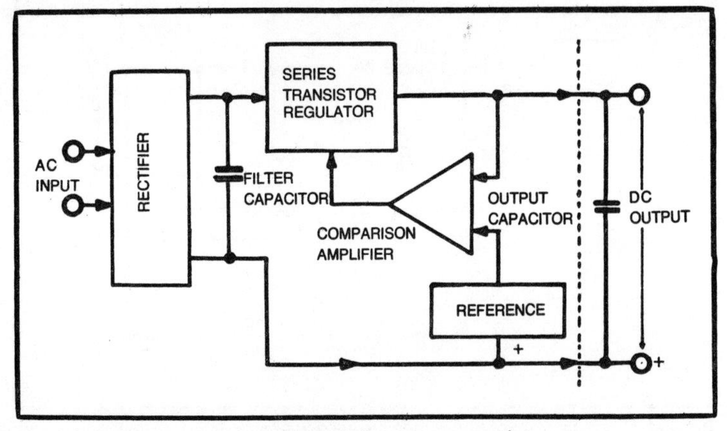

Fig. 9-3. Block diagram of a constant voltage power supply.

demanded by the load. Since it is possible that the load resistance connected to a constant current power supply may vary, the ideal constant current source must have an infinite source impedance at all frequencies.

Figure 9-4 shows the block diagram of a constant current regulated power supply. This supply resembles, in many respects, the block diagram of a constant voltage regulated power supply. However, instead of comparing the reference voltage with the output voltage, the comparison amplifier of a constant current power supply, compares the reference voltage with an IR drop caused by the output current flowing through a fixed resistor. The action of the feedback loop is such as to adjust the conductance of the series regulating element, so as to maintain the IR drop across the series monitoring resistor constant, and equal to the reference voltage, and thereby holding the output current to some constant value as desired.

It is desirable for many reasons to limit the maximum instantaneous current which the series transistors can pass to some predetermined value. The primary reason for doing so is to protect the series transistors themselves from damage due to excessive heating and to be able to charge a large load capacitor without blowing a fuse on the power supply. Consequently, a protection circuit is provided to many power supplies to limit the maximum output current under any load condition. This is achieved by having an automatic crossover between the two modes of operation, namely constant voltage and constant current.

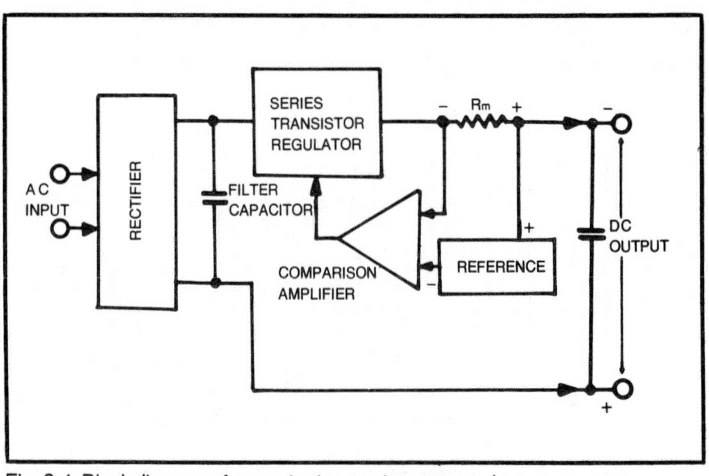

Fig. 9-4. Block diagram of a constant current power supply.

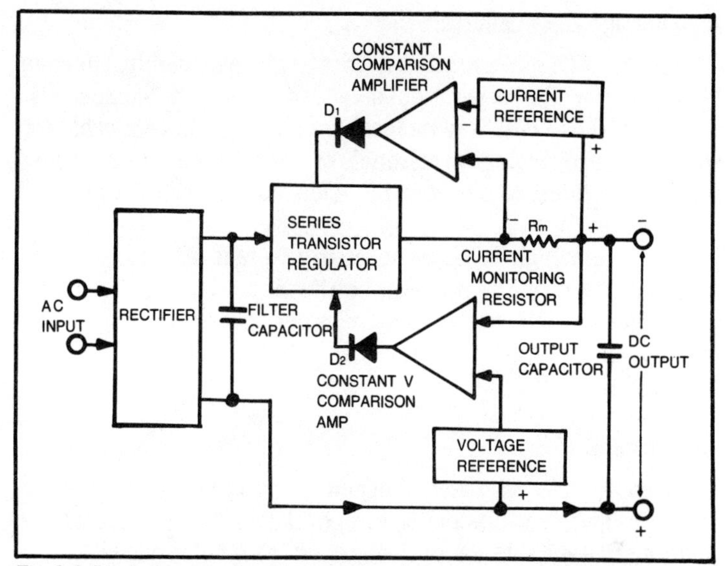

Fig. 9-5. Block diagram of a constant voltage/constant current power supply.

Constant Voltage-Constant Current Regulated Power Supply

Figure 9-5 shows the block diagram of CV/CC power supply. The diodes are so connected in the circuit that when the supply is in constant voltage operation, the upper diode is forward biased or shorted while the lower diode is reverse biased or open. Conversely, when the supply is in constant current operation, the upper diode is reverse biased and the lower diode is forward biased. Thus, the series regulator is only called upon to respond to either the constant voltage comparison amplifier or the constant current amplifier. The effectiveness of one amplifier is not diluted by the shunt presence of the other. In this circuit the same output terminals are used for both constant voltage and constant current operation. The supply delivers constant voltage with a continuously variable current limit or of constant current operation with a continuously variable voltage limit.

Whether the supply is in constant voltage or constant current operation at any instant depends upon the relationship between the dc load resistance and the critical value of the load resistance—defined as the ratio of the front panel voltage control setting to the front panel current control setting. If the load resistance is greater than the critical load resistance, the supply will be in constant voltage operation. If the load resistance is less than this critical load resistance, the supply will be in constant current operation.

High Voltage Electrophoresis

High voltage electrophoresis is a rapid and highly effective technique for the separation of a wide range of compounds, especially those with low molecular weights. It is invaluable for work with many biological materials of which amines, amino acids, indoles, peptides, sugars, organic acids, dansyl derivatives, etc., are only a few.

The very high voltages necessary to take advantage of the technique gives rise to problems like heating, which are overcome by a water cooling system or some other special technique. The high voltages normally employed are 10,000 volts at 100 mA or 5,000 volts at 200 mA.

Densitometer

Quantitative estimates obtained by elution of the dye from 5 mm strips and measurement of the optical densities of the resulting solutions in a suitable absorptiometer offer a lengthy and laborious method and gives only approximate results. The method has been simplified and made more accurate by the direct measurement of the optical density on the dyed paper. In this method the stained electrophoretic strip is made transluscent by prolonged immersion in a 1:1 mixture of paraffin oil and bromonaphthalene and then placed between two parallel glass plates and moved before the slit of a photometer and the reading of the optical densities of the sections of strips are measured. While taking the readings, it is advisable to use a light filter in doing the colorimetry, since the wavelength of light used has a rather large effect on the readings obtained.

For making direct photometric recordings, a number of scanners are commercially available. All of them work more or less on the same principle. Figure 9-6 shows the typical arrangement. The stained strip is placed between two glass plates and made to move in front of a 1 mm slit. Light from a constant source after passing through the slit falls on the translucent paper strip mounted on a slide. A photoelectric cell is placed in line with the light source and the slit, but on the other side of the paper having a suitable optical filter in front of it. The photocell produces a small, but measurable electrical current in direct proportion to the intensity of light falling on it, which can be directly read from a microammeter.

At right angles to the paper holding slide is attached a transparent rule, suitably graduated so that readings from the

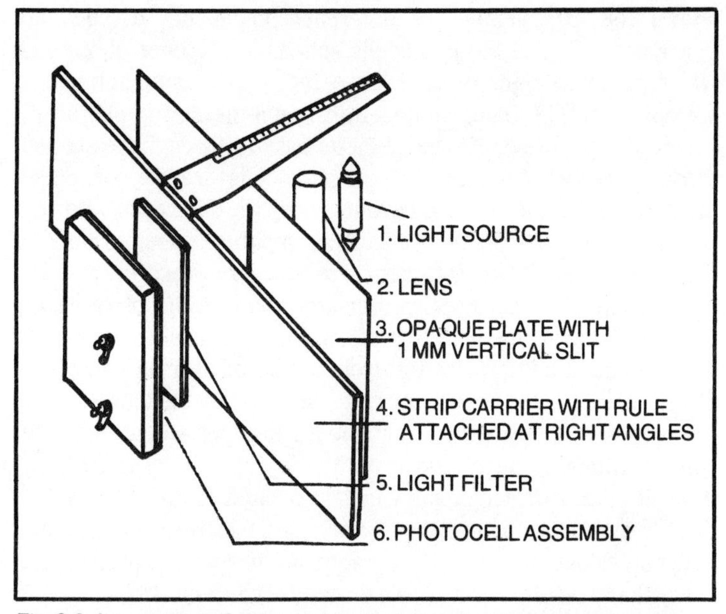

Fig. 9-6. Arrangement of various components in a single beam densitometer.

microammeter may be plotted at required intervals. The operator only has to move the transparent rule and, therefore, the slide, and at each movement plot along the rule the microammeter reading, which records the intensity of dye at that point on the strip. By joining the points so obtained, a clear picture of the pattern of fractionation is immediately obtained.

When light from the source lamp passes through a blank portion of the oily filter paper, the meter is set to read zero corresponding to zero OD and 100% transmission. When the paper moves before the light path, the portion which is richest in protein concentration will offer maximum opacity to the light path and thus the photocell will receive minimum light corresponding to which the recorder will trace a large area curve on OD scale.

A serious problem with densitometers using a single photocell is their instability against line fluctuations. With the changes in the input voltage, the source lamp voltage will change and affect the amount of light falling on the photocell. Hence, the interpretations of the concentrations from the graph will not be accurate. The stabilization of the source lamp can be achieved either by using a saturation core transformer or incorporating an electronic regular circuit. However, the preferred method is to have a double beam densitometer having facilities for automatic recording. Figure 9-7 shows the arrangement of different components of such an instrument. One of the photocells acts as a reference photocell, which receives light directly from the source lamp through a variable slit. The other photocell, called a measuring photocell, receives light through a fixed slit after passing through the stained electrophoretic paper strip. The photocells are so connected electrically that the net output from them is the difference between the photovoltage of the two cells. A dc amplifier is used to amplify this difference to a level sufficient to drive a servo-recorder.

Initially, the zero optical density is set on the recorder or meter by allowing the light to pass through the translucent portion of the stained paper placed in front of the measuring photocell. Then, as the stained paper is made to move across the light source, the recorder will trace curves on the graph paper according to the concentrations of the protein samples. Any fluctuations in the lamp intensity are received simultaneously by both the photocells and the effect is automatically nullified. Some instruments include attachments like electronic integrators for measuring a peak height in which the results are available directly in concentration.

SPECTRODENSITOMETERS

The introduction of agar gel and subsequently of poly-acrilamide gel electrophoresis for fractionation of ribonucleic acids, and the need of a direct ultraviolet densitometry of the gels, has stimulated the development of electrophoretogram scanning techniques. Staynov and Stainov (1969) describe a double beam recording spectrodensitometer for measuring the optical density of dry agar electrophoretograms.

The Schoeffal Instrument Company manufactures a double beam ratio type scanning spectrodensitometer, incorporating a high intensity continuous xenon light source and a low stray light quartz prism monochromator. Thin layers, as well as gels and paper strips of all types, and preparative forms can be rapidly and very accurately scanned for quantitative determinations (Fig. 9-8).

The instrument also facilitates liquid samples in static or flow cells for a great number of investigations in the field of column chromatography, spectrodensitometry, and kinetic and denaturation research. Figure 9-9 shows the optical system of the instrument. In the instrument, a 150 watt xenon high pressure short arc lamp or 200 watt xenon-mercury lamp (1) is fitted in a special air-cooled housing which is equipped with high transmission suprasil optics (2) and a front surface spherical reflector (3) for

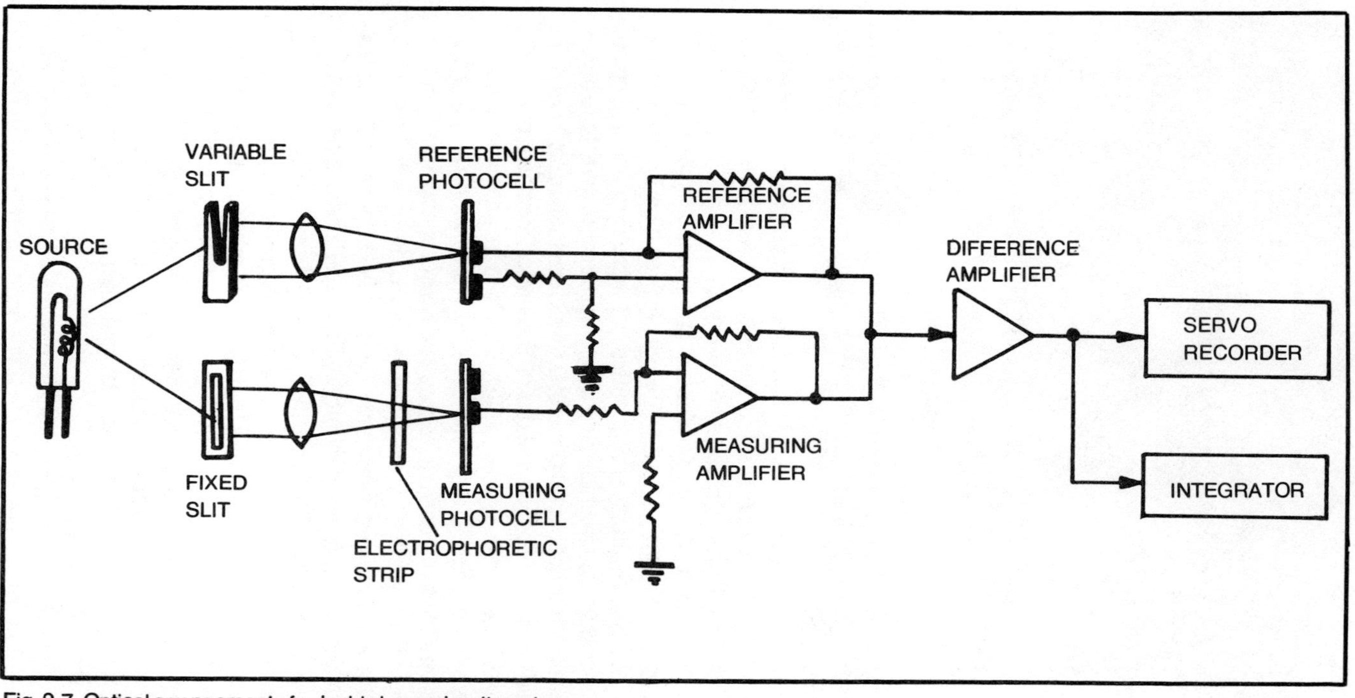

Fig. 9-7. Optical arrangement of a double beam densitometer.

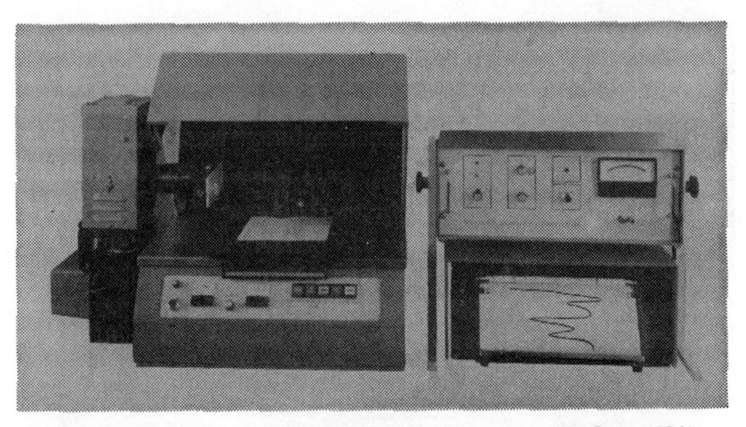

Fig. 9-8. Spectrodensitometer (courtesy of Schoeffel Instruments Corp., USA).

greater efficiency of operation. The reflector is adjustable so that the primary light image and reflected image can be aligned to be coincident.

A light tube attached to the lamp housing focuses the light beam onto the entrance slit of the monochromator. A slot located just past the focusing adjustment is provided for the insertion of an interference or other type of filter, and a special split ring which serves to cover this slot while not in use. The light tube also contains the beam diverging optics, all necessary beam aperture and balance adjustments, and a deflector mirror to divert the beams down to the exit post (8) adjustment which includes a calibrated slit width control (9). The details of the optical components of the monochromator are given in (17). The light beam emanating from the source strikes the TLC plate or other media. The media scatters the light, making it act as a pair of secondary light sources. These light beams then pass through protective quartz lenses (10) and are maintained separately by a baffle between the lenses, which is also in close proximity to the plate. These lenses collect the light and focus it onto the individual photomultiplier tubes (11). The lenses are located in such a way that only the incident beams or the scattered light resulting from the incident beams are directed onto the photomultiplier tubes. The photomultipliers are equipped with individual dynode chain divider circuits and are mounted in a sturdy cast aluminum housing inside the main instrument. They are arranged so that both angular and lateral adjustments can be made at the factory.

The photomultiplier tubes are supplied with a highly regulated, low ripple power supply. The gain control is adjusted by varying the amount of high voltage being applied to both photomultiplier tubes. The balance control changes the voltage on the signal photomultiplier with respect to the voltage applied to the reference photomultiplier. The balance control does not affect the reference signal. The amplifiers consist of high gain amplifiers which act as current to voltage converters. They are adjusted to have equal output from the sample channel and reference channel with a zero balance setting on the controls. This means the ratio would be 1, i.e., 100% T or zero optical density. Changes in the sample channel will affect this ratio and can, therefore, be measured. The ratio system compensates for any changes in the light due both to the lamp power supply and characteristics of the lamp itself. In case

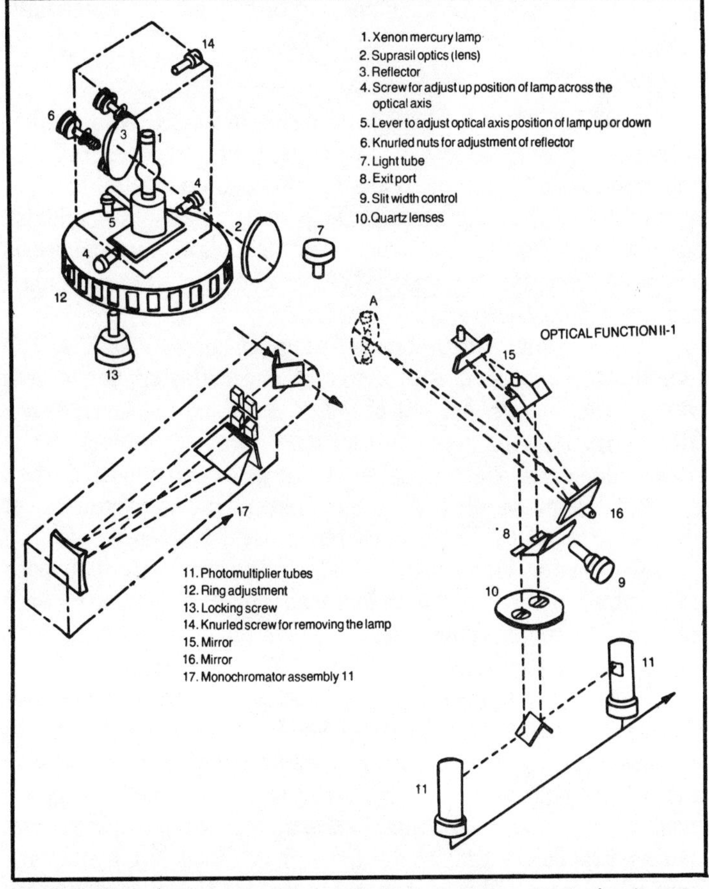

Fig. 9-9. Optical arrangement in a ratio type scanning spectrodensitometer (courtesy of Schoeffel Instrument Corp., USA).

optical density is to be obtained, the log function of the ratio of the two input signals coming from the reference and sample photomultiplier is obtained by using a log amplifier. This gives a linear output for recording and integration.

The monochromator used is calibrated in nm from 200 to 700 in 10 nm increments, linearly. It incorporates high transmission quartz optics in a modified-Littrow mount. The wavelength cam drive can linearize non-linear dispersion over the spectral range. The monochromator can be fitted with a motor drive assembly for spectral scanning. The scanning is done at a speed of 200 nm per minute. The wavelength drive should have zero backlash design. A slip clutch permits manual override of the motor drive, permitting any selected wavelength to be manually set. The drive will automatically take over upon release of the manual wavelength control.

Densiscan KS 3

DENSISCAN type KS 3 from KIPP and ZONEN is another automatic density scanner with a built-in computer unit for collecting quantitative data obtained from micro, semi-micro or macro bands. The instrument is particularly suitable for cellulose acetate strips, polyacrylamide gels and all kinds of electrophoresis strips. A complete microzone electrophoresis scan, including printout of data, takes only 25 seconds.

The instrument incorporates a tungsten lamp 4 V/0.75A. The light beam is focused onto a mirror and reflected downwards through the sample, the slit of 0.1 × 3mm and the interference filter. The light intensity is measured with a phototube. The background density of the membrane or gel is compensated when the carrier is in the idle position, and for this the membrane or gel should be clear at the beginning to obtain a true reference zero.

The carrier with a sample holder is moved horizontally back and forth across the slit. The actual scan is made during the forward motion. The motor control enables selection of scan length, scan speed and automatic return of carrier at the end of a scan.

The peaks can be selected automatically by the instrument. It detects the minima point between a falling and rising slope of two successive peaks. The first peak is integrated from the start pulse and the last peak until the stop. When peaks are overlapping, the peak selector fails and manual selection is advisable. The area under each peak is computed and is printed in relative percentage on a paper strip. The intensity patterns of the electrophoresis strips are recorded with an x-y recorder.

Modern densitometric investigations require a high optical resolution. Also, the output measured should preferably allow a choice between linear and logarithmic response. The logarithmic response is of great advantage in the evaluation of electrophoretic samples and in all cases where a linear relationship may be assumed between concentration and extinction.

KIPP Micro-Densitometer DD2

In the KIPP micro-densitometer DD2, the light transmitted through the sample reaches a vacuum phototube with a strictly linear response. The phototube is coupled to an operational amplifier which may be switched to linear or logarithmic operation. The circuit gives an output of 10 mV full scale on the recorder.

The sample carriage is moved by a micrometer spindle driven by a reversible synchronous motor. The sample may be scanned with speeds of 5, 10 and 20 mm per minute in both directions. The maximum movement is 140 mm. For line distance measurements, the spindle is provided with a digital readout giving an accuracy of 0.1 mm.

The permissible scanning speed is determined by the width of the absorbing lines to be measured and the response time of the recorder. The response time of the recorder should be approximately 1 second. Table 9-1 suggests the scanning speeds to be available in relation to the width of the narrowest line in the sample.

When too high a scanning speed is selected, the recorder will not be able to reach the true deflection while passing a narrow line. The lowest speed is chosen in connection with the resolving power of the optical system and the response time of the recorder. In most cases a chart speed of 50 mm/min will be found suitable.

When the wavelength of maximum absorption in the sample is known, the most suitable filter is selected. The filter disc is provided with three filters having transmission maxima at 450, 525 and 600 mμ. The fourth position allows measurement with white light. For grey absorbing materials, such as X-ray films, no filter is

Table 9-1. Available Scanning Speeds in Relation to Line Width.

Line width	Scanning Speed
above 1 mm	20 mm/min.
1 - 0.4 mm	10 mm/min.
0.4 - 0.1 mm	5 mm/min.

required. If the wavelength of maximum absorption in the sample is not known, the most suitable filter is selected by scanning one or two lines using the three different filters. The filter that gives the lowest transmission or highest peak for a certain line is chosen for the measurement.

Samples may be fixed in different ways. Glass plates are best held in position between the two magnets placed on the carriage. Photographic films, cellogel strips and the like may be fixed and flattened by partly placing the magnets over them. Since the light spot is clearly visible, samples are easily aligned parallel to the scanning direction. For the measurement of acrylamide rods, the glass tubes containing these rods are placed in the slot of the carriage, with their axis approximately in the plane of the latter.

MICROELECTROPHORESIS

The electrophoresis cell used in normal applications requires a relatively large volume of solution about 20 ml, which might mean as much as 100 to 200 mg of material if accurate measurements are desired. Such large material requirements are not easily arrangeable in some biochemical investigations. This requirement has led to the development of methods for microelectrophoresis which have the advantage of reducing the volume and concentration of the solution required.

The simplest method for reduction in the volume of solution required by reducing the dimensions of the cell was achieved by Labhart and Staub (1947) and Lotmar (1953). The cell in this case is only 30 mm in height and 1.5 × 5mm in cross section. For this cell, about 0.5 ml of solution and 100 ml of buffer are required.

Use of sensitive optical methods can also aid in achieving a reduction in the volume of solution required. One of the methods used for the purpose is the Jamin interferometric system used by Labhart and Staub and improved by Lotmar. The method is based on the formation of interference fringes, which are produced by the interference of two beams, one of which has passed through some homogeneous reference medium and the other through the cell under investigation. In the Jamin interferometric system, an image of the cell is obtained which is crossed by fringes. The variation in spacing of the fringes represents the variation in concentration along the cell. The fringes will be crowded at a boundary and, in effect, each component of a mixture will make itself seen by a corresponding group of fringes. A simple counting of the fringes gives an approximation of relative concentrations. A slight variation in the measurement technique using an interferometer is incorporated in the microelectrophoresis apparatus manufactured by Shandon Antweiler. In this method (Fig. 9-10) the light from the source (A) is split by the first mirror (B) into two coherent light beams which pass through the measuring and reference channels of the cell (C). The two light beams are recombined in the rotatable mirror (D). Whenever the two coherent beams encounter variations in optical density in the cell because of the migration of the proteins into the buffer solution in the measuring channel and the presence of buffer in the reference channel, they pass through optically unequal paths. This difference in optical path length can be compensated for by rotating the mirror (D). Correct compensation can be checked by means of a spectrum and an index line in the eye piece. The mirror is rotated by hand through a micrometer screw, whose movement can be read off and recorded. Rotation of the micrometer screw corresponds directly to a difference in the concentration of the solutions in the two channels. The channels are measured out at intervals of 0.1 mm from the top downwards, along the protein migration distance.

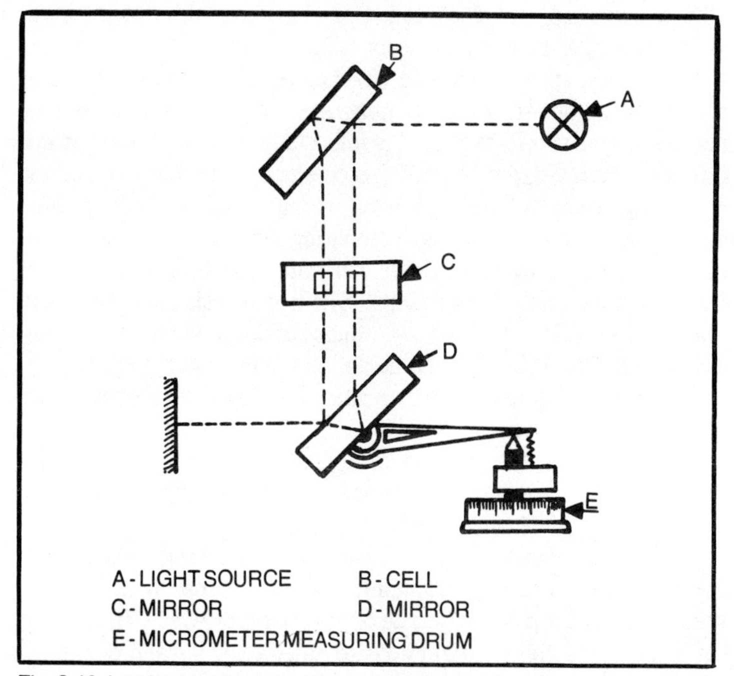

Fig. 9-10. Interferometric method for measurement of density on electrophoretic strips.

STEPS IN THE ELECTROPHORESIS PROCEDURE REQUIRING QUALITY CONTROL

With appropriate knowledge of sample handling and with a useable control specimen, the following points should be borne in mind to reduce errors and improve reproducibility. Sample application is the most critical portion of the electrophoresis procedure. Overloading of the samples should be avoided. This can result in an inability of the densitometer to read the most intense peaks due to the optical density of the dye protein complex being in a region of non-linearity with respect to concentration. This is readily checked by making 5%, 10% and 25% dilutions of the sample and carrying out electrophoresis on undiluted and diluted samples.

Sample applicator repeatability is likewise important to assess. This is especially true in those laboratories where several applicators are used interchangeably.

Buffer pH and composition should be carefully checked. Ideally, conductivity measurements should be made on each new batch of buffer made. Variations in buffer conductivity as large as ± 10% can generally be accepted. Buffer pH should be maintained within ± 0.10 of the given value for buffer.

Power supplies should be checked regularly to determine if the actual output to the electrophoresis cell is the same as the reading on the supply meter. It is important to monitor and record initial and final current levels for each electrophoresis run. Buffer and media changes, open or short circuits and unstable power supplies can be evaluated by comparing current levels between runs. Applied voltages should be controlled to within ± 5% of the recommended voltage conditions. Electrophoresis running times must be controlled to within ± 1 minute of the recommended time for a given procedure. The combination of buffer voltage, running time and medium interact to determine the degree of sharpness of the separated fractions.

Invariably, it is not possible to control the properties of the electrophoretic medium. Therefore, within the lot and lot-to-lot variations in performance must be assessed.

Post-electrophoresis conditions such as staining, rinsing and clearing times should be maintained within ± 30 seconds. These steps directly control the degree of transparency which, in turn, is reflected in the uniformity of densitometer scanning and resultant values.

Chapter 10

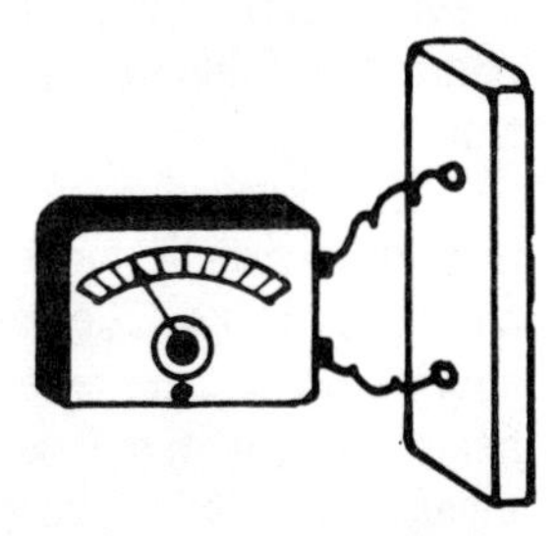

Mass Spectrometers

In a mass spectrometer, the sample to be analyzed is first bombarded with an electron beam to produce ionic fragments of the original molecule. These ions are then sorted out by accelerating them through electric and magnetic fields according to their mass/charge ratio. A record of the numbers of different kinds of ions is called the ***mass spectrum***. The uniqueness of the molecular fragmentation is the basis for identification of different molecules in a complex mixture as no two molecules will be fragmented and ionized in exactly the same manner. Very small sample sizes, which may be of the order of a few tenths of a milligram, are generally sufficient as long as the material is able to exist in the gaseous state at the temperature and pressure existing in the ion source. Several other useful applications of mass spectrometry include the direct determination of the molecular weight, the placement of functional groups into certain areas of the molecules, and their interconnection and investigation of reaction mechanism. One of the main advantages of mass spectrometry over other spectroscopic techniques is its sensitivity. It is possible, on some instruments, to obtain full mass spectra with 1 ng of material in about a second. Compounds now amenable to mass spectrometry vary from low molecular weight gas mixtures to high molecular weight, natural products. Various ionization techniques are used to provide complimentary structural information and molecular weight, and elemental composition data is readily obtainable. A significant advantage of mass spectra is its suitability for data storage and library retrieval since the positions (masses) of the peaks in the spectrum of a given compound are fixed. Another asset of mass spectrometry is its capability in handling complex mixtures currently via gas chromatography-mass spectrometer (Brooks & Middleditch 1977) and high resolution gas chromatograph-mass spectrometer (Meili, 1975). Burlingame (1976) illustrates the enormous fields of application presently covered by mass spectrometry.

The procedure for analyzing a substance by mass spectrometry starts by converting the substance into a gaseous state by chemical means. The gas is introduced into the highly evacuated spectrometer tube where it is ionized by means of an electron beam. The positive ions thus formed are deflected and focused by means of suitable magnetic and electric fields. For a given accelerating voltage, only positive ions of a specific mass pass through a slit and reach the collecting plate. The ion currents thus produced are measured by using a sensitive electrometer tube. By varying the accelerating voltage, ions from other mass species may be collected and the ion currents measured in such cases would be proportional to the amount of the given mass species present. A great advantage of mass spectrometry over other direct methods is that the impurities of masses different than the one being analyzed do not intefere with the results.

Figure 10-1 shows the principle of operation of a mass spectrometer. The molecules in the gas sample (A) to be analyzed are bombarded with electrons to produce ions (B). These ions are accelerated in a high vacuum into a magnetic field (C) which deflects them into circular paths (D). Since the deflection for light ions is greater than that for heavy ions, the ion stream separates into beams of different molecular weight. A suitably placed slit (F) allows a beam of a selected molecular weight to pass through to a collection electrode (E). As the accelerating voltage on the ion source is gradually reduced, ion beams of successively greater mass pass through the slit. When these ions fall on the collector electrode, they produce minute electric currents, which after suitable amplification may be measured. Their amplitude will indicate the number of ions in each beam. The proportion of molecules of different masses in the gas sample may be found, and a complete analysis of the gas sample may be made, provided all the constituent gases have a different molecular weight. This is usually the case in respiratory gas analysis work.

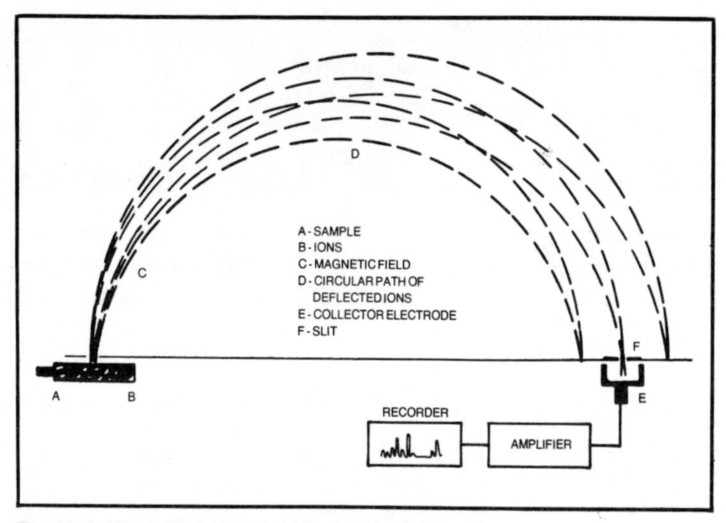

Fig. 10-1. Principle of operation of a mass spectrometer.

TYPES OF MASS SPECTROMETERS

Several methods of producing mass spectra have been devised. However, the principal difference between the various types of spectrometers lies in the means for separating the ions according to their mass to charge ratio. The important types are described below.

Magnetic Deflection Mass Spectrometer

Figure 10-2 shows the essential parts of a typical magnetic deflection mass spectrometer tube due to Nier (1940). The heated tungsten filament produces an electron beam which passes between plates A and B. A difference in electrical potential between A and B pulls ions out of the beam so that they pass through the slit in B into the region between B and C. The potential difference between B and C is adjustable from 0 up to several thousand volts. The ion beam then enters the space between two trapezoid-shaped magnet poles where it is deflected through an angle of 60°, 90°, 120° or 180°.

When ions of mass "m" and charge "e" pass through an accelerating electric field, they would attain a velocity v which can be expressed in terms of the accelerating voltage as follows:

$$\frac{1}{2} mv^2 = eV \qquad \textbf{Equation 10-1}$$

where $\frac{1}{2} mv^2$ is the kinetic energy of the ion as it leaves the electric field. From Equation 10-1, v can be written as

$$v = \left[\frac{2eV}{m}\right]^{\frac{1}{2}} \qquad \text{Equation 10-2}$$

If the ions next enter a magnetic field of constant intensity h which is applied at right angles to their direction of motion, the ions would be diverted into circular orbits. Equating the centripetal and the centrifugal forces:

$$\frac{mv^2}{v} = hev \qquad \text{Equation 10-3}$$

The radius of curvature of the trajectory is given by:

$$r = \frac{mv}{eh} \qquad \text{Equation 10-4}$$

Substituting for v from Equation 10-2, we get

$$r = \left[\frac{2eV}{m}\right]^{\frac{1}{2}} \times \frac{m}{eh}$$

$$= \left[\frac{2eV}{m} \times \frac{m^2}{e^2h^2}\right]^{\frac{1}{2}}$$

$$= \left[\frac{2V}{h^2} \times \frac{m}{e}\right]^{\frac{1}{2}} \qquad \text{Equation 10-5}$$

This equation shows that the radius of the orbit is a function of the mass/charge ratio of the particles.

In practice, all the quantities of equation (v) are kept constant with the exception of m and V. By varying the accelerating voltage, V, it is possible to cause an ion of any mass to follow the path which may coincide with the arc of the analyzer tube in the magnetic field. Ions of different m/e ratio will strike the tube at some point and would get grounded.

Under specified conditions, the ions which will be collected would follow the expression:

$$\frac{m}{e} = \frac{h^2r^2}{2V} \qquad \text{Equation 10-6}$$

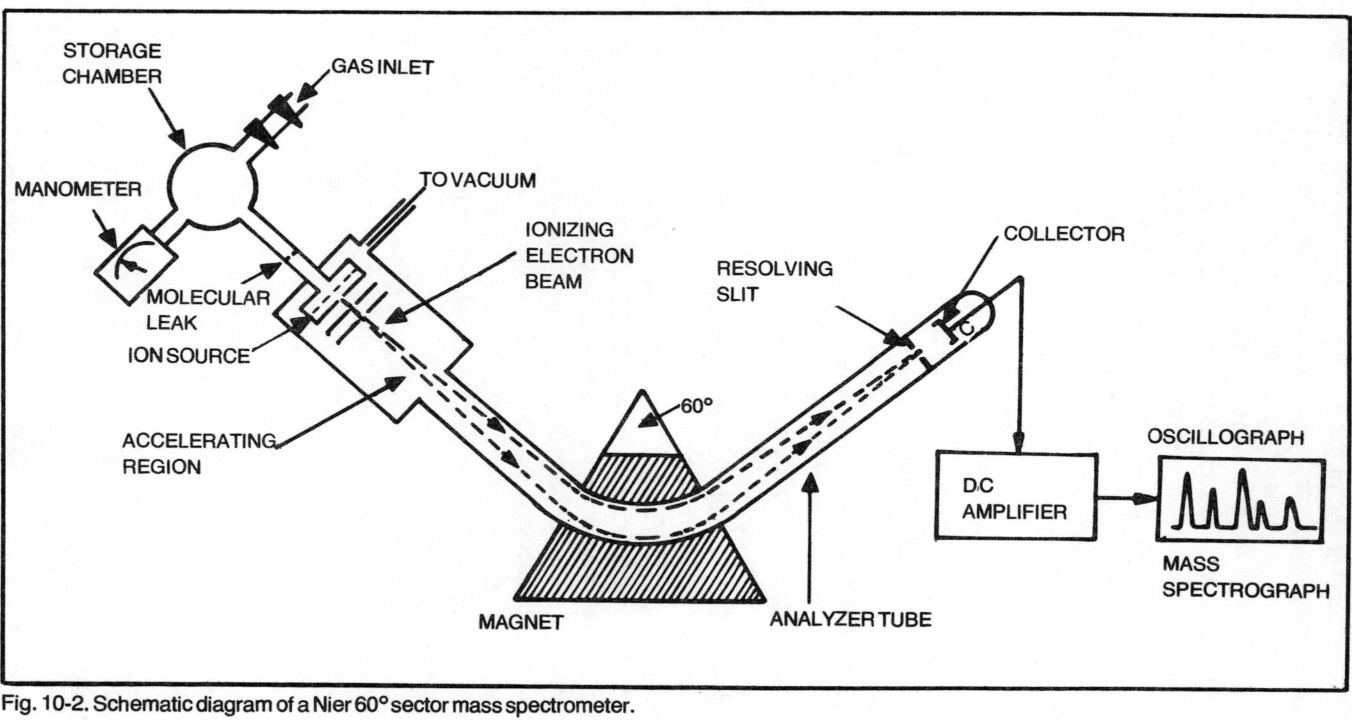

Fig. 10-2. Schematic diagram of a Nier 60° sector mass spectrometer.

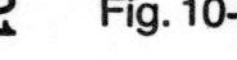

The relationship shows that for obtaining a mass spectrum, the accelerating voltage or the magnetic field strength can be varied. Usually, it is the magnetic field which is kept constant and the voltage is adjusted to bring to a focus specific m/e. The magnetic field must be very uniform over a large area. If electromagnets are employed, they would require large power supplies capable of providing several kilowatts of power and must be highly regulated. For a particular instrument, the angle of deflection (radius) is fixed for a given analyzer tube (180°, 90° and 60°). Therefore, Equation 10-6 can be written as

$$mV = \text{Constant}$$

If this constant is, say, 2000, then as V is set at 100 volts, particles of mass 20 can reach the collector plate, while to direct particles of mass 25 to it, V must be set at 80 volts. Dampster (1918) introduced direction focusing by deflecting the ion beam along a 180° trajectory through the magnetic field. A large magnet is required for a 180° mass spectrometer. When the ion source is in the uniform magnetic field, the gap between the poles must be large enough to contain the ion source. To overcome this difficulty, Nier (1940) employed the direction focusing properties of a 60° sector magnetic field. In this arrangement, the magnetic field does not envelop the ion source, a separate source magnet is required. With sector type instruments, a mass resolution of 200 to 600 mass units can be obtained. Bartky and Dempster (1929) suggested the use of simultaneous direction and velocity focusing to achieve high resolution mass spectrograph. Mattauch and Herzog (1936) achieved decisive improvement in resolution by using the double focusing principle.

The instrument focuses for both direction and velocity. This is done by placing an electrostatic analyzer between the ion accelerating slits and the magnetic field. After the usual acceleration of the ion beams in an electrostatic field, the ions are deflected through a tandem arrangement of an electrostatic analyzer and then a magnetic analyzer. The arrangement enables the focusing of ions having the same mass/charge but different initial velocities and directions. Resolving power of 8500 with a 0.05 mm entry aperture was achieved by Newmann and Ewald (1962) by utilizing the focusing properties of electric and magnetic sector fields in a parabolic configuration.

Commercial mass spectrometers appeared in the United States in the early 1940s. In these instruments, ions were separated according to the quotient mass/charge, and the separated ion beams were recorded directly on a photographic plate.

Later, the design of a mass spectrometer for precision mass measurements was developed. The instrument incorporated the arrangement, known as *Nier-Johnson geometry*, which involves a deflection of $\pi/2$ radians in a radial electrostatic field analyzer, followed by a magnetic deflection of $\pi/3$ radians. One ion-beam at a time is brought to a focus on an exit slit and measured electrically. These types of instruments became commercially available in the early 1960s.

About the same time, commercial mass spectrographs using *Mattauch-Herzog geometry* also became readily available. In this arrangement, a deflection of $\pi/4\sqrt{2}$ radians in a radial electrostatic field is followed by a magnetic deflection of $\pi/2$ radians, and all ions of different mass can be simultaneously focused on a photographic plate. The spectrographs with photographic recording are used for analysis of solids and for the recording of organic mass spectra. An accuracy of 1 part in 10^9 has been obtained in precision mass measurements.

The Time-of-Flight Mass Spectrometer

In a time-of-flight mass spectrometer, ions of different mass/charge ratio are separated by the difference in time they take to travel over an identical path from the ion source to the collector. Stephens (1946) described the first pulsed mass spectrometer in which ion packets of a few microseconds duration are emitted at intervals of a few milliseconds from a voltage source. The ions transverse an evacuated tube called the drift tube to reach the detector. The detector is sensitized for a brief instant to register their arrival. Since ions of different masses arrive at the detector at different times, the accurate measurement of the time between activating the source and sensitizing the detector gives information concerning the mass of the ions. The signal from the ions reaching the detector is amplified and applied to the vertical deflection plates of an oscilloscope. The horizontal axis deflection of the oscilloscope commences as the ion packets start out. This produces a mass spectrum on the screen of the oscilloscope. The device gives a mass spectrum in a very short time. The essential parts of a time-of-flight instrument are shown in Fig. 10-3. It consists of:

—An electron gun for the production of ions.

—A grid system for accelerating ions to uniform velocities in a pulsed mode.

—An evacuated tube, called the drift tube.

—An ion detector and suitable electronic circuitry for translating the time-dependent arrival of ions of different velocities into a time base that is related to mass number.

If L is the length of the drift tube in centimeters and t is transit time in microseconds, for singly charged ions of mass m and constant energy Ve, then

$$t = L\left[\frac{m}{2Ve}\right]^{1/2}$$

$$= L\left[\frac{m}{e} \times \frac{1}{2V}\right]^{1/2}$$

$$\frac{m}{e} = \frac{2V \times t^2}{L^2}$$

If the detector is sensitized for a period Δt at time t, the resolution for constant energy is given by

$$\text{Resolution} = \frac{\Delta m}{m} = 2\frac{\Delta t}{t}$$

Equation 10-1 shows that the time resolution will increase with increased drift tube length and will decrease with increasing accelerating voltage.

The current produced by the ions arriving at the collector may have a very short duration which necessitates the use of a wide-band amplifier. Goodrich and Wiley (1961) designed a special magnetic electron multiplier for this purpose. This multiplier uses a strip of semiconducting material for the multiplying surface instead of dynodes. A gain of 10^7 is attained with a dark current of only 3×10^{-21}A.

The main advantages of the *time of flight* (TOF) spectrometers include their speed and ability to record the entire mass spectrum at one time. A conventional spectrometer detects only one peak at a time. Its accuracy depends on electronic circuits, rather than on extremely critical mechanical alignment and on the production of highly stable and uniform magnetic fields. The main disadvantage of the TOF spectrometers is their poor resolution due to display on an oscilloscope screen.

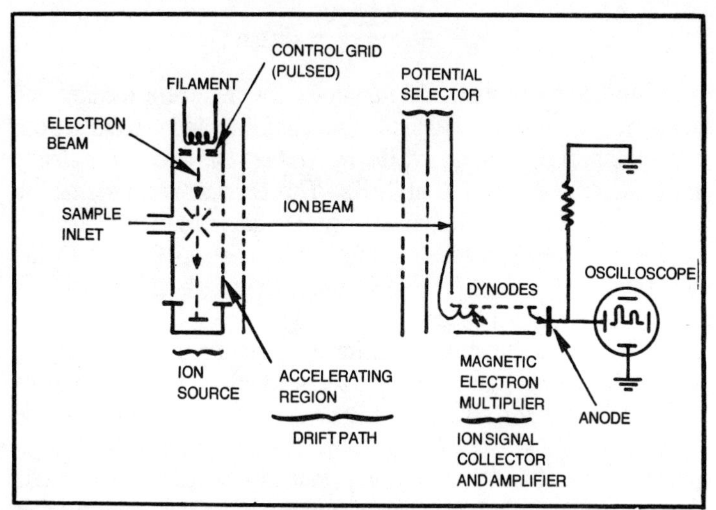

Fig. 10-3. Schematic diagram of a time-of-flight mass spectrometer.

Radio Frequency Mass Spectrometer

The most popular radiofrequency mass spectrometers make use of the Bennett (1950) tube. The arrangement in the instrument is such that the charged particles emerging from the ion source are all accelerated to the same energy in an electrostatic field, and then they pass through a system of radiofrequency electrodes. The energy acquired by the ions in this process is a function of their specific mass/charge ratio. Maximum energy increment would be acquired by these ions only if the ions start with the correct velocity at the optimum phase of the RF field. A potential energy selector is placed before the detector, which balances out the energy of the ion beam, and the mass spectrum is recorded by detection of the ions with the highest energies as the frequency of the alternating RF voltage is varied.

The RF field is applied in one or more RF stages. Each stage is a series of three equally spaced parallel grids. An alternating RF voltage is applied to the central grid with respect to the outer grids which are kept at ground potential.

If v is the velocity attained by the ions in phase with the radiofrequency field, then

$$v = df$$

where f is the frequency of the RF field in MHz and d is the spacing between adjacent grids in cms. The mass/charge ratio of the ion beam reaching the detector is given by

$$\frac{m}{e} = \frac{0.266V}{d^2f^2} \text{ (in cgs units)}$$

The RF mass spectrometer does not require a magnet and therefore is comparatively lightweight and simple in construction. In the Bennett spectrometer, the RF voltage has a fixed frequency and is modulated at 10% at 1kHz. The current received at the detector is amplified with an ac amplifier tuned to the modulation frequency. The dc ion-accelerating voltage is swept from 50 to 250 volts twice per second. The spectrum in the range M = 10 to 50 is reproduced twice per second.

It has been shown through investigations that the resolving power of a Bennett tube is primarily determined by the distances between the individual RF accelerating stages. A two stage tube would be the most favorable, but the instrument tends to provide spurious lines. Separation of the principal line from the spurious lines with a retarding field would greatly affect the sensitivity of the spectrometer. The spurious lines may be suppressed in a three stage tube, although resolving power is slightly reduced.

Henson (1950) improved upon the resolution of the Bennett tube by using a square wave radio frequency signal in place of sinusoidal RF voltage. Several improvements have been incorporated over the original design and mass spectrometers for special applications have been built up.

Quadrupole Mass Filter

Figure 10-4 shows a path stability spectrometer without a magnetic field, in which the stability of the ion paths is determined by the specific charge e/m. The arrangement consisting of four cylindrical rod-shaped electrodes provides a potential field distribution, periodic in time and symmetric with respect to the axis, which will transmit a select mass group and cause ions of improper mass to be deflected away from the axis. This mass selection scheme uses a combination of dc potentials plus a radio frequency potential. By proper selection of potentials and frequency, an ion of desired mass can be made to pass through the system while unwanted masses will be collected on one of the electrodes.

COMPONENTS OF A MASS SPECTROMETER

Common to most mass spectrometer instruments, there are the following units: the inlet sample system, the ion source, the electrostatic accelerating system, the detector, amplifier and display system and auxiliary equipment (pumping system).

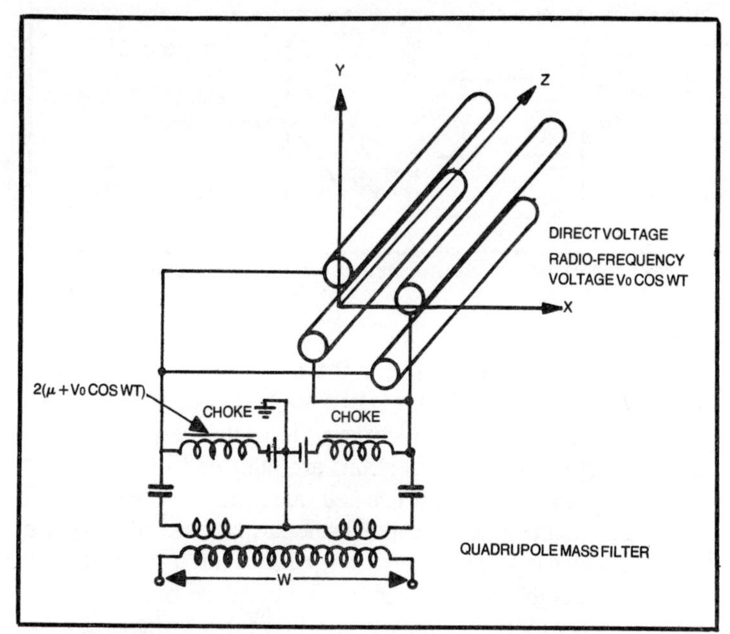

Fig. 10-4. Quadrupole mass filter.

The Inlet Sample System

Introduction of *gases* involves merely transfer of the sample from a gas bulb into the metering volume. The arrangement is a small glass manifold of known volume attached to a mercury manometer. The pressure range is generally 30 to 50 mm of Hg.

Liquid samples may be introduced by a hypodermic needle and injected through a silicone rubber dam or by a break off device which consists in touching a micropipette to a sintered glass disc under mercury. The low pressure in the reservoir draws in the liquid and vaporizes it instantly.

Solid samples can be vaporized to gaseous ions by instantaneous discharges with a power up to 100 KW by using a radio frequency (1 MHz) spark. Under these conditions, all constituents of the sample are converted to gaseous form at an equal rate without regard to their vapor pressure, thereby eliminating the possibility of preferential vaporation.

The gas sample is introduced into the mass spectrometer ion source through a leak of some kind. Generally, the leak is a pinhole in metal foil. Hogg (1969) explains the construction of a variable leak inlet system used in high resolution mass measurement.

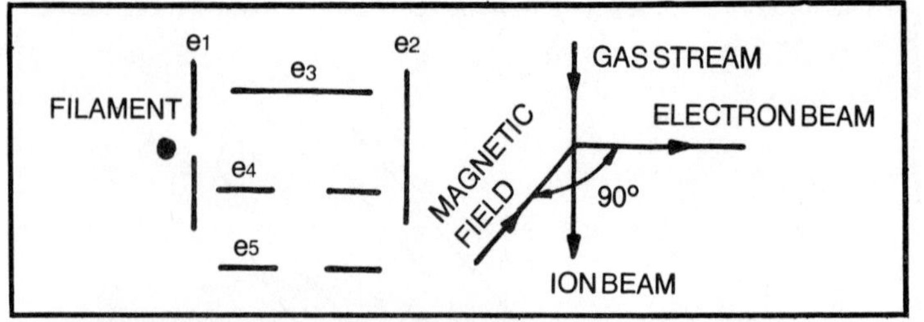

Fig. 10-5. Schematic diagram of electron gun and ionization chamber.

Ion Sources

Following the leak is the ionization chamber which is maintained at a low pressure (10^{-4} to 10^{-7} mm Hg) and at a temperature of 200°C. The electron gun is located perpendicular to the incoming gas stream. Electrons are emitted from a filament (Fig. 10-5) normally of carbonized tungsten, but for special purposes tantalum or oxide coated filaments may be used. They are drawn off by a pair of positively charged slits through which they pass into the body of the chamber. The potential present on the slits control the electron emission and the energy of the electrons. An electric field applied between these slits accelerates the electrons which, on subsequent collisions with molecules of the passing gas stream, produce ionization and fragmentation. To obtain a mass spectrum, the electric field is kept between 50 to 70 V. The electron beam is usually collimated by a magnetic field which is confined to the ionization region.

Electrostatic Acceleration System

Positive ions, which are separated from electrons by a weak electric field, are accelerated in a strong electrostatic field between the first and second accelerating slit. Voltages of the order of 400-4000 volts accelerate the ions to their final velocities of up to 150,000 miles/second, and they acquire a kinetic energy of a few thousand electron volts. Such a relatively high kinetic energy is imparted to the ions to produce an almost monoenergetic beam when it finally emerges out of the final accelerating slit which is approximately 0.076 mm in width. The electrostatic voltages are highly stabilized to an accuracy of better than 0.01%.

Ion Collecting System and Recording of Mass Spectrograph

The ion beam passing through the exit slit of the analyzer tube is normally collected in a cylinder (Faraday cage) which is connected to the grid of an electrometer tube whose output is, in turn, amplified. The use of an electrometer tube is necessitated because of an extremely low magnitude of the ion current (10^{-6} to 10^{-10}amp). Vibrating electrometers have also been used in order to convert the dc current output to an ac signal. The amplified ion current is recorded as a function of the ratio m/e on an oscillograph or pen and ink strip chart recorder. The recorder must have a provision for automatically recording peaks of widely varying amplitudes. This is achieved by using five separate galvanometers with relative sensitivities of 1, 3, 10, 30 and 100. This arrangement enables the height of any peak to be recorded with better than 1% accuracy over a range of magnitude of 1 to nearly 1000.

Small, low cost commercially available mass spectrometers have become common analytical instruments in several branches of the natural sciences. Most of these instruments are not equipped with sufficient stability for continuous high-sensitivity measurements of the maximum of a specific peak in the spectrum. In such cases, a simple circuit arrangement described by Lundsgaard and Petersen (1974) can be useful. This circuit facilitates continuous oscillatory scanning about a peak and readout of the maximum peak height. The principle of the circuit is illustrated with the block diagram in Fig. 10-6.

A triangular pulse generator feeds the mass spectrometer at the low voltage end of the mass selector potentiometer. With no signal from the generator applied to this potentiometer, the

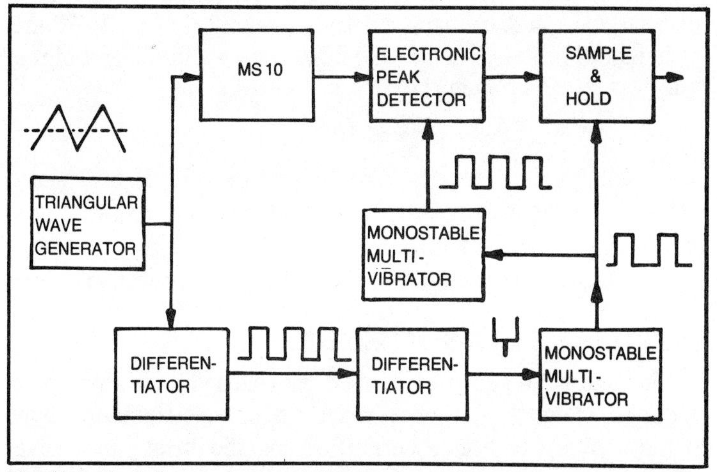

Fig. 10-6. Block diagram for a peak detector for continuous mass spectrometric measurements (after Lundsgaard and Petersen, 1974).

desired mass peak can be selected. An oscillatory scan about this peak is then obtained by application of the triangular signal from the generator. At the end of each scan, the sample and hold circuit reads the maximum signal attained under the scan from the output of the electronic peak detector. The peak detector is then reset and ready to accept the value from the next scan. Thus, the readout from the sample and hold circuit will be maximum value read during the previous scan. The generator signal is differentiated twice and fed into a monostable multivibrator to form a trigger-pulse of sufficient duration at the end of each sweep. This pulse controls the sample and hold circuit.

Vacuum System

To prevent undue scattering by collision of ions with residual gas molecules, the mass spectrometer requires a good vacuum system. Generally, separate mercury or oil diffusion pumps are employed in the source and analyzing regions of the spectrometers. Extreme cleanliness must be maintained on the surfaces in all regions of the evacuated system. The hands should not touch any interior surface; nor should any volatile lubricant be used.

Resolution

Resolution of a mass spectrometer is the mass divided by the difference in mass number between two distinguishable neighboring lines of equal height in the mass spectrum.

If two ions of mass M_1 and M_2 differing in mass by ΔM give adjacent peaks in their mass spectrum as shown in Fig. 10-7, and the height of peak is H above the base line, and the height of the valley h is less than or equal to 10% of the peak H, i.e.,

$$\frac{h}{H} \leq 10,$$

the resolution is, then, $M_1/\Delta M$. For example, if the peaks representing two masses 100 .000 and 100 .005 are separated by a 10% valley, the resolution is 100 .000/0.005, i.e., 20,000. Instruments having resolution of 150,000 are readily available.

APPLICATIONS OF MASS SPECTROMETRY

In addition to general inorganic analysis and trace analysis in inorganic chemistry, rocket-borne mass spectrometers now analyze the upper atmosphere. Perhaps the most impressive experimental achievement in mass spectroscopy is the successful mass spectrometric analysis carried out on the surface of Mars.

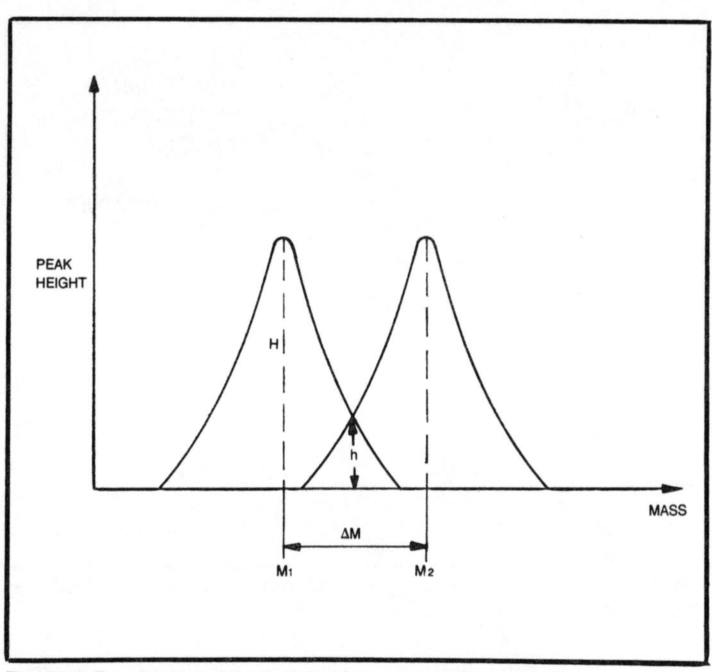

Fig. 10-7. Resolving power calculation for mass spectrometer.

Process monitor mass spectrometers with the idea of direct control of industrial processes have been constructed.

An important requirement for studying respiratory physiology in routine clinical investigations is of continuously analyzing the gas flow in a patient's breathing cycle. A mass spectrometer would perform such an analysis in an elegant form. The instrument should be capable of analyzing gases in the mass range 15 to 50 which covers the full range of respiratory gases from water to carbon dioxide. A mass spectrometer would require very little volume of gas for analysis which may be only about 15 milliliters per minute. This is a negligible fraction of the patient's respiratory minute-volume. The instrument provides a rapid response, with an overall accuracy of 2 to 3%. For respiratory work, mass spectrometer with an analyzer tube of 180° in a permanent magnetic field is quite suitable. This would provide a resolution of about 50. Figure 10-8 shows a typical mass spectrograph of exhaled air as seen on oscilloscope display. Mass spectrometers are widely used in biochemical analysis in medicine and other fields.

In the early 1950s, mass spectrometers were being used for the quantitative analysis of mixtures of the lighter hydrocarbons

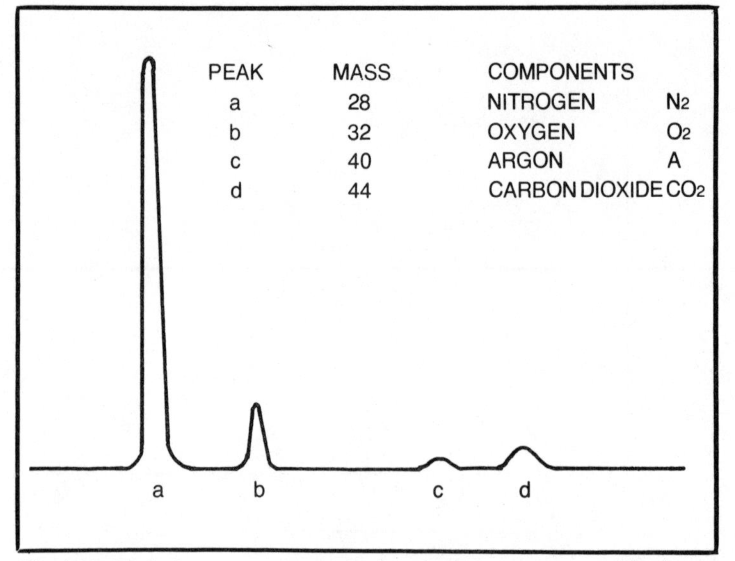

Fig. 10-8. Typical mass spectrograph of expired air as seen on cathode ray oscilloscope.

containing molecules with up to six or seven carbon atoms. The mass spectra of the various components in a mixture were taken to be linearly additive, and the analysis was worked out by setting up a set of simultaneous linear equations and solving them, usually by matrix inversion, using desk calculators. The advances in electronic digital computers have considerably transformed the situation concerning calculations of such equations. Mass spectrometry is presently used to make analysis of mixtures into different hydrocarbon types even when many hydrocarbons are present with 20 to 30 carbon atoms in the molecule. Therefore, they are used in refineries for trace element investigations, analysis of lubricating oils and quantifying the substances in mixtures of organic compounds. In addition, they are used in detecting and measuring the concentrations of pollutants in air and water.

GAS CHROMATOGRAPH-MASS SPECTROMETER GC-MS

Gas chromatography provides an excellent method for separation of a mixture's components. However, the technique does not provide direct identification of the separated components. The only information presented for each component is its retention behavior in comparison with that of the other constituents under a given set of GC operating conditions. The most usual and reliable method of identification is to isolate the compound using spectral methods like IR, UV, NMR and mass spectrometry. Mass spectrometry is the most sensitive of the spectral methods and permits the direct introduction of a gas chromatograph's effluent stream into it. Moreover, both require samples in the vapor state. GC-MS can be thus recognized as an entity in itself rather than just a combination of the two.

Problems

One of the most obvious problems encountered in combining the GC and MS is the considerable difference in operating pressures. The pressure at the exit of the column in GC is atmospheric. The flow rate of the carrier gas is 15-50 ml/min, depending upon the type of column used. Mass spectrometers, on the other hand, accept ion source pressures generally no higher than 10^{-4} torr (1 torr = 1 mm Hg). Therefore, the total flow of the GC instruments cannot be introduced into any MS of commercial type. In the earlier days some sort of splitting arrangement was used which permitted to feed only a portion of the effluent into the mass spectrometer. More efficient linkages have been reported in literature which, in principle, strip the sample from the carrier gas, discard the carriers and introduce the sample into the mass spectrometer. These are known as enrichment devices or molecular separators, since the sample gets concentrated by passage of the effluent stream through the separator.

Molecular Separator

Figure 10-9 shows the principle of a jet type molecular separator which provides a pressure dip from approximately

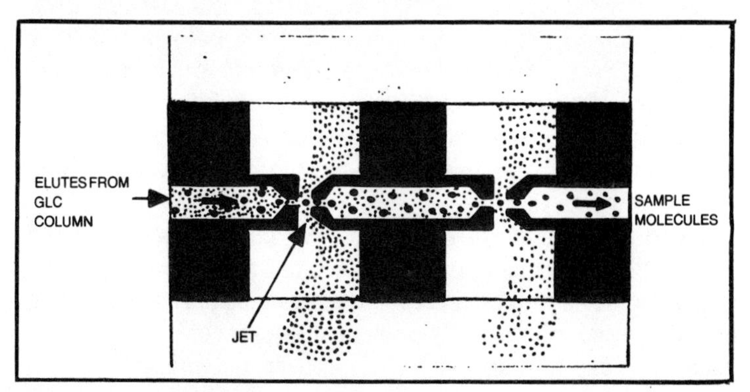

Fig. 10-9. Principle of a jet type molecular separator.

atmospheric to about 10^{-4} mm Hg. Substances eluted from the GLC column are carried along with helium carrier gas into the system which, in two stages, removes most of the helium (carrier gas) while the sample passes into the ion source chamber in a highly concentrated molecular beam. Each stage consists of an evacuated chamber connected to a pumping system. The column effluent enters the system through a very fine jet which is aligned with a small exit orifice positioned a short distance from the entrance jet. The column effluent would enter the separator as a very high speed stream of gas with the entrance jet acting as a restrictor between the GC and the separator. The carrier gas with the low-molecular weight would diffuse at a higher rate than the higher molecular weight sample. Therefore, the carrier gas would diffuse away from the line of flow and would be pumped away. The sample molecules along with the remaining carrier gas then pass into the second stage, when the process repeats itself. This type of separator is used in most of the commercial GC-MS systems. Other methods for separating out the carrier from the sample include porous tubes, Teflon tube and selective membranes.

Description of GC-MS Combination

Figure 10-10 shows a simplified block diagram of a gas chromatograph-mass spectrometer which is self-explanatory. The GC system incorporates a coiled column housed in a compact oven. The effluent stream of the GC enters the molecular separator. The separator is maintained at a controlled temperature to prevent sample condensation. Its output is connected to the MS system through a three-way valve system. The mass spectrometer is usually a single focusing instrument equipped with a 60°sector, 20 cm radius magnetic analyzer. The system for measurement of ion intensity consists of an electrometer and a wide band amplifier feeding a direct writing UV recorder. The high capacity vacuum system consists of two isolated pumping assemblies, one for the analyzer tube and one for the inlet systems. A cold trap is provided for liquid N_2 or CO_2, but this is used only for mass spectral analysis of extremely small samples.

The optimization of the parameters controlling ion source sensitivity, mass spectral scan cycle time, and chromatographic elution profile as well on-line computer systems to record, display in real time and subsequently evaluate these sets of GC-MS data is mandatory for full utilization of present potentiality of this combination.

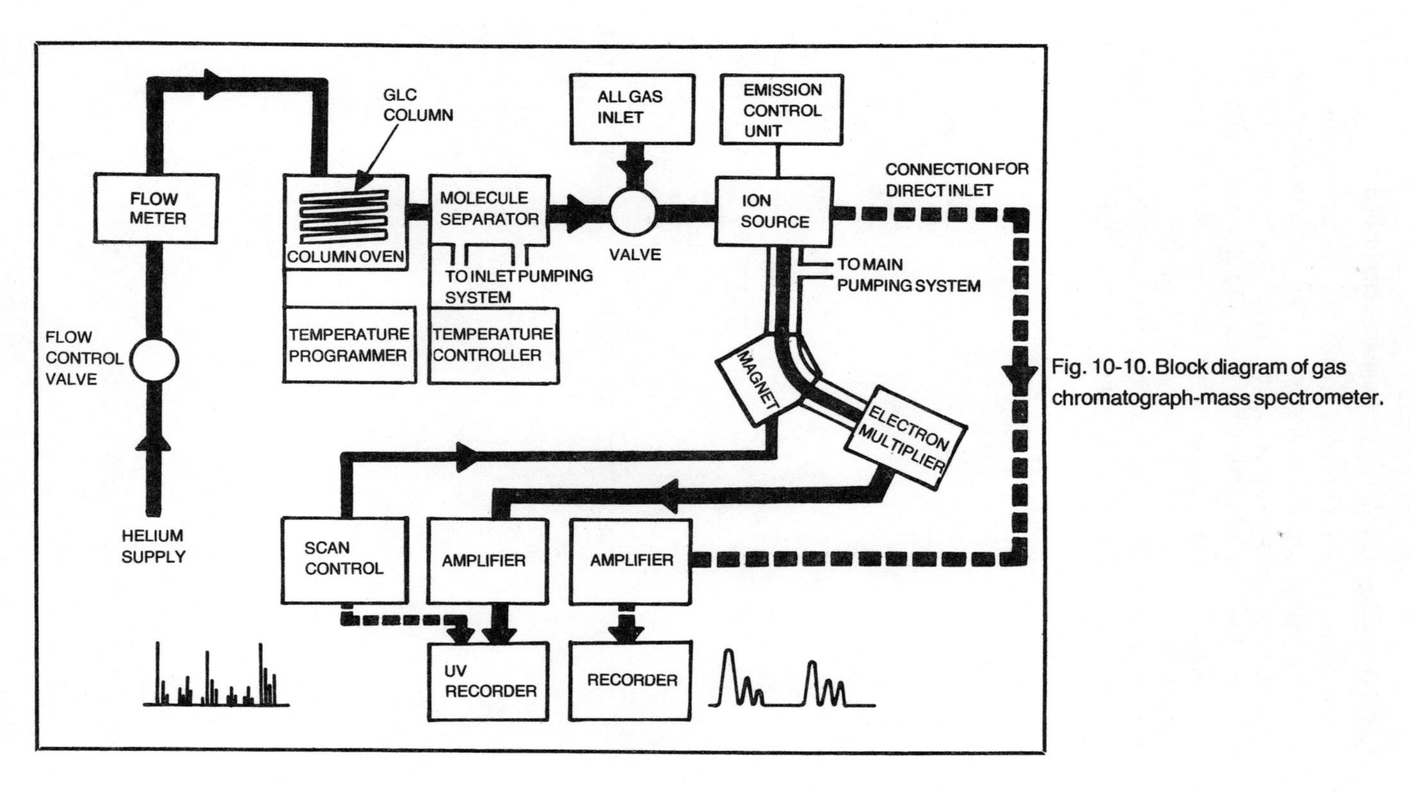

Fig. 10-10. Block diagram of gas chromatograph-mass spectrometer.

Liquid Chromatograph-Mass Spectrometer Combination

A potentially advantageous area is the effort being devoted toward development of combinations of liquid chromatographic systems with mass spectrometers. However, the levels of utilization from the point of view of the types of molecules which are amenable to HPLC separation versus interface and mass spectral characterization are still so significantly disparate that these developments are just exploratory from an analytical point of view. Burlinagame (1978) reviews the progress in this field.

Chapter 11

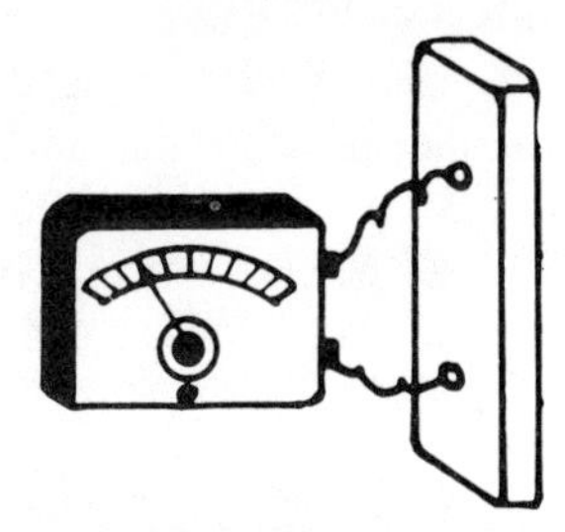

Nuclear Magnetic Resonance Spectrometers

Nuclear magnetic resonance (NMR) spectroscopy is recognized as one of the most powerful techniques for chemical analysis. The importance of this technique is reflected in the efforts that have been made to extend its applicability to smaller and smaller sample sizes. The NMR spectrometer provides an accurate and non-destructive method of determining structure in liquids and soluble chemical compounds.

The study of absorption of radio frequency radiation by nuclei in a magnetic field is called *nuclear magnetic resonance*. For a particular nucleus, an NMR absorption spectrum may consist of one to several groups of absorption lines in the radio frequency region of the electromagnetic spectrum. They indicate the chemical nature of the nucleus and the spatial positions of neighboring nuclei.

PRINCIPLE OF NMR

Elementary particles such as electrons or a nucleus are known to behave as if they rotate about an axis and thus have the property of *spin*. The angular momentum associated with the spin of the particle would be an integral or a half-integral multiple of $\frac{h}{2\pi}$ where h is *Planck's constant*. The maximum spin component for a particular particle is its spin quantum number I.

Based on the property of spin, nuclei may be classified into three types:

■ If the number of neutrons and the number of protons is even, the spin would be zero. Nuclei of this type do not give rise to an NMR signal; neither do they interfere with an NMR signal from other nuclei. Examples are C^{12}, O^{16}.

■ Nuclei that have either the number of protons or the number of neutrons as odd have half integral spin. Examples are H^1, B^{11}, P^{31}, etc.

■ Nuclei which have both the number of neutrons and the number of protons as odd, would have integral spin—for example, H^2 and N^{14}.

Nuclear Energy Levels

Since a nucleus possesses a charge, its spin gives rise to a magnetic field that is analogous to the field produced when an electric current is passed through a coil of wire. The resulting magnetic dipole or nuclear magnetic moment μ is oriented along the axis of spin and has a value that is characteristic for each kind of particle.

When a spinning nucleus is placed in a strong uniform magnetic field (H) (Fig. 11-1), the field exerts a torque upon the nuclear magnet. This would make the nucleus assume a definite orientation with respect to the external field. The torque is a vector with its direction at right angles to the plane of μ and H. This results in a rotation of the nuclear axis around the direction of the external field. This is called *precessional motion.*

Each orientation of the nucleus corresponds to a different energy level or state. The interrelation between particle spin and magnetic moment leads to a set of observable magnetic quantum states given by

$$m = I, I\text{-}1, I\text{-}2, \text{------------} - (I\text{-}2), -(I\text{-}1), -1$$

The particle will thus have $(2I + 1)$ discrete possible energy states.

The spin number for both the electron and proton is ½. Thus, each has two spin states, corresponding to I = ½ and I=– ½. The spin number of a nucleus is related to the relative number of protons and neutrons it contains. Therefore, for heavier nuclei, the spin number range from zero to at least 9/2. When brought into the influence of an external magnetic field, a particle that possesses a magnetic moment tends to become oriented such that its magnetic dipole and hence its spin axis is parallel to the field. The energy levels are a function of the magnitude of the nuclear magnetic moment and the strength of the applied magnetic field.

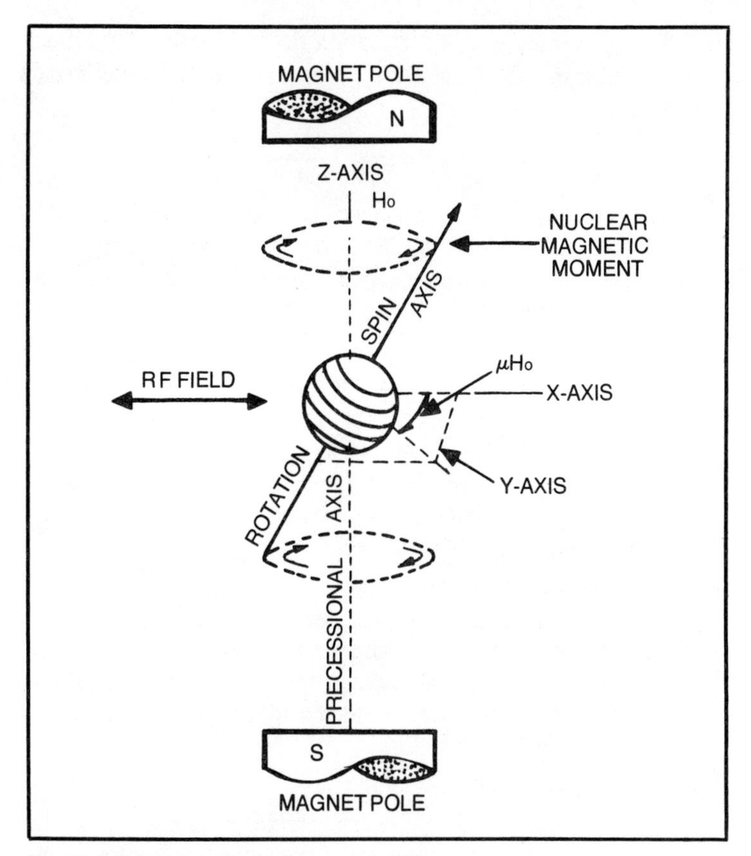

Fig. 11-1. Spinning nucleus in a magnetic field.

Resonance Conditions

When an alternating radio frequency field, superimposed over the stationary magnetic field, rotates at exactly the frequency of an energy level, the nuclei will be provided enough energy to undergo a transition from a lower energy level to a higher energy level. In general, the energy difference between two states is given by

$$\Delta E = \mu\beta \quad \frac{H_0}{I}$$

where H_0 is the strength of the external magnetic field in gauss, β is a constant called the nuclear magneton, 5.049×10^{-24} ergs gauss ^{1}, and μ is the magnetic moment of the particle expressed in units of nuclear magnetons. The value of μ for the proton is 2.797 nuclear magnetons.

The frequency v of the radiation that will effect transitions between energy levels can be determined from the Planck's equation.

$$\Delta E = hv = \mu\beta \quad \frac{H_0}{I}$$

The frequency of the resonance absorption can be adjusted by varying the value of the applied magnetic field. Difficulties in construction of large magnets limit the field to approximately 23,000 gauss. In that case,

$$v = \frac{\mu}{h} \times \beta \times \frac{H_0}{I} \quad h = 6.625 \times 10^{-27} \text{ erg sec.}$$

$$= \frac{2.7927 \times 5.05 \times 10^{-24} \times 23,000}{(6.6256 \times 10^{-27})\ (\tfrac{1}{2})}$$

$$= 95 \times 10^{6} \text{ Hz} = 95 \text{ MHz}$$

Therefore, proton will precess 95 million times per second in a fixed field of 23,000 gauss. The frequency 95 MHz lies in the radio frequency range of the electromagnetic spectrum.

With a field strength of 14092 gauss, the frequency would be 60 MHz. Similarly, for a fixed field strength of 10,000 gauss, the frequency is 40 MHz.

NMR Absorption Spectra

The NMR absorption spectra can be obtained either by changing the frequency of the radio frequency oscillator or by changing the spacings of the energy levels by varying the magnetic field. Constructing a highly stable oscillator whose frequency can be varied continuously is a difficult job. Also, there are no dispersing elements analogous to a prism or a grating for radio frequency radiation. Therefore, it is more practical to hold the oscillator frequency constant and vary the magnetic field continuously. Since for a given nucleus, frequency and field strength are directly proportional, the magnetic field (H_0) can be used equally well as abscissa for recording an NMR absorption spectrum.

The Chemical Shift

The difference between the field necessary for resonance in the sample and in some arbitrarily chosen reference compound is called the *chemical shift*. For protons, it is usual to refer spectra to tetramethyl silane (TMS_i) with extrapolation to infinite dilution in an inert solvent such as CCl_4. TMS_i gives sharp resonance line at the high field end of the range of observed proton shifts and, therefore, it does not obscure any other proton lines arising from the sample.

The chemical shift is expressed as

$$\delta = \frac{(H_{sample} - H_{TMS})}{H_1} \times 10^6$$

where H_{sample} and H_{TMS} are the positions of the absorption peaks for the sample and reference material respectively in Hz and H_1 is the radiofrequency of the signal used. The chemical shift δ units is expressed in parts per million.

CONSTRUCTION DETAILS

Figure 11-2 shows the block diagram of a nuclear magnetic resonance spectrometer. The instrument consists of the following basic components.

- A magnet which produces magnetic field in the range 10,000 to 25,000 gauss.
- Radio frequency transmitting system.

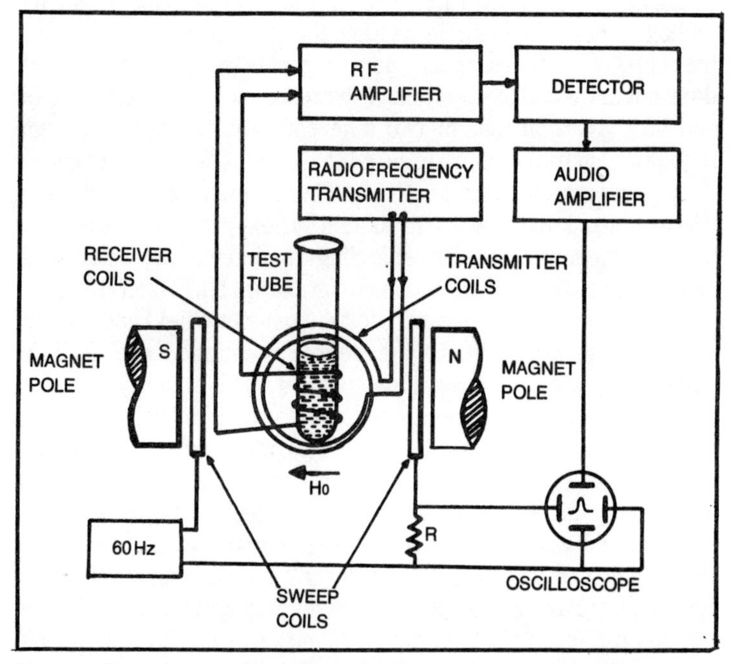

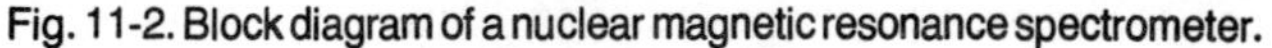
Fig. 11-2. Block diagram of a nuclear magnetic resonance spectrometer.

- The signal amplifier and detector.
- A display device, which may be a recorder or an oscilloscope.
- A non-magnetic sample holder which holds the sample.

Magnetic Field

The magnet used in these instruments may be a permanent magnet or electromagnet. Alternately, the magnetic field may be produced from super-conducting solenoids. Important requirements of the magnet are that it should be stable and homogeneous. Stability of the magnetic field is achieved by continuously compensating for small rapid fluctuations in the magnetic field with coils would around the pole faces and by controlling the temperature of the magnet to minimize thermal fluctuations. Inhomogeneity is compensated also with small magnetic fields produced, by passing dc current through small electrical coils located on the faces of the magnet, and spinning the sample to effectively average out field gradients in the direction perpendicular to the spin axis. This, however, produces spinning side bands due to modulation of the magnetic field. They can be identified from the main signal by changing the spinning rate. The magnetic field must be highly homogeneous in the sample area and the order of homogeneity is 1 part in 10^8. Decidedly, a permanent magnet is less expensive, but it does not allow the observation of the resonance of different nuclei and of a given nucleus at two different field strengths. In high resolution instruments, to ensure the field stability of the magnet, it is temperature compensated. The magnet is placed in a thermostatted oven and is surrounded by heavy thermal insulation. Special magnetic and RF shielding minimizes the effects of environmental perturbations which can disrupt NMR operation.

It is not usually necessary for the magnetic field and the RF frequency to be stable to the degree mentioned above; it is sufficient if their ratio remains constant. The stability in field-frequency can be conveniently achieved by continuously adjusting the magnetic field or the frequency to maintain the resonance condition for a particular sample. A pair of coils located parallel to the magnet faces permit alteration of effective field by a few hundred milligauss without loss of field homogeneity. Generally, the field strength is changed linearly and automatically with time. For a 60 MHz instrument, the sweep generator periodically sweeps the magnetic field in the vicinity of 14,092 gauss with a sweep of 1000 Hz or some integral fraction thereof.

For the study of nuclear magnetic resonance, there are great advantages in working at the highest possible applied magnetic field strength, especially for the resonances of nuclei other than hydrogen and fluorine (Halliday et. al., 1969). The expected signal to noise ratio increases rapidly as the strength is increased (Abragam, 1961), provided that the resonant frequency does not become so high as to make the problem of coupling radio frequency power in and out of the sample very difficult. For higher field strengths, it is better to use a superconducting solenoid. Magnets of this kind also have the advantage of quite remarkable stability when operated in the persistent mode. The superconducting solenoid used by Halliday is 26 cm long, with a 3 cm bore and outside diameter 8.36 cm. It is wound with Nb-Zr 25% wire and produces a field of 5 with a current of about 20 A.

Permanent magnets can yield maximum magnetic fields approximately 14,000 gauss whereas electromagnets can produce fields up to about 24,000 gauss, Superconducting solenoids have been used to give approximately 70,000 gauss. In case of electromagnets, the pole pieces are about 12 inches in diameter and are spaced about 1.75 inches apart.

Sthanapati et al. (1977) describe a current regulated magnet power supply for magnetic resonance and susceptibility studies. The power supply system provides a current of 0-5A into a 40 ohm magnet coil together with linear current sweep facilities. The dc drift is about 1 part in 10^5 per hour and low frequency peak to peak noise is less than 1 part in 10^6.

The Radio-Frequency Transmitter

The *radio-frequency transmitter* is a 60 MHz crystal controlled oscillator. The RF signal is fed into a pair of coils mounted at right angles to the path of the field. The coil that transmits the radio-frequency field is made in two halves to allow insertion of the sample holder. The two halves are placed in the magnetic gap. For high resolution work, the transmitted frequency must be highly constant to about 1 part in 10^8. The oscillator is of low power, generally of less than 1 watt. The basic oscillator is (Fig. 11-3) usually crystal controlled at a fundamental frequency of 15 MHz. It is followed by a buffer doubler, the frequency being doubled by tuning the variable inductance to the second (30 MHz). It is further connected to another buffer-doubler with the collector inductance, tuned to 60 MHz. A buffer amplifier T4 is provided to avoid circuit loading on the tuned doubler. Precision resistors are used in the transmitter for their low noise characteristics.

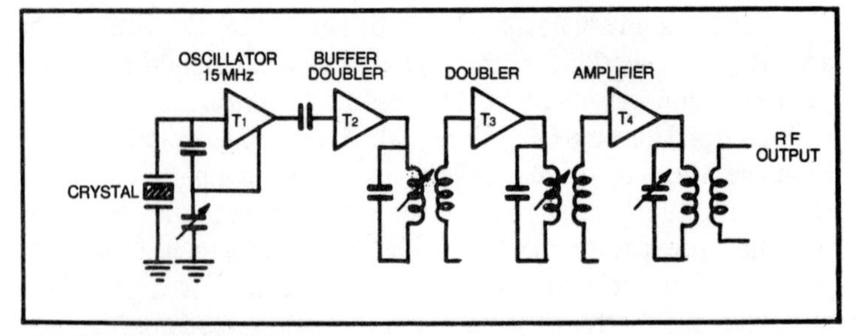

Fig. 11-3. Transmitter circuit.

The Signal Amplifier and Detector

The radio frequency signal produced by the resonating nuclei is detected by means of a coil that surrounds the sample holder. This coil consists of a few turns of wire and is placed at right angles to the source coil and the stationary field to minimize pickup from these fields. Even so, the coupling between transmitting and receiving coils cannot be completely eliminated and some leakage is always known to be present. The problem of this coupling, is solved to some extent by using devices called *paddles* which act as inductors mutually coupled to both the receiver and transmitter.

The signal results from the absorption of energy from the receiver coil when nuclear transitions are induced and the voltage across the receiver coil drops. This voltage change is quite small and must be amplified in a radio frequency amplifier before it can be displayed.

Referring to Fig. 11-4, after impedance matching in a transformer T_1, the input RF signal is amplified in a three stage low noise a cascode amplifier. The output of the last RF stage is applied to the AGC and spectrum detectors. The AGC voltage is fed back to the previous stage through an AGC amplifier, which is kept cutoff due to a delaying bias until a pre-determined signal level is reached. This delayed AGC allows maximum gain on weak signals and prevents overdriving on strong signals.

The Display System

The detected signal is applied to the vertical plates of an oscilloscope to produce the NMR spectrum. The total amplification required is usually of the order of 10^5. The spectrum can also be recorded on a chart recorder. The NMR spectrometers have built-in electronic integrators for measuring the relative areas under the peaks. The integrated spectrum is a step function with the height of each step being directly proportional to the area under the peak corresponding to that step.

Recording Unit

The recorders with NMR spectrometers are usually self-balancing null type potentiometric recorders having an FET chopper amplifier in the input stage. Phase detected NMR signals are given to the amplifier. The chopper is driven at the line frequency and it produces a square wave output signal which is phase compared to the input signal and applied to a source follower. The signal is amplified and applied to the drivers and power amplifiers. The power amplifier stage obtains filtered voltage from a rectifier circuit to establish a phase relationship with the chopper output signal and the power line frequency. The output of the drivers is 0 when no NMR signals are present. A signal input causes a square wave at the chopper and, depending upon the polarity, it results in a lead or lag voltage with reference to the power line frequency. The servometer then rotates in a direction so as to cause the difference voltage between NMR and the reference signal to become zero. The rotation of the servometer causes vertical movement of the pen. The horizontal movement of the pen carriage is controlled by a synchronous sweep motor. A sweep potentiometer is turned by the gear train in synchronism with the horizontal movement of the pen carriage. Variable sweep width can be selected with a manual switch by controlling the amplitude of the dc voltage applied to the probe sweep coils from the sweep potentiometer.

The chart size in an NMR spectrometer is usually around 8 × 11 inch with its absicissa calibrated in Hertz. The recorder has

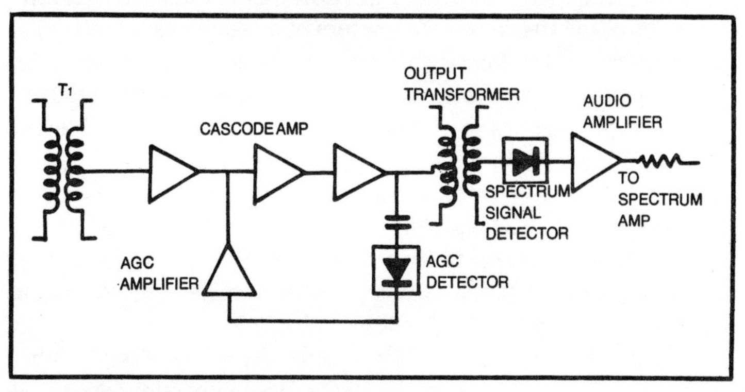

Fig. 11-4. Receiver and detector circuit.

calibrated sweep width ranges and sweep speeds selected on choice.

Sample Holder

The *sample holder* for NMR studies consists of a glass tube, generally of 0.5 mm outer diameter. Microtubes for smaller sample volumes are also available. The sample is invariably in the liquid form. In case studies are to be made on solid or gaseous samples, the solids may be studied above their melting point and gaseous samples below their liquefaction point.

The sample holder is placed in a sample probe which also contains the sweep source and detector coils. This ensures reproducible positioning of the sample with respect to these components. The sample probe is also provided with an air driven turbine for rotating the sample tube along its longitudinal axis at several hundred rpm. This rotation averages out the effects of inhomogeneities in the field and provides better resolution.

For continuous monitoring of a chemical process, it is necessary to use a special sample cell situated in the detector coil of the spectrometer through which the solution is made to flow continuously. McIvor (1969) designed a flow probe for NMR spectroscopy which could be readily fitted to the Perkin Elmer model R.10NMR spectrometer. It incorporates a flow system which uses the magnet of the spectrometer as the polarizing field. The sample cell is made from a soda glass tube of a diameter 0.180 inch and is locked at the center of the polarizing coil. The latter is made of nylon tube within inner diameter of 0.20 inch. The detector coil made from four turns of 40 SWG copper wire is wound directly on the outside of the cell.

It is frequently necessary in NMR studies to determine the behavior of the measured parameters of a sample as a function of temperature. Often the range from room temperature down to the boiling point of liquid nitrogen is appropriate for these studies. This necessitates sample temperatures to be kept stable for several hours, while allowing rapid changes to be made when passing a cooled gas over the sample or using a controlled heat leak from the sample to a liquid nitrogen reservoir. Norris and Strange (1969) describe a cryostat which employs the direct injection of liquid nitrogen into it. This leads to a considerable improvement in the efficiency of the cooling system.

In the Varian T-60A NMR spectrometer, the sample tube, with spinner attached, is simply dropped into a chimney-like outlet and floats down on to a cushion of compressed air, eliminating the possibility of damage to inner components of the probe. Spinning speed is adjustable and the speed can be optically monitored and displayed on a directly calibrated meter.

VARIAN T-60A NMR SPECTROMETER

The Varian T-60A NMR spectrometer (Fig. 11-5) is designed for use with protons only over the limited chemical shift range of 2000 Hz below and 500 HZ above tetramethyl silane at 60 MHz. The crystal-controlled 60 MHz transmitter is coupled to the probe and sample receiver through the directional coupler. The sine wave from the 5 kHz oscillator is applied through the field modulator to the ac sweep coils, thereby modulating the H_0 field at a 5 kHz rate. The signal in the sample receiver coil would have a frequency of 60.005 MHz, which is both amplitude and phase modulated by tne nuclear resonance being observed. After detection in the receiver, which removes the 60 MHz component, the signal is amplified in the spectrum amplifier and then compared in the phase detector with a reference signal derived from the 5 kHz oscillator. Since increases in RF power result in phase shift on NMR displays, the reference signal is made continuously variable in the phase shifter, which supplies a defined baseline for display.

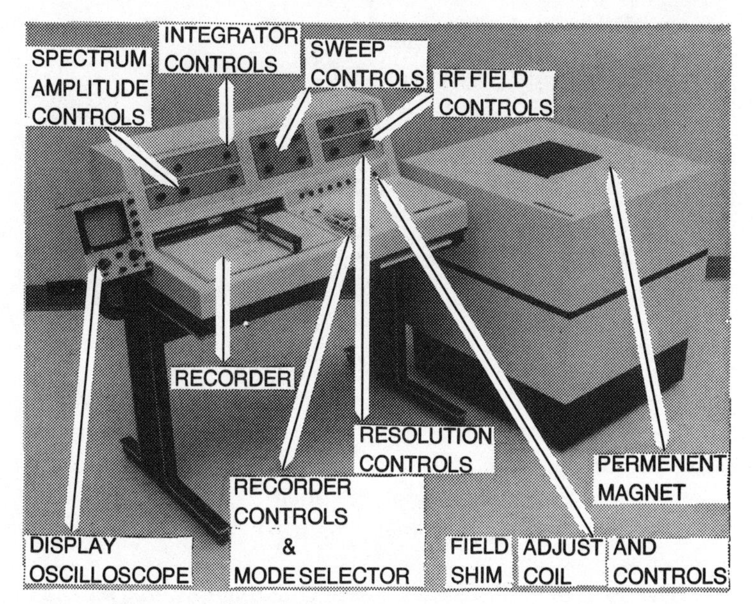

Fig. 11-5. Varian 60 MHz NMR spectrometer.

The resulting phase corrected absorption signal is applied to the dc amplifier, noise filter and recorder for spectrum display or to the integrator and recorder to obtain the corresponding integral.

An air turbine sample spinner is used to average the effects of field inhomogeneities which may exist in the plane perpendicular to the tube axis, thereby improving the system resolution. The spinner speed tachometer samples the spinner speed and the rate is indicated on a front panel meter.

NUCLEAR SIDEBAND OSCILLATOR OPERATION (NSBO)

In the previously described system, the 5 kHz signal which emerges from the receiver-detector can be amplified and phased so that, when applied to the modulation coils in the probe, a self sustaining oscillation is established in the loop comprised of ac sweep coils, sample tube, receiver coil, receiver and field modulator. Such a closed feedback loop is called a *nuclear sideband oscillator* and the spectrometer is said to be locked to a nuclear resonance line in the sample. It is convenient to switch to NSBO operation when adjusting the magnetic field homogeneity for optimum resolution. After homogeneity has been optimized, the NSBO loop is opened and the spectrometer returned to normal operation.

SENSITIVITY ENHANCEMENT FOR ANALYTICAL NMR SPECTROSCOPY

The sensitivity of an instrument is a measure of the ability to differentiate signal from the surrounding noise. It is usually measured as the ratio of peak signal amplitude to rms noise.

$$\text{Sensitivity} = \frac{S}{N} = \frac{\text{Peak signal height}}{\text{rms noise}}$$

Equation 11-1

$$\text{where rms noise} \approx \frac{\text{peak to peak noise}}{2.5}$$

Equation 11-1 shows that the sensitivity may be improved either by increasing the signal amplitude or by reducing the observed noise. Both of these are considered in the initial design of a NMR spectrometer. However, special operating techniques allow the NMR operator to optimize them for a particular sample and spectrometer system.

Erwine (1975) discusses three operating techniques for sensitivity enhancement: optimization of sample volume, optimization of instrumental parameters and time averaging. Erwine studied the spectrum of a 1 mg sample of Coral (a commercial phosphorus insecticide with molecular weight 363) by investigating the effects of optimizing various parameters on the EM-360, a low cost analytical NMR spectrometer. He showed that the first two enhancement techniques (optimization of sample volume and spectral parameters) can often enhance the sensitivity by a factor of 10 over the normal operating conditions. Further use of signal averaging with a computer can achieve a further increase of 10 within reasonable time limits. This can significantly increase the applicability of analytical NMR to the investigation of small samples to give the NMR operator the best possible use of his spectrometer.

Optimization of Sample Volume

Better sensitivity can be achieved with good resolution by concentrating the sample inside a specially designed NMR microcell with a volume which approximates the active volume of the receiver coil and with a shape which eliminates the effects of the magnetic field distortion. The most common shapes used are a spherical cell with a 4 mm diameter and a 1 mm capillary cell centered along the axis of the coil. These cells can be positioned inside the standard 5 mm cylindrical tube. Preferably, a liquid of similar susceptibility is placed outside the cell to minimize movement of the microcell within the sample tube.

Optimization of Instrumental Parameters

It is obvious that noise may be reduced by proper choice of filter setting. Filtering operates on the basis that the frequency spectrum of the noise generally occupies a wider band than the signal spectrum. Therefore, suppression of frequencies, which contain no signal, increases the effective S/N ratio.

The resonant nuclei in the active region of the receiver coil absorb RF energy over a small range of applied field-to-frequency ratios because of the slight differences in their molecular magnetic environments. In a swept NMR experiment, either the applied magnetic field or frequency is varied to make the different types of protons resonate sequentially. The rate of change of the field or frequency may be adjusted by the operator for maximum sensitivity.

Sensitivity can be enhanced by maximizing signal amplitude. The amplitude of the NMR signal caused by the absorption of RF energy at the resonance frequency depends upon the power of the RF energy applied. It will, however, depend upon the sweep rate selected. Therefore, optimization of signal amplitude with RF power should only be considered for a given sweep rate.

Use of Signal Averaging Technique

The signal averaging technique is based on the principle that the signal, if added coherently, will increase linearly with the number of scans, N, while the noise, being random, will add as the square root of the number of the scans. Thus, the sensitivity increases by the square root of the number of scans ($N^{½}$). To have a long measuring time for increasing sensitivity, it is better to have several fast scans than a slow signal scan.

Signal averaging would involve a system, which must repetitively scan the spectral region of interest; some storage device for storing the spectral information; and a system, which would coherently add the individual spectra. These are usually accomplished by a small computer (multichannel pulse height analyzer) suitably modified to provide an MNR sweep and coherent spectral addition. These special purpose computers are commercially available, such as the Varian 547 Signal Averager.

SPIN DECOUPLER

A spin decoupler permits spin-coupled interactions between nuclei to be eliminated through the use of a technique called nuclear magnetic double resonance. In this technique, specific regions of the spectrum are irradiated by a strong component of radio frequency power or RF sideband power, causing the multiplet patterns from spin coupled resonances in other regions to coalesce into less complex patterns that are more easily assignable. In principle, double irradiation helps to solve the problem introduced by complicated proton couplings. The second radio frequency field is adjusted to the resonant condition for the group whose coupling is to be eliminated. Under these circumstances, the proton being split sees only one equivalent state and single peak results.

In addition to the usual H_1 RF field used to examine the nuclear magnetic resonance response to the sample, a second RF field, H_2, is introduced simultaneously into another region of the proton spectrum. The spectrum is recorded by sweeping the main magnetic field H_0 and the H_2 field while the H_1 field is held constant. H_1 modulation of the H_0 field generates the sidebands for the observing field. For frequency sweep, a voltage proportional to the magnetic field is applied to a voltage controlled oscillator which supplies the decoupling frequency. The magnitude and frequency of the H_2 are controllable.

Figure 11-6 shows a block diagram of spin decoupler arrangement. The decoupling frequency to the field modulator is supplied by a 5 kHz voltage controlled oscillator. This oscillator is a Wien bridge oscillator whose gain stability is provided by a thermistor in the feedback loop. This frequency is amplified and applied to the ac sweep coils to generate the H_2 or decoupling field. The decoupling power level is adjusted at the oscillator output. In a frequency sweep operation, tracking of the oscillator with the sweep is accomplished through the frequency control circuit. As the dc component of the field is swept, a differential voltage is developed across the resistor R placed in series with the sweep coils. The differential voltage is amplified in the differential amplifier and applied to the error amplifier. The error amplifier receives input from three different sources—the frequency control setting, the differential voltage from the dc sweep coils, and the feedback voltage from the frequency to voltage converter.

The frequency control sets the dc operating level that produces the desired frequency from the voltage controlled

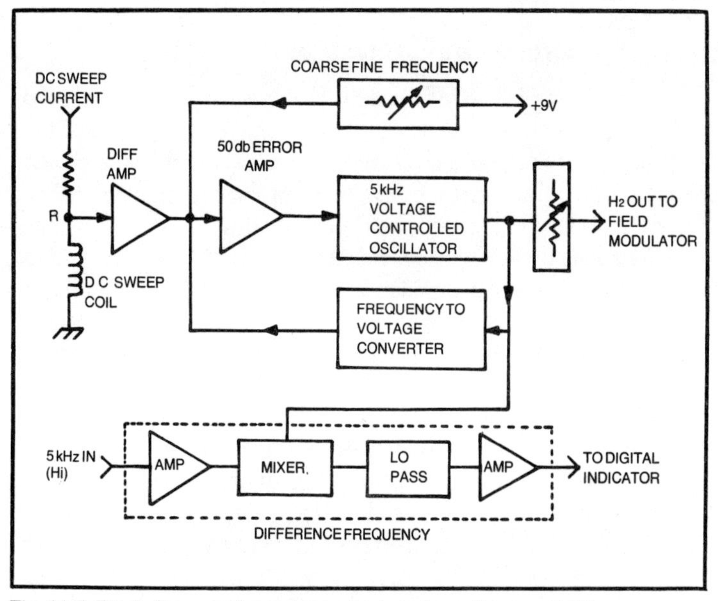

Fig. 11-6. Block diagram of spin decoupler (courtesy of Varian, USA).

oscillator. The dc sweep voltage from the differential amplifiers adds to the operating level and the oscillator follows the sweep. The frequency of the oscillator is sensed by the frequency to voltage converter which applies an error signal to the error amplifier when the oscillator drifts or overshoots. The frequency to voltage converter provides a direct conversion of a sinusoidal signal to a dc signal. In this circuit, the input signal is squared by a Schmidt trigger circuit and differentiated to trigger a monostable multivibrator. The pulses from the monostable multivibrator are applied to a low-pass RC filter which smoothes the pulses to a dc voltage.

A difference frequency circuit is included to provide a readout frequency for the digital indicator. This circuit provides a means of mixing two audio signals and a filter to recover the low frequency component. The readout frequency is the difference frequency between the 5 kHz spectrometer control oscillator (H_1) and the spin decoupler oscillator frequency (H_2). The spin decoupler module temperature is maintained to ± 1°C by a temperature controller circuit. This circuit is a dc amplifier which compares a dc reference voltage to the voltage change produced by a thermistor mounted in the spin decoupler oven. The output of the dc amplifier is applied to a power transistor which controls the power supplied to the oven heaters.

FOURIER TRANSFORM NMR SPECTROSCOPY

In conventional NMR spectrometers for high resolution studies, the spectra are scanned by sweeping the frequency or the field through the region of NMR absorption. From the point of view of sensitivity, this is an inefficient mode of excitation, since only a very narrow band of frequencies is contributing a signal at any one time. Multichannel excitation and detection would be expected to improve the sensitivity considerably. A spectrometer consisting of a large number (M) of transmitter frequencies, each matched by a suitable receiver channel, can be built. However, the method proves rapidly uneconomical as M increases. The Fourier transform accomplishes the same thing (Ernst and Anderson, 1966) in a much more satisfactory manner. In this method, a strong radio frequency pulse excites the entire range of precision frequencies and these frequencies can be detected in the receiver. Under certain conditions, the Fourier transform of this precision signal is identical with the steady-state slow passage spectrum. The transformation process plays the role of a multichannel receiver, eliminating the need for a cumbersome array of a narrow band filters and detectors.

Any complex waveform can be converted to a spectrum of frequencies by Fourier transformation. In NMR, the waveform in question is the superposition of a set of nuclear precision frequencies with amplitudes decaying due to relaxation and field inhomogeneity. Fourier transformation can be accomplished by using a spectrum analyzer. However, the spectrum would show only the absolute value of the XY component of magnetization, with no account taken of the phase information which often results in distortion of the lines. The transformation is carried out most satisfactorily on a digital computer. Usually, dedicated computers are used for this purpose.

Sensitivity enhancement is the major attraction of the Fourier transform technique. Thus, for proton NMR, it is possible to consider samples that are an order of magnitude less concentrated than the previous limit, and this has great importance in the biochemical field. An XL-100 Pulsed-Fourier transform NMR spectrometer from Varian is a high resolution 23.5 KG spectrometer, which can be operated in the frequency range from 6 to 100 MHz. This instrument is designed around a fundamental clock frequency. A 15.4 MHz crystal oscillator generates the deutrium resonance frequency at 23.5 kilogauss. The deutrium resonance signal is then used to lock the magnetic field to the clock frequency. The choice of a 15 inch or 12 inch magnet system permits to accommodate a wide size range of sample tubes. This permits you to optimize sensitivity versus availability of samples. For Fourier transform, the spectrometer is provided with programmable RF pulse hardware and a digital computer programmed to control, acquire and transform spectral data.

USE OF COMPUTERS WITH NMR SPECTROMETERS

Computer equipped NMR spectrometers have been used for a variety of tasks. Simple time-averaging for signal-to-noise enhancement has been extremely helpful in continuous wave NMR studies. It only requires relatively simple hardware. With Fourier transform NMR becoming more common, general purpose digital computers are required for data acquisition and data display. It is possible to place under computer control such functions as pulse timing, delay and acquisition timing, digitization rate, filter bandwidth, transmitter and decoupler offsets, receiver gain, noise bandwidth, and plotter and pulse sequence. This leaves only a few essential manual adjustments such as establishing the NMR lock for field/frequency ratio stabilization and trimming up the field homogeneity controls. Varian CFT-20 Fourier transform NMR spectrometer has 16K software which provides automated unattended performance. This increases productivity and also improves the signal-to-noise ratio.

Chapter 12

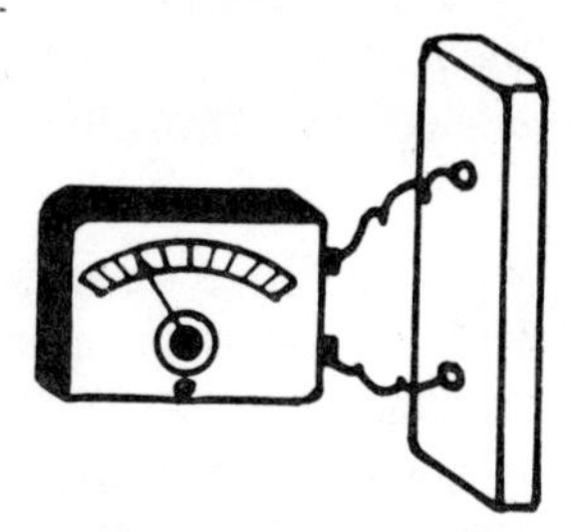

Electron Spin Resonance Spectrometers

Electron spin resonance (ESR) is the study of magnetic dipoles of electronic origin by applying, usually, fixed microwave frequencies to a sample residing in a varying magnetic field. It is also known as *electron paramagnetic resonance* (EPR) spectroscopy and is a valuable research and analytical tool in chemistry, physics, biology and medicine. It is used in the study of molecular structure, reaction kinetics, molecular motion, crystal structure, electron transport and relaxation properties. Wasson and Corvan (1978) have presented a detailed review of ESR applications and spectral analysis techniques.

MAGNETIC MOMENT AND SPIN OF ELECTRONS

The phenomenon of electron spin resonance is based on the fact that an electron possesses a spin, and associated with the spin there is a magnetic moment, the value of which is called the *Bohar magneton.* The atoms, ions or molecules having an odd number of electrons exhibit characteristic magnetic properties which arise from the spinning or orbiting action of the unpaired electrons about the nucleus. When a strong magnetic field is applied to the unpaired spins of an electron, the electrons will be split into two groups. In one group, the electron dipoles or magnetic moments of the electrons are aligned either parallel or anti-parallel to the direction of the external magnetic field. The electrons will precess about the axis of the magnetic field at a frequency proportional to both the applied magnetic field and the electron magnetic moment.

If a second weaker radio frequency alternating magnetic field having the frequency of precession of the electron is applied at right angles to the fixed magnetic field, the resonance occurs. At resonance, the absorption of energy from the rotating field causes the spin of the electrons to flip from the lower energy level to the higher level. The two levels are separated by

$$E = h\nu = 2\mu H$$

when h is Planck's constant and ν is the frequency. In comparison to NMR, the electron has a much smaller mass and larger magnetic moment than a proton. For a given magnetic field, the precession frequency is, therefore, much higher. For a free electron, the frequency of absorption is given by

$$\nu = \frac{2\mu H}{h}$$

$$= \left(\frac{2\mu}{h}\right) \times H = (2.8026 \times 10^6) \times H$$

In a field of 3400 gauss, the precession frequency is approximately 9500 MHz.

In actual practice, the radio frequency is in the microwave region and is held at a certain constant value. The magnetic field strength is varied to obtain conditions where resonance occurs.

The incident radiation is absorbed by the electrons in the lower energy level and they jump into higher energy state.

In general, if n_1, the population of the ground state, exceeds n_2, the population of the excited state, a net absorption of microwave radiation takes place. The signal would be proportional to the population difference $(n_1 - n_2)$.

By the Boltzmann distribution law, the population ratio in the two states is given by

$$\frac{n_1}{n_2} = e^{-2}\mu H/Kt$$

where K is the Boltzmann constant. The sensitivity of measurement is greatly enhanced by using a high magnetic field.

In most of the substances, chemical bonding produces paired electrons as they would be either transferred from one atom to another to form an ionic bond, or electrons are shared between different atoms to form covalent bonds. The magnetic moments and spins of paired electrons point in opposing directions with the result that there is no external spin paramagnetism. However, in a paramagnetic substance having an unpaired electron, resonance occurs at definite values of the applied magnetic field and incoming microwave radiation. The deviation from the standard behavior of the unpaired electron due to the presence of magnetic fields in its surroundings gives knowledge about the structure of the substances under study.

The ESR measurements are made by placing the sample under study in a resonant cavity positioned between the pole pieces of an electromagnet. The microwave frequency is set to a matched condition with the help of a tuning device. The magnetic field is varied to bring about resonant conditions and the microwave energy absorbed by the sample is plotted on a recorder. The spectrum is then analyzed to determine the behavior and mechanisms associated with the unpaired electrons' interaction with the external magnetic field and its environment.

For routine investigations with the ESR spectrometer, it is general practice to employ an X-band frequency (8.5-10 GHz). This cannot be generated by using ordinary vacuum tubes, and special tubes capable of operating at microwave frequencies are employed.

The limitations of triode, pentode and similar tubes that arise at very high frequencies as a result of transit time effects are avoided by employing types of tubes which make the use of transit time in achieving their normal operation. One such tube is the ***reflex klystron*** or reflex oscillator, as it is sometimes called, and it requires only a single resonant cavity. Since it has an efficiency of only a few percent, the reflex klystron is essentially a low-power device, typically being used to generate 10 to 500 mW. The reflex klystron is particularly satisfactory for use in the frequency range 1000 to 25000 MHz. Pierce (1945) describes the theory of reflex klystron in detail.

CONSTRUCTION OF AN ESR SPECTROMETER

Figure 12-1 shows the block diagram of an ESR spectrometer. The sample is irradiated by microwave energy from the klystron in the microwave bridge. A klystron normally operated at 9.5 GHz generates a microwave field. The magnetic field at the sample is modulated at 100 kHz. The klystron output passes through an isolator, a power leveler and a directional coupler. The field is applied to the resonant cavity, which is connected to one arm of the microwave bridge. During the magnetic field scan, when field

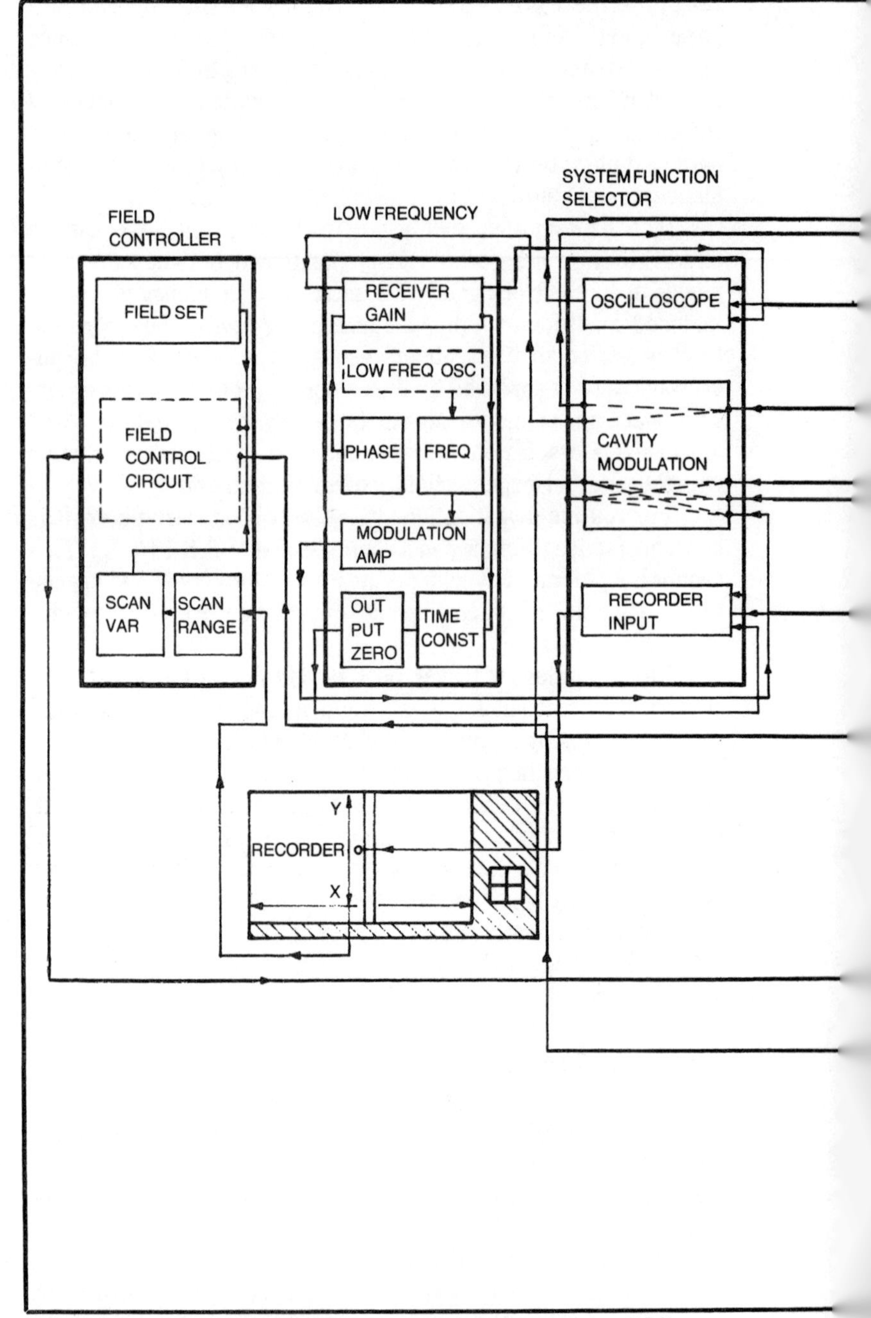

Fig. 12-1. Block diagram of an ESR spectrometer (courtesy of Varian, USA).

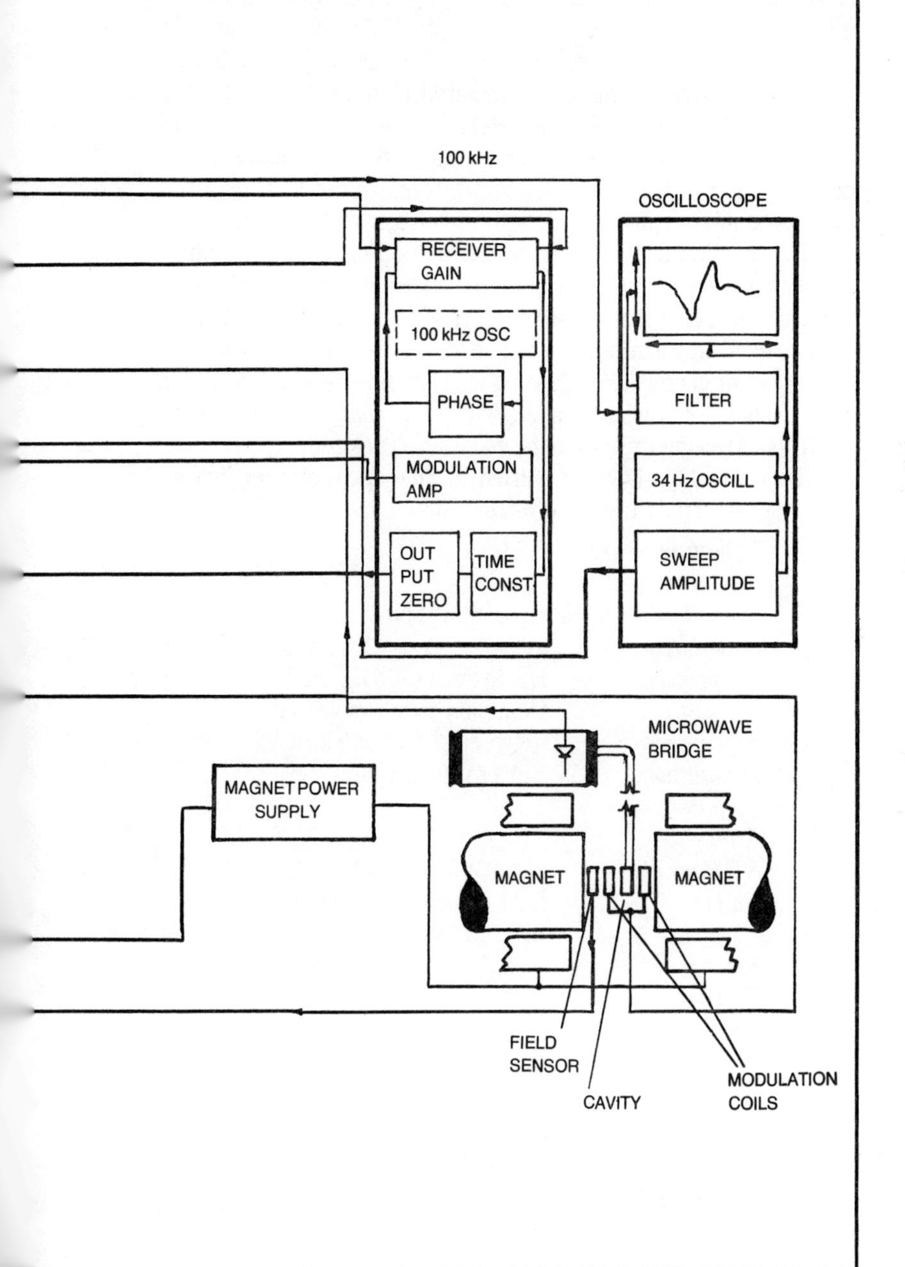
100 kHz
OSCILLOSCOPE
RECEIVER GAIN
100 kHz OSC
PHASE
MODULATION AMP
OUT PUT ZERO
TIME CONST
FILTER
34 Hz OSCILL
SWEEP AMPLITUDE
MICROWAVE BRIDGE
MAGNET POWER SUPPLY
MAGNET
MAGNET
FIELD SENSOR
CAVITY
MODULATION COILS

intensity reaches the value required to induce electron spin resonance in the sample, a change occurs in the amount of microwave energy absorbed by the sample. This causes a change in the microwave energy reflected from the cavity. The reflected microwave energy, which is modulated at the field modulation frequency, is directed to the detector crystal. The detector is usually a silicon tungsten crystal rectifier. After detection at the crystal, the resulting 100 kHz, which contains the ESR information, is amplified in the preamplifier circuit and applied to the receiver section of the 100 kHz modulation unit. The amplified signal is phase detected to obtain the spectrum which appears as a deflection on the Y-axis of the recorder. The field scan potentiometer is linked mechanically to the recorder X-axis. The X-axis is calibrated in gauss. The ESR signal may also be applied to the oscilloscope for visual display. The field controller accurately controls the magnetic field to the value set as desired.

Figure 12-2 shows a Varian E-line spectrometer whose important specifications are given below:

Operating Frequency	8.8 to 9.6 GHz
Sensitivity	5×10^{10} H* Spins
Modulation Frequencies	100 kHz, 1 kHz, 270 Hz, 35 Hz. In lieu of 270 Hz, another frequency may be selected between 35 Hz and 3 kHz.
Modulation Amplitude Range	5 mG to 40 G in steps.
RF Power to Cavity	200 mW calibrated over full frequency range. ± 15 mW 8.9 to 9.5 GHz.
Field Scanning Rates	½, 1, 2, 4, 8, 16 min; ½, 1, 2, 4, 8, 16 hrs.
Field Scanning Ranges	200 mG to 10 kG or 20 kG.
Receiver Gain Range	5 decades, 10 settings per decade.
Receiver Time Constants	100 kHz; 0.003 to 100 sec. Low frequencies: Up to 100 sec.

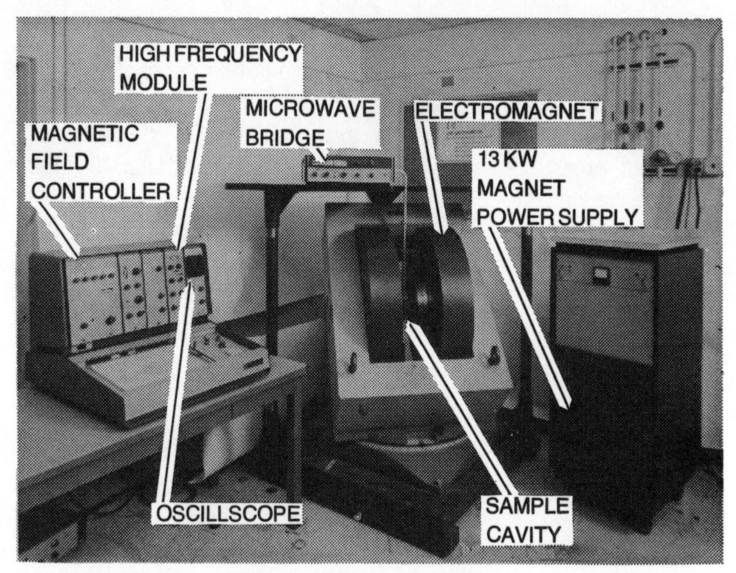

Fig. 12-2. ESR spectrometer (courtesy of Varian, USA).

Stability of Field	Within 1 ppm of set field or 3 mG whichever is greater.
Resolution	Depends on sample size, modulation frequency and magnet configuration. At 100 kHz limited by 100-kHz sidebands (70 mG).
Oscilloscope Sweep Width	0.2 to 40 G (34 Hz sawtooth).
Magnetic Field Homogeneity	6-inch magnets: within 50 mG at 3,400 G. 9-inch and larger magnets: 14 mG at 3,400 G.
Magnet Air Gap	2.625 inches on 6-inch magnet, 3 inches on 9-, 12- and 15-inch magnets.

* H is defined as the signal linewidth in gauss at half maximum absorption with 1-second integration time, the sample having negligible dielectric loss.

MAGNET AND THE MAGNETIC FIELD CONTROLLER

The magnet used in the ESR spectrometers is usually of the electromagnet type. It provides a homogeneous magnetic field which can be varied from 200 milligauss to 20 kilogauss in calibrated steps. Stability of 1 part in 10^6 is satisfactory for adequate resolution of ESR spectra.

The *magnetic field controller* provides direct control and regulation of the magnetic field in the air gap. It is an ac carrier type servo system that accurately controls the magnetic field. Figure 12-3 shows the schematic diagram of the magnetic field controller in an E-line spectrometer.

The driver amplifier supplies a 30 mA rms, 1230 Hz exciting current through the field set reference resistor (R), the Hall effect magnetic field sensor element and the primary of the scan voltage transformer (T_4). A field reference voltage is developed across R. Similarly, a field scan reference voltage is developed across the constant input impedance of scan range attenuator by the transformed reference current in the secondary winding of T_4.

The output of the summation circuit (i.e. input to the error amplifier) is the algebraic sum of the output of the Hall-effect field sensor (transformed by T_5) added to the field set and *field scan voltage*. The sum of these three voltages is amplified by the error amplifier and applied to the input of the phase sensitive detector. When the output of the Hall-effect field sensor is less than the sum of the field set and field scan voltages, the output of the phase sensitive detector is of a polarity which "turns on" the magnet power supply to increase the magnet current and thus the magnetic field. In a similar manner, when the output of the Hall-effect field sensor exceeds the sum of the field set and field scan voltages, the current through the magnet winding is decreased and the magnetic field is decreased.

The field scan potentiometer is center-tapped and the wiper is driven by the recorder's horizontal axis drive mechanism. The voltage output of the scan potentiometer acts to oppose the field set voltage when the recorder pen is in the left-half of its operating range, but adds to the field set voltage when the recorder pen is moved into the right half of the recorder chart.

—The 1230 Hz oscillator is a stable Wein bridge oscillator whose output level is internally regulated.

—The error amplifier is a high gain stable amplifier with a pass band centered near 1230 Hz. It has a phase shifting potentiometer which permits the phase to be adjusted so that the

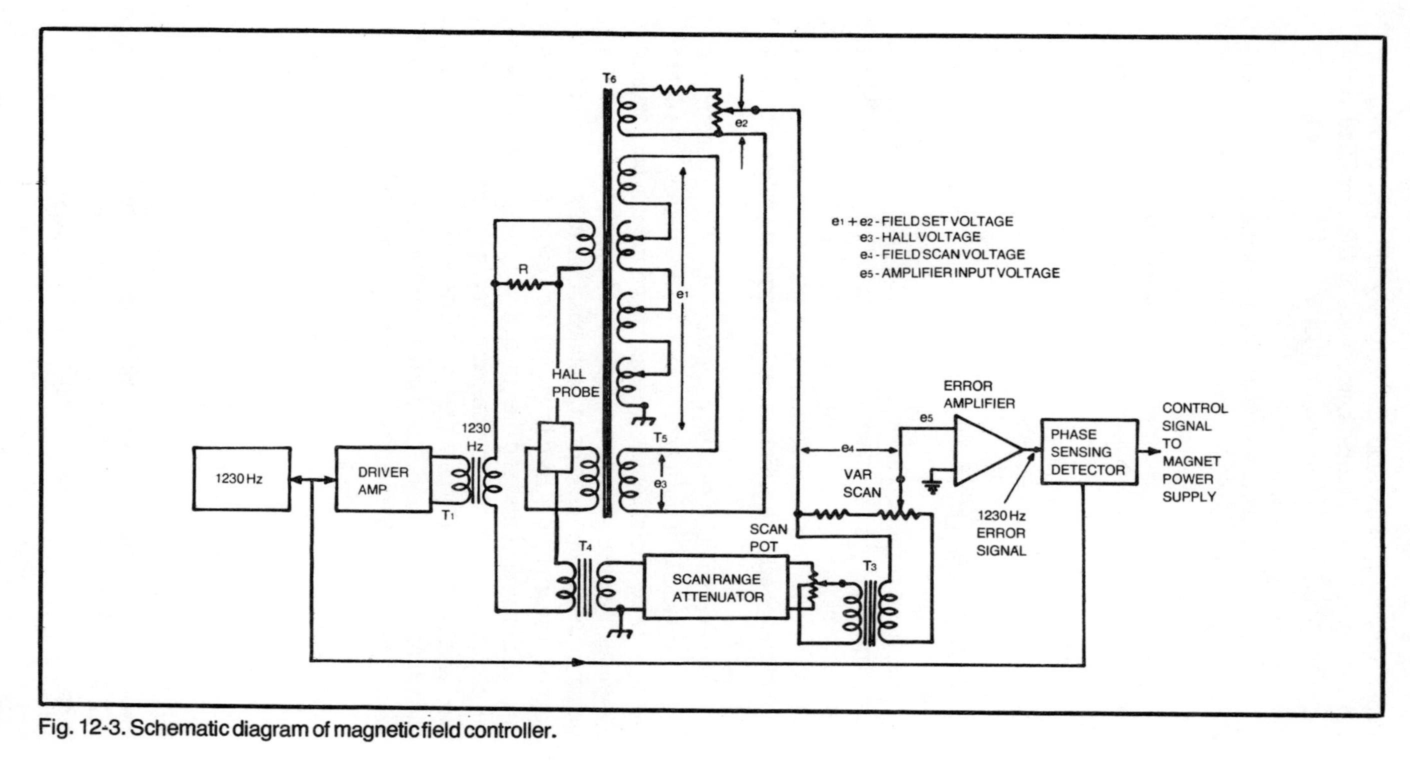

Fig. 12-3. Schematic diagram of magnetic field controller.

signal component of the amplified error signal is exactly in phase and the spurious quadrature component of the amplified error signal is exactly out of phase, so that the latter is completely rejected by the phase sensitive detector.

—The phase sensitive detector converts the amplified 1230 Hz output signal of the summation circuit into dc control voltages to drive the magnet power supply.

—The Hall probe is maintained at a constant temperature. The temperature control is maintained by controlling current through a heater, sensed by a thermistor.

—The magnet power supply provides controlled dc to the low impedance electromagnet. The magnet current is supplied by the main transformer and rectifiers.

The magnets used could be 9, 12 or 15 inches. They are generally floor mounted. They require 2.5 to 22.5 KW power supplies to drive them. The magnets are equipped with ring-shim pole caps designed for maximum field homogeneity. A 3-inch clearance (air gap) provides ease of access. The pole caps are made compatible with tapered ring-shim tips which reduce the air gap to 1.75 inches to achieve optimum field performance when operating at 35 GHz.

Goldbert and Crowe (1975) report an inexpensive method of improving magnet performance in ESR spectrometers. In order to dissipate heat, coolant lines are run around the magnet, magnet power supply and microwave bridge. If the temperature of the raw water is below the dew point, condensation may occur in the magnet power supply. In this event, the raw water can be run through the magnet first and then through the power supply. Also, the coolant water flow should be correct for the magnet system in use. This is ensured by checking the difference between inlet and outlet pressures. If the difference is more than that specified for the magnet system, the cooling system may be clogged.

If the ambient room air temperature or magnet cooling water temperature changes, the setting of the Hall probe temperature should be examined to insure that the temperature control system is regulating, i.e., it is not saturated or cut off. This is particularly important in those installations where large annual variations in room or water temperatures take place. Noise and jitter can be induced into the field control system by transients on the power lines that are coupled into the Hall probe circuit. As much as possible, the Hall probe cable should be isolated from power lines connecting the console and the magnet power supply.

Instabilities of the magnetic field are often difficult to detect and troubleshoot since there is normally no convenient external field measuring device available to monitor the magnetic field with sufficient stability or accuracy. However, the manufacturers of instruments suggest several test techniques that may be employed with easily obtainable auxiliary test equipment. If the magnet current rises beyond normal, the Hall sensor probe may be improperly phased. This is corrected by removing power from the system and reversing power leads to magnet coils.

MICROWAVE BRIDGE

The *microwave bridge* detects the ESR signal reflected from the sample cavity. The bridge contains microwave circuitry, a klystron power supply, a preamplifier and automatic frequency control circuit (Fig. 12-4).

The klystron generates microwave energy at 9.5 GHz which is used to irradiate the sample. The kystron output is applied to an isolator which allows signal flow only in one direction. The klystron is water-cooled through its mounting flange. The 35 GHz systems are also available.

The output of the isolator is given to a 4-port circulator where the microwave power is directed to the sample cavity. The circulator then directs reflected power from the sample cavity back to the microwave detector. The crystal dc bias current is indicated by the detector level meter.

When all the power is absorbed and none reflected, the detector receives no signal. At resonance, the sample absorbs microwave energy and thus unbalances the impedance of the cavity. The absorption in the sample is detected as a dip in the output of the crystal detector. The crystal output containing the ESR signal and the 70 kHz AFC signal is amplified by the preamplifier. It is then coupled to the receiver and to the AFC amplifier.

The automatic frequency control circuit contains a 70 kHz crystal controlled oscillator which generates the AFC carrier and the AFC phase detector reference voltage. The AFC carrier is superimposed on the klystron reflector voltage which results in a 70 kHz frequency modulation of the klystron output. The frequency modulated RF is applied to the sample cavity, resulting in an FM to AM conversion. The resultant AM modulated microwave power is detected and amplified in a preamplifier, followed by a 70 kHz tuned amplifier. The output of this amplifier is phase detected, amplified, filtered and applied to the klystron reflector tracking network. Another parallel output is given to an integrator which quickly returns the phase detector dc output to zero. This arrangement permits a very accurate lock, over a range of at least ± 15 V reflector voltage correction.

Noise elimination is an important factor in enhancing the sensitivity of plotting ESR spectrum. Most of the noise is eliminated in the phase-sensitive detector because only that part of the noise which is at the same frequency and in-phase with the reference signal is allowed to pass through it. The noise is further eliminated by using a long time constant filter which averages out noise at frequencies greater than the reciprocal of its time constant. A better method of noise elimination is through continuous averaging by a computer which employs a multichannel pulse-height analyzer. This technique of using computer for average transients (CAT) efficiently removes both low and high frequency noise.

Sthanapati et. al. (1977) describe a phase sensitive detector for wide line NMR and ESR spectroscopy. This detector has an input dynamic range better than 90 db over the frequency range 20 Hz-1 MHz.

To achieve advantages of higher frequency operation such as increased sensitivity for small samples, minimization of second order shifts and increased resolution for powder samples with different g-values, the klystron frequency is set at 35 GHz instead of 9.5 GHz. The higher frequency also permits the observation of transitions that require higher energy.

As the klystron requires + 650V for the klystron beam and up to −400 V for the klystron reflector. The supply voltages are obtained from the 20 V dc inputs in inverter circuits, working at a frequency of 35 kHz. This frequency is obtained by dividing the 70 kHz AFC modulation frequency by 2. The klystron body or tuning shaft should not be touched when the bridge is in tune or operate position because the klystron body is at + 650 volts when operating. It is insulated from ground by an insulating gasket placed between the klystron flange and the water cooling flange.

100 kHz MODULATION UNIT

The 100 kHz modulation unit acts as a transmitter and receiver. The transmitter provides the power to drive the cavity modulation coils. The receiver processes the ESR signal from the

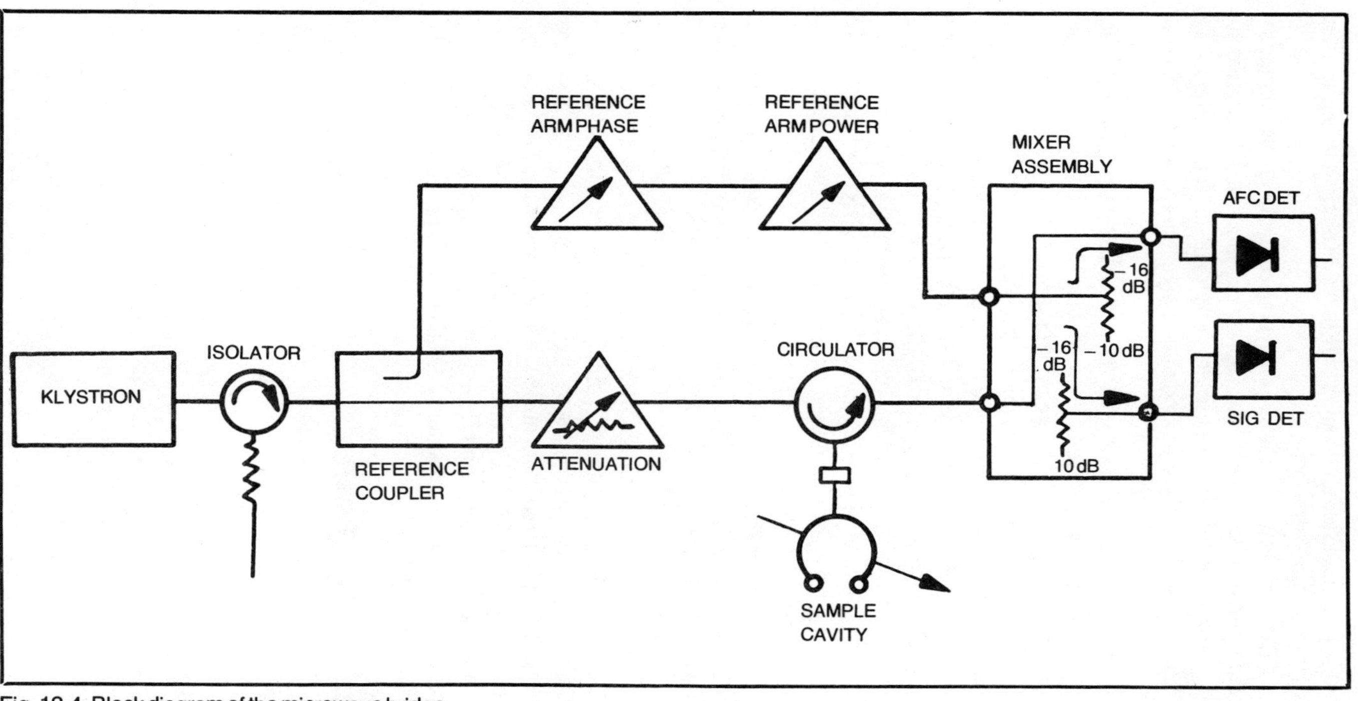

Fig. 12-4. Block diagram of the microwave bridge.

bridge pre-amplifier and converts it to a dc voltage for application to the recorder Y-axis (Fig. 12-5).

The 100 kHz oscillator is crystal controlled. Its output is attenuated by the modulation amplitude control and is applied to the output modulation amplifier. The gain of the amplifier is controllable. The 100 kHz modulation signal is transmitted to the cavity modulation coils.

The 100 kHz ESR signal from the detector crystal in the bridge is filtered and amplified in the 100 kHz receiver. The receiver gain control can be set in 1 db steps.

The amplified 100 kHz ESR signal is applied to the input of the phase detector. The amplitude of the dc phase detector output is proportional to the amplitude of the 100 kHz signal input. A low pass filter removes any residual 100 kHz or harmonics from the phase-detected signal. Additional noise filtering is provided by an RC network which provides variable TIME CONSTANT. This is followed by a buffer amplifier which provides low output impedance stage connection to the recorder.

Modulation may also be carried out at low frequencies like 10 kHz, 1 kHz, 270 Hz or 35 Hz. Low frequency module shown on the block diagram serves to accomplish this function.

Smith et. al. (1976) reports the conversion of a Varian E-3 spectrometer 100 kHz to 1 MHz modulation (Kim and Seissman, 1977) describes a modification of this spectrometer for detection of transient species generated by light flashes e.g. from a pulsed nitrogen laser. Non-uniformities of the modulation field and the radio frequency field alter the CW saturation behavior of inhomogeneously broadened ESR lines. Corrections to ESR saturation data have been considered (Mailer, 1977) both theoretically and experimentally.

SYSTEM FUNCTION SELECTOR

The system function selector provides the selection of signal inputs to the recorder and oscilloscope and cavity modulation to the cavity. When the oscilloscope monitor and the recorder input switches are in the 100 kHz or low frequency position, the phase detected EPR signal passes from the receiver to the oscilloscope and recorder respectively. In the external position, the recorder input may be supplied from an external source and the oscilloscope may be fed an external signal.

The cavity modulation switch allows the operator to apply either 100 kHz or a low frequency modulation to the cavity coils. If

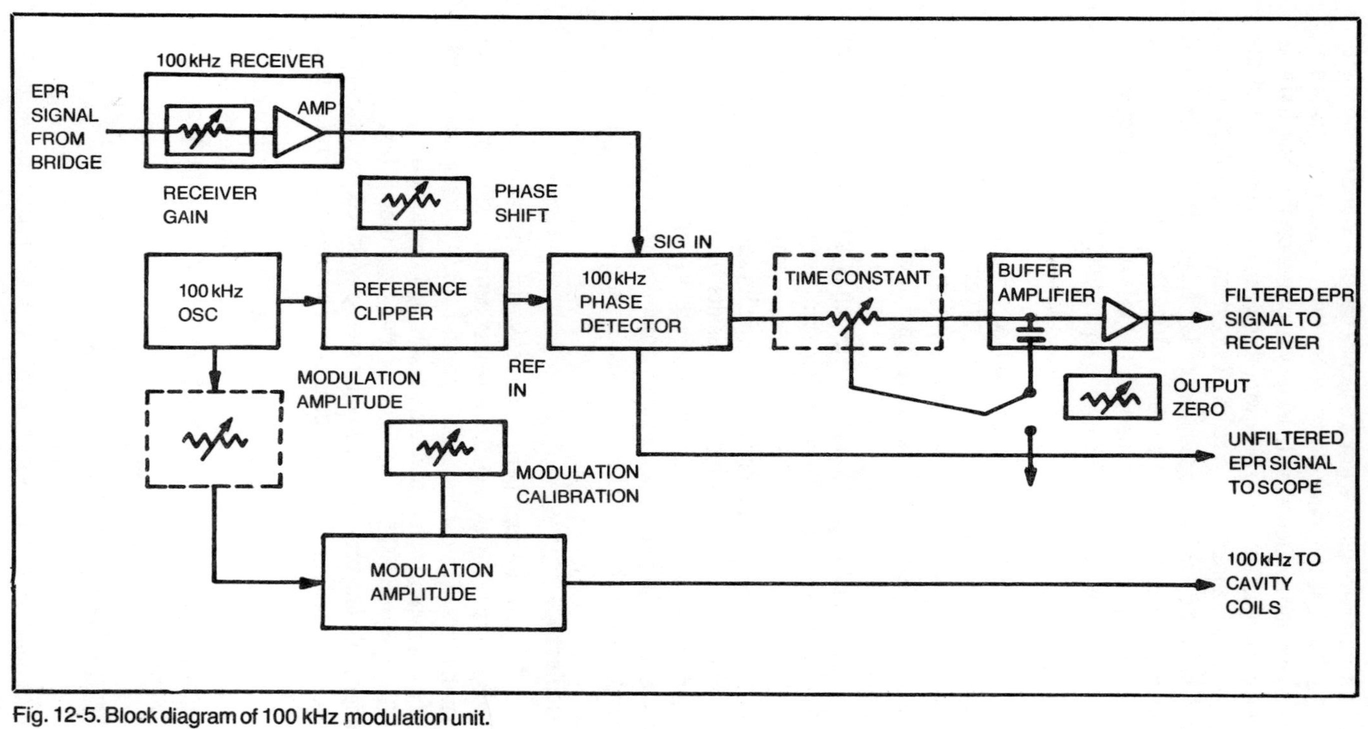

Fig. 12-5. Block diagram of 100 kHz modulation unit.

two cavities are used, a low frequency can be applied to one cavity and 100 kHz to the other cavity, or the two modulation signals may be added and applied simultaneously to either cavity. The 100 kHz modulation and the 34 Hz sawtooth sweep are combined in the modulation summation circuit. Two decade capacitors series resonate the cavity modulation coils at 100 kHz for the most efficient energy transfer to the cavity. The EPR signal can be observed on the oscilloscope with 100 kHz or 10 kHz modulation, but not with any of the other lower frequencies.

RECORDER

The recorder is used to produce a hard copy of the ESR spectrum as a function of time on a 30 × 40 cm flat chart. By selecting the synchronized frequency to the stepper motor, the recording time can be selected from a few seconds to several hours.

Horizontal travel (X-axis) of the pen is achieved by a pulsed stepper motor and belt driven capstan, which moves the pen carriage horizontally. Vertical (Y-axis) travel of the pen is controlled through a second system of pulleys and wire cable from a servomotor. The Y-axis movement results when the servo loop, of which the motor is a part, is unbalanced by a signal applied to the recorder amplifier. The recorder amplifier is turned on by the servomotor to balance the servo loop.

Oscilloscope

The oscilloscope permits direct visual observation of rapidly changing or decaying signals as well as optimization of instrument parameters. The oscilloscope supplies 34 Hz sawtooth modulation to the sample cavity modulation coils and displays the resulting ESR spectrum. The amplitude of the field modulation is kept adjustable and may be selected by sweep width control. Noise on the displayed spectrum may be reduced by the addition of an input filter.

The oscilloscope is also used to display the klystron mode. It is accomplished by applying the same 34 Hz sawtooth modulation to the klystron reflector and displaying the output of the microwave detector on CRT.

Sample Cavities

Signal intensity from a particular sample can vary greatly depending on the microwave cavity configuration. This is important when signal-to-noise is a serious problem. The two main effects arise from the difference between cylindrical (TE_{011}) and rectangular cavities (TE_{012}) and the effect of quartz such as dewar inserts on the microwave field distribution. The use of a cylindrical cavity results in a net improvement over the rectangular cavity. Both the types of cavities are, however, transverse electric (TE) modes which means that the electric field lines are confined to the plane perpendicular to the longitudinal axis. No such restriction exists for the magnetic field lines. The sample cavity should be so designed that it can be held in the volume of maximum homogeneity of the magnetic field. The cavities are available for a wide variety of ESR signals ranging from solids to liquids. They are designed to be compatible with other accessories like aqueous solution sample cell, flow mixing chamber, electrolytic cell, liquid nitrogen Dewar and the variable temperature accessory.

The ESR studies involving comparative measurements are made by using dual cavity. This cavity does this by allowing a reference sample of known magnetic resonance characteristics to be exposed to microwave energy simultaneous with, and in the same cavity as, the sample being studied.

Rotating cavity is designed to facilitate ESR studies of crystal anisotropy. The cylindrical rotating cavity has exterior modulation coils which are free to turn with the rotating electromagnet and thus provide a modulation field which is always parallel to the dc magnetic field. The sample tube remains fixed in the cavity which keeps the Q constant and eliminates the need for retuning after rotation.

SAMPLE CELLS

Aqueous solution sample cells are specially designed to carry out ESR studies in lossy or aqueous solutions. Further, as optical irradiation experiments are often performed with these cells, the quartz is selected to give a maximum optical transmission in the uv-visible region of the spectrum. Special types of flow cells, tissue cells and electrolytic sample cells are also used when required. Generally, the standard sample tube is made of high purity quartz and measures 3 mm ID and 4 mm OD Glass which is not used because it contains traces of Fe^{3+}. For maximum sensitivity, the tube may be filled to a height of 2.5 cm for a rectangular cavity and 5 cm for a cylindrical cavity. Allendoerfer (1975) describes a new design for a cell to make simultaneous electrochemical and ESR measurements.

The single crystal method of ESR analysis requires a large number of measurements with precise orientations of the sample. Such measurements are often difficult because of the small size of microwave cavities. Some *goniometers* for ESR studies have been described in literature (Batley et. al. 1972), but either they only permit 360° rotation about a single axis or the angular variation of the signal/noise ratio (modulation coils not aligned with H_0) prevents the spectrum from being recorded for some orientations. Berelaz et. al. (1977) describe a device which allows the precise orientation of a single crystal in any given direction with respect to the static magnetic field. This goniometer was designed for an X-band spectrometer using 100 kHz field modulation.

All sample tubes should be cleaned before inserting them into the cavity. Only standard EPR sample tubes should be used to prevent breakage and to minimize contamination signals. Sources of contamination such as cigarette ashes or smoke should be kept away from the immediate vicinity of the cavity whether the cavity is installed in the air gap of not. Loose magnetic materials must be kept away from the gap of the magnet to prevent damage to the cavity when it is in the air gap. Watches should be kept away from the magnet gap.

Sample materials or solvents (water-acetone) with high dielectric loss requires the flat cell. For materials with low dielectric, the regular 3 mm sample tube is used.

Chapter 13

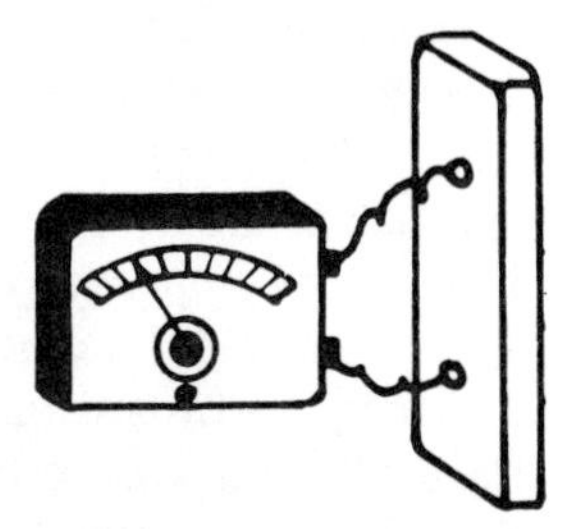

Electrochemical Instruments

Electrochemical methods are characterized by a high sensitivity, selectivity and accuracy. Analytical sensitivity attainable even exceeds the 10^{-10} molar level and analyzes at the subnanogram range are possible. Electrochemical methods have been extensively developed and each basic electrical parameter—namely current, resistance and voltage—has been utilized alone or in combination for analytical purposes. The simplest and the most commonly used electrochemical technique consists in passing current through a solution for exhaustive electrolysis. The solution gets electrolyzed until all of the product is deposited on a mercury electrode or platinum gauze. The electrode is weighed to complete the analysis. Besides direct analysis, electrical measurements are excellent indicators in all areas of titremetry. Reagents are added volumetrically by automatic or manual means to an end point that is conveniently detected electrometrically. Sometimes, even the reagent can be coulometrically generated within the sample to obtain a complete electrochemical system. Electrochemical sensors are based on EMF (potentiometric), resistive (conductometric) or current (amperometric) principles, each having its own use in a given situation.

When using electrochemical techniques for chemical analysis, the potential at an electrode is a function of the concentration of some analyte, the current that flows during the measurement, a time function, any mass transport rate that might exist at the electrode and the area of the electrode.

ELECTROCHEMICAL CELL

An electrochemical cell (Fig. 13-1) consists of two electrodes of the same metal or different metals 1 and 2 immersed in the same or in different electrolytes. Each compartment is a half-cell. Electrodes 1 and 2 are connected to an electrical instrument P, which can exchange electrical energy with the cell. The electrical energy exchanged between the cell and instrument P is liberated or consumed by reactions involving transfer of charge at the electrodes. Electrons are consumed at the cathode, where reduction occurs, and are supplied by the anode 1, where oxidation occurs. Free electrons generally do not exist in electrolytes, and electricity is transported through the cell by migration of ions. Positive ions migrate toward the negative electrode, and negative ions toward the positive electrode.

There are two types of electrochemical cells: *galvanic* (voltaic) and *electrolytic*. A galvanic cell consists of two electrodes and one or more solutions (two half-cells). In these cells, a chemical reaction involving an oxidation at one electrode and a reduction at the other electrode occurs. An electrolytic cell is one in which the electrical energy is supplied from an external source, the cell through which the current is forced to flow.

In an electrochemical cell, the potential at which the current is half its limiting value is known as the *half-wave potential* ($E_{1/2}$), which is one of the most useful properties of an electroactive substance. When fast electron transfer rates are involved, $E_{1/2}$ is equal to E_0 of the Nernst equation. For slow rates, more voltage

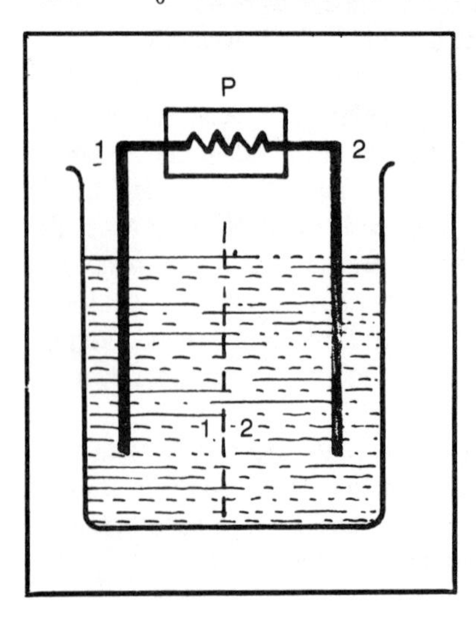

Fig. 13-1. Schematic diagram of an electrochemical cell.

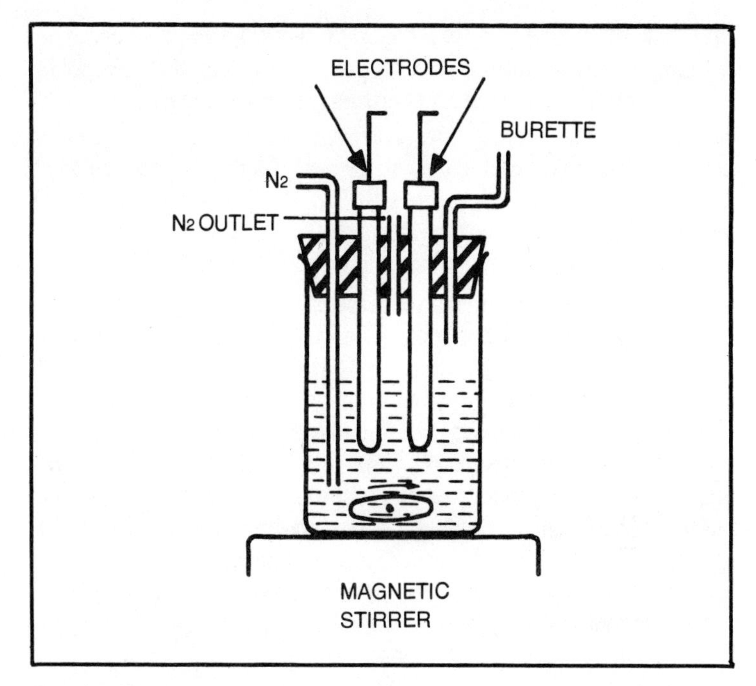

Fig. 13-2. Titration cell.

than predicted by the Nernst equation is required to reach the $E_{½}$ point.

Figure 13-2 shows the components of a titration cell. This cell consists of an indicator electrode and a reference electrode. The reference electrode is usually the commercial calomel electrode. Continuous stirring of the solution is done with a magnetic stirrer. It is usual to carry out many oxidation-reduction titrations in an air-free medium. Nitrogen is then bubbled through the solution before and during titration.

TYPES OF ELECTRODES

There are a wide variety of electrodes which are used for different electrochemical techniques. Basically, they can be classified as working, auxiliary and reference electrodes.

Working Electrode

The working electrode is the controlled electrode at which the desired reaction occurs. Platinum disc, carbon paste and dropping mercury electrode are examples of working electrodes. The primary reaction takes place at the surface and the change in current or potential resulting from this reaction is measured.

Solid surface electrodes and mercury electrodes normally cover different voltage ranges. In general, metal electrodes cover the positive voltage range (+ 1.0 V to −0.5 V versus SCE) and mercury the negative voltage range (0.0 V to −2.0 V versus SCE). Carbon electrodes cover a much wider range.

Mercury is used in various forms. The dropping mercury electrode is the most common. The mercury pool offers about a tenfold increase in sensitivity over the DME but requires the use of a large pool (2 cm diameter) to minimize changes in curvature and area with change in applied potential due to interfacial tension. The mercury plated electrode provides a very high resolution and operation over extended negative potentials. Nickel, silver and platinum are most often used as the plating surface. The solid electrode is usually in the form of a wire, disc or wire mesh. Both the platinum wire and disc are often rotated to obtain increased current signal-to-background ratio obtained by the enhanced mass transport or convection of the ions.

Much work has been done with platinum than any other solid electrode. However, a platinum surface is subject to formation of exide films on its surface by either chemical or electrolytic oxidation. A gold surface is less susceptible than platinum to attack by some chemical oxidizing agents. Gold is probably a better electrode material than platinum for general utility.

The carbon paste electrode is one of the most practical of solid electrodes and, for routine applications, the carbon paste surface is superior to gold or platinum. Vitreous carbon is highly resistant to chemical attack and is relatively insensitive to changes in pH. It has, however, higher residual current than the carbon paste electrode.

Reference Electrode

This electrode provides a stable potential which may be taken as a reference. Most literature states the potential measurements in relation to a saturated calomel electrode, which is a convenient electrode to use. Another common reference electrode is the silver-silver chloride electrode. Reference to hydrogen electrode, the difference in potential between a calomel (−0.244 V) and silver-silver chloride (−0.200 V) electrode is 44 mV.

Auxiliary Electrode

The electrode required for completing the electrolysis cell is called the auxiliary electrode. It performs the reverse reaction of that which takes place at the working electrode. Usually, a platinum disc may be used as an auxiliary electrode. A *frit* separates the auxiliary compartment solution and sample compartment containing the working electrode. This frit is normally sufficient to prevent contamination in quiet solution techniques. However, an agar plug can be used for stirred solutions such as are required in coulometry.

POTENTIOMETERS

Potentiometry involves the measurement of the difference in potential between an indicating electrode and a reference electrode immersed in a solution of the ions to be determined. The potential E (half cell potential) of any electrode is given by the generalized form of the Nernst equation.

$$E = E^{o} + \frac{RT}{nF} \log \frac{a_{ox}}{a_{red}}$$

where E^{0} = Reduction potential of the half-cell under standard conditions.
E = The potential of the half-cell.
R = a constant, 8.314 J/deg.
T = Absolute temperature.
F = Faraday number (96,494 C).
n = Number of electrons transferred in the electrode.

Note that a_{ox} and a_{red} are the activities of the oxidized and reduced forms, respectively, of the electrode action. Substituting concentrations for activities and various other constants and assuming the temperature to be 25°C, the Nernst equation becomes

$$E = E^{o} + \frac{0.0591}{n} \log \frac{(ox)}{(red)}$$

where (ox) is the concentration of the oxidized form of the ion and (red) is the concentration of the reduced form of the ion.

From this equation, it is obvious that measurement of the potential developed by a half-cell serves as a measure of the concentration of the components in a solution. In practice, the half cell to be measured is connected to a standard or reference half-cell to form a complete cell. The method of potentiometry is used to measure the voltage of the complete cell, and the potential of the half-cell is calculated from the relationship

E (half-cell) – E (Standard half-cell) = EMF (observed)

The EMF of the standard half-cell being known, the unknown potential can be determined. A potentiometer is used to measure the voltages instead of a voltmeter as it draws negligible current and hence does not produce any depolarization during the measurement.

Although potentiometry is useful in several analytical areas, its most common application lies in the determination of pH. Potentiometric measurements are also applied to the detection of the end point of titrations. Another area where potentiometry is being increasingly applied is that of the use of ion-selective electrodes for the direct measurement of cations and anions.

Principle of a Potentiometer

Although the voltage of an electrochemical cell can by measured by connecting a voltmeter to the cell, it has the disadvantage that it is not very precise and draws a relatively large current from the cell. With the flow of current through the cell, reaction may occur at the electrodes and alter the composition of the solution. The problem can be solved to a certain extent by using FET input amplifiers having very high input impedance so that they draw very small current from the cell. They, too, may lack the precision due to a rather short scale length. Very accurate measurements can be made by the potentiometric method. This method is a comparison technique in which the unknown cell emf to be measured is compared with a known emf source.

The principle of a simple potentiometer is explained in Chapter 1. The standard voltage across the slide wire is provided by using a standard cell which is usually a Weston cadmium cell. The positive terminal in this cell is a mercury electrode whereas the negative terminal is a cadmium amalgam with an excess of cadmium to maintain saturation at all temperatures. Voltage recording is common in potentiometric methods and instrumental analysis in general. Self-balancing null type potentiometric recorders are used for this purpose. When potentiometry is used as an end point detection system, the following techniques can be employed.

Zero Current Potentiometry

No net electrochemical reaction occurs at the indicating electrode, and the measurement is independent of mass transport.

In this method, the course of a titration is followed by measuring the change in potential of an indicator electrode at zero current.

Constant Current Potentiometry

In this method a small but constant current flows through the indicator electrode system during the titration. The potential change during the titration is monitored as usual. In a stirred solution, the current is sufficiently small so that a stable potential is achieved due to diffusion of the electroactive species present.

Null-Point Potentiometry

In this method, the potential difference between the two electrodes is brought to zero by having same concentration of the investigated substance in both the half-cells. This is done on adding titrant to the unknown sample in order to bring the concentration of the species of interest to the same concentration of the same species in the other half of the cell system. Identical indicator electrodes are used in both sides of this two chamber cell.

Cyclic Chronopotentiometry

In *chronopotentiometry,* a constant current is passed through an electrolytic cell. The potential of the working electrode is monitored. The variation of this potential with time takes the form of a wave, having a characteristic transition time (Fig. 13-3).

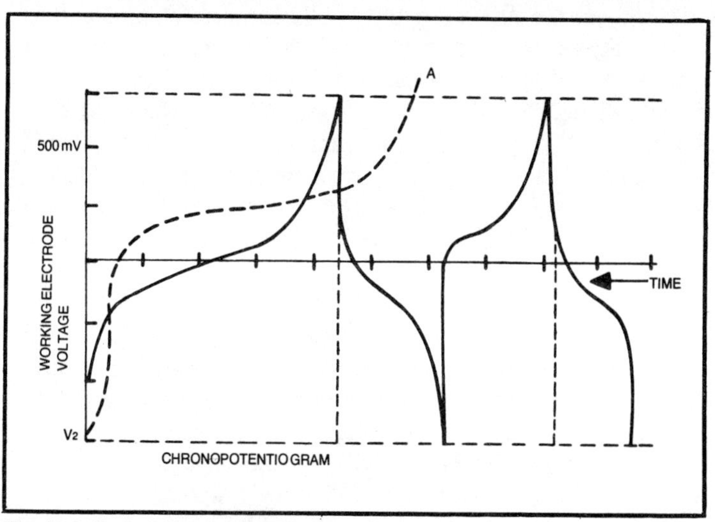

Fig. 13-3. Typical chronopotentiogram.

Information about the reactions occurring at the electrode can be obtained from the shape and duration of this wave.

Cyclic-chronopotentiometry is a useful extension of the technique in which current is reversed at the end of each transition time, so that a sequence of forward and reverse transitions is produced. The current reversal is made, when the electrode potential reaches either of two preset valves V_1 and V_2, so that potential oscillates between these values. The current is reversed at the appropriate electrode potentials using a relay.

If the potential drop across the cell is small, a series resistor can be used to control the current adequately, using a stable high voltage supply. If the cell resistance is high (e.g. with non-aqueous electrolytes), a high value of series resistor and a high voltage would be needed. To avoid this, an electronic current stabilizer (galvanostat) is generally used. The potential of the working electrode is measured against a suitable reference electrode. Ideally, negligible current should be drawn from this type of electrode. The emf between the reference and the working electrodes is measured using a high input impedance, fast response amplifier. The design of such an amplifier is simplified by holding the working electrode at earth potential.

Figure 13-4 shows the circuit arrangement used for cyclic chronopotentiometry. A stabilized voltage supply is connected to the working electrode through a resistance R. Amplifier 1 amplifies the voltage difference between the working electrode and earth, and the output is applied to the auxiliary electrode, thus maintaining the working electrode at earth potential. Since the amplifier draws negligible current in the input circuit, all the current passing through the resistor also passes through the cell. The cell emf is supplied by the amplifier. The cell current is therefore determined by the supply voltage and the series resistor. The supply to the resistor is supplied by amplifier 2 which is an inverter following the output of the galvanostat voltage drive generator. The cell current is reversed at the desired switching potentials (V_1 and V_2) by reversing the potential which controls the amplifier 2. The switching potentials are controlled by reference potential amplifier 3. The voltage swing is determined by the ratio at P_2, and it can be extended over a wide range by adjusting P_3. With a 1 μf capacitor connected between the auxiliary electrode and the working electrode terminals and a proportion of the auxiliary electrode voltage fed back to the reference electrode terminal, a steady triangular wave of good form can be obtained. English

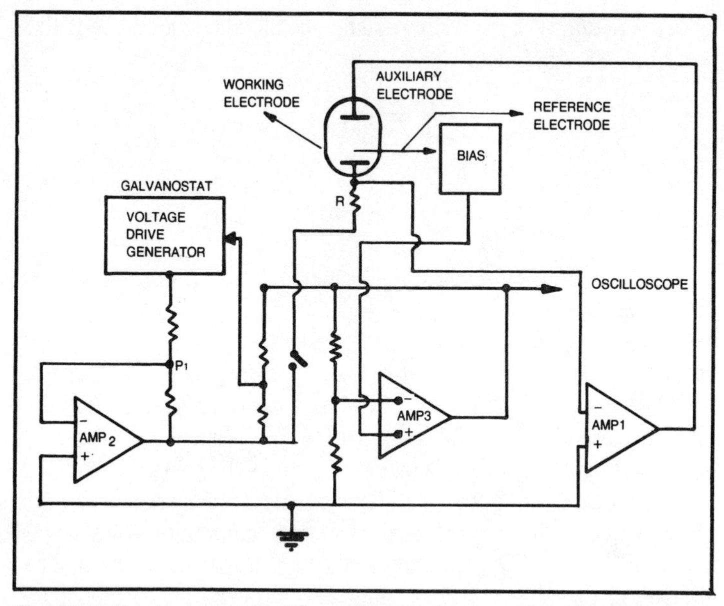

Fig. 13-4. Circuit arrangement for making cyclic chronopotentiometric measurements (after English, 1970).

(1970) describes the circuit details of an apparatus for cyclic-chronopotentiometry in non-aqueous solvents.

CONDUCTIVITY METERS

The conductivity of an electrolyte is a measure of the ability of the solution to carry electric current. The current through the solution takes place through the movement of electrically charged particles called ions. When a potential difference is applied to two electrodes immersed in the solution, ions are almost instantaneously accelerated towards the electrodes. Since the conductance of a solution of electrolyte is related to the concentration of electrolyte, analytical applications of conductance are possible.

Like a metallic conductor, electrolyte solutions obey Ohm's Law. The reciprocal of the resistance R of the electrolytic solution (1/R) is called the conductance. It is expressed in reciprocal ohms or mhos.

The resistance of a solution depends upon the length (l) area (a) and the intrinsic properties of the solution. It can be expressed as:

$$R = \frac{pl}{a}$$

where p is known as specific resistance. Since conductance is the reciprocal of resistance:

$$\therefore \frac{1}{R} = \frac{1}{p}\left(\frac{a}{l}\right)$$

$$= K\left(\frac{a}{l}\right)$$

The constant K is called the specific conductance. It is expressed in ohm^{-1} $centimeters^{-1}$.

The specific conductance of an electrolyte is a function of concentration. As the solution is diluted, the specific conductance will decrease. This is because fewer ions are present to carry the electric current in each cubic centimeter of solution. The ability of individual ions to conduct is usually expressed by a function called the *equivalent conductance.* The *equivalent conductance* is the conductance of a hypothetical solution containing one gram equivalent of an electrolyte per cubic centimeter of solution. The equivalent conductance Λ is connected with the specific conductance and concentration (in gm-equivalent per 1000 cm^3) as follows:

$$\Lambda = \frac{1000K}{C} \qquad K = \frac{\Lambda C}{1000}$$

The conductance of an electrolyte between two electrodes, when expressed in terms of equivalent conductance and concentration is given by:

$$\frac{1}{R} = K\left(\frac{a}{l}\right)$$

$$= \frac{\Lambda C.a}{1000l}$$

The equivalent conductance of a salt is the sum of the equivalent ionic conductances of its ions:

$$\Lambda = \lambda_+ + \lambda_-$$

The total conductance of a solution at infinite dilution is:

$$\Lambda_\infty = \Sigma\lambda_+ + \Sigma\lambda_-$$

which shows that the migration of ions are theoretically independent of each other and are not affected by other ions in the solution. This is, however, not strictly true. There are very slight differences in ionic conductance of an ion in the presence of various other ions.

MEASUREMENT OF CONDUCTANCE

Conductivity is usually determined by measuring the resistance of a column of solution. This is done by using a Wheatstone bridge in which the conductivity cell forms one arm of the bridge (Fig. 13-5). In order to avoid changes in ionic concentrations due to net chemical reactions at the electrodes, alternating current rather than direct current devices are employed. The choice of frequency is not critical and may be anything between 50 to 10,000 Hz. However, most commonly employed frequency is 1000 Hz. This ac source may be low voltage tapping on the 50 Hz transformer or transistor oscillator.

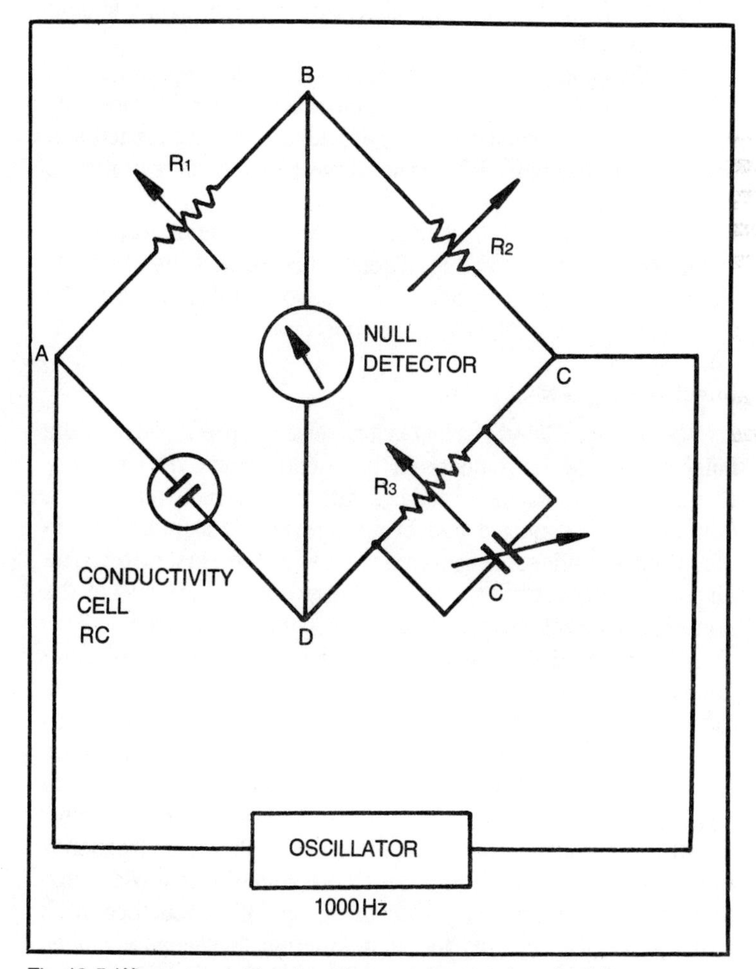

Fig. 13-5. Wheatstone bridge circuit for measurement of conductivity.

Null Method

When the bridge is balanced, and it is assumed that the conductivity cell behaves as a pure resistance, then the voltage between B and D is equal to zero.

$$R_c = \frac{R_1}{R_2} \times R_3$$

By adjustment of the ratio R_1/R_2, a wide range of resistance can be measured. However, whenever possible, this ratio is kept unity. This condition is the most favorable to precise measurements. The arms AB and BC represented by resistance R_1 and R_2 and usually in the form of a single calibrated slide wire resistor with a sliding contact connected to the null detector.

As the conductivity cell contains electrodes separated by a dielectric, an appreciable cell capacitance is invariably present. This capacitance is balanced out by providing a variable capacitor in parallel with resistance R_3. It is so adjusted that the detector gives a sharply defined balance point.

The null detector is not an ordinary galvanometer as the same is not sensitive to alternating current at the frequencies which are employed to excite the bridge. The most popular detector in use is the magic eye or the cathode ray oscilloscope.

Direct Reading Method

In practice, direct reading instruments are preferred over the null balance type instruments. In these instruments, the necessity of converting resistance readings into conductance readings is eliminated. The unbalanced bridge current is amplified in an electronic amplifier and displayed on a calibrated panel meter. The display in conductance units is achieved by making use of the non-linear transfer characteristics of a specific electron tube.

Kinetic studies of reactions in solution can also be advantageously made by conductance measurements. Direct reading instruments are preferred for this purpose. Knipe et. al. (1974) describe a fast response conductivity amplifier which employs integrated circuits (Fig. 13-6). The excitation for the conductivity cell is a square wave generated by a simple astable multivibrator constructed around a μ A 709 operational amplifier. Its frequency is adjusted to 10 kHz. The conductance amplifier is a wide band high gain operational amplifier in which difference feed back resistors can be selected to obtain different conductance ranges between $0.2 - 10^{-7}$ ohm^{-1}. This is followed by a full wave rectifier

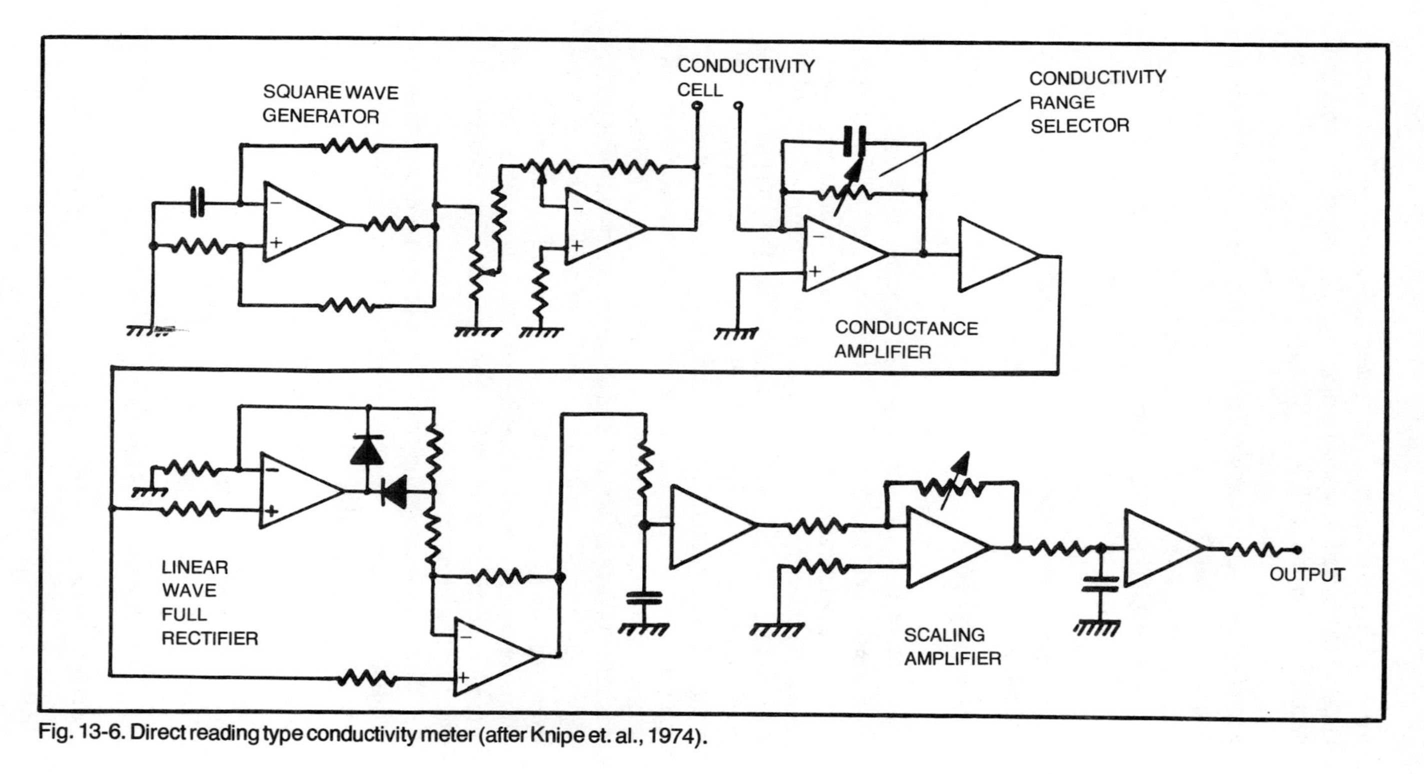

Fig. 13-6. Direct reading type conductivity meter (after Knipe et. al., 1974).

for rectifying the output of the conductance amplifier. The output stage of the circuit is a scaling amplifier. The dc signal from the rectifier is, in volts, 0.5×10^n times the conductance of the solution, where n is a scale factor introduced by the conductance amplifier feedback resistors. This is multiplied by 2 to obtain an output signal that is a simple factor of 10 of the conductance. The original paper describes the performance of the circuit in respect of linearity, stability and sensitivity.

The output of the amplifier is fed to the oscilloscope, via a logarithmic amplifier. The arrangement gives a straight line with a slope equal to the rate constant of the reaction with reasonably good linearity.

Conductivity Cells

A conductivity cell consists of two electrodes which may be two parallel sheets of platinum fixed in position by sealing the connecting tubes into the sides of the measuring cell. In order to reduce the polarization effects which produce a large cell capacitance, the effective area of the electrode is greatly increased by coating the electrode with platinum black. This deposit can be obtained by immersing the electrodes in a solution containing 7.5×10^{-2} M chloroplatinic acid and about 8×10^{-4}M lead acetate and applying a direct current, reversing the direction every half-minute. When a black deposit is obtained, the electrodes are washed in water and the occluded gases are removed by electrolyzing in dilute H_2SO_4 for about half an hour with current reversal every minute. Electrodes are washed again and stored in distilled water overnight before using. Also, the electrodes should be stored in distilled water when not in use. This prevents the platinum black from drying out.

Most of the conductivity cells are of such a design that the solution completely surrounds the electrodes. In such cases, the conductance of the cell is given by

$$\frac{1}{R} = \rho \left[\frac{a}{l}\right]$$

$$\rho = \frac{1}{R} \left[\frac{l}{a}\right]$$

The term $\frac{l}{a}$ is called the cell constant and may be denoted by θ.

$$\therefore \rho = \frac{\theta}{R}$$

The effective value of θ for a cell is not simply related to the cell geometry. However, it has a constant value for electrolytes, measured in that particular cell. The cell constant can be determined by measuring R for a solution of known specific conductance.

Solutions of potassium chloride of known concentrations are invariably employed for this purpose. The specific conductance of these solutions is determined once and for all with a cell in which the cross sectional area is uniform and known with accuracy.

Conductivity cells are available in different types, sizes and shapes. The simplest is the dip type which is immersed in the liquid to be tested. The solution may be in an open container and have volumes in the range of 5 ml. Pipette cells permit measurements of conductivity with small volumes of solution which may be as small as 0.01 ml. Epoxy cells are employed for high temperature use.

For the majority of applications, the cells used are made from a specially developed high density carbon that has the same desirable quality as platinized platinum of eliminating electrochemical errors but without the need for frequent replatinization and recalibration. The annular carbon electrodes are fitted within the tubular bore of the cell. The cell body is moulded from an epoxy resin. The cells are now available for screw in, flow line and dip type installations.

Two terminal conductivity cells are commonly used. These cells are quite satisfactory in many applications, but with dirty solutions, fatty acids or other sticky deposits, fouling takes place. This modifies the surface area and thus results in change of the cell constant, resulting in incorrect readings. This problem has been largely overcome by the four terminal conductivity cell. Here a four electrode cell is used, two outer electrodes being for current and the two inner ones as voltage electrodes.

Temperature Compensation In Conductivity Measurements

The conductivity of electrolytic solutions varies with temperature. This is because the ionic mobilities are temperature dependent. The temperature coefficient is usually of the order of 1.5 to 2% per degree centigrade at room temperature. Control of temperature is thus very essential in precision work. This is usually done by introducing into the bridge circuit a resistive element which will change with temperature at the same rate as the solution under test. The temperature compensating resistor may be a rheostat calibrated in temperature which can be manually adjusted. Automatic temperature compensation can be provided by using thermistor and resistance combination in contact with the solution which would automatically offset the effect of changes of temperature of the solution under test.

An important accessory to a precision type conductivity meter is a thermostatic bath capable of providing very high long term temperature stability. A proportional controller is employed in preference over the conventional on-off methods. With this method, a temperature stability of 0.02°C may be achieved. Thermostatic baths are very essential for continuous measurement of electrical conductivity in liquid streams, especially where very small changes in salt concentrations take place.

Conductivity Measurements Using High Frequency Methods

A high frequency method of measuring the conductivity of solutions offers the advantage of placing the electrodes outside the solution container and out of direct contact with it. This eliminates the possibility and danger of electrolysis or electrode polarization.

The method consists in placing the container with the sample to be analyzed between the plates of a capacitor, which forms a part of the high frequency generator circuit functioning at a frequency of a few megacycles per second. Since the capacitor is a part of the oscillator circuit, any changes in the composition of the solution will result in the changes in the plate and grid currents and voltages due to change in the conductance and capacitance of the cell. The frequency of a parallel resonance circuit is given by:

$$f = \frac{1}{2\pi\sqrt{LC}}$$

The sample cell is usually placed in parallel with a calibrated capacitor. In order to achieve the resonant frequency, the exact amount of capacitance which is added by the sample is removed by adjustment. This gives a measure of the conductivity.

A beat frequency method can also be used to measure the output frequency of two oscillator circuits. One of these circuits (Fig. 13-7) contains the sample cell as a part of the oscillator capacitance and the other is of fixed frequency. The output of the two when given to a mixer unit would be the difference of the two frequencies (f-f_0) which would be directly proportional to the

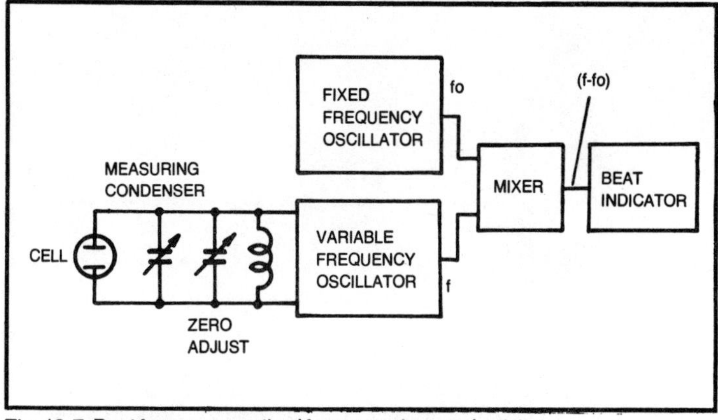

Fig. 13-7. Beat frequency method for measuring conductivity.

changes in high frequency capacitance of the cell. The difference frequency is measured directly with a beat indicator.

A typical sample cell for high frequency conductivity measurement may be composed of two metallic plates sealed on to the wall of a rectangular container. When the solution is put in the container, the metal plates act as a condenser with solution and glass as the dielectric. The equivalent circuit of the cell is as shown in Fig. 13-8. Cg represents the capacitance of the glass walls of the cell, C_s is the capacitance of the sample and R_p is the resistance in parallel with C_s. The resistive component is very high and offers negligible contribution. Capacitive effect is the major factor in high frequency measurements whereas resistive balance is more important in low frequency measurements.

POLAROGRAPHS

When a voltage is applied to a pair of inert electrodes, placed in a solution, specific relationships exist between current and voltage which depend upon the electroactive species present in solution. Current-potential curves can be plotted which prove useful for chemical analysis. These curves can be plotted by varying the voltage applied to a cell and measuring the current flowing through it. Usually, it is assumed that the ohmic drop in the cell is negligible and the potential of one electrode is independent of current. This electrode is said to be unpolarized and the other electrode as polarized.

Polarography is the name given to the technique in which a dropping mercury electrode (DME) is used as an indicator electrode. The basic procedure deals with the measurement of

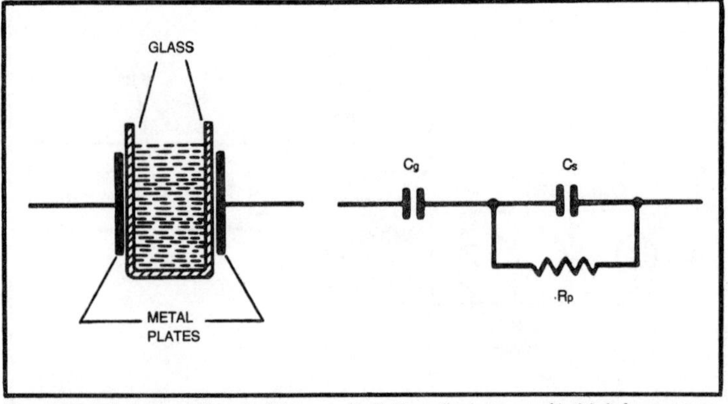

Fig. 13-8. Equivalent circuit of the conductivity cell when used in high frequency method.

current as a function of electrode potential. Recording instruments are available for directly having a plot of these characteristics.

Basic Polarographic Instrument

The essential instrumental requirements of a polarograph are few and simple. The arrangement must have a means of applying a variable but known voltage ranging from 0 to ± 3 V to the cell. Also needed is a method for measuring the resultant current, which is usually in the range of 0.1-100 μa. A block diagram of a polarographic setup is shown in Fig. 13-9.

The electrolysis cell is shown in the diagram as G. The beaker contains the test solution. K represents a dropping mercury electrode consisting of a narrow capillary from which mercury emerges at a rate of 20-30 drops per minute. A is a non-polarizable electrode, e.g. a mercury pool at the bottom of the beaker. Instead of this, it is common to use a calomel half-cell.

The polarographic method is based on recording of the variations in the current flowing through the electrolysis cell as the potential between the electrodes is gradually increased from one value to another, say from 0 to 2 volts. This process is achieved through a motor which simultaneously drives the potentiometer and feeds the chart on which the electrolysis current is recorded as a function of the electrolysis voltage.

A large variety of polarographic instruments are commercially available. The simpler instruments require manual control and point by point plotting of current voltage curves. They are less expensive as compared to those featuring graphic recording of the current-voltage traces. Typically, sensitivities of these instruments range from 0.025 to 0.003 μa per division on a galvanometer scale.

Linearly increasing polarizing voltage can be conveniently obtained by using an operational amplifier as an integrator. This enables the potential scan-rate to be varied over a wide range. Also, small currents encountered in polarography are advantageously amplified in a high input impedance operational amplifier without loading the system with an undesirable voltage drop, when connected with grounded positive input and negative feedback. The input bias current of the operational amplifier used as an integrator must be, at the most, of the order of 10^{-11} A if integration times of the order of minutes are to be achieved. Figure 13-10 shows a simple two electrode polarographic system with amplifier 1 connected as an integrator to the output of which the dropping mercury electrode is connected. The counter electrode is connected to the input of amplifier 2. The integrator output is monitored on voltmeter V. The output of current amplifier is connected to a recorder.

The polarographic current pulsates between zero and a maximum value during the growth and fall of the mercury drops.

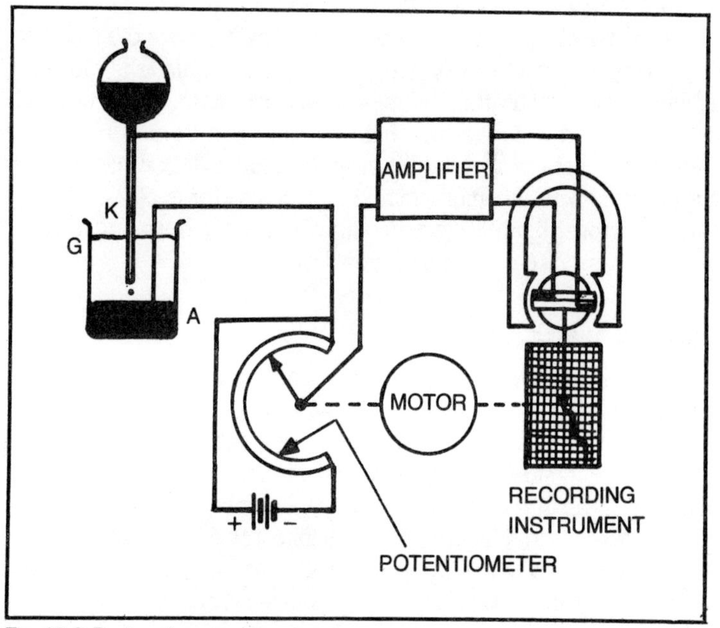

Fig. 13-9. Basic polarographic setup.

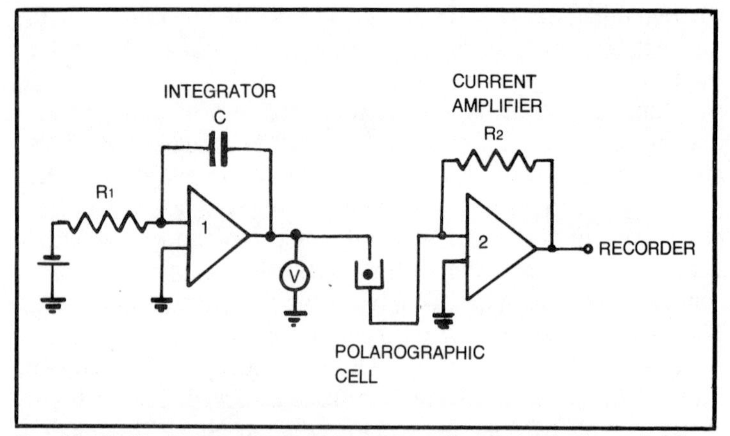

Fig. 13-10. Circuit arrangement for obtaining linearly increasing polarizing voltage.

Rapid oscillations are not followed by the moving coil meters. The electromagnetic currents are generated which cause overswing of the needle. This effect is controlled by connecting a suitable resistor across the meter which is adjusted to provide slight overdamping. To smooth out excessive fluctuations which may still be present, a capacitor resistance circuit is usually connected in parallel with the microammeter.

Commercial polarographs are often provided with the facility of plotting derivate polarograms. If the gradient di/du is plotted against applied potential, the peaks become sharp. In recording instruments, the derivative of the current can be recorded by means of a slight circuit modification (Fig. 13-11). A capacitor is put in series with the microammeter and a resistor is connected in parallel across them. Several modifications have been suggested over this basic circuit to improve resolution.

Current Compensation

It is obvious that the maximum sensitivity cannot be achieved for measuring the diffusion current of a very small wave, which follows a large one, unless the first wave can be canceled out by passage of a counter current through the microammeter, equal and opposite to the first diffusion current. Counter currents are applied from a compensating potentiometer connected directly across the microammeter. Zero setting controls are supplied on some instruments which may be used to obtain a balance at any desired voltage.

Differential current compensation is another technique for applying counter currents. In this method, two polarographic cells with identical electrodes and solutions are connected in series with their like poles connected, the opposing currents would balance each other giving zero current. However, it is difficult to obtain and maintain perfect synchronization of drop fall in two DMEs. This is achieved preferably by a suitable smoothing circuit to clamp out the oscillating current.

Dropping Mercury Electrode

The dropping mercury electrode is exactly as its name implies. It consists of a length of marine barometer tubing with a fine capillary and a head of mercury above it. Mercury, usually under force of gravity, is forced through a section of very fine glass capillary. A mercury drop starts, grows and finally falls off as

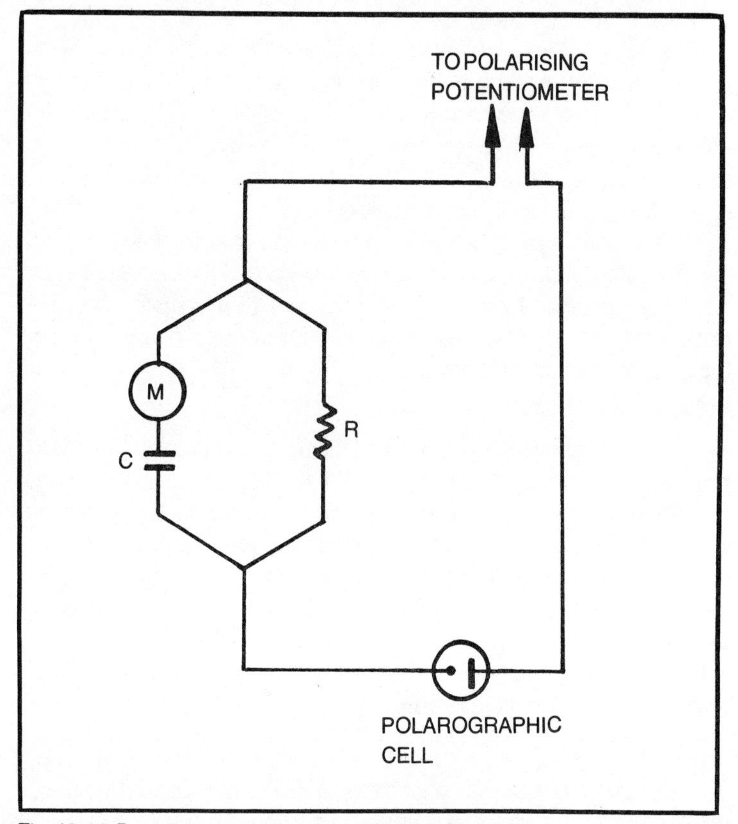

Fig. 13-11. Damping arrangement in recording polarograms.

another drop starts. The measured current will naturally tend to follow this process of increasing steadily, dropping sharply and finally increasing again. The mercury head is so adjusted that it gives a drop time of 2-5 seconds. The head is generally kept between 40 to 80 cm. The internal diameter of the capillary tube is of the order of 0.03 to 0.05 mm, and the length of the capillary is about 8 cm. A platinum wire is immersed in the mercury reservoir and the dropping mercury electrode is coupled with an unpolarized electrode. This electrode is useful over the range 0.4 to 2.8 volts referred to the normal hydrogen electrode. Above 0.4 volts, mercury dissolves and gives an anodic wave. At potentials more negative than -1.5 volts, the electrolytes begin to discharge.

The dropping mercury electrode is a truly elegant electrode and has the following advantages over other types of electrodes.

- Hydrogen has a very large over-potential, more so than mercury. It renders possible the deposition of substances difficult to reduce, as for example, the alkali ions.
- It provides very nearly ideal conditions for obtaining a diffusion controlled limiting current which is reproducible.
- It provides a continuously refreshed surface which is conducive to a high degree of reproducibility for the current measurements. The constant renewal of the electrode surface eliminates passivity or poisoning effects.

The dropping mercury electrode, however, cannot be used for dilutions less than 10^{-5} M due to the presence of a relatively large charging current. Nevertheless, high sensitivity derivative instruments may be 200 times more sensitive as they compensate for the effect of charging current.

Reference Electrode

A mercury pool at the bottom of the polarographic cell acts as a reference electrode. It has a large area and, therefore, the current is generally very small. The concentration over-potential at this electrode is negligible and its potential may be regarded as constant. Though convenient, the mercury pool never possesses a definite known potential. Therefore, the reference electrode is usually a saturated calomel electrode (SCE). It is almost a universal practice in polarography to express half-wave potentials with reference to this electrode.

Typical Polarogram

Figure 13-12 shows a polarogram of a solution containing copper, cadmium and zinc. Initially, the electrolysis current is

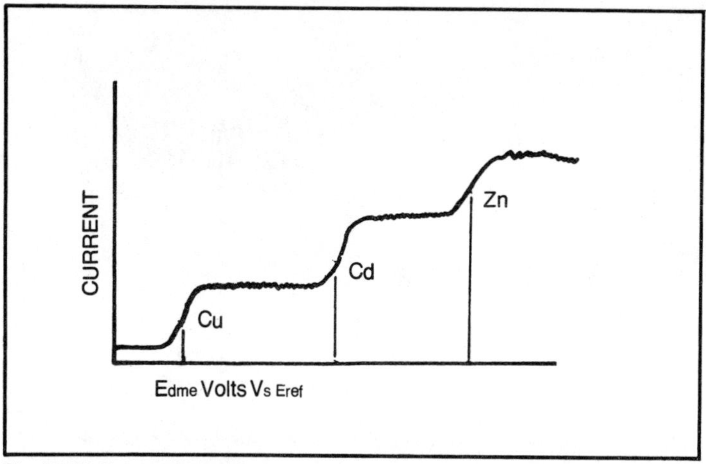

Fig. 13-12. Typical polarogram.

practically zero. The current then starts rising rapidly with the voltage. The increase is brought about as an increasing number of Cu^{2+} ions are discharged at the dropping mercury electrode. The current soon reaches a constant value which is governed by the velocity of diffusion at which the Cu^{2+} ions are transferred to the cathode. This difference would reach the maximum value when the Cu^{2+} ions are discharged as fast as they appear at the mercury cathode. This means that the height of the copper wave is proportional to the Cu^{2+} concentration. As the potential is increased further, the Cd^{2+} ions will begin to discharge at the cathode, which would result in a wave corresponding to the Cd^{2+} concentration. The Zn^{2+} is recorded at a still higher potential. This figure shows that the height of the wave corresponds to the concentration of the ion concerned, where as the potential at which the wave is produced is characteristic of the ion.

The spiked oscillations (Fig. 13-13) seen on the recorded polarograms are due to the growth and fall of the mercury drops. As shown in the inset, the current increases as the drop grows and then drops sharply when the mercury drop detaches itself and falls.

Quantitative Aspects Of Polarography

The quantitative aspects of polarography are based on the equation given by Ilkovic and is as follows:

$$(Id)_{max} = 708\,(n)(m^{2/3})(t^{1/6})(D^{1/2})\ C \text{ at } 25°C$$

where (I_d)max is the maximum diffusion current.

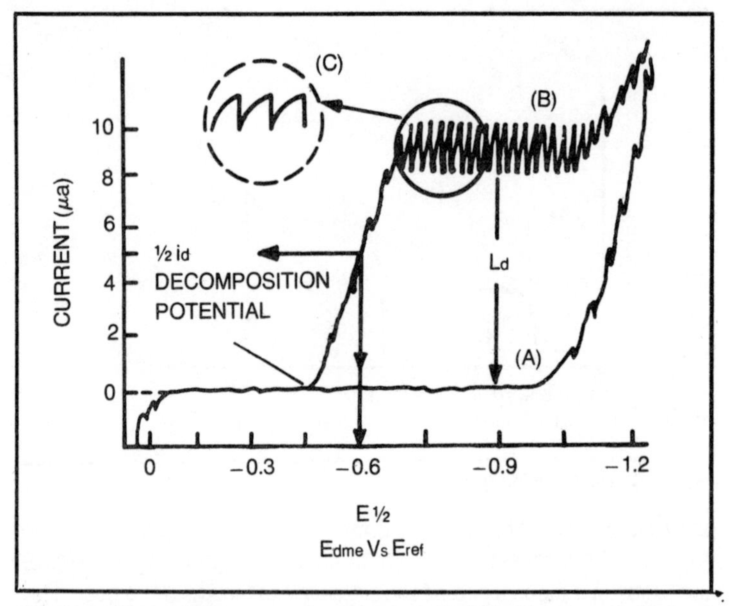

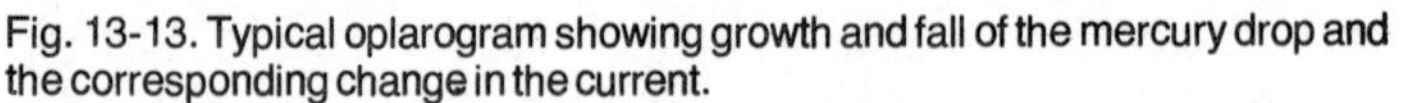

Fig. 13-13. Typical oplarogram showing growth and fall of the mercury drop and the corresponding change in the current.

The rate of diffusion of the substance being reduced or oxidized at the electrode controls the current given by the above equation which is therefore called the diffusion current.

n = Number of electrons in the electron transfer reaction.
m = average mercury flow rate in mg/sec. through the capillary.
t = drop life (sec.)
D = Diffusion coefficient of the electroactive species.
C = Concentration of the analyte in the solution (mM).

The linear dependence between diffusion current and concentration forms the basis for quantitative polarography. Ilkovic's equation also shows quantitatively the influence of other factors which also directly affect the diffusion current.

When a fast recording system is used having less than 1 second full scale response, the peaks of the oscillations equal $(I_d)_{max}$. When the recording system is damped the average diffusion current $(Id)_{ave}$ can be determined from the average of the recorded oscillations. The relationship between average and maximum diffusion currents is given by

$$(Id)_{max} = \frac{7}{6}(Id)_{ave}$$

While the electroactive material in polarography is quantitatively related to the limiting diffusion current, it is quantitatively characterized by the half-wave potential (Fig. 13-14). This is the potential at the point of inflection of the current-voltage curve (one-half the distance between the residual current and the final limiting current plateau). Thus, an $E_{1/2}$ value from the polarogram of an unknown sample can be indicative of possible substances.

Quantitative polarography has been most successfully applied in the 10^{-2} to 10^{-5} formal region. At still lower concentrations, the residual current is often as large as the diffusion current. The selectivity depends greatly upon the solvent used. When analyzing a multi-component system, species with $E_{1/2}$ values differing by 150 mV or greater can easily be resolved. Usually, the first species to be electrolyzed can be determined with the greatest reliability.

Polarograms often exhibit current maxima as shown in Fig. 13-15. This is due to the streaming of solutions past the mercury surface at certain potentials. These maxima can often obscure the wave of interest. They can be often eliminated by adding a small amount of surface active material such as gelatin.

Cathode Ray Polarograph and ac Polarography

Polarography has advantages over non-electrochemical methods in speed, selectivity and sensitivity. An earlier method of

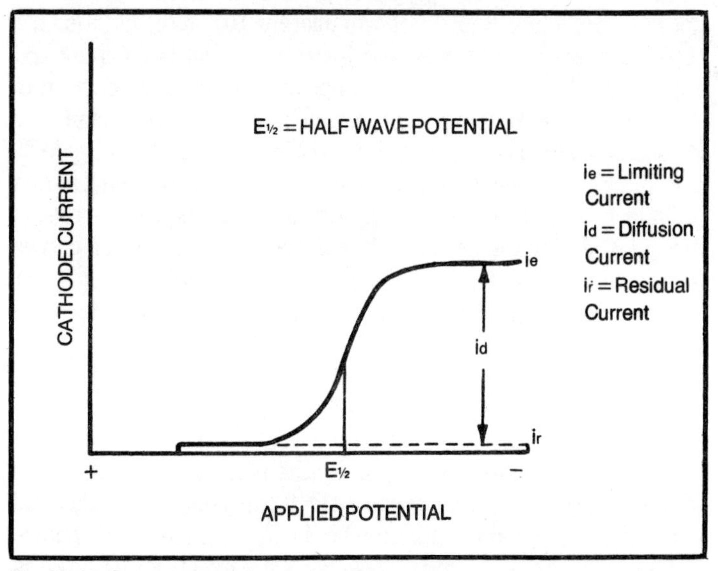

Fig. 13-14. Current potential curve in a polarogram.

achieving these was by the single sweep cathode ray polarograph in which the scan of a quickly increasing dc potential applied across the polarographic cell, in the last quarter of the life of every mercury drop, led to sensitivity improvement by avoiding the serrations caused by scanning over the lives of several drops. The curves do not resemble a conventional polarogram, but possess peaks of characteristic shape with the value of the current at the summit.

Another technique known as *ac polarography* was later developed. A small ac voltage is superimposed upon the potential scan of conventional polarography. The superimposed voltage may be sinusoidal. The disadvantage is that the small polarographic waves are obscured by the capacity current with consequent sensitivity restrictions. The undesirable effect of capacity current on the sensitivity of ac polarography is eliminated by applying a square wave voltage in place of the sinusoidal voltage, in addition to the linearly increasing voltage which is effectively applied to each drop. The amplitude of the ac component of the cell current is measured shortly before each sudden change in the applied voltage.

Pulse Polarography

The method of square wave polarography gave derivative type polarograms with peaks proportional to concentration down to a possible detection limit of approximately 10^{-8} mol dm^{-3} for the reactive species. However, the instrumentation is complex and expensive. This technique was improved by the introduction of derivative pulse polarography in which one pulse of rectangular voltage, usually having a duration of 1/25 second and amplitude 30 mV, is applied to the slowly and linearly increasing potential after a definite time in the life of each mercury drop and the current is measured during the second half of the pulse, the recorded curves having the shape of peaks. Pulse synchronization permits very short time intervals between sample points in drop lifetimes and fast scan rates are also possible.

In the integral variant of *pulse polarography*, the chopping mercury electrode is polarized with a pulse of rectangular voltage having linearly increasing amplitude and the current measured during the second half of the pulse. The polarograms obtained in this way are identical in shape with those obtained in classical polarography, but since the time between the moment of application of pulse to the moment at which the current is measured is

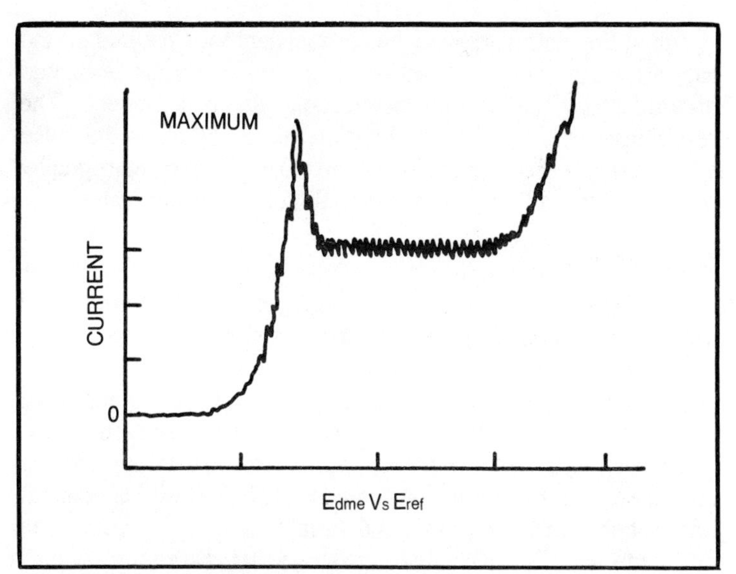

Fig. 13-15. Polarogram exhibiting maxima.

much shorter than the droptime, the limiting value of the measured current is much higher than that observed in classical polarography (Thomas, 1977).

Pulse polarography can be carried out in two modes: *normal pulse polarography* and *differential pulse polarography.* Normal pulse polarography involves the imposition of square wave voltage impulses of increasing magnitude upon a constant dc voltage. In the differential pulse mode, fixed magnitude pulses superimposed on a dc voltage ramp are applied to the working electrode; the currents are measured prior to pulse application and just before termination of the voltage pulse. For a given pulse, the output is recorded as the difference between the two current flows. Burge (1970) describes the details of pulse polarographic technique.

Figure 13-16 shows the various types of pulses involved in pulse polarography. The cell current in pulse polarography would consist of the following:

■ A background current due to reductions occuring at the applied ac voltage.

■ A current due to the charging and discharging of the double layer capacity around the mercury drop when the pulse is applied. If the cell impedance is low enough, this falls to zero within 20 ms.

■ A faradic current associated with the change of reaction rate as the pulse is applied.

For the polarogram to be independent of background and capacitive currents, the effect of background current is estimated instrumentally and is subtracted from the cell current. The remaining current is measured during the last 20 ms of the pulse when the capacitive current has fallen to zero. Pulsed polarography is useful in determining trace metal pollutants in air and water.

COULOMETERS

Ordinary titrations are carried out by measuring the volume of a standard solution that is required to react with the substance to be determined. The process involves the use of a burette and the preparation and storage of standard solutions. On the other hand, coulometric methods of analysis depend on the exact measurement of the quantity of electricity that passes through a solution during the course of an electrochemical reaction. The quantity of reactant formed between the beginning and the interruption of current at the end of the process is directly related to the net charge transferred, Q. Analytical methods based on the measurement of a quantity of electricity are designated by the generic term of *coulometry,* a term derived from coulomb.

According to Faraday's law, the quantity of electricity involved in the electrolysis of one equivalent of substance is one faraday or 96,494 coulombs. The weight W of a substance consumed or produced in an electrolysis involving Q coulombs is

$$W = \frac{W_m Q}{96{,}494\,n}$$

where W_m is the gram atomic weight or gram molecular weight of the substance being electrolyzed. Note that, n is the number of electrons involved in the electrode reaction.

The equation can also be expressed as

$$Q = nFVC_b$$

where F equals 96,490 coulombs, V equals volume of solution in liters and C_b equals bulk concentration of the electrolyzed analyte in moles/liter.

The total amount of electricity (Q) which is required to electrolyze a certain species is the current time integral

$$Q = \int_0^t i\,dt$$

This integral is equal to the area under the i-t curve.

Coulometric methods employ two techniques: *potentiostatic coulometry* or controlled potential coulometry and *amperostatic*

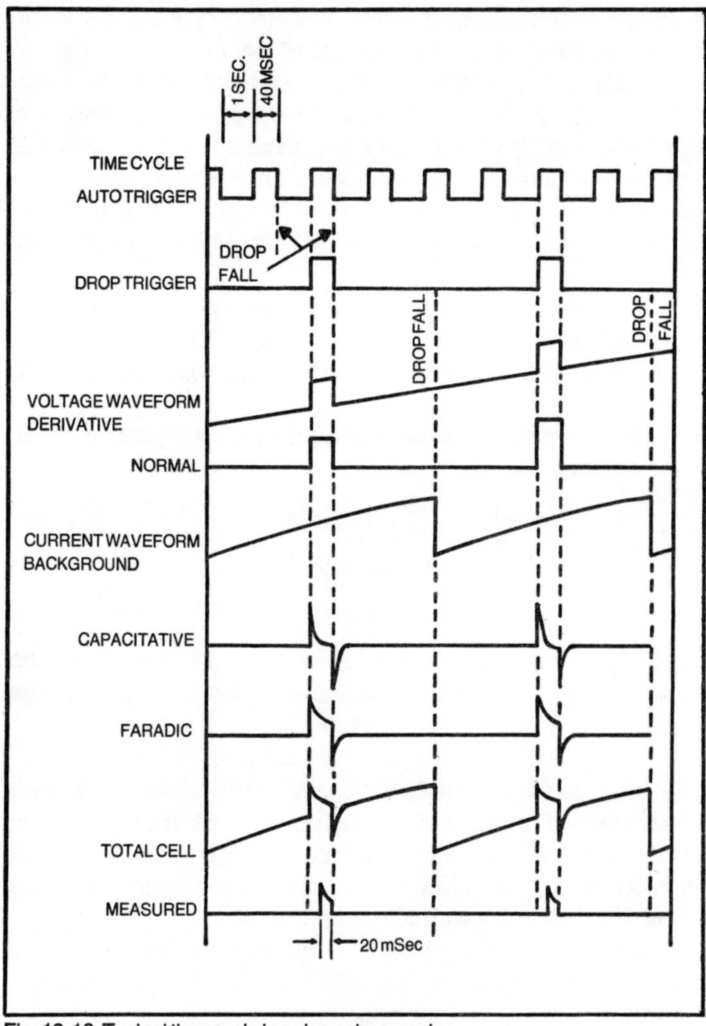

Fig. 13-16. Typical time cycle in pulse polarography.

coulometry or controlled current coulometry. In controlled potential coulometry, the potential of the working electrode is controlled at a constant value and the electrolysis current is measured against time. Completion of the electrolysis is indicated by decay of the current to a negligibly small value. Typical working electrodes used in this analysis are platinum, silver, silver-chloride and mercury.

Figure 13-17 shows a typical current time curve for controlled potential coulometry. Before starting the electrolysis, the working electrode potential is chosen which will electrolyze the species of interest. If this potential is not known before hand, it can be determined from a polarogram of a standard solution of the given analyte. When the electrolysis is begun, the current increases to a high value initially. It then falls exponentially as the analyte is consumed. Obviously, it is possible to carry out the electrolytic generation to infinite time. In practical analysis and the electrolysis is complete when the current has decayed to less than 0.1% of the initial current.

Integration of the i-t curve may be done by graphic, mechanical, electromechanical or electronic means. The integration unit may be attached to the potentiometric recorder which draws the i-t curve.

The current decay in controlled potential coulometry is given by the equation

$$i_t = i_o \times e^{-kt}$$

$$\text{or } 2.3 \log \frac{(i_o)}{i_t} = kt$$

where i_o is the initial current and i_t the current at time t. When the logarithm of the current is plotted as a function of time, the intercept at $t = 0$ is i_o and the slope is $-\frac{k}{2.3}$

In a controlled current titration, the current is set at a working electrode and maintained throughout the titration, with the potential at the indicating electrode measured against time. A mercury or platinum electrode is used as the working electrode along with a reference and indicating electrode.

The success of most coulometric titrations depends on the ability to attain 100% current efficiency. This implies that the amount of titrant produced in the electrolysis is exactly equal to that predicted by Faraday's law. Judicious selection of the current density (current/unit electrode area) can mean the difference between success and failure of a coulometric titration. A value of 0.5 mA, cm^{-2}, mN^{-1} is commonly stated as near the maximum limiting current density for many substances. Coulometry involves only the fundamental quantities of current and time. It is thus free from many of the uncertainties and errors associated with standard solutions. It offers excellent precision and accuracy and is useful for analysis of very small amounts ranging as low as a few hundredths of a microgram in volumes of 5-50 ml.

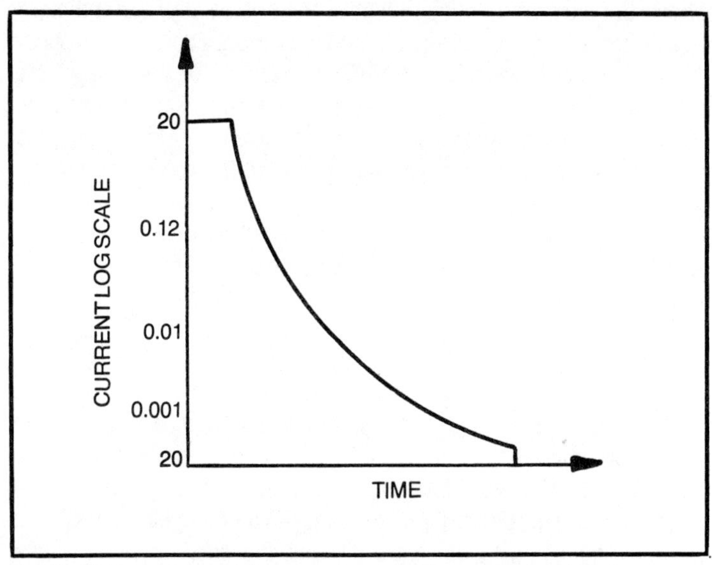

Fig. 13-17. Typical current-time curve for controlled potential coulometry.

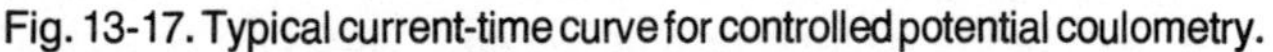

A coulometric cell is an electrochemical cell in which the two electrodes can be a platinum pair. As electrochemical reactions can occur at either electrode, the two electrode compartments must be separated by a suitable membrane to prevent interaction between the products. A sintered glass disc or an agar-gel plug is often used for this purpose. The electrode compartment of interest must be suitably stirred. The end points are usually detected with a microammeter.

Coulometric experiments range from chronocoulometry (Anson 1966, Christie et. al. 1964 a, b) whereby steps or sweeps of electrode potential produce current transients of less than 10^{-3} seconds duration, to the recording of charge-time relations or total final charge values in electrosynthesis, electrodeposition or controlled potential coulometry (Bard and Santhanam 1970), operations which last several minutes or hours.

AMPEROMETERS

Amperometry is the method of determining the concentration of an electroactive substance by applying a fixed voltage across an indicator and reference electrode and then measuring the current passing through the cell. This technique is particularly well suited to trace analysis. The current measured is generally on the diffusion-current plateau of the current-voltage curve, a region where the current is independent of the potential of the indicator electrode. The rate of diffusion and hence the current is proportional to the concentration of diffusing material in the bulk of the solution. The most common use of amperometry is in titrations where the current is measured as a function of the volume of titrant added. Concentration changes during a titration are reflected in a change in the current.

Amperometric technique may be used with either a single polarized electrode or two polarized electrodes. The electrodes in either case are usually small with surface areas of a few tenths of a square centimeter.

In the single electrode method, a polarized electrode, coupled with an unpolarized electrode (saturated calomel electrode), is immersed in the solution being titrated. The polarized electrode may be a dropping mercury electrode, a rotated electrode or a stationary electrode. The cell is connected to a manual type polarograph. The potential of the polarized electrode is held at a constant value during the titration and the current which flows through the system is observed. The second electrode acts as a reference electrode. Typical titration curves in amperometric titrations with one polarized electrode may take a variety of forms depending on whether the electroactive species is the titrate, titrant or a product of the titration reaction.

In the two electrode system, two stationary platinum wire electrodes are immersed in the titration cell. The potential of both electrodes vary during the titration, but the potential difference between them is kept constant. The current through the cell is measured during titration. The equivalence point is deduced from the plot of current against the volume of titrant. The applied potential difference is relatively small, 0.01 to 0.1 volt, and the current is generally not as large as in single indicator electrode amperometry. The two electrode system is also called *dead stop end point* and is particularly applicable when a reversible oxidation-reduction system is present either before or after the end point.

Amperometric titrations are carried out in a polarographic cell suitably modified to permit entry of a burette and stirrer. H-type cells are also convenient, but a wide-mouthed 100 ml flask or beaker fitted with a suitably pierced cover is mostly employed. When oxygen is known to interfere, it may be removed in the usual way. For stirring, gas bubbling is frequently employed. Any polarographic instrument may be employed to carry out amperometric and titrations.

Advantages of amperometric titrations are that the electrode characteristics are unimportant and the method offers greater sensitivity than conductance and potentiometric titrations. The method is applicable in very dilute solutions even down to 10^{-4} or 10^{-5} M, according to the type of electrode used.

AQUAMETERS

Aquametry consists of determining small amounts of water in solids, liquids and gases. This is done by titration with Karl Fischer Reagent (KFR) and the end point is determined colorimetrically or electrometrically. The electrometric method is generally preferred because highly colored samples obscure the colorimetric end point. Mitchel and Smith (1948) describe the technique and applications using KFR.

Automated titration equipment has been designed in which the titrant delivery is controlled by an automatic correcting circuit that varies the rate of titrant addition to minimize titration time while maintaining high precision and repeatability Sensitivity as great as 1 ppm of water in 100 ml of sample is attainable.

A Karl Fischer titrator essentially consists of two parts. The burette assembly consists of a piston traveling in precision bore glass tubing. The inside diameter of the tube is very accurately

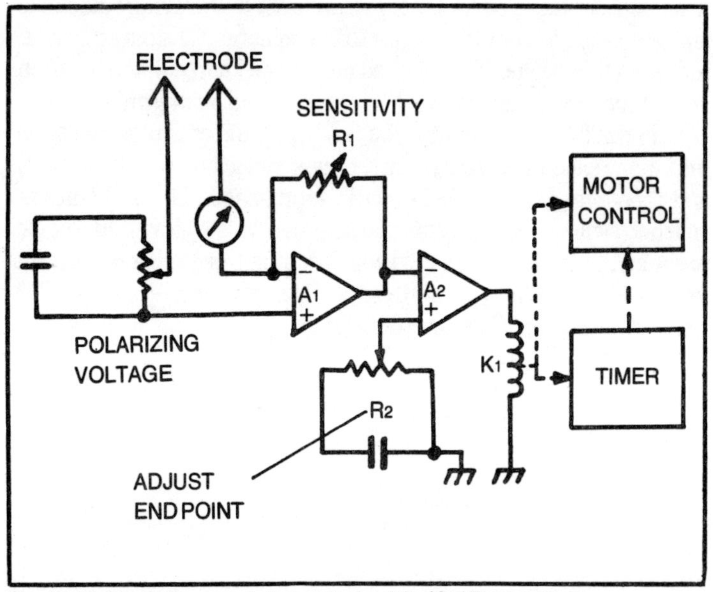

Fig. 13-18. Block diagram of electronic circuit for Karl Fischer titrator.

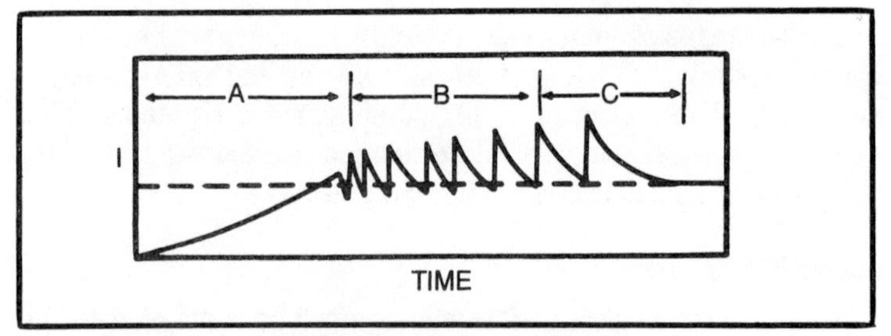

Fig. 13-19. Electrode current versus time for a simulated titration.

controlled with a tolerance of + 0.0002 inch. The pistons are coupled with the counters which are driven from the lead screw by a toothed belt. Backlash is held to less than 0.002 ml. A small dc motor is connected to the lead screw by means of a friction drive and gear set with a speed reduction of 200:1. Limit switches are used to automatically stop the motor at both ends of the piston stroke.

Figure 13-18 shows the block diagram of the Karl Fischer titrator KFR-4 from Beckman. A voltage is applied to two platinum electrodes immersed in the sample solution. The resultant current is measured by amplifier A_1 and compared to a reference voltage by amplifier A_2. When the output of A_1 exceeds the reference voltage set at R_2, relay K_1 is energized. The relay contacts turn the burette drive motor on and off. A timer circuit is used to provide selectable and point delays from 10 seconds to two minutes. The burette drive motor should be speed controlled during the course of the titration so that the time required to reach a precise end point is minimized.

In the Karl Fischer titration, each addition of titrant causes an initial increase in electrode current which decays, rapidly at first, and then slowly as the end point is approached. The end point is reached when the current stays above a set level for some preselected time. Figure 13-19 shows a plot of electrode current versus time for a typical titration. Haagen-Smit et. al. (1971) describe the sample handling techniques.

Chapter 14

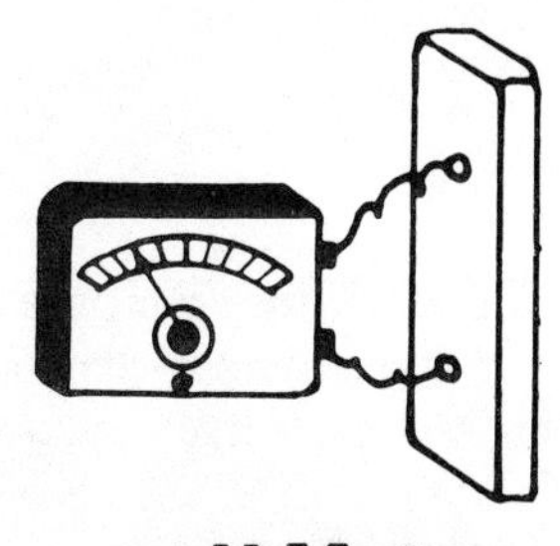

pH Meters

The concept of pH was introduced by Sorensen in 1909. He recognized that hydrogen ion concentrations, as distinct from total acidities, are frequently of importance in chemical processes. While studying enzymatic reactions, he found it convenient to define a symbol which could represent the concentration of hydrogen ions and called this symbol pH. It is defined by the following equation:

$$pH = -\log_{10} C_H$$

where C_H is the hydrogen ion concentration.

$$\therefore C_H = 10^{-pH} \qquad \textbf{Equation 14-1}$$

Pure water is known to be a weak electrolyte, and it dissociates to form hydrogen ions and hydroxyl ions as follows:

$$H_2O \quad H^+ + OH^-$$

Assuming that activity coefficients are unity, the dissociation constant K_W of pure water is given by:

$$K_W = C_{H^+} \times C_{OH^-} \qquad \textbf{Equation 14-2}$$

The product of hydrogen and hydroxyl ions in water at 25°C is 1.008×10^{-14} moles^2 liters^{-2}, and the concentrations of hydrogen and hydroxyl ions will of necessity be equal. Since the positive and negative electric charges in the solution must balance, each of these concentrations is given by

$$C_{H^+} = C_{OH^-} = \sqrt{K_W}$$

$$= \sqrt{1.008 \times 10^{-14}}$$

$$= 1.004 \times 10^{-7}$$

$$C_{H^+} \approx 10^{-7} \qquad \textbf{Equation 14-3}$$

Therefore, the pH of pure water = 7

It is obvious that the neutral point or point at which the hydrogen and hydroxyl ions are present in equal concentrations is located at pH 7. The pH of an acidic solution [i.e. $(H^+) > (OH^-)$] at 25°C will be less than 7, and that of an alkaline solution will be greater than 7. The peculiarity of the logarithmic scale is exemplified by the fact that a tenfold decrease in hydrogen-ion concentration corresponds to an increase of one pH unit, whereas a twofold increase of the concentration means pH will decrease by log 2, i.e., by 0.301 units.

The dissociation constant K_W is a function of temperature and, therefore, the neutral point will vary as the temperature is changed. The pH for neutral conditions decreases to 6.5 at 60°C from the value of 7 at 25°C and increases to 7.5 at 0°C. The range of the pH scale also depends upon the magnitude of K_W.

The approximate practical range of the pH scale is from −1 to 15 at room temperature, although most of the commercial instruments are designed to measure 0 to 14 pH.

There is, at present, a wide utilization of pH measurements in the chemical laboratories, industries and the clinics. This has been made possible by the discovery of the hydrogen ion function of glass membranes which led to the development of convenient, practical glass electrodes, pH meters and controllers that allow the pH of process solutions to be adjusted automatically. The technology of pH instrumentation has been greatly developed and extremely sensitive instruments are now commercially available.

With the developments in chemical thermodynamics, it has gradually become clear that Sorensen's experiments did not, in reality, yield hydrogen ion concentrations. No doubt the numbers obtained depended in a complex manner on the activity of the electrolytes in the solution under investigation. They were not an exact measure of the hydrogen ion activity and indeed could never be made so. Sorensen's measured values of ph are not, therefore, values of $C_H^{\cdot}$ + (hydrogen ion concentration) as he originally considered them to be. However, the methods of pH measurement have been standardized by convention in a manner which allows a maximum of theoretical significance to be place upon the experimental results.

PRINCIPLE OF pH MEASUREMENT

The measurement of hydrogen ion concentration (pH) in a test solution is made by measuring the potential developed in an electrochemical cell. The electrochemical pH cell consists of a measuring electrode and a reference electrode, both immersed in the solution under investigation. The two electrodes are connected to a measuring instrument and the emf between these two electrodes is measured. The measuring electrode is pH sensitive and its potential is proportional to the pH of the solution in which it is immersed, while the reference electrode would always develop a constant electrical potential against which the potential of the glass electrode is measured.

The potential of the measuring electrode may be written by means of the Nernst equation:

$$E = E_o + 2.3026\,RT/F \quad \log C_H$$

$$E = E_o - 2.3026\,RT/F \quad pH_c$$

when

E_o = Standard potential

R = Gas constant

T = Absolute temperature

F = Faraday constant

pH_c = pH value deviation from 7

The equation shows that the emf developed in the electrochemical pH cell is a linear function of pH_c. Figure 14-1 shows the relationship between pH and temperature of a typical glass electrode.

Change of pH of one unit = 58.2 mV at 20°C
= 62.2 mV at 40°C

The factor −2.3026 RT/F is called the *slope factor* and is obviously dependent upon the solution temperature. It is clear that with 1° change in temperature, the emf changes by 0.2 mV. Further, pH measurement is essentially a measurement of millivolt signals by special methods.

For measurement of pH, the electrodes are first immersed in a buffer solution of known pH. The pH meter zero reading is adjusted by the standardization control until the pH value of the buffer is indicated by the meter. This standardization automatically compensates for the various potentials in the electrode system. Subsequently, immersion of the electrodes in a test solution produces a potential that is proportional to the pH of the solution. This potential registers directly as pH on the scale of the pH meter. The temperature compensation knob is set at the temperature of the solution.

ELECTRODES FOR pH MEASUREMENT

The *hydrogen electrode* is the primary electrode to which all electrochemical measurements are referred. However, owing to the experimental difficulties associated with it, other electrodes are commonly employed for routine pH measurements. Nevertheless, the performance of all other electrodes is always evaluated in terms of the hydrogen electrode.

Hydrogen Electrode

The *hydrogen electrode* consists of an inert but catalytically active metal surface, most frequently platinum, over which hydrogen is bubbled to achieve electrochemical equilibrium with the hydrogen ions in the solution. The following redox reaction takes place:

$$H^+ + \bar{e} \rightleftharpoons 1/2\, H_2$$

The electrode is immersed in the solution under investigation and electrolytic hydrogen gas at 1 atmosphere pressure is bubbled through the solution and over the electrode in such a way that the electrode surface and the adjacent solution gets saturated with the gas at all times. Electrode life is 7-20 days before its response becomes sluggish (Perley, 1948). The potential setup at the hydrogen electrode by a given activity of hydrogen ions is governed by the Nernst equation. When the partial pressure of the hydrogen is other than 1 atmosphere pressure correction would have to be applied.

Since the hydrogen electrode is essentially a redox system and as such is affected by the presence of oxidizing and reducing

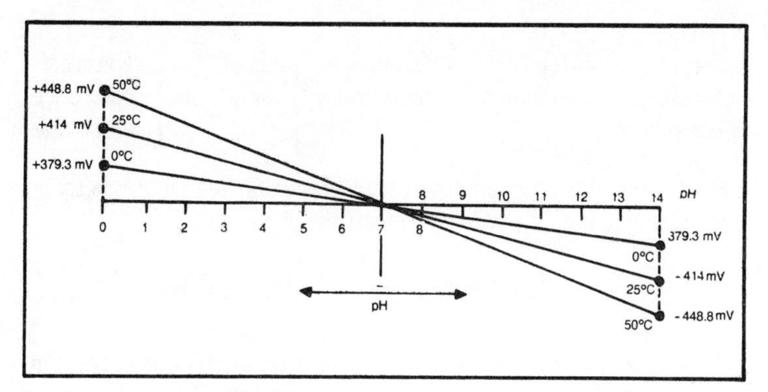

Fig. 14-1. Relationship between pH and emf at different temperatures (courtesy of Beckman Instruments, USA).

agents, it is, therefore, subject to a number of limitations in its application (Mattock 1961).

Glass Electrode

The widespread application of pH measurements in the control of industrial processes as well as in research can be largely attributed to the development of the convenient and versatile glass electrode. Its action is based on the principle that when a thin membrane of glass is interposed between two solutions, potential difference is observed across the glass membrane, which depends on the ions present in the solutions. Depending on the composition of the glass, the response may be to H^+ ion or it may be to other cations. The selective response of certain glass compositions to H^+ has led to the development of pH-responsive glass electrodes.

In construction, the glass electrode consists of a thin walled bulb of pH-sensitive glass sealed to a stem of non-pH sensitive high resistance glass. The pH response is limited entirely to the area of the special glass membrane, thus making the response independent of the depth of immersion. The membrane normally has a thickness of the order of 0.05 to 0.15 mm, and the bulbs are of the order of 10 mm in diameter. Figure 14-2 shows typical construction of a glass electrode. Both surfaces of the membrane are pH sensitive. On the inside of the membrane is a system of effectively constant pH. It is composed of a silver-silver chloride or calomel electrode dipped in hydrochloric acid. Changes in electrical potential of the outer membrane surface are measured by means of an external reference electrode and its associated salt bridge. The complete pH cell is represented as follows:

Internal reference electrode	Internal electrolyte	Glass membrane	Test solution	External reference electrode

The ideal pH response of a glass electrode behaving exactly in the same manner as a hydrogen electrode is given by:

$$E_2 - E_1 = 2.3026 \frac{RT}{F} (pH_2 - pH_1)$$

where E_1 and E_2 are the values of the electromotive force of cell 1 in test solutions of pH equal to pH_1 and pH_2 respectively.

This equation shows that ideal pH response is 54.2 mV at 0°C, 59.16 mV at 25°C and 73.04 mV at 95°C. Unfortunately, no glass electrode yet constructed has the theoretical response in all types of test solutions and over the entire pH range.

The most important characteristics of a glass electrode are low melting point, high hygroscopicity and relatively high electical conductivity. For many years, the best pH sensitive glass available was CORNING 015 or SCHOTT 4073 glass. However, this glass gives a reasonably good response only in range 3 to 9 pH. Outside this range, the glass membrane is subject to errors, tending to give a high reading below a pH of 3, and too low above pH 9. The errors also depend upon the ions which may be present. The error in the alkaline range is pronounced especially in the presence of sodium ions. Other types of glasses with considerably smaller alkaline errors have been developed.

A glass membrane exhibits quite a high electrical resistance and consequently the internal resistance of the cell with a glass electrode is of the order of 50-1000 MΩ. The emf measurement, therefore, necessitates the use of measuring circuits with high input impedance. Furthermore, the electrodes and the leads from them must be supported on holders made of a good insulating material in order to eliminate electrical leakage across the outside surface of the glass bulk. Sometimes the upper part of the outside of the glass electrode is rendered water repellent by the application of a silicone oil or paraffin wax. The high resistance of glass electrodes renders them very susceptible to capacitive pickup from ac mains or charged bodies. In order to minimize such effects, it is necessary to screen the electrode cable. The screen may be connected directly to earth or is grounded to the case of the instrument.

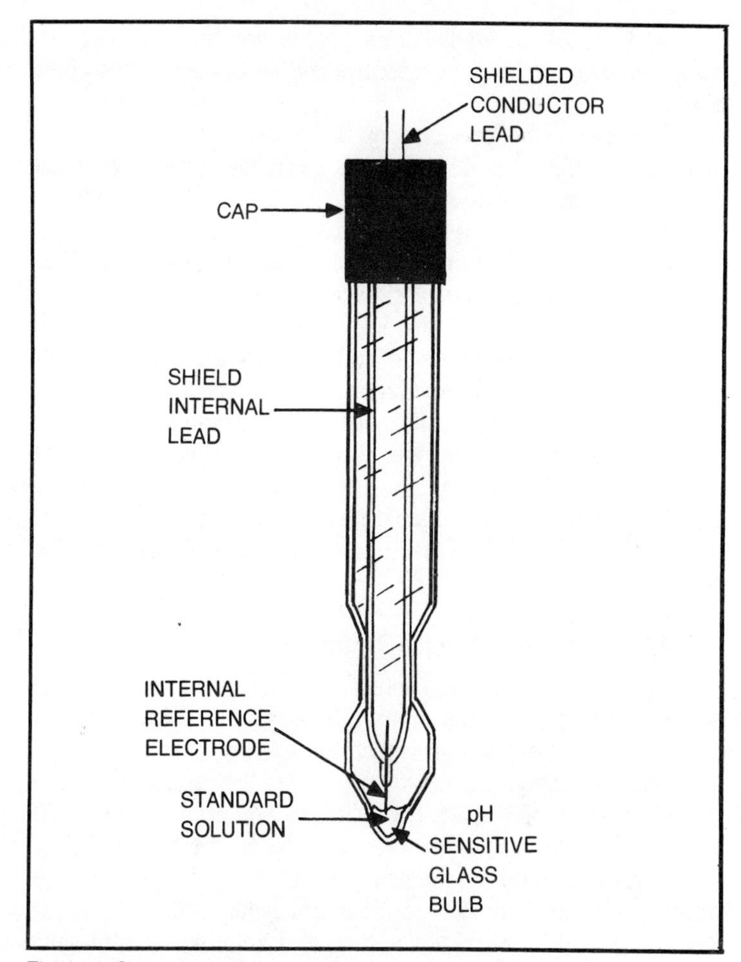

Fig. 14-2. Construction detals of Beckman glass electrode.

Commercial glass electrodes are available in a wide variety of sizes and shapes. They are designed to operate with samples as small as one drop; others require at least 5 ml of solution. Special microcells for the pH measurment of blood are supplied by several manufacturers.

New electrodes or those that have been stored dry should be conditioned or activated before use by soaking the bulb for a period of 12-24 hours in 0.1 N hydrochloric acid. The relative fragility of the glass membrane demands reasonable care in its handling. The pH sensitive tip of the electrode should not become scratched or cracked through contact with the sample or container. Therefore, it should not be allowed to rest on the bottom of the sample container. The tip should be dried by gentle rubbing with absorbent tissue.

Sometimes the presence of an inhibiting film can have a deleterious effect on pH response. Even the extraction of the soluble components of the glass from the surface of the membrane may cause a type of deterioration marked by sluggish behavior. Sluggishness caused by immersion in alkalis can often be eliminated by soaking in hydrochloric acid, and that caused by strong acids is best overcome by standing the electrode in water. If this treatment fails to correct the difficulty, it is sometimes possible to rejuvenate the electrode by immersing it for about 1 minute in a 20% solution of ammonium bifluoride at room temperature. For the fluoride solution, a waxed paper cup or a waxed beaker should be used. This treatment should be carried out very carefully since attack can cause puncture of the thin membrane. It should be tried only when other measures fail to satisfactory performance of the electrode.

Calomel Electrode or Reference Electrode

In order to measure the potential changes of the pH sensitive electrode directly, it is necessary that the ph cell be completed by means of a stable reference electrode whose potential remains unaffected by changes in the composition of the cell solution. The reference electrode against which the potential of the glass electrode is measured is the *calomel electrode*. It consists of (Fig. 14-3) a metallic internal element typically of mercury-mercurous chloride (calomel) or silver-silver chloride, immersed in an electrolyte which is usually a saturated solution of potassium chloride. The electrolyte solution forms a conductive salt bridge between the metallic element and the sample solution in which the measuring and reference electrodes are emplaced. For a stable electrical connection between the internal metallic element and the sample solution, a small but constant flow of electrolyte solution is maintained through a liquid junction in the tip of the outer body of the reference electrode. Depending upon the nature of the application, this junction may be formed in several ways. For example, the tip of the electrode could be formed by an embedded linen or asbestos fiber or by a permeable composition of pressed-sintered Carborundum and glass pellets.

$$1/2\,Hg_2\,Cl_2 + e \rightleftharpoons H_g + Cl^-$$

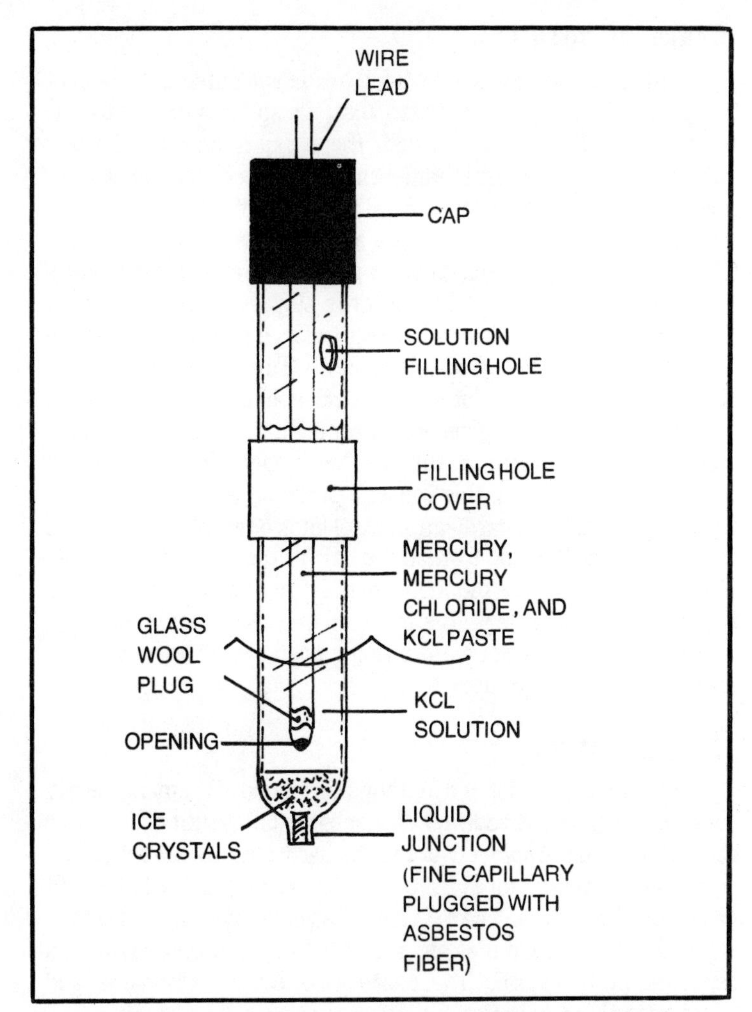

Fig. 14-3. Calomel electrode (courtesy of Beckman Instruments, USA).

Since at a given temperature the activity of the mercurous chloride is constant and that of the mercury is unit, by definition, it is the chloride ion activity which is potential determining. When this is fixed, the electrode has a fixed potential at a fixed temperature. The most commonly used source of chloride ion is potassium chloride at saturated 3.8 M, 3.5 M or 0.1 M concentrations. Saturated electrodes have largely replaced electrodes with other KCl concentrations. They are used in practical pH measurements at high as well as low temperatures. However, their useful life is known to be short at temperatures above 70°C.

ASYMMETRY POTENTIAL

If identical solutions are placed inside and outside the bulb of the glass electrode, it is found that, in spite of the apparent symmetry of the cell so formed, there exists an emf of a few millivolts. This potential difference is called the *asymmetry potential* and is thought to arise from the different states of strain on the inside and outside surfaces of the glass electrode bulb (Yoshimura, 1937). The asymmetry potential of a glass electrode is not absolutely constant but may drift slightly from day-to-day. However, it is not ordinarily subject to large and sudden fluctuations particularly in good electrodes. The existence of this potential is, therefore, of no practical consequence in the pH measurement. It may be regarded, for a short series of measurements, as a constant of the cell assembly. In most of the commercial pH meters, this constant potential is compensated by adjustment of the standardizing control knob or zero adjuster when the instrument is made to read the pH for the standard solution. The standardization of the instrument is carried out immediately before the measurements are made on unknown solutions. As an extra precaution, the standardization procedure can be repeated at the end of the measurement.

BUFFER SOLUTIONS

A *buffer* may be defined as a solution whose pH remains nearly constant despite the addition of a substantial quantity of acid or base. Buffers are employed for the standardization of pH cells. The cell emf is measured for given buffers and related to the known pH values for the buffers. The pH value of an unknown solution is then derived from this calibration. Many buffer solutions have been reported in literature. There are also British standards and standards of NBS on buffers.

Buffer tablets are commercially available which, when dissolved in an appropriate quantity of fresh distilled water, give buffer solutions. These tablets may contain the buffer material admixed with substances which do not significantly affect the pH. Buffer tablets for pH 4, 7 and 9.2 are available commercially.

DESIGN OF pH METERS

In the earlier days, the pH of a solution was determined with the change in color observed on pH paper when dipped in the solution of interest. The color was compared with the color chart. It is possible to read pH to an accuracy of 0.1 pH by this method.

With the advent of glass and calomel electrodes and with the development of very stable, drift free dc amplifiers with extremely high input impedance, pH measurements to much better accuracy have become possible.

The following considerations govern the design of pH meters:

■ The internal resistance of the glass electrode is very high. It is of the order of 1000 M. Therefore, the input impedance must be at least 1000 times more than the resistance of the pH cell. Also, the measurement should be unaffected by large changes in magnitude of this resistance.

■ No current should either be drawn by the pH meter from the solution or should flow on the electrode which might result in the polarization of the electrode. Such polarized electrodes will give rise to erroneous results.

■ The meter must have a provision for compensating changes in pH readings due to changes in temperature.

In general, commercial pH meters can be categorized broadly into two main types: the *null detector* type or the *potentiometer* type and the *direct reading* type. In the instruments of the first type, the procedure followed is essentially that used in potentiometers. An emf equal and opposite to that of the pH cell is applied so as to give zero reading on a galvanometer. Direct reading instruments are similar to voltmeters of the deflection type. The current signal available after the amplifier is used to operate a meter, suitably calibrated in pH units. Of the two types, the null-detector type of instrument is inherently capable of greater accuracy than the direct reading type. However, the convenience and the usefulness of the direct reading type meters in following the changes of pH that occur during the course of a reaction have made them more popular.

pH meters invariably make use of some amplifying device to amplify the emf produced in the pH cell. As the amplifier input stage must have an extremely high input impedance, a number of commercial pH meters have been making use of vacuum tubes like 12A × 7 as amplifiers. They were specially selected for low grid currents. However, special electrometer tubes have been designed in order to minimize the flow of grid current and the stray currents emanating from surface leakage, emission of photoelectrons by the grid under the action of light and soft X-rays. These tubes are operated at lower potentials as compared to other vacuum tubes to minimize the ionization of gas in the tube. Quite recently, electrometer tubes have been replaced by solid state devices like field-effect transistors, MOSFETs and integrated circuits having high input impedance. IC 8007 is an example of the integrated circuit having a FET input stage which is quite suitable to be used in the input stage of a pH meter.

For making pH measurements, the emf of a pH cell is amplified in a direct coupled amplifier by giving it directly to the grid of a suitable electrometer tube. Note that dc amplifiers give rise to zero drift errors which must be eliminated for getting accurate results. Improvements in zero stability are possible by the use of low drift and low input offset voltage integrated circuits and highly stable power sources. Zero drift can also be reduced by the use of balanced and differential amplifiers. They are so constructed that their responses to external signals are additive while those to internal noise or drift are subtractive. Several other methods are available by which zero stability can be achieved to a great extent by using zero-corrected dc amplifiers, a contact-modulated amplifier and vibrating capacitor modulated amplifiers. These methods are discussed in the subsequent sections.

The Nernst equation on which pH measurements are based contains a temperature dependent component. Therefore, arrangements are invariably made for automatic or manual compensation in changes of pH due to changes in temperature in the commercial pH meters. The instrument is calibrated at one temperature (say 25°C) and compensation is applied by suitable adjustments of the output current of the amplifier. The current is adjusted by incorporation of a variable resistance in the output circuit so that the calibration point may then correspond to the desired temperature. This ensures that for a given pH, the current to the meter is constant.

Automatic temperature compensation is achieved by using a resistance thermometer or thermistor in the output circuit. The thermistor is mounted on the electrode holder and is immersed in the test solution along with the electrodes. As the temperature of the solution changes, the circuit constants are altered accordingly.

NULL DETECTOR TYPE pH METERS

The simplest form of schematic for pH measurements with glass electrodes employing an ordinary potentiometric circuit with an electrometer tube for signal amplification is shown in Fig. 14-4. The glass electrode lead is connected to the grid of the electrometer so that changes in the electrode potential causes changes in the anode current. A standard cell sends a constant current through the

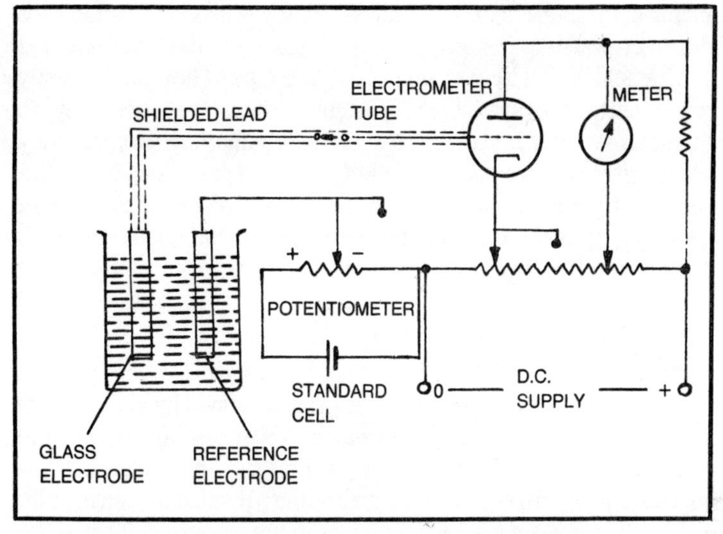

Fig. 14-4. Null detector type pH meter using an electrometer tube.

slide wire and establishes a voltage difference across it as in any potentiometer.

The potentiometer circuit is connected to the cathode of an electrometer value. A sensitive galvanometer is placed to indicate zero, usually at center scale, for a particular value of anode current corresponding to zero difference between the potentials of electrode and potentiometer. This requires the provision of an auxiliary control which is adjusted when the electrodes are placed in a buffer solution and the potentiometer dial set to the known pH value of the buffer. The electrodes are then placed in the solution of unknown pH and the potentiometer is adjusted to reestablish balance with the emf of the pH cell. The pH value is read from the calibrated potentiometer dial marked off in pH units. The initial standardization of the potentiometer is carried out by replacing the electrodes with a standard cell and making the adjustment by varying the current in the potentiometer by means of a variable series resistor. Also, the point on the potentiometer against which the standard cell is balanced is usually made adjustable by a control calibrated in temperature. This permits the pH calibration of the potentiometer to be made correct for any desired solution temperature. It is necessary to repeat the standardization procedure at any new temperature setting. Instruments of this type are fundamentally simple in principle as the electrode potential is determined by direct comparison with a standard cell. The limiting factor in measurement with null-detector circuits is the slide wire accuracy which is generally 0.1%. Accuracy better than 0.01 pH can be achieved with instruments of this type. They can be easily made battery operated and portable. Also, they are easy to maintain as the electronic circuit is very simple. However, they would often need replacement of the standard cell. To improve the sensitivity of detection, and additional stage of amplification may be used. As these circuits require readjustment frequently, they are unsuited to long time unattended operations in the industrial field.

DIRECT READING pH METERS

Direct reading instruments are characterized by simplicity of operation and speed of measurement which are of considerable importance for making routine laboratory and industrial pH measurements. In these instrumenty, the pH value is indicated by the display system (meter pointer or in the numerical form in the digital display system) without any balancing process by the operator. After making an adjustment on one buffer solution only, samples may be measured in rapid succession without having to operate any control on the instrument. Also, if employed to measure a solution in which the pH value is changing, the direct reading meter will give a continuous indication of the value which could even be recorded for subsequent control of a process.

CHOPPER AMPLIFIER TYPE pH METERS

In the *chopper amplifiers*, the direct voltage is chopped at the mains frequency and amplified as alternating voltage. The output is rectified by a phase-sensitive circuit and indicated by a suitable meter.

Basically, a chopper is a form of relay consisting of very light moving parts so that they may be operated repetitively at a fairly high frequency. The moving contact vibrates in response to the presence of an alternating current electromagnet and alternately makes and breaks contact with a pair of contactors connected to the ends of the center tapped primary winding of a transformer. The secondary winding is connected to the input of the amplifier.

Figure 14-5 shows the block diagram of a chopper amplifier type pH meter. The oscillating contact of the chopper, in one position, short-circuits the input of the amplifier while in the other it is connected to the glass electrode. The reference electrode is connected back to the amplifier through a negative feedback

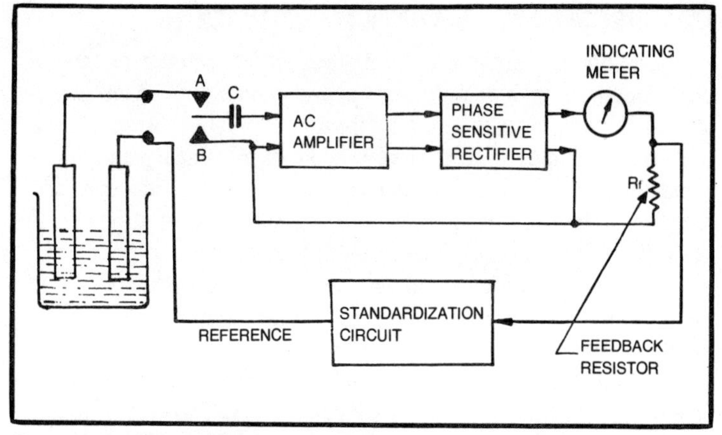

Fig. 14-5. Chopper amplifier type pH meter.

resistor R_f. If the current in the feedback resistor is zero and the chopper contact is at position A, the glass electrode potential will be applied to the amplifier input. When the relay changes to position B, the amplifier input is brought to zero abruptly. When the contact moves at a fast rate between the positions A and B, a square wave voltage waveform is produced which can be capacitively coupled to the input amplifier.

Capacitive coupling isolates the electrodes from any grid potential which may be present. The alternating voltage is amplified by an ac amplifier and given to a phase-sensitive rectifier. The polarity of the dc output of the phase-sensitive rectifier reverses with the changes in the phase of the ac input by 180°. A resistance R_f is included in the feedback circuit and the current flowing through it produces a potential across it which is in opposition to the electrode potential, indicating that the meter indication gives the difference voltage.

The use of chopper amplifiers for high input resistance signals presents problems not encountered with low input resistance signals. For example, the making and breaking of the contact to the glass electrode leads changes the capacity at the surfaces very abruptly and gives rise to a sudden voltage surge which is transmitted to the amplifier as a series of *spikes* on the waveform. This results in some degree of zero instability in the dc output of the phase-sensitive rectifier. Also, with aging the surface conditions change and the magnitude of the surges increase. These factors demand special care in design and construction of pH meters based on this principle.

VIBRATING CONDENSER AMPLIFIER TYPE pH METER

In a chopper amplifier, the main problem is the presence of spikes due to transient potentials at the contractors. This disadvantage is overcome by using a vibrating condenser in place of a mechanical chopper. The dc signal is converted to ac by applying it to one plate of a condenser whose capacity is changed at a constant frequency. The capacity is changed by vibrating one of its plates. If the distance between the plates is varied sinusoidally, capacity will vary in an inversely sinusoidal way. So, for a given value of charge q, the voltage V across the condenser will vary sinusoidally. The ac signal is processed in the same manner as in a chopper amplifier. The method offers infinite input resistance which is limited only bt the insulating material used in its construction. The zero stability is much better than a direct coupled amplifier. For a good performance, the frequency and amplitude of the vibrator should be stable and constant. Figure 14-6 shows the block diagram of a vibrating capacitor amplifier type pH meter.

ZERO CORRECTED dc AMPLIFIER TYPE pH METER

The zero stability of a *direct-coupled amplifier* can be improved by incorporating an additional circuit for measuring the amount of drift present in the amplifier and making a corresponding correction automatically. This type of amplifier is used in the Beckman Zeromatic pH meter. In this instrument, automatic zero correction is employed at intervals of one second without disturbing the reading. The pH meter is provided with a switch between the glass electrode and the input valve which makes it possible to disconnect the electrode and short circuit the amplifier input. This is used to

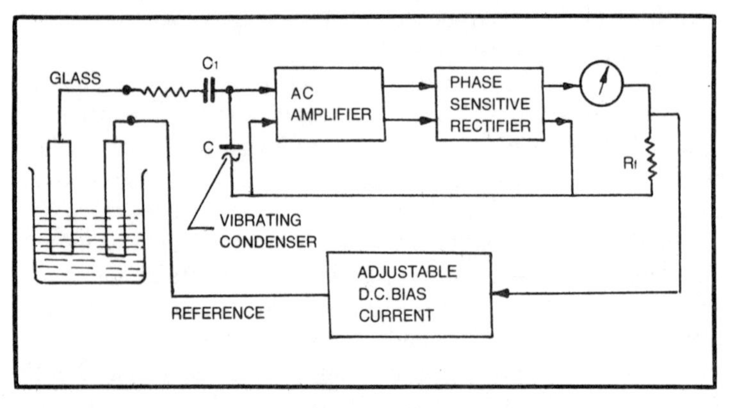

Fig. 14-6. Vibrating capacitor amplifier type pH meter.

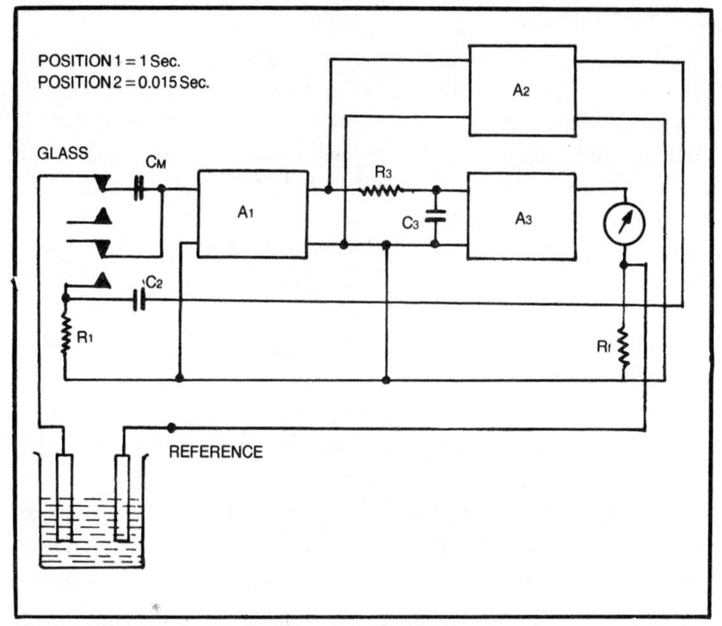

Fig. 14-7. Schematic diagram of Beckman Zeromatic pH meter.

measure the amplifier zero drift periodically, making an automatic correction at once. The correction is made by charging a memory capacitor in series with the amplifier input to a voltage equal and opposite to the amplifier drift.

Referring to Fig. 14-7, the pH measuring circuit consists of a two-stage amplifier A_1 followed by a cathode follower stage A_2 connected to the meter and feedback resistor R_f so that the whole system acts as a negative feedback amplifier. Capacitor C_M acts as a memory capacitor. The relay contacts in position 1 remain there for the measuring period of 1 second. The correcting action is applied when the contacts are moved to position 2, where they remain for 15 msec. On change of the contacts to position 2, the input voltage V to amplifier A_1 would normally decay to zero and, if no drift is present, the output would also fall to zero. If drift has occurred in A_1, the resulting output is amplified by A_2 and returned through C_2 to produce a voltage across R equal to the amount of drift. Since C_M and R are parallel in position 2, this voltage is stored on C_M and constitutes a correcting voltage when the contacts are brought back to position 1. R_3 and C_3 work as a smoothing circuit for the disturbances which are produced during the zero correction and prevent them from reaching the meter.

CURRENT DESIGNS

Inherently, bipolar transistors are current amplifying devices with a low input impedance. For these reasons, conventional transistor amplifiers are seldom employed for pH measurements. Bipolar transistors may be replaced at the input stage with field-effect transistors to achieve a higher input impedance. Metal-oxide silicon field-effect transistors of the insulated gate type have very small input leakage currents. A pair of these can be used to construct a pH meter. Figure 14-8 shows one such circuit. This circuit makes use of two MOSFETs in a differential amplifier configuration. This circuit is suitable to be used with a combination type pH electrode.

The differential input cancels the common mode errors such as effects of temperature and supply voltage variations. The meter used is of center zero type. Initially the input is grounded and the potentiometer R_7 is adjusted to bring the meter pointer to center scale (pH=7). R_3 is adjusted to read 0 pH after a simulated electrical signal (for 25°C) is given to the input corresponding to a change of 7 pH units.

Before the pH of a solution is measured, the pH meter is standardized by dipping the electrode in a solution of known pH and adjusting R_6 (standardize control) until the meter indicates the known pH. After the meter is thus standardized, the probe is rinsed in distilled water, wiped dry and placed in the solution whose pH is to be measured.

The pH readings can be corrected if the measurements are made at temperatures other than 25°C by the following formula:

$$\text{pH error} = \frac{(T-25)(pH_1 - pH_2)}{T+273}$$

where pH_1 is the instrument reading and pH_2 is the pH of the buffer.

Care should be taken to wire the circuit on high quality glass epoxy base printed circuit board. As MOSFETs are easily damaged by static electric charges, care should be taken to short their leads with wires during assembly and wiring of the circuit. This circuit can be employed in pocket pH meters.

Figure 14-9 shows another circuit arrangement which uses a matched pair of field-effect transistors housed in a single can. The circuit would provide input impedance greater than 10^{12}. The emf produced by the measuring electrode is given to the gate of transistor T_1. The potential applied to the input 1 of the operational

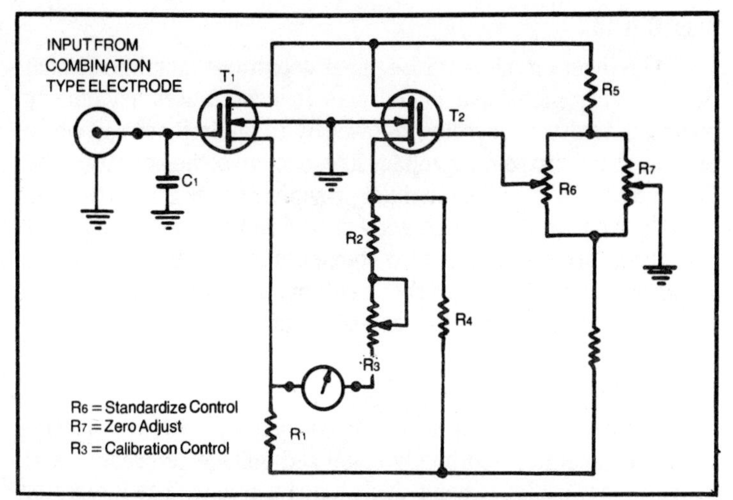

Fig. 14-8. Circuit arrangement of a pH meter using MOSFETs.

amplifier would depend upon current which flows through the transistor and its corresponding resistance R_1. The potential applied to the gate of transistor T_2 is set by the buffer bias adjustment which is fed from a zener-stabilized potential supply. The potential developed across R_2 would depend on the current through T_2. Resistances R_1 and R_2 are kept equal. Therefore, the output of the operational amplifier will depend upon the difference in potential developed on the measuring electrode and the potential setup in the instrument. The current flowing through the indicator also flows through the manual or automatic temperature compensating resistor. Thus, potential applied to the reference electrode can be arranged to compensate for the change in slope of the pH/temperature relationship, i.e., the gain of the system can be changed by the negative feedback across the temperature compensator so as to match the slope of the pH/temperature relationship.

DIGITAL pH METERS

Operational amplifiers with FET input stage are available in the integrated form. They can be directly coupled to the electrodes to amplify signals from the pH cell. Also, the current trend is to use digital display of the pH value in preference to the conventional moving coil meters. Digital display offers better resolution and readings without any ambiguity.

Figure 14-10 shows the block diagram of a digital pH meter. It consists of the following main parts:

Input Circuit

The input circuit is a FET input operational amplifier which offers a very high input impedance. It incorporates asymmetry control for electrode zeroing followed by a unipolarity circuit which provides a positive output voltage irrespective of the polarity of the input voltage. The buffer amplifier provides further gain so that the pH value can be displayed directly on an analog meter if desired. It also incorporates temperature compensation control. Basically, this manual control changes the sensitivity of the amplifier in such a way that its output becomes independent of temperature.

A-D Converter

The *analog-to-digital converter* consists of a switch, an integrator, comparator and highly stabilized voltage reference. A-D conversion is achieved using dual slope technique. The integrator uses an operational amplifier having a capacitor in the feedback circuit.

At the beginning of a measurement, the input to the integrator is switched to input signal, the capacitor has zero charge and the counter is set to zero. V_{in} is integrated and the integrator output appears at the input of the capacitor with an average slope of V_{in}/RC. When the counter accumulates its maximum count (say 2000 count) and the capacitor charges for a constant time, the switch changes over the integrator input to V_{ref}. V_{ref} being opposite to V_{in} in polarity causes the integrator capacitor to start discharging. At the same time the counter starts counting. The input to the comparator changes at a rate V_{ref}/RC. When the charge on the integrating capacitor reaches zero, the comparator triggers the control circuit to stop the counter. The counts N accumulated up to that time are proportional to the input voltage and are displayed on the digital readout. Since the time integral of the signal during the 2000 counts is equal and of polarity opposite to that of the reference:

$$\therefore 2000 \frac{V_{in}}{RC} = N \frac{V_{ref}}{RC}$$

$$N = 2000 \times \frac{V_{in}}{V_{ref}}$$

Control and Display Circuit

This part consists of reset, transfer polarity control, 20 kHz clock, bidirectional counters, latches, decoder driver and display

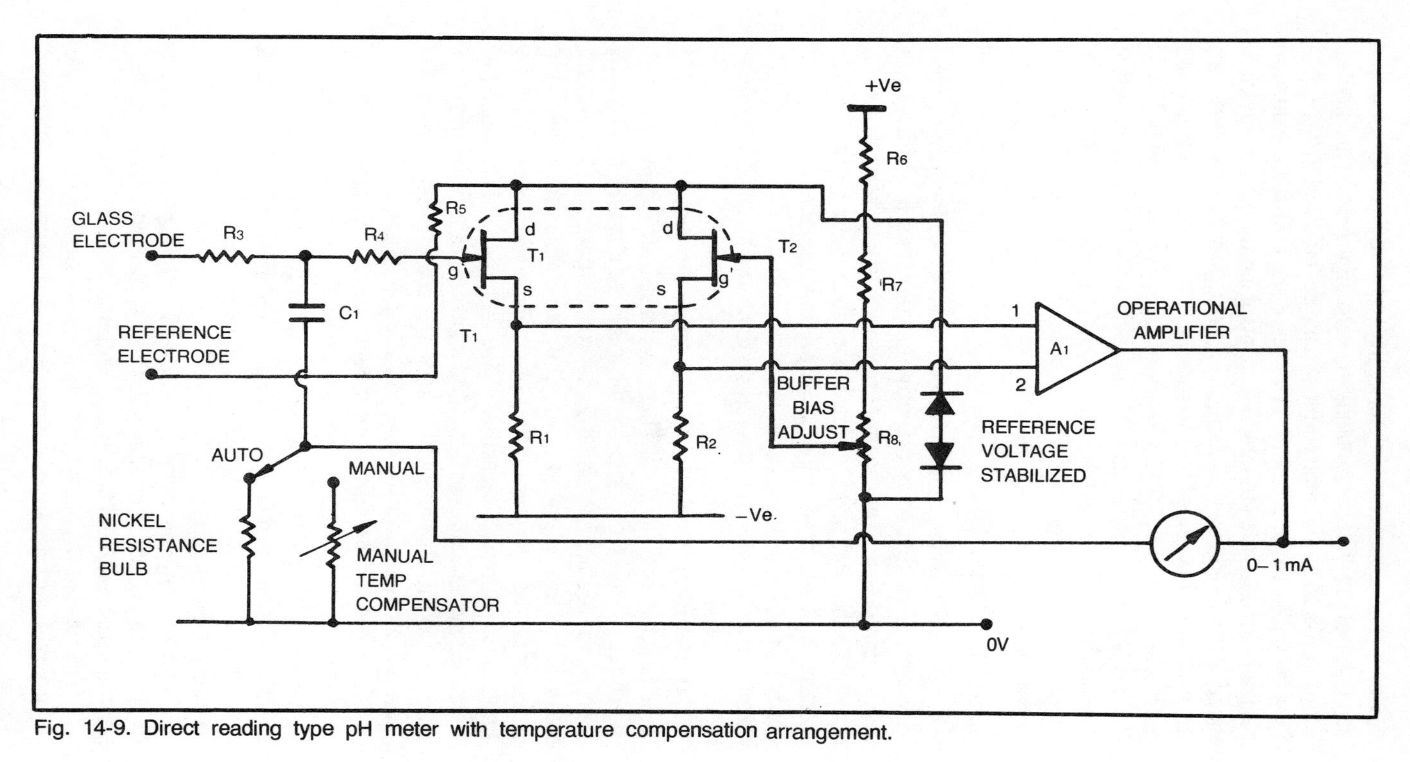

Fig. 14-9. Direct reading type pH meter with temperature compensation arrangement.

devices. The counters are normally reset to 700. This would display a pH of 7 corresponding to zero signal from the neutral solution. As such, the bidirectional counters are used to have both forward and reverse counting. The reset and transfer circuits are basically monostable circuits. The control circuit mainly does two functions; as soon as the counter is reset, it connects integrator input to the input signal and as soon as the counter reaches to its maximum count, i.e. 2000, integrator input is connected to V_{ref}. If the pH value is more than 7, the electrode develops negative potential and the counter starts counting in forward direction. When the pH is less than 7, the counter counts in the reverse direction.

INDUSTRIAL pH METERS

pH as a measurement has been carried out for many years. However, it is relatively recently that its value to industry has been realized either for the proper reaction to take place or for the quality of the end product, thus contributing to better quality products at lower production costs. pH measurements and control normally need to be done either in a tank or on stream in a pipe. For these examples, the electrode systems could be dip type or mounted on line.

Quite often, the electrode system and the controls unit could be as far apart as 50 meters. There would be a strong possibility of interference due to pickup by the cable connecting the two. A special type of shielded cables is used for this purpose. The second factor is the loss of signal strength in the connecting cable. This problem can be overcome either by compensating the loss by the amplifier circuit or, alternatively, the amplifier is fitted into the electrode system and the amplified signal is transmitted through the cable to the display system kept at a convenient place. This, however, introduces problems of on-site maintenance.

There could be situations when the measurement is to be carried out in an explosive atmosphere. In these cases, the design should be such as not to ignite the combustible gases. Normally, the amplifier part is housed in an explosive proof chamber and it is operated preferably on batteries, thereby minimizing chances of sparking in the circuit.

Industrial pH measuring instruments are fitted with contacts for operating visual or audible alarms or activating control valves, providing pH control of the process within preselected limits.

The output from the industrial pH amplifier is fed into recorders, controllers, indicators, etc. There is usually a choice of

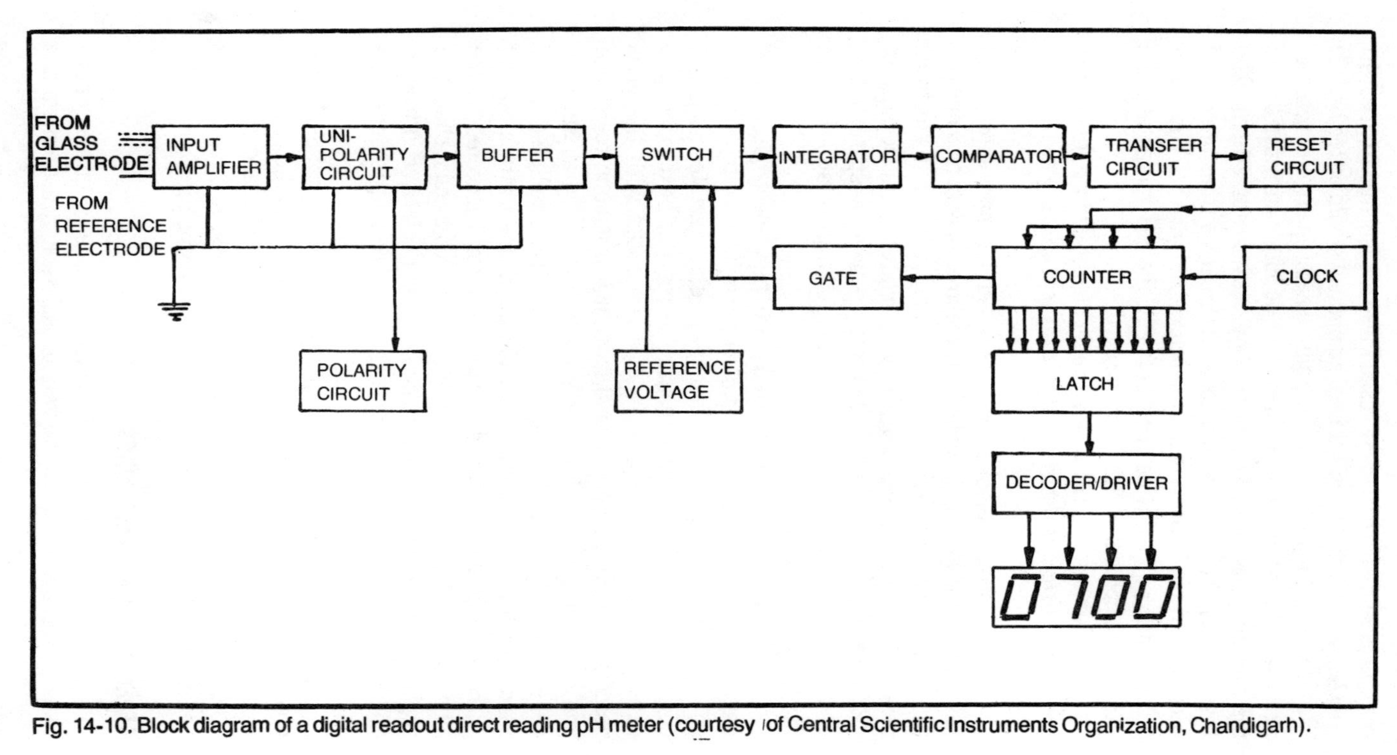

Fig. 14-10. Block diagram of a digital readout direct reading pH meter (courtesy of Central Scientific Instruments Organization, Chandigarh).

outputs provided and in many cases these are isolated from earth. Any leakage current from external equipment would be fed back to the circuit via the reference electrode causing an error in reading. An isolator eliminates this problem. The dc output current from the pH meter is periodically reversed by a transistor switch. The resultant square wave is then coupled through a transformer and reconverted to dc by a second synchronized switch. The transformer interrupts the dc path between the pH meter and the external equipment so that it may be earthed.

The failures in a pH meter can be classified into three main categories: *defective electrodes, defective input circuitry* and *defective electronic circuitry*.

If the instrument balances at zero with the input shorted and fails at pH, it is logical to replace the electrodes with new ones. If the instrument works with new electrodes, a defective electrode must be replaced. Since a majority of failures occur because of defective electrodes, it is advisable to have a spare set of electrodes with every instrument and use them by rotation to avoid drying up due to long storage.

If the instrument fails even with new electrodes, input circuitry must be checked. The failures in input circuitry would be due to poor insulation between glass electrode terminal and common terminal, excessive leakage currents and excessive grid current. Because of high resistance associated with the glass electrode, input circuitry should have high insulation resistance and low leakage currents. These may be impaired at times by collection of dust and other vapors. In such cases, input terminals, band switch and electrometer should be cleaned with a solvent such as carbon tetrachloride. The capacitor connected at the amplifier input should also be checked for leakage and it should be replaced if defective. If the instrument is still defective, it may be due to excessive grid current. In such a case, the electrometer tube should be replaced.

If the instrument fails to balance at zero, it is most likely that electronic circuitry is defective. Before attempting the repairs, it is advisable to short circuit the input externally and note the position of the pointer in all measuring ranges. If the pointer is within the scale, coarse buffer adjustments may be tried.

SELECTIVE ION ELECTRODES

Over the past decade, the pH meter has been at the center of a most important change in the field of analytical measurement due to the introduction of selective ion electrodes. As their name implies, these electrodes are sensitive to the activity of a particular ion in solution and quite insensitive to the other ions present. As the electrode is sensitive to only one ion, a different electrode is needed for each ion to be studied. Approximately 20 types of selective ion electrodes are presently available.

The development of a range of ion-selective electrodes has stimulated interest in the solid-state chemistry of glass and crystals and in the specificity for metal ions of synthetic and natural organic and complexing agents. Following developments in cation-responsive glasses (Eisenman et. al. 1957) and precipitate-impregnated silicone rubber membranes, the real impact of ion-selective electrodes was made by the fluoride ion-selective solid state lanthanum fluoride crystal membrane electrode (Frant and Ross 1966). The fluoride electrode was quickly followed by a calcium ion-selective liquid ion-exchanger membrane electrode (Ross, 1967). These and a whole host of other ion-selective electrodes in various styles have been applied in fields like environmental and industrial monitoring, reaction rate studies, enzyme reactions, general analysis and non-aqueous media studies. Ion-selective electrodes are classified into four major groups.

Glass Electrodes

The first glass ion-selective electrode developed is the one sensitive to hydrogen ions. Glasses containing less than 1% Al_2O_3 are sensitive to hydrogen ions (H^+) but almost insensitive to other ions present. Glasses of which the composition is Na_2O 11%, Al_2O_3 18% and SiO_2 71% is highly selective toward sodium, even in the presence of other alkali metals. Glass electrodes have been made that are selectively sensitive to sodium, potassium, ammonium and silver.

Solid State Electrodes

These electrodes use single crystals of inorganic material doped with a rare earth. Such electrodes are particularly useful for fluoride, chloride, bromide and iodide ion analysis.

Liquid-Liquid Membrane Electrodes

These electrodes are essentially liquid ion exchangers separated from the liquid sample by means of a permeable membrane. This membrane allows the liquids to come in contact with each other, but prevents their mixing. Based on this principle, cells have been developed that are selective to calcium and magnesium. These cells are used for measuring water hardeners.

Gas Sensing Electrodes

These electrodes respond to the partial pressure of the gases in the sample. The most recent of these to be developed are the gas sensing electrodes for ammonia and sulfur dioxide. Ammonia or sulfur dioxide is transferred across a gas permeable membrane until the partial pressure in the thin film of filling solution between the glass electrode membrane and the probe membrane equals that in the sample. The resultant pH change is measured by a combination pH electrode, a potential is developed related to the partial pressure and hence the ammonia or sulfur dioxide concentration.

Applications using ion-selective electrodes are many. Most are time-saving and simple to use. The electrodes are now used in continuous monitoring of ammonia, nitrate, fluoride, chloride, cyanide, sodium etc., providing vital information for the power industry, environmental work and process control industries.

CHEMICALLY SENSITIVE SEMICONDUCTOR DEVICES

Considerable effort has recently been directed toward the development of ion-sensitive electrodes based on a modification of the metal-oxide semiconductor field effect transistor. In these devices, the chemical sensitivity is obtained by fabricating the gate insulation of the FET out of ion-sensitive materials, usually a polymer or SiO_2. These devices are called ISFETs (ion-selective field-effect transistors). A simple review which discusses the chemistry and physics of chemically sensitive semiconductor devices is due to Janata and Moss (1976).

In these devices, the ion-sensitive material is bonded to the FET itself. This requires the material and its method of fabrication to be compatible with the substrate (high purity silicon). This very significant requirement puts a severe limitation on the use of some of the best-characterized membrane materials including ion-sensitive glasses. Martin and Sinclair (1976) report the development of a pH-sensitive electrode by means of thick film screening techniques. This electrode retains the advantages of ion-sensitive FET transducers, but eliminates the restrictions on membrane selection and fabrication. Here the ion-sensitive structure is physically separated from the FET. In this way, the ion-sensitive membrane can be fabricated on a compatible substrate and the FET can then be attached appropriately and placed in close proximity to the ion-sensitive membrane. A hybrid electrode structure permits the incorporation of a source follower FET amplifier directly adjacent to the pH membrane, significantly reducing response time and noise pickup.

Chapter 15

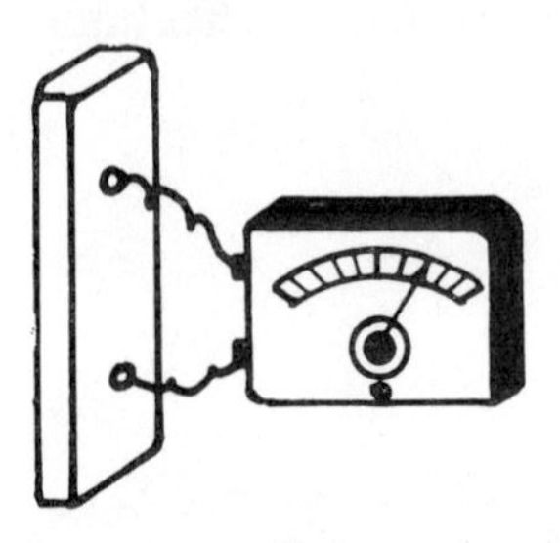

Blood Gas Analyzers

A blood gas analyzer is essentially used to measure the pH, partial pressure of carbon dioxide (pCO_2) and oxygen (pO_2) of the body fluids with special reference to human blood. The measurement of these parameters throws considerable light on the acid-base balance in the body. Disturbances in the acid-base equilibrium may develop at any time in patients undergoing major surgical interventions. This may occur during long operations on the heart and major blood vessels especially employing hypothermia and cardiopulmonary bypass. Sudden changes in blood pH and/or pCO_2 may also have serious consequences during or after anaesthesia. Brown and Miller (1952) were the first to report that a sudden change in pH and pCO_2 could precipitate cardiac arrhythmia and death. This shows that the maintenance of physiological neutrality in blood is of utmost importance. An increase in H-ion concentration, i.e., a decrease in pH is termed as *acidosis*. Likewise, a decrease in H-ion concentration, that is, an increase in pH is termed an *alkalosis*. The acid-base equilibrium of a patient may be upset by disturbances of metabolic or respiratory origin. Acidosis of respiratory origin follows the rapid accumulation of CO_2 in the blood, lungs and tissues. This may result from alveolar ventilation which may be inadequate for the CO_2 production. It may occur in the presence of lung disease or even with structurally normal lungs having central depression of respiration caused by cerebral damage or drugs. Acidosis of metabolic origin may complicate renal or hepatic disease, poisoning by external agents or acute tissue anoxia of any cause.

At a partial pressure of oxygen of about 100 mm of mercury, in the blood, all the blood hemoglobin is in the oxygenated form, i.e., the blood is 100% saturated with oxygen. With lower oxygen pressure, the percentage saturation falls off. The percent of oxygen saturation at a given partial pressure is less if the blood is more acidic. When the blood reaches the pulmonary capillaries, the oxygen pressure in the blood in a normal subject is about 40 mm corresponding to 70% saturation. While passing through the pulmonary capillaries, carbon dioxide is lost, the blood becomes slightly more alkaline and this aids oxygenation of hemoglobin. However, hemoglobin becomes more acidic in changing from reduced to oxygenated form. Thus, the changes in blood pH are greatly reduced in the transition from the venous blood to the arterial blood. The oxygen saturation of the arterial blood is about 96% corresponding to a partial pressure of about 95 mm Hg. The difference between alveolar and arterial oxygen partial pressure varies in individual normal human subjects between 5 and 11 mm.

The mechanism of exchange and transport of carbon dioxide is quite different from that of oxygen. The blood carries carbon dioxide chiefly in the form of bicarbonate—sodium bicarbonate in the plasma and potassium bicarbonate in the cells. Carbon dioxide is produced continuously and its concentration is regulated by the rate of respiration. Normally, a primary fall in bicarbonate concentration is accompanied by a secondary fall in plasma carbon dioxide brought about by increased ventilation which washes out more carbon dioxide from the alveolar air. Similarly, primary increase in bicarbonate is followed by a reduced rate of breathing, thus increasing carbon dioxide concentration in the alveolar air and in the plasma. In this way, a metabolic acidosis or alkalosis is compensated by a change in the rate of respiration. If the change is adequately compensated, plasma pH remains within normal limits. However, as the disturbance becomes more severe, compensation may be only partial and the pH may become abnormal.

Table 15-1 gives the normal range for pH, pCO_2, pO_2, total CO_2 and bicarbonate when the measurements are taken at 37°C.

BLOOD pH

When making pH measurements, the Sorenson's definition for pH of blood and plasma is adopted which is given by

$$pH = -\log_{10}(H^+)$$

where H^+ designates the concentration of hydrogen ions.

According to this relation, whole blood with a (H^+) of 4×10^{-8} moles per liter would have a pH of 7.4. This implies that an increase in the (H^+) to 1×10^{-7} moles per liter would correspond to a decrease in pH to 7.0.

Blood or plasma pH is a fairly constant number. The range of pH in health and disease is relatively narrow, in health 7.36 to 7.42, in disease from about 7.00 to 7.80. Any greater fluctuations of pH may result in acidosis or alkalosis and death may occur on either side of variation.

	7.00		7.36		7.44		7.88	
DEATH	|	ACIDOSIS	|	NORMAL	|	ALKALOSIS	|	DEATH

Electrochemical pH determination utilizes the general method of measuring the difference in potential occurring between two solutions of different pH using the glass electrode. However, there are special problems associated with the measurement of the pH of blood. Particular attention must, therefore, be given to the problems that will be mentioned now.

Temperature Control

The pH of blood changes linearly with temperature in the range 18 to 38°C. The temperature coefficient for the pH of the blood is 0.0147 pH unit per degree centigrade. This calls for the necessity of using a highly accurately temperature controlled bath to keep the electrodes with the blood sample at 37°C ± 0.01°C.

Because of the possible individual variations in the temperature coefficient of blood pH, the method of measuring at some temperature other than 37°C followed by correction is not recommended. Moreover, although the plasma may be separated from the whole blood at the temperature of measurement without the two solutions showing markedly different pH values, measurements made or plasma separated at a temperature different from subsequent measurements could be misleading.

This is because the red cells are in equilibrium with the plasma solution, whose separate pH temperature characteristics are quite different from those of whole blood (Rosenthal, 1948). Therefore, maintenance of constant temperature is very essential. It is advisable to keep both the glass as well as the reference electrode at the temperature of measurement.

Type of Electrodes

Several types of electrode systems have been described in literature for the measurement of blood pH. They are all of the

Table 15-1. Normal Range for pH, pCO_2, pO_2, Total Co_2 and Bicarbonate When Measurements Are Done at 37°C. (courtesy of Corning Medical Instruments).

		Arterial or Arterialized Capillary (Whole Blood)	Venous Plasma Separated at 37° C
pH		7.37 to 7.44	7.35 to 7.45
PCO_2	Men	34 to 36 mm Hg	36 to 50 mm Hg
	Women	31 to 42 mm Hg	34 to 50 mm Hg
pO_2	Resting adult-	80 to 90 mm Hg	25 to 40 mm Hg
	Resting adult-over 65	75 to 85 mm Hg	
Bicarbonate	Men	23 to 30 m mol/1	25 to 30 m mol/1
	Women	20 to 30 m mol/1	23 to 28 m mol/1
Total CO_2 (plasma)	Men	24 to 30 m mol/1	26 to 31 m mol/1
	Women	20 to 30 m mol/1	24 to 29 m mol/1

glass electrode type made in different shapes and arrangements so that high accuracy and best reproducibility can be obtained. The most common type is the *syringe electrode* which is convenient for taking small samples of blood anaerobically. The plunger of the syring constitutes the glass electrode. A small dead space exists between the electrode bulb and the inner surface of the syringe barrel. Fill up this space with dilute *heparin* solution (about 0.2 mg per 10 ml blood) before blood is drawn. Heparin is used mainly to prevent blood coagulation. Before the measurement is made, the syringe is rolled between the hands to ensure thorough mixing.

Blood pH measurements can also be carried out by using microcapillary glass electrodes. They can be used for continuous monitoring of pH during surgery. This type of electrode requires very small volumes of the sample, which may be of the order of 0.1 ml.

Sanz (1957) describes a micro-pH electrode in which the blood sample is held inside a fine polyethylene tipped capillary tube to ensure anaerobic measurements. The accuracy of the system was reported to be 0.01 pH unit. A warm water jacket was provided around all susceptible to shield against static charges.

Astrup (1956) constructed a micro-electrode for clinical applications which corresponds to the principle of Sanz. It requires only 20-25 μl of capillary blood for determination of pH. The electrode is enclosed in a water jacket with thermostated circulating water at 38°C. The water contains 1% NaCl for shielding against static interference. The capillary is protected with a polyethylene tubing. The internal reference electrode is Ag/AgCl and the calomel reference electrode is connected to a small pool of saturated KCl, through a porous pin. It is possible to obtain an accuracy of ± 0.001 pH against a constant buffer. Figure 15-1 shows the construction details of a blood pH electrode and the measurement setup used in practice.

Aging Effect on pH Electrode

It is often noticed that when a glass electrode has been in use for a couple of months or more, its sensitivity and speed of response may diminish. This situation may arise from the fact that the surface of the glass membrane has become inactive due to a depletion of certain ions. A screen is thus formed which partly covers an active layer of the glass. The manufacturers provide instructions on how to etch the electrode surface clean in order to restore the sensitivity.

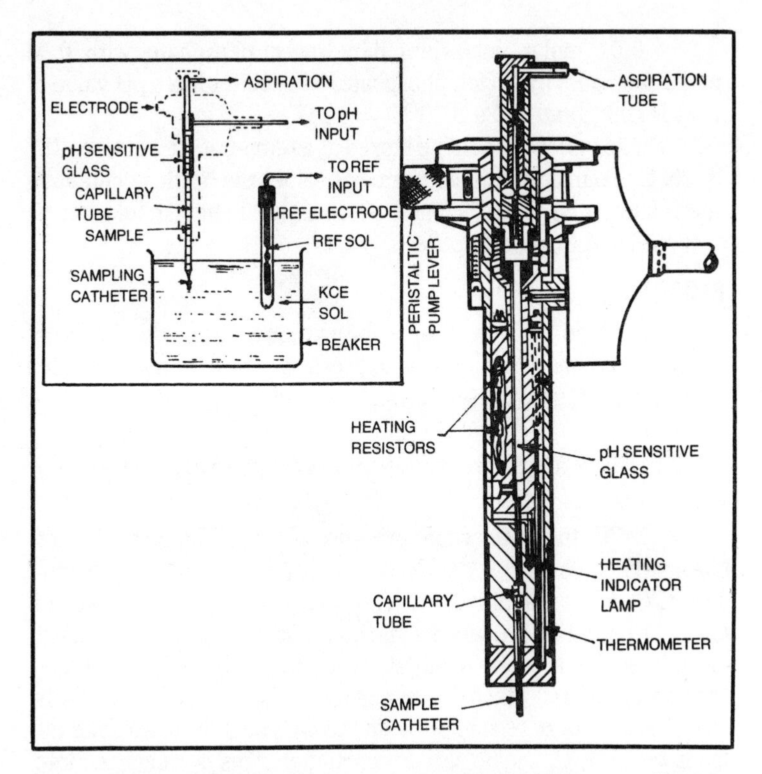

Fig. 15-1. Construction details of blood pH electrode (courtesy of Corning, USA).

Glass electrodes deteriorate if they are allowed to remain in contact with blood for a considerable length of time. The poisoning effect appears to be due to protein deposition. As a precautionary measure, in an apparatus where blood necessarily remains in contact around the electrode for long periods (more than 30 minutes), the response must be checked frequently against buffer solutions. The poisoning effect can be removed by standing the electrode in pepsin and 0.1 N Hydrochloric acid, followed by careful wiping with tissue paper.

Buffer Solutions

Buffer solutions used in blood pH measurements are as follows:

—0.025 molar potassium dihydrogen phosphate with 0.025 molar disodium hydrogen phosphate. This solution has a pH value of 6.840 at 38°C and 6.881 at 20°C (Bates and Acree, 1945).

—0.01 molar potassium dihydrogen phosphate with 0.04 molar disodium hydrogen phosphate. This buffer has a pH value of 7.416 at 38°C and 7.429 at 20°C.

These buffers should be stored at a temperature between 18° to 25°C. To maintain an accurate pH value, the bottles containing them should be tightly closed. A decanted buffer should not be returned to the original bottle.

BLOOD pCO_2

The blood pCO_2 is the partial pressure of carbon dioxide of blood taken anaerobically. It is expressed in mm Hg and is related to the percentage CO_2 as follows:

$$pCO_2 = \text{Barometric pressure-water vapor pressure} \times \frac{\% \, CO_2}{100}$$

At 37°C, the water vapor pressure is 47mm Hg, so at 750 mm barometric pressure, 5.7% CO_2 corresponds to a pCO_2 of 40 mm. The pCO_2 of arterial blood is directly proportional to the amount of CO_2 which is being produced in the body and inversely proportional to the rate of alveolar ventilation in the lungs. Also, the CO_2 tension of arterial blood corresponds to that of alveolar air. It is thus possible to determine it from the CO_2 content of alveolar air from which pCO_2 can be simply calculated. This is converted into millimoles per liter by multiplying the pCO_2 by 0.030.

In clinical practice, most of the methods used for estimating pCO_2 in the arterial blood are based on the well known Henderson-Hasselbalch equation which relates the partial pressure of carbon dioxide to the pH as follows:

$$pH = pK' + \log\left[\frac{(HCO_3^-)}{(H_2CO_3)}\right] \qquad \textbf{Equation 15-1}$$

where pK′ is the logarithm of the reciprocal of the fist dissociation constant of carbonic acid ($pK' = -\log K'$) and (HCO_3^-) and (H_2CO_3) are concentrations expressed in molar concentrations. (H_2CO_3) exists almost entirely in the form of dissolved carbon dioxide in the normal physiological pH range. Equation 15-1 can be written as

$$pH = pK' + \log\left[\frac{(HCO_3^-)}{(CO_2 \text{ dissolved})}\right] \qquad \textbf{Equation 15-2}$$

Since the dissolved carbon dioxide is equal to the partial pressure of gaseous carbon dioxide (pCO_2) times the solubility coefficient of CO_2 in serum (a), and (HCO_3^-) equals the total CO_2 minus CO_2 dissolved, Equation 15-2 can be written as

$$pH = pK' + \log \left[\frac{\text{total } CO_2 - \alpha pCO_2}{\alpha pCO_2} \right] \quad \textbf{Equation 15-3}$$

CO_2 is usually expressed in millimoles/liter and pCO_2 in millimeters of mercury.

pK′ for the bicarbonate - CO_2 buffer system = 6.10

$$\alpha = 0.0301$$

Substituting in Equation 15-3

$$pH = pK' + \log \left[\frac{\text{total } CO_2 - 0.0301\, pCO_2}{0.0301\, pCO_2} \right] \quad \textbf{Equation 15-4}$$

$$\therefore pCO_2 = \frac{\text{total } CO_2}{0.0301\,[1 + \text{antilog}(pH - pK')]} \quad \textbf{Equation 15-5}$$

This equation relates the three quantities—pH, pCO_2 and total carbon dioxide. If any two of the three are known, the third can be calculated. In particular, the equation is used to determine pCO_2 from values obtained for pH and total CO_2. In clinical practice, pH is usually measured on whole blood whereas total CO_2 is derived from plasma. The latter is usually measured by gasometric techniques.

The relationship between pH and pCO_2 is logarithmic and therefore, emphasizes the need for careful pH measurement. Approximately, measuring error of 0.02 pH will correspond to 4.5% error in the pCO_2 determination.

In practice, it is usual to assume constant values for pK′ and α in spite of the fact that wide variations in the value of pK′ have been reported in literature. Deave and Smith (1957) obtained a spread of values from 6.07 to 6.25. Dill et.al. (1937) gave pK′ for oxygenated cells as 6.04 and for reduced cells as 5.98.

Severinghaus et. al. (1956) reported that a high degree of accuracy can be achieved in the estimation of pCO_2 using the Henderson-Hasselbalch equation. They compared and computed the known pCO_2 values in 20 samples of plasma obtained from normal subjects. The average error was less than 1%. Ludbrook (1959) studied the factors which may affect the accuracy of pCO_2 estimation based on this equation and concluded that the method yields satisfactory results which are quite acceptable for clinical purposes, provided the sources of error are recognized and either corrected or avoided.

Effect of Temperature on the Determination of pCO_2 Using Henderson-Hasselbalch Equation

Changes in temperature cause many important alterations in the biophysical and biochemical status of the blood. It has been investigated by Severinghaus et. al. (1956) and several other research workers that the value of the dissociation constant pK′ changes considerably with the blood temperature. If the patient's blood temperature is significantly altered, as with high fever or during hypothermia for surgical procedures, the Henderson-Hasselbalch equation cannot be directly used for calculation of pCO_2 unless necessary corrections are applied. The measurements are, therefore, made at the recommended temperature of 37°C. Lenfant (1961) developed a nomogram which permits rapid determination of the pCO_2 and of other factors having a role in acid-base equilibrium regardless of the temperature of the blood at the time of sampling. The initial measurement of blood pH is made at 37°C along with the measurement of the blood temperature at the time of sampling and a separate determination of the total plasma CO_2 content.

The Astrup Technique

Astrup (1956) suggested a method for the determination of pCO_2 which eliminated many of the difficulties associated with variations in pK′ from one sample to another. Also, he along with his co-workers [(Astrup et. al. (1969), Astrup and Siggaard-Andersen (1963)] have introduced an apparatus by which it is possible to estimate all the required values of acid-base equilibrium without the need for separate determination of the total CO_2 content. The method is based on observations made by Astrup (1956) that the relation between the pH and the logarithm of the pCO_2 is very nearly linear in the physiological range of pH. In fact, the method enables determination of pH, pCO_2, standard bicarbonate and base excess or deficit by measuring the pH of the blood at its actual pCO_2 and at two other accurately known pCO_2 values, one higher (about 60 mm) and one lower (about 20 mm Hg) than the normal pCO_2. The latter measurements are made by equilibrating an oxygenated whole blood sample at two known CO_2 tensions. If log pCO_2 is plotted against pH for a specimen of blood, a straight line is obtained. On this basis, a nomogram has been constructed from which once the log pCO_2, pH line has been drawn from the pH values for the specimens equilibrated at a high and low pCO_2, the unknown pCO_2 of blood and the standard bicarbonate and base excess can be read.

The technique developed by Astrup and his co-workers is suitable for both micro and macro samples. Using the pH micro-electrode requiring only about 100 μl of capillary blood, the error in pCO_2 measurement is about ± 2%.

DIRECT MEASUREMENT OF pCO_2

Stow et. al. (1957) described the use of a pH sensitive glass electrode for the direct estimation of pCO_2. The method is based on the fact that the dissolved CO_2 changes the pH of an aqueous solution. The pCO_2 electrode developed by Stow consisted of a thin rubber membrane stretched over a glass electrode with a thin layer of water separating the membrane from the electrode surface. The CO_2 from the blood sample diffuses through the membrane to form H_2CO_3 which dissociates into $H^+ + HCO_3^-$ ions. The change in the pH of the water as measured by the glass electrode is a function of the CO_2 concentration of the medium. After these measurements, the voltage response of the electrode showed a linear relationship between the pH and the negative logarithm of pCO_2. The electrode, however, could not provide sensitivity and stability required for clinical purposes.

Severinghaus and Bradley (1958) modified this electrode to a degree that made it suitable for routine laboratory use. In the construction suggested by them, water layer was replaced by a thin film of an aqueous sodium bicarbonate solution. The bicarbonate solution is interposed between the electrode and a Teflon membrane which is permeable to CO_2 but not to any ions which might affect the pH of the bicarbonate solution. The Teflon membrane used is backed with a layer of cellophane 0.002 inch thick. The CO_2 from the blood diffuses into the bicarbonate solution. There will be a drop in pH due to CO_2 reacting with water forming carbonic acid. For this measurement, a reference electrode such as silver-silver chloride is also required.

The pH falls by almost one pH unit for a tenfold increase in the CO_2 tension of the sample. Hence, the pH change is a linear function of the logarithm of the CO_2 tension. The optimum sensitivity in terms of pH change for a given change in CO_2 tension is obtained by using bicarbonate solution of a concentration of about 0.01 mole/liter. The response time of the CO_2 electrode is of the order of 0.5 to 3 minutes. The electrode must be calibrated with the known concentrations of CO_2. A standardizing gas with pCO_2 close to the blood sample may be used before and after determinations. Figure 15-2 shows the construction of a typical pCO_2 electrode. This electrode is twice as sensitive and drifts much less than the Stow's electrode.

Hertz and Siesjo (1959) achieved further improvements in stability and response time. They used a dilute solution of $NaHCO_3$ (0.0001 N) which reduced the response time but introduced drift problems. Ag-AgCl reference electrode was replaced by a calomel cell which was made an integral part of the electrode. The compromise between response time and drift was achieved by using 0.001 N solution of $NaHCO_3$.

Severinghaus (1962) improved upon the earlier Severinghaus-Bradley electrode in the low pCO_2 range by replacing the cellophane spacer with a very thin nylon mesh. Glass fibers or powdered glass wool were also found to be good separators. Using a membrane of ⅜ mil Teflon and glass wool for the separator, electrodes with 95% response in 20 seconds were constructed. Severinghaus also reported that response time could be further reduced by the addition of hemolyzed blood to the electrolyte. Reyes and Nevillé (1967) constructed a pCO_2 electrode using 0.5 mil polyethylene as a membrane and used no separator between the glass surface and this membrane. They added commercial preparation of carbonic anhydrase to the electrolyte. The response time was 6 seconds for 90% of a step change from 2% to 5% CO_2.

Performance Requirements of pH Meters Required for pCO_2 Measurement

The potential developed by a pCO_2 electrode is a direct logarithmic function of pCO_2. It is found that a tenfold change in pCO_2 causes the potential to change by 58 ± 2 mV. Also, the pH versus log pCO_2 relationship is linear within ± .002 pH unit from 1 to 100% carbon dioxide. As 0.01 unit pH change corresponds to a 2.5% change in pCO_2 or 1 mm Hg in 40 mm Hg., for obtaining an accuracy of 0.1 mm Hg, it is desirable to read 0.01 pH unit (60 μV). This order of accuracy is not attainable in the ordinary pH meters. Therefore, it is necessary to use either an expanded scale pH meter or a digital readout type pH meter. The instrument must have a very high degree of stability and a very low drift amplifier. The electrode virtually has no inherent drift. The input impedance of the electronic circuit must be at least 10^{12} ohms.

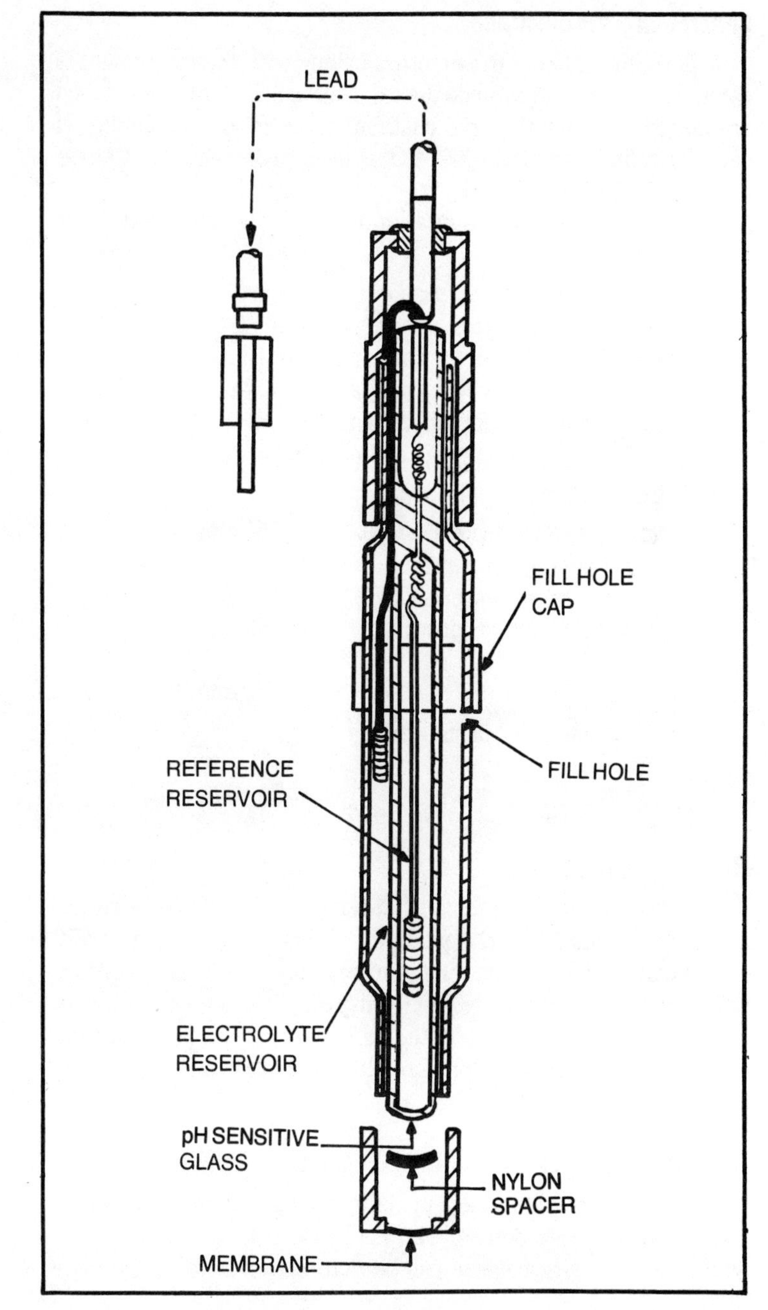

Fig. 15-2. Construction details of pCO_2 glass electrode (courtesy of Corning, USA).

Calibration of the Electrode

The calibration of the electrode along with the instrument is carried out by using gas since the calibration lines for gas and blood are identical. In practice, gas containing 6 or 8% carbon dioxide is used for calibration, and 100% CO_2 is used to determine the slope of the calibration line.

The procedure for calibration of the pCO_2 electrode is as follows:

- Measure the ambient barometric pressure.
- Subtract the water vapor pressure at the humidifying temperature (3.70°C) from the barometric pressure.
- Multiply the correct dry gas pressure by the percentage of gas being determined in the calibrating standard.

For example: to calculate the pCO_2 of the 5% gas standard, the following figures would apply:

Barometric pressure	= 760 mm Hg
Humidifying temperature	= 37°C
Water vapor pressure at 37°C	= 47 mm Hg
Corrected dry gas pressure	= 760 − 47 = 713 mm Hg
Calculate pCO_2 of standard gas	$= 713 \frac{\times 5}{100}$
	= 35.65 mm Hg

Effect of Temperature

It is essential to maintain the temperature of the electrode assembly constant within close limits. It has been experimentally shown that variation in the temperature of ± 1°C produces an error ± 1.5 mm Hg or about ± 3% at 50 mm pCO_2. The combined effects of temperature changes upon the sensitivity of the pH electrode and upon the pCO_2 of the blood sample amount to a total variation in sensitivity of 8% per °C.

Calculations for Bicarbonate, Total CO_2 and Base Excess

Calculations are usually made in conjunction with blood pH and gas analysis for determining acid base balance. An accurate picture of acid-base balance can be determined from the equilibrium

$$CO_2 + H_2O \rightarrow H_2CO_3 \rightarrow H^+ + HCO_3^-$$

which for bicarbonate has an equilibrim constant

$$\frac{K_{H_2CO_3}}{HCO_3^-} \qquad \frac{(K^+)(HCO_3^-)}{(H_2CO_3)}$$

where (H^+), (HCO_3^-) and (H_2CO_3) refer to the concentration of these substances.

Since $(H_2CO_3) = 0.03\, pCO_2$
and since $pH = -\log (H^+)$, we can write

$$pH = pK + \log \frac{(HCO_3^-)}{0.03\, pCO_2}$$

where pK equals 6.11 for normal plasma at 37°C. This formula is used for calculating actual bicarbonate.

Total CO_2 is calculated from the relationship

$$(HCO_3^-) + (0.03 \times pCO_2) = \text{Total } CO_2 \text{ (millimoles/liter)}$$

Base excess is calculated from the formula described by Siggaard-Andersen

$$\text{Base Excess} = (1 - 0.0143 \times Hb)(HCO_3^-) - (9.5 + 1.63\,Hb) \times (7.4 - pH) - 24$$

where Hb represents the patient's hemoglobin value.

Cleaning Procedure for pCO_2 Electrode

After continuous use of pCO_2 electrode, it is sometimes observed that the response of the electrode begins to slow down as in a pH electrode. The usual cause is some substance which begins to form a thin film on the surface of the sensing glass. This film of material accumulates in varying degrees and progressively makes the electrode sluggish. To restore the electrode's normal response time, the following procedure may be adopted:

■ **Procedure 1.** Remove the membrane without touching its face and place it on a tissue. Empty the filling solution. Add enough cleaning solution (cleaning solution contains pepsin 9.6 g/l, hydrochloric acid 0.1 N and a preservative agent) to a beaker padded with a folded paper towel to cover the sensing bulb. Soak for at least two hours. Flush with tap water, distilled water and filling solution.

■ **Procedure 2.** Repeat the first two steps from procedure 1. Wash the electrode tip with 6 N HCl for several minutes. Flush with tap water, distilled water and filling solution.

■ **Procedure 3.** If the above two procedures fail, repeat the first two steps from procedure 1. Immerse the sensing bulb of the electrode into a 20% by weight solution of ammonium bifluoride for 60 seconds. Wash tip immediately with 0.1 N HCl. Flush with tap water, distilled water and filling solution.

BLOOD pO_2

The partial pressure of oxygen in blood or plasma reflects the extent of oxygen exchange between the lungs and the blood and, normally, the ability of blood to adequately perfuse the body tissues with oxygen. For studying the oxygen concentration levels in biological systems, use is made of an oxygen electrode which measures the partial pressure of oxygen directly at the point of insertion into the tissue. The electrode and O_2 measuring arrangement in its simplest form is shown in Fig. 15-3. The electrode is a piece of platinum wire embedded in a glass holder with the end of the wire exposed to the solution under measurement. When the platinum electrode is made negative with reference to Ag-AgCl reference electrode, oxygen reaching the surface of the platinum is reduced electrolytically. There is a characteristic polarizing voltage at which any element in solution is predominantly reduced and, in the case of oxygen, it is 0.6 volts to 0.9 volts. In this voltage range, it is found that the current flowing is proportional to the oxygen concentration in the solution.

The Clark Electrode

The most commonly used oxygen electrode for biological investigation is the *Clark* (1956) type electrode. This type of electrode consists of a nearly invisible platinum cathode, a Ag/AgCl anode, in electrolytic filling solution and a polypropylene membrane. The electrode is of single unit construction and contains the reference electrode also in its assembly. Figure 15-4 shows the construction of the Clark type pO_2 electrode. The entire device is isolated from the solution under measurement by the membrane. Oxygen from the blood diffuses across the membrane into the electrode filling solution and is reduced at the cathode. The circuit is completed at the anode, where silver is oxidized, and the magnitude of the resulting current indicates partial pressure of oxygen. The reactions occurring at the anode and cathode are:

$$O_2 + 2H_2O + 4\bar{e} \longrightarrow 4OH^-$$

(cathode reaction)

$$4Ag \longleftarrow 4Ag^+ + 4\bar{e}$$

(anode reaction)

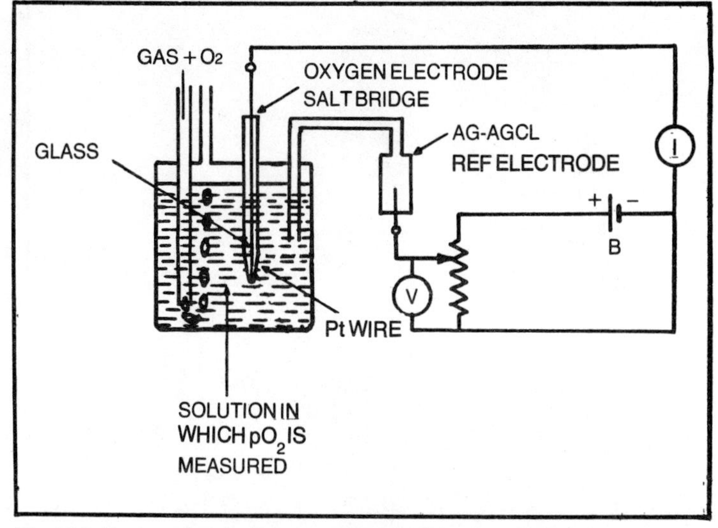

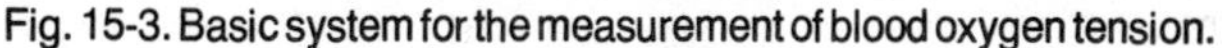
Fig. 15-3. Basic system for the measurement of blood oxygen tension.

Use of the Clark electrode for measuring pO_2 has been extensively studied and found to be of particular advantage for measuring blood samples. The advantages are:

—The electrode can be made small enough to measure oxygen concentrations in highly localized areas.

—Sample size required for the measurement can be small.

—The measurement can be made in seconds whereas it takes much longer time if determinations are made by chemical methods.

Measurement of Signal Generated by a pO_2 Electrode

The measurement of current developed at the pO_2 electrode due to the partial pressure of oxygen presents special problems. The difficulty arises because of the extremely small size of the electrical signal. An idea of the size of the signal can be had from the work done by Davies and Brink (1942) who employed the total range of the indicator as 0.05 μa for an oxygen tension of 180 mm Hg. The current measured by Tobias (1949) was 0.650 μa for a 500 mm Hg oxygen tension. The recorder sensitivity employed by Clark et. al. (1958) produced a full scale deflection for 0.98 μa.

The sensitivity (current per torr of oxygen tension) is typically of the order of 20 pA per torr for most commercial instruments. It is subject to constant drift and is not independent of the sample characteristics.

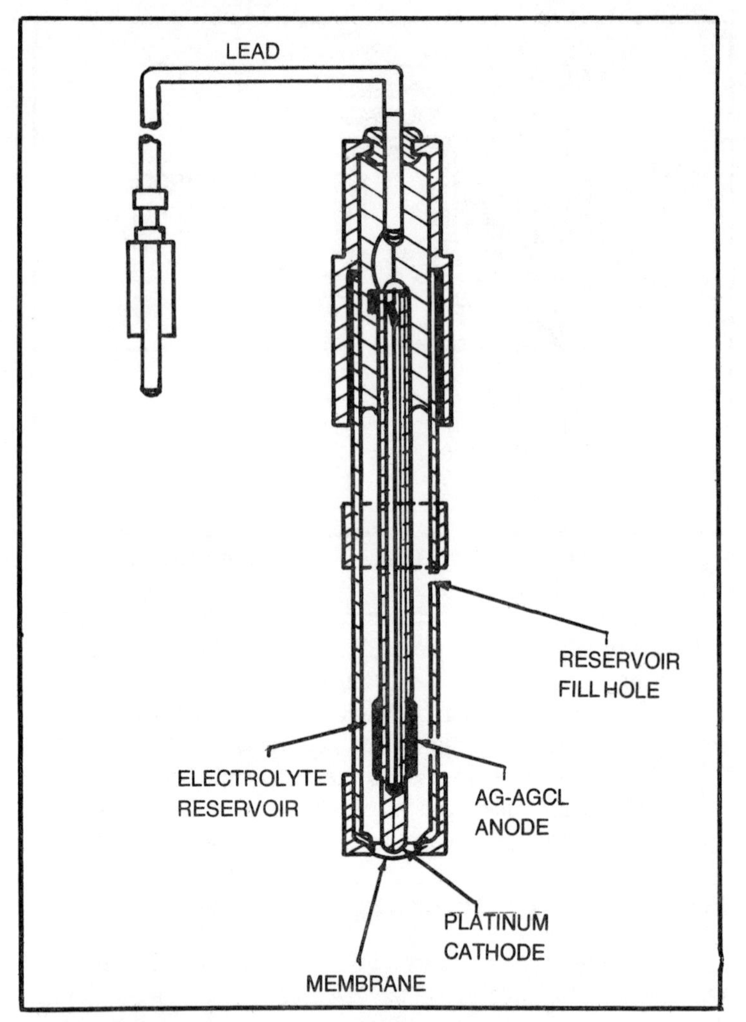

Fig. 15-4. Clark type pO_2 electrode (courtesy of Corning USA).

Measurement of oxygen electrode currents is made by using high input impedance, low noise and low current amplifiers. Field effect transistors usually form the input stage of the preamplifiers. Clark and Becattini (1967) describe a circuit using a MOSFET amplifier for use with a Clark type oxygen electrode.

Hahn (1969) describes the use of a field-effect transistor operational voltage amplifier to measure small currents. The op amp is connected as a transresistance converter, the output of which can be read directly by a digital voltmeter.

The circuit (Fig. 15-5) shows that cell B_1 (1.3 V) and variable resistance VR_1 constitute the polarizing source for the electrode. The zero current for the cell which is present even if the gas or liquid sample is free of oxygen is canceled by means of VR_2 and battery B_2 and 1 G ohm resistance. Capacitor C(100 Pf) is included to limit the bandwidth of the amplifier to reduce noise and to ensure good dynamic stability.

McConn and Robinson (1963) found that zero electrode current was not given by a solution having zero oxygen tension but occurred at a definite oxygen tension which they called the *electrode constant*. From a knowledge of this constant, for a particular electrode, it is possible to calibrate the electrode. They showed that when the straight line calibration curves (Fig. 15-6) were extended backwards, they did not pass through the origin but intersected the oxygen tension axis at a negative value. A true zero current of less than 10 nA could be obtained by deoxygenating the electrolyte of the electrode by bubbling nitrogen through it for 30 minutes and then placing the electrode in water redistilled from alkaline *pyrogallol*.

Measurement of Oxygen Tension of Whole Blood Using Oxygen Electrode

For measurement of the oxygen content of whole blood, 0.5 ml of blood is diluted with 20 ml of a solution of *saponin* and *potassium ferricyanide*. This solution contains 6 grams of potassium ferricyanide and 3 grams of saponin per liter of distilled water. This solution is stored in a 500 ml bottle partially immersed in a water bath at 38°C. Addition of the solution to the blood sample produces

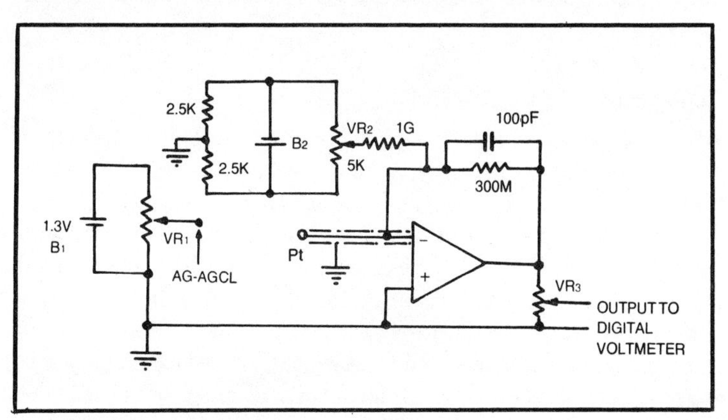

Fig. 15-5. pO_2 current amplifier.

hemolysis and releases oxygen from combination with the hemoglobin. The oxygen remains in a dissolved state in the solution and its tension can then be measured using polarographic electrode.

Measurement of Tissue Oxygen Tension Using Polarographic Electrode

Tissue oxygen tension can be determined by using a platinum wire 0.004 inch in diameter covered in epoxy resin except for the tip and mounted inside a hypodermic needle [Montgomery (1957), Johnson and Krog (1959)]. The reference electrode would consist of a chlorided silver plate covered with ECG electrode paste and strapped on to an arm or leg. Each electrode can be supplied with polarizing voltage and the polarographic current recorded with an electrometer amplifier. These electrodes may be calibrated by putting them in a physiological saline solution equilibrated with air at the start and finish of the experiment. Several other types of tissue oxygen electrodes are described in literature.

Catheter Type Oxygen Electrode

Beneken Kolmer and Kreuzer (1968) have described an oxygen polarographic electrode whose dimensions are small enough to be mounted at the tip of a catheter for in vivo work. The total length of the electrode is 10 mm and has an outer diameter of 2 mm. The cathode consists of a platinum wire of 300 μm in diameter and the anode is of silver which supports a Teflon membrane of 3 to 6 μm in thickness. The current output of the electrode is 4 μa in 100% oxygen at 37°C. It has a response time of 0.20-0.25 seconds for a 95% deflection for 100% change in concentration in both the directions.

Aging Effect

The polarographic electrode exhibits a slow reduction in current over a period of time, even though the O_2 tension in the test solution is maintained at a constant level. Therefore, it requires frequent calibration of the electrode. It is perhaps associated with material attaching itself to the electrode surface. The effect due to aging can be avoided by covering the electrode with a protective film of polyethylene. However, it has the undesirable effect of increasing the response time. The second procedure employed is to reverse the flow of current frequently to lower or reverse the accumulation of surface contaminants (Geddes and Baker, 1968).

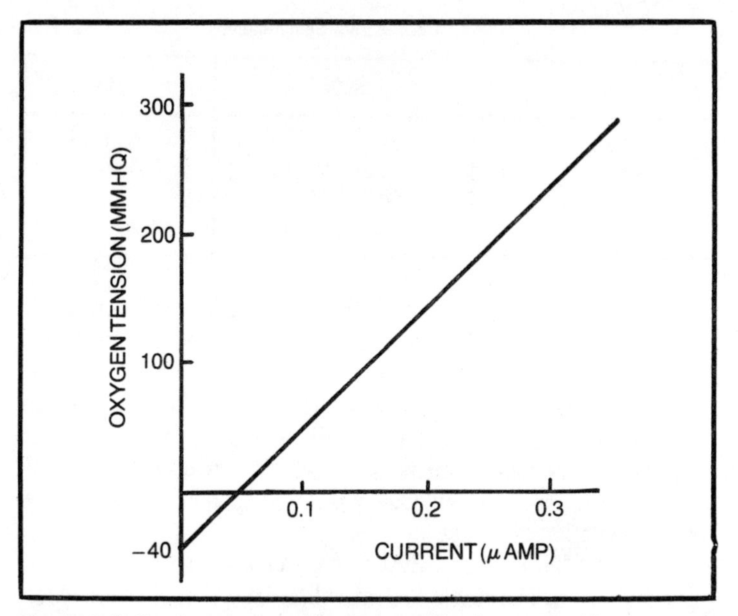

Fig. 15-6. Calibration curve used to obtain a calibration constant for an oxygen electrode.

Cleaning Procedure for the pO_2 Electrode

The platinum cathode of the oxygen electrode tends to become contaminated or dimensionally unstable with time and use. The result is usually an inability to calibrate and slope the electrode. When this happens, the electrode may be cleaned as follows.

■ **Procedure 1.** Remove membrane and O-ring and discard. Empty the filling solution. Flush with tap water. Gently work pumice on the tip of the electrode giving a rotary motion using a towel. Flush with tap water, distilled water and filling solution.

■ **Procedure 2.** Remove membrane and O-ring and discard. Empty the filling solution. Apply the ammonium hydroxide on the tip of the electrode (10% solution) with a gentle rotary motion using a swab. The AgCl will dissolve in ammonium hydroxide. Flush with distilled water and pO_2 filling solution.

COMPLETE BLOOD GAS ANALYZER AND SPECIFICATIONS

From the practical view point, blood gas analysis is mainly the function of the diagnostic chemical pathologist. He would require pH and pCO_2 to give him base deficiency or excess so that the

Table 15-2. Electrode Specifications.

pH	pCO2	pO2
Micro-capillary flow through with Ag/AgC1 Reference	Modified Stow - Severinghaus	Modified Clark

doctor can treat the patient by intra-oral chemical addition. Such requirements exist in intensive care units where any minor changes in pH are sought. The anesthesiologist wants to know pCO_2 and pO_2 to start with so as to control his gas flows in respiration. He may also require pH as a check with pCO_2 or respiratory acidosis or alkalosis which he can control with his anesthetizing gases. Knowledge of pCO_2 and pO_2 is also necessary for the inhalation therapist.

Modern blood gas analyzers are designed to measure pH, pCO_2 and pO_2 from a single 175 microliter insertion of whole blood in approximately 90 seconds (Corning Model 165 blood gas analyzer). With a built-in calculator, this instrument can also compute total CO_2, HCO_3 and base excess. Separate sensors are used for pH, pCO_2 and pO_2. The instrument contains three separate high input impedance amplifiers designed to operate in the specific range of each sensor electrode. A separate module houses and thermostatically controls the pO_2, pCO_2 and pH electrodes. It also provides thermostatic control for the humidification of calibrating gases. A vacuum system provides aspiration and flushing service for all three electrodes. Calibrating gases are selected by a special push button control and passed through the sample chamber. Two gases of accurately known O_2 and CO_2 percentages are required for

Table 15-3. Range, Resolution and Repeatability.

	pH	pCO2	pO2
Operating Range	0.000 to 14.000	1.0 to 240.00mm Hg	0.0 to 1600.00mm Hg
Resolution	0.001 pH	0.1 mm Hg	0.1 mm Hg
Repeatability	0.002 pH	±0.2 mm Hg	±0.2 mm Hg
The accuracy of HCO3, TCO2 and Base Excess depends upon pH, pCO2 and Hb accuracy.			

Table 15-4. Electrical Specification.

Input Impedance	10^{13} ohms
Offset current	5×10^{-12} amps
Operating ambient temperature	18° - 32° C
Output	Analog or digital

calibrating the analyzer in the pO_2 and pCO_2 modes. These gases are used with precision regulators for flow and pressure control. For calibrating the analyzer in the pH modes, two standard buffers of known pH are required. Buffers that bracket the usual blood range (6.838-7.382) must be used. See Tables 15-2 through 15-5 for specifications of a typical blood gas analyzer (Corning Model 165).

BLOOD GAS ANALYZER WITH ANALOG DISPLAY

The signal from each of the three electrodes is a low level dc signal which requires amplification preferably by a chopper amplifier before it can be displayed on a properly calibrated analog meter. Figure 15-7 shows a block diagram of the amplifying arrangement used in a blood gas analyzer. The dc voltage with respect to a reference is connected to the input. This signal is modulated by a photochopper and passes through a preamplification stage. This stage has an electrometer tube in the input to provide a very high input impedance. This is followed with an ac amplifier having gain of about 5000. The signal is fed back to the screen of the preamplifier electrometer tube for stabilization of operating point and gain.

Table 15-5. Sample Chamber.

Chamber construction	Teflon coated stainless steel
Temperature range	36° - 38° C
Temperature response time	90 sec. for 0.5 ml samples at 4°C
Temperature stability	±0.05°C
Sample size	Syringe or vacuum blood collection tube 500 μl nominal Capillary 250 μl nominal (125 μl minimum)

The output of the ac amplifier is fed to a synchronous demodulator operating from the same oscillator as the chopper circuit. A dc amplifier stage with a gain of approximately 2000 amplifies the demodulated signal to the required level for the measuring circuit and meter. The modulator and demodulator are driven at approximately 50 Hz.

The meter can be calibrated for each of the three inputs with three color coded scales. Two controls are provided for each parameter. A calibration control varies the amplifier reference input and a *slope* control varies the gain. The calibration and slope controls are used to adjust the instrument to values of known standards. The details of the aspiration system are given later in the chapter.

DIGITAL READOUT TYPE BLOOD GAS ANALYZER

Figure 5-8 shows a block diagram of a complete blood gas analysis system which gives digital readout. Basically, the system consists of the following parts.

Electrode and Amplifier System

Regarding the pH channel a sample is drawn into a capillary of pH sensitive glass where an electrical potential is developed at the glass/liquid interface. The resulting potential is dependent upon the hydrogen ion activity in the sample. To complete the cell, the sample also contacts a reference electrode which develops a stable half-cell potential independent of sample pH. The signal from the two electrodes is given to a floating differential input amplifier consisting of a dual FET stage. The amplifier is provided for a high common mode rejection ratio. The amplified signal is given to the A-D converter. Calibration and slope adjustments are incorporated.

Let's turn our attention to the pO_2 channel. The pO_2 electrode can measure the partial pressure of oxygen in both gaseous and liquid samples by the reduction of oxygen. The Clark electrode is used for this purpose. The current corresponding to the amount of diffused oxygen which is proportional to the pO_2 of the sample is converted to voltage in the input amplifier and subsequently a voltage gain of about 50 is introduced. An additional current offset is provided to balance out initial electrode offset occurring at the low gas calibration point, usually zero mm. The signal is given to A-D converter with slope and calibration adjustment provisions.

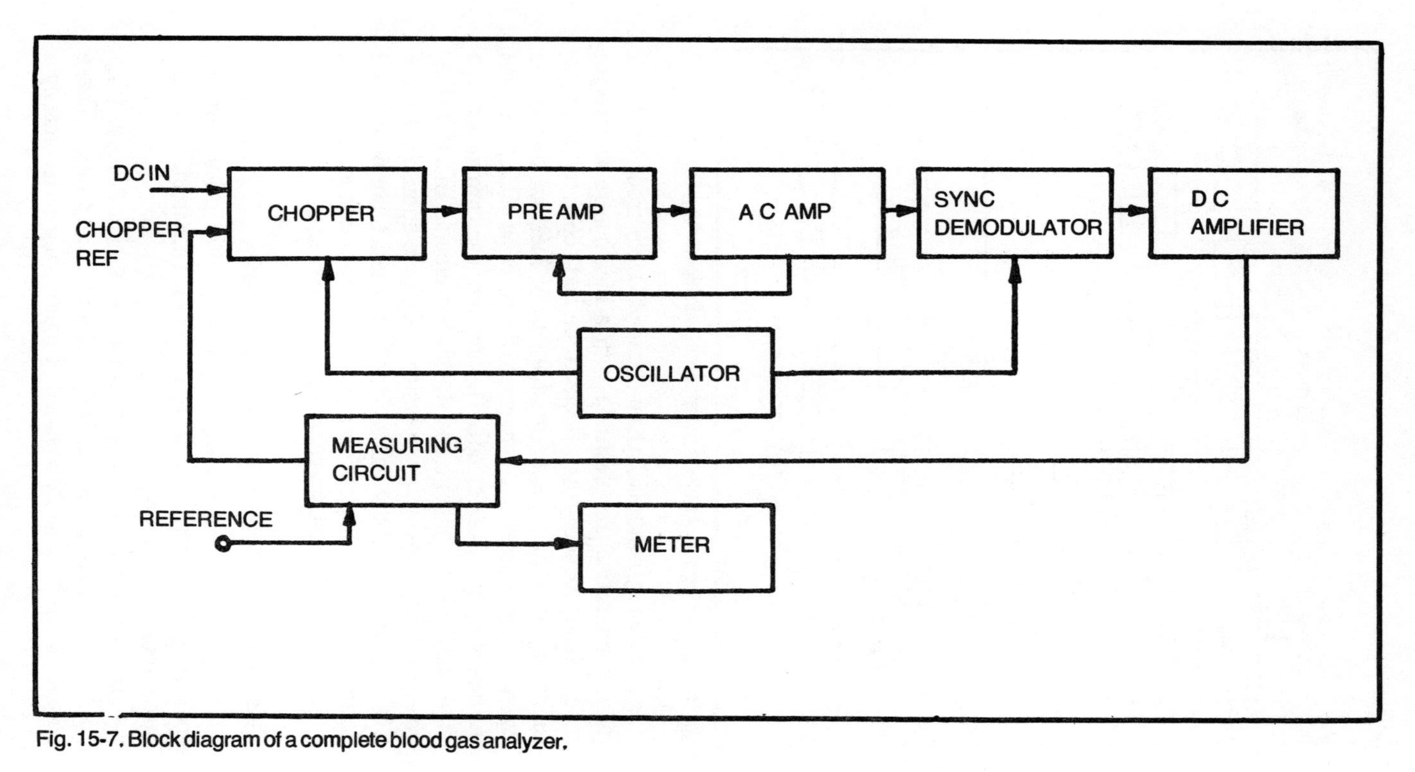

Fig. 15-7. Block diagram of a complete blood gas analyzer.

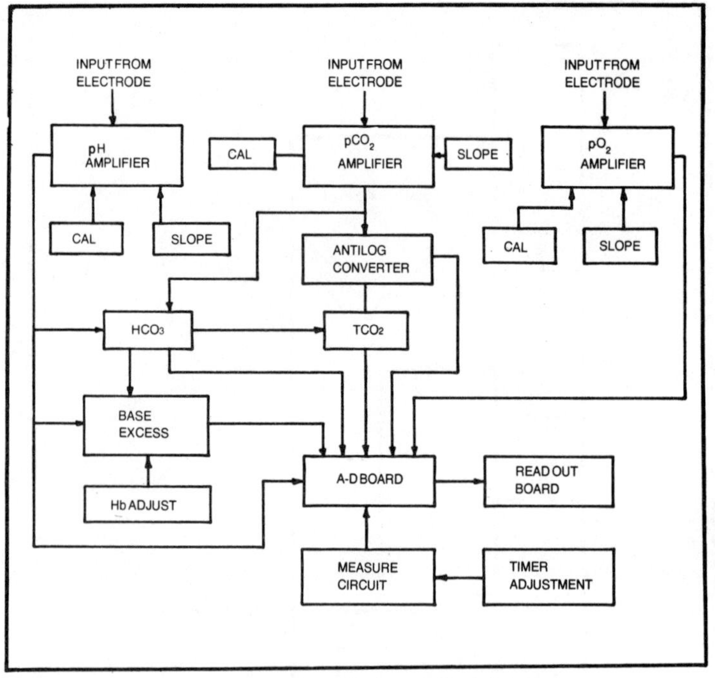

Fig. 15-8. Block diagram of a blood gas analyzer with digital display.

Pertaining to the pCO_2 channel, the pCO_2 electrode employs a glass combination pH electrode with a bicarbonate electrolyte reservoir surrounding the pH electrode and separated from the sample solution by a gas permeable silastic membrane cup. The change in pH of the electrolyte due to the reaction of CO_2 with it is measured by the pH electrode. The amplifier used provides a high impedance input, calibrate and slope adjustments. This is further given to an anti-log converter, which uses the relationship that the CO_2 electrode normally has a slope characteristic of 60 mV per decade of CO_2 concentration and thus gives a direct reading of concentration. The output goes to the A-D converter.

Calculators for HCO_3 Total CO_2 and Base Excess

Modern blood gas analysers have built-in calculators for computing HCO_3, total CO_2 and base excess. The circuit blocks for these would be as follows.

HCO_3 Calculator. The input signal to this calculator (Fig. 15-9) comes from the outputs of pH and pCO_2 amplifiers. These outputs are appropriately adjusted by multiplying each signal by a

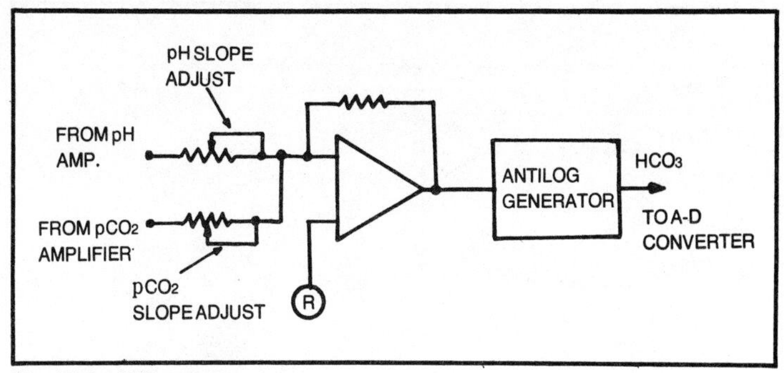

Fig. 15-9. HCO_3 calculator.

constant and are given to a summing amplifier. The following stage is an antilog generator whose output is connected to the A-D converter for readout. The output is adjusted to zero by having a variable potential dividing network.

Total CO_2 Calculator. Total CO_2 is calculated (Fig. 15-10) by summing the output signals of the HCO_3 calculator and the antilog output of the pCO_2 amplifier. The output of the summing amplifier is switched to the input of the A-D converter for readout when the appropriate switch is operated.

Base Excess Calculation. The output of the pH amplifier is inverted (Fig. 15-11) in an operational amplifier whose gain is controlled with a potentiometer (hemoglobin) placed on the front panel. The output of the HCO_3 calculator is also inverted. The two inverted signals are given to a summing amplifier whose output is switched to the A-D converter.

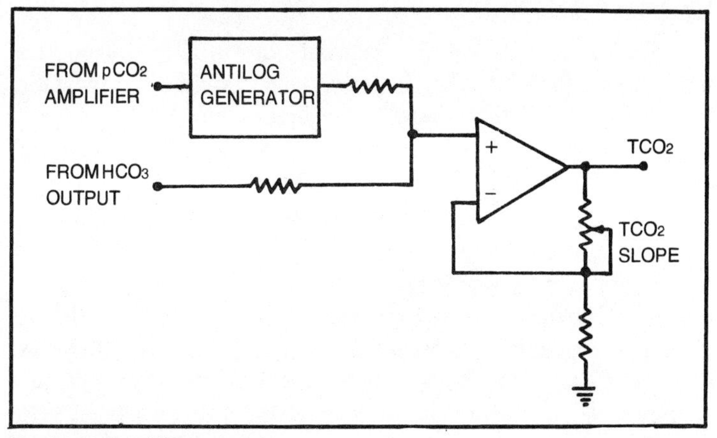

Fig. 15-10. Total CO_2 calculator.

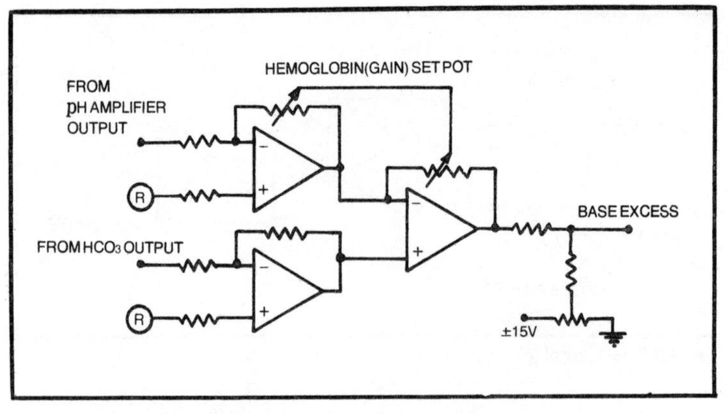

Fig. 15-11. Schematic of base excess calculator.

A-D Converter and Digital Readout

The analog output of the selected amplifier drives the input to an A-D converter. The dual slope integration method is used to convert the analog signal to digital form. The digital output is given to the readout circuit which performs the following functions: a four stage decade counter which supplies timing signal to the A-D converter, a set of storage latches to hold data from the decade counters while the counters are still running, a set of decoder/drivers which convert the 4 line BCD data stored in the latches to 10 line decimal data to drive the display devices, and a clock supplying a series of pulses to the input of the decade counter. Figure 15-12 shows a digital display type blood gas analyzer.

Electrode Control Module

The pCO_2, pO_2 and pH electrodes are usually housed in a thermostatically controlled chamber. It also provides thermostatic control for the humidification of the calibrating gases.

Heating Control

The pH electrode is kept at a constant temperature within the range of 36 to 38°C with a temperature stability of ± 0.01°C. For this purpose, a proportional temperature controlling circuit is utilized. Once the required temperature is set, any variation in electrode temperature is detected by a thermistor which provides an error signal at the input to the amplifier. The error signal is amplified and fed to the heater control circuit in such a manner as to counteract the change in temperature.

Membrane Check Circuit

Many of the blood gas analyzers have a provision for checking the membrane of a pO_2 or pCO_2 electrode. In the check position, a potential is placed across that membrane. Should there be a leak in the membrane of sufficient magnitude, it will result in a considerable lowering of the resistance, maybe from 100 megohms to 500 K. This fall in resistance can be used to have a change of potential to switch on a transistor which would cause a lamp to light. This indicates that a new membrane is required.

Aspiration Control

Figure 15-3 shows a typical aspiration flow chart. The aspiration control is provided by a peristaltic pump. The operation of the FLUSH key initiates the cleaning cycle and the flush solution flows through the sample chamber and the tubing. The ASPIRATE key completes the cleaning cycle by causing the vacuum in the instrument to draw the waste out of the lines and sample chamber and into the waste bottle. The protein trap bottle is filled with

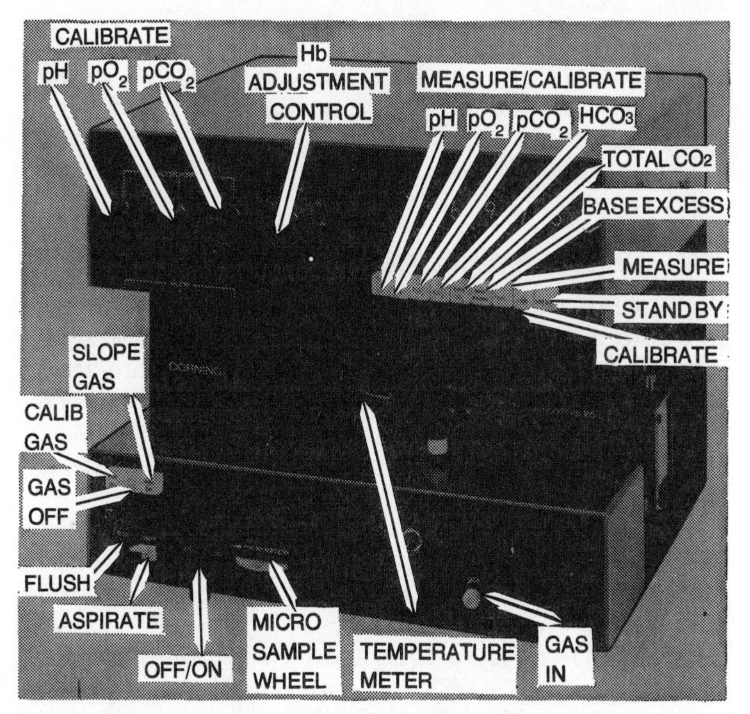

Fig. 15-12. Blood gas analyzer Model 165 (courtesy of Corning, USA).

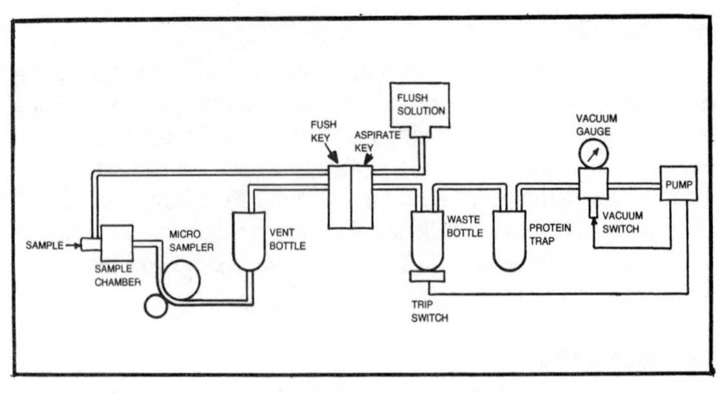

Fig. 15-13. Typical aspiration flow chart.

loosely packed cotton. The vacuum system provides and maintains a vacuum between 10 and 12 inches of mercury which is indicated by a vacuum gauge. The pump automatically shuts off when the desired vacuum is achieved.

Gas Flow System

The calibrating gas cylinders are provided with pressure regulators. They are connected to the humidifier bottles through needle valves. The humidifer bottles are filled with distilled water. The two humidifiers are connected to the sample chamber through a manifold and solenoid valves. The operation is self-explanatory from Fig. 15-14.

SPECIMEN COLLECTION AND STORAGE

Blood sample collection must be done under medical supervision. Arterial blood has been recommended for use in blood gas studies. A sterile technique is required for arterial puncture, which may be performed in the brachial, femoral or radial artery. Capillary blood may also be used and can be obtained from the heel, finger or ear lobe. The area can be prewarmed to promote arterial circulation.

Properly drawn venous blood may be used for the routine chemical estimation of acid-base parameters. No hand movement such as clenching the fist should be permitted during withdrawal of blood.

Heparin is mostly used as *anticoagulant.* Other common anticoagulants such as EDTA, citrate or oxalate may have a significant effect on blood pH and, therefore, should not be used.

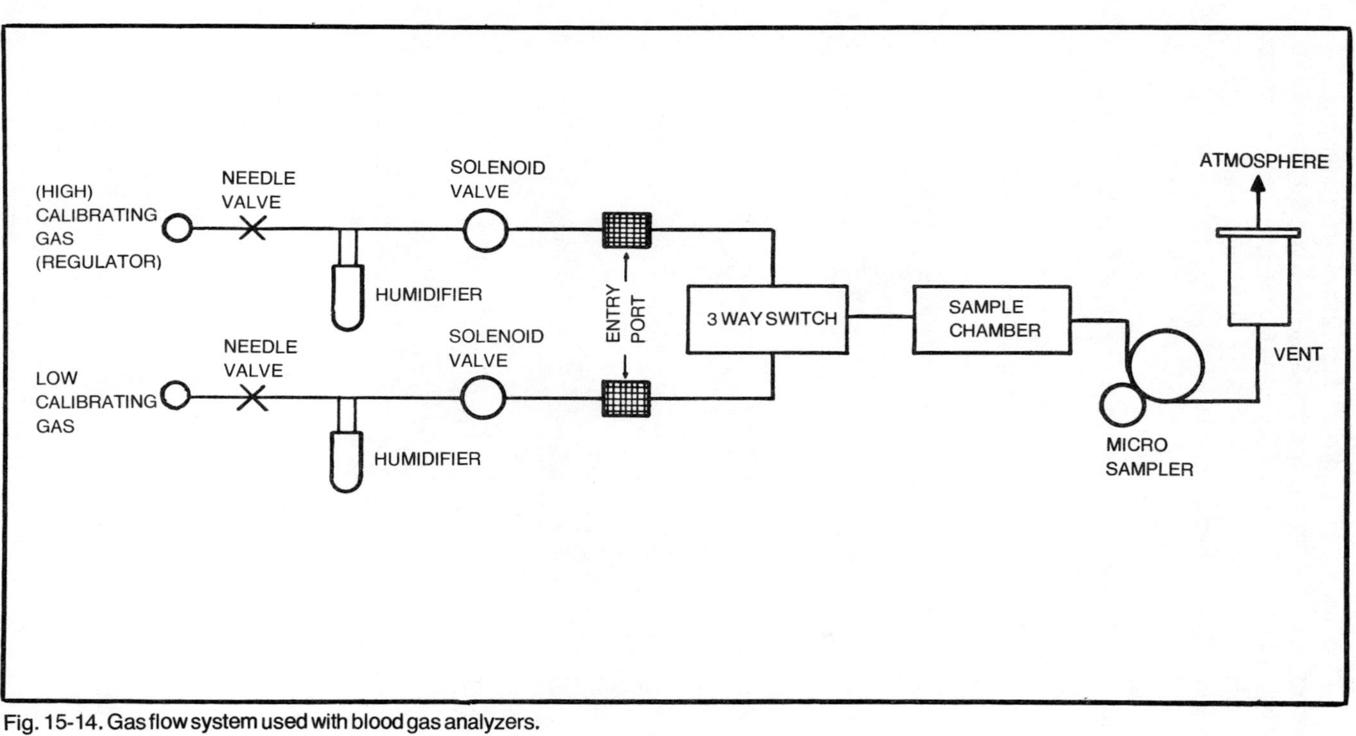

Fig. 15-14. Gas flow system used with blood gas analyzers.

Blood collected in airtight heparinized glass syringes will satisfy most of the requirements. Anaerobic, heparinized vacuum tubes are commercially available and may also be used for routine acid-base studies. For micro-sample collection, capillary tubes coated with dried heparin may be used.

Whole blood pH falls and pCO_2 rises with time. The actual change depends upon time and white blood cell count. When the utmost precision and accuracy is required, pH and blood gas analysis should be done within five minutes of sampling. If samples cannot be measured within 20 minutes of withdrawal, they should be immediately placed in ice water. Samples must be stored anaerobically.

Chapter 16

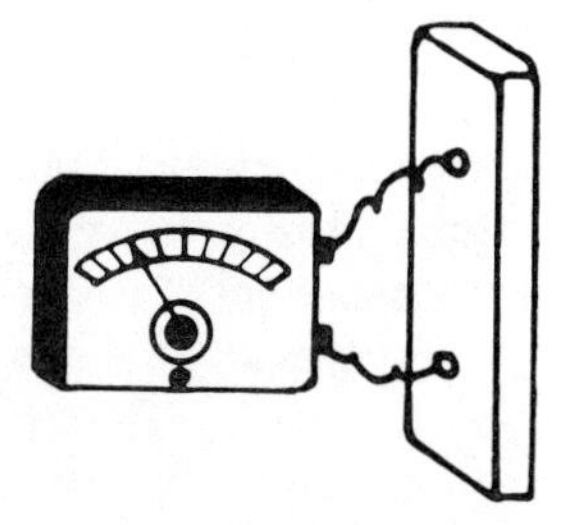

Industrial Gas Analyzers

Several physicochemical properties have been utilized for analysis of gases in simple or multi-component mixtures. However, most of the commercially available analyzers work on the measurement of quantities such as infrared absorption, paramagnetism, thermal conductivity and the gas density. In addition to these, many of the methods described in other chapters can also be used for the determination of gaseous components. These methods include visual and ultraviolet spectrophotometry, infrared spectrophotometry, mass spectrometry and various electrochemical methods.

All molecules, with the exception of the noble gases, consist of several atoms which exhibit a regular three-dimensional structure by chemical forces and whose valence electrons can attain defined energy states. The molecules are primarily in rotational and translational motions relative to their surroundings, thus giving *Brownian molecular motion*. Also, the single atoms within a molecule can vibrate mutually and the shells of the valence electrons within the molecule can reach different states of energy. Eventually, analyzers can be designed so that the molecules of measuring gases may be made to give a physical or chemical reaction which may reveal their nature and extent. Three different interactions are generally utilized.

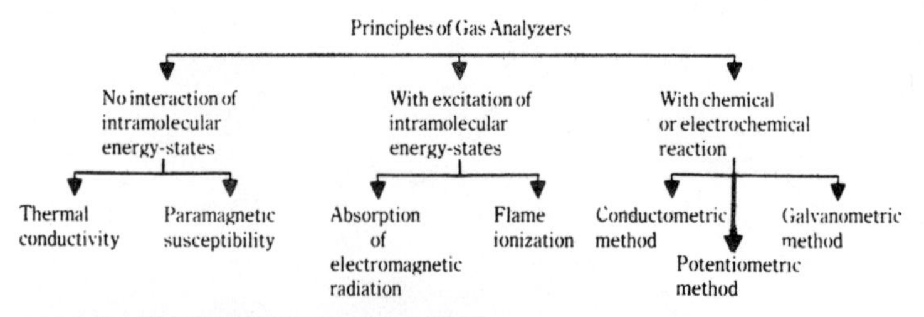

PARAMAGNETIC OXYGEN ANALYZER

Oxygen has the property of being paramagnetic in nature, i.e., it does not have as strong magnetism as permanent magnets, but at the same time it is attracted into a magnetic field. Nitric oxide and nitrogen dioxide are other two gases which are paramagnetic in nature. Most gases are, however, slightly diamagnetic, i.e., they are repelled out of a magnetic field.

Figure 16-1 shows relative paramagnetism exhibited by different gases. The magnetic susceptibility of oxygen can be regarded as a measure of the tendency of an oxygen molecule to become temporarily magnetized when placed in a magnetic field. Such magnetization is analogous to that of a piece of soft iron in a field of this type. Similarly, diamagnetic gases are comparable to non-magnetic substances. The paramagnetic property of oxygen has been utilized in constructing oxygen analyzers.

A paramagnetic oxygen analyzer was first described by Pauling et. al. (1946). Their simple dumbbell type of instrument has formed the basis of more modern instruments. Figure 16-2 shows the schematic of a paramagnetic analyzer from Beckman. The arranagement incorporates a small glass dumbbell suspended from a quartz thread between the poles of a permanent magnet. The pole pieces are wedge-shaped in order to produce a non-uniform field.

Referring to Fig. 16-3, when a small sphere is suspended in a strong non-uniform magnetic field, it is subject to a force proportional to the difference between the magnetic susceptibility of this sphere and that of the surrounding gas. The magnitude of this force can be expressed as:

$$F = C(K - K_0).$$

where C = A function of the magnetic field strength and gradient.

K_0 = Magnetic susceptibility of the shpere.

K = Magnetic susceptibility of the surrounding gas.

The forces exerted on the two spheres of the test body are thus a measure of the magnetic susceptibility of the sample and therefore of its oxygen content.

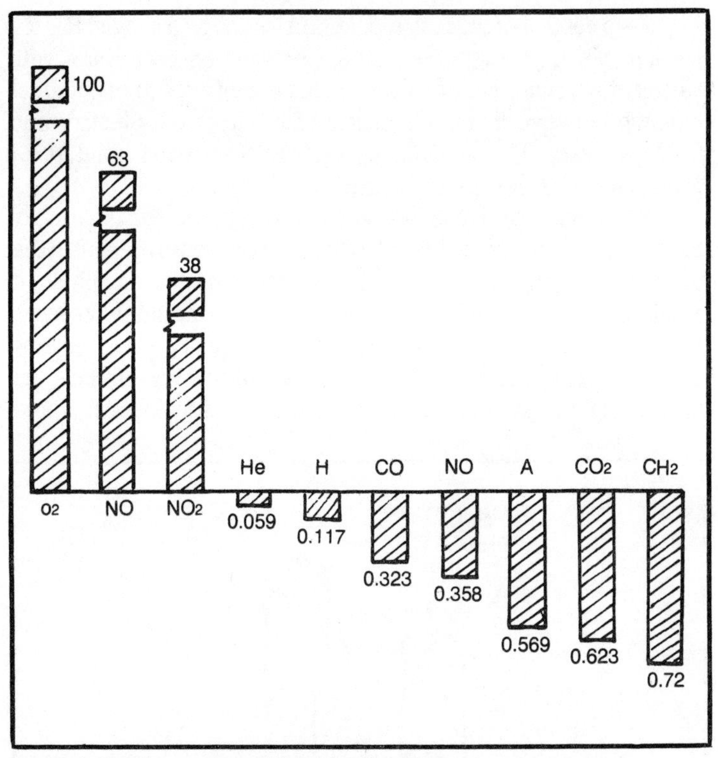

Fig. 16-1. Relative paramagnetism of various gases.

The magnetic forces are measured by applying to one sphere an electrostatic force equal and opposite to the magnetic forces. The electrostatic force is exerted by an electrostatic field established by two charged vanes mounted adjacent to the sphere (Fig. 16-4). One vane is held at a higher potential than the test body, the other at a lower potential. Since the glass test body must be electrically conductive, it is sputtered with an inert metal.

The test body is connected electrically to the slider of null adjust potentiometer R_{20}. This potentiometer is part of a voltage-dividing resistor network connected between ground and B^+. Potential to the test body can be adjusted over a large range.

An exciter lamp directs a light beam on to the small mirror attached to the test body. From the mirror, the beam is reflected to a stationary mirror and then on to a translucent screen mounted on the front panel of the instrument. The geometry of the optical system is so arranged that a very small rotation of the test body causes an appreciable deflection of the image cast by the beam.

Zero control of the instrument is provided by a ganged R_{13}-R_{15} setting which changes the voltage present on each vane with respect to ground, but does not change the difference in potential existing between them. This adjustment alters the electrostatic field. Rheostat R_{19} sets the upscale standarization point, i.e., provides span or sensitivity control.

When no oxygen is present, the magnetic forces exactly balance the torque of the fiber. However, if oxygen is present in the gas sample drawn in the chamber surrounding the dumbbell, it would displace the dumbbell spheres and they would move away from the region of maximum magnetic flux density. The resulting rotation of the suspension turns the small mirror and deflects the beam of light over a scale of the instrument. The scale is calibrated

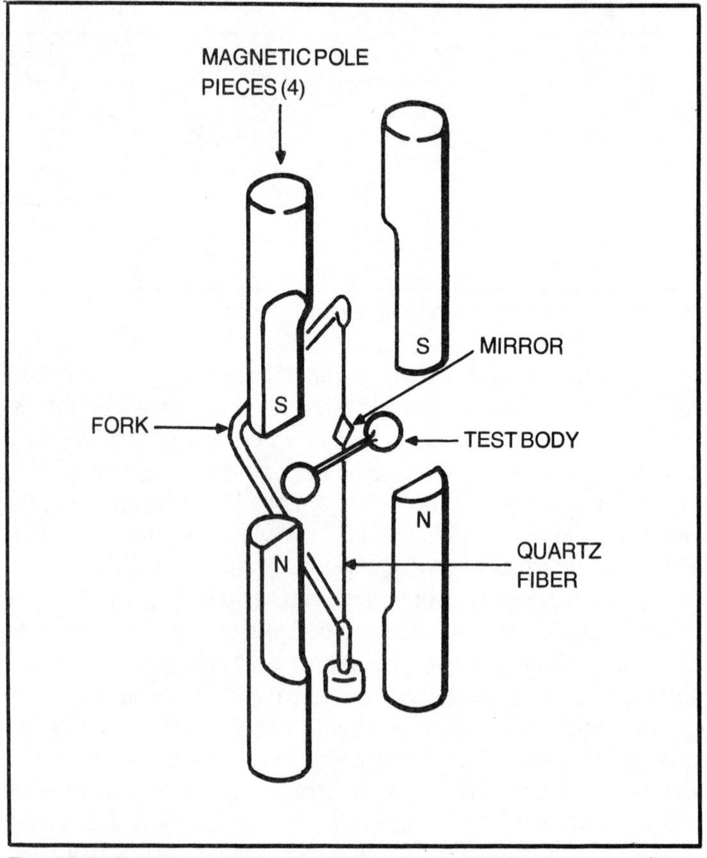

Fig. 16-2. Arrangement of magnets in a paramagnetic oxygen analyzer. (courtesy of Beckman, USA).

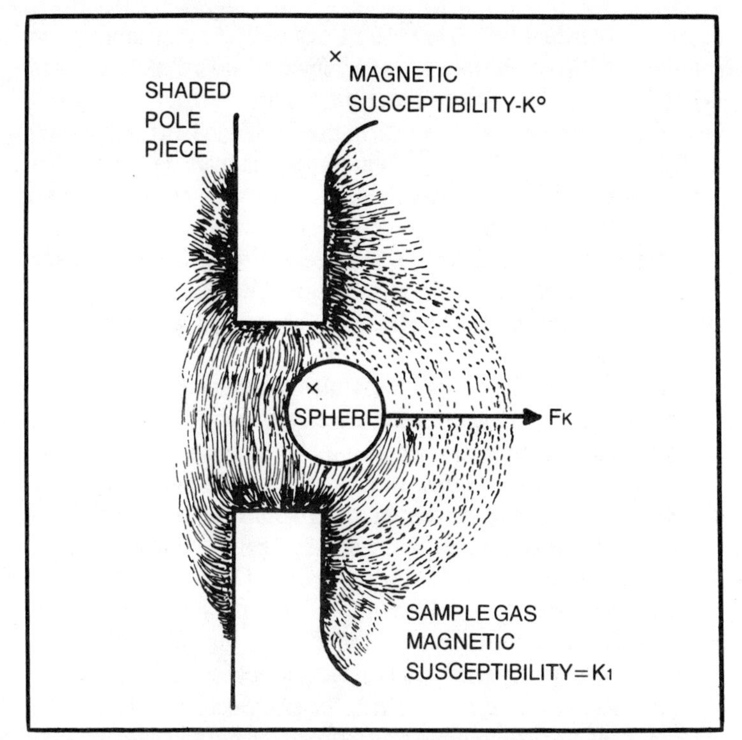

Fig. 16-3. Sphere in non-uniform magnetic field (courtesy of Beckman, USA).

in percentages by volume of oxygen or partial pressure of oxygen. Paramagnetic oxygen analyzers are capable of sampling static or flowing gas samples.

Only a few improvements have been suggested and carried out in the development of oxygen analyzers based on paramagnetism. The original quartz suspension has been replaced largely with a platinum-iridium suspension which is more robust. Instead of measuring the deflection of the dumbbell, a null-balance system is preferred wherein the deflection is offset by passing a current through a coil of wire attached to the dumbbell. The current required is proportional to the deflecting couple and thus to the oxygen tension of the gas. The control of current is carried out by a helical potentiometer which is duly calibrated.

Displacement of the dumbbell results in unbalancing the output from a pair of photocells. The difference in their output signals is fed to a differential amplifier which supplies its output current to the dumbbell coil to null the deflection. The coil current is indicated on a meter. Oxygen analyzers are available with continuous readout 0-25% or 0-100% oxygen. The instruments are calibrated with the reference gas specified. Standard cell volume is 0-10 ml and response time is about 10 seconds.

The recommended flow rate in the Beckman instrument if 50 to 250cc/min when the sample enters the analysis cell through a porous diffusion disc. If the sample enters directly for rapid response, the flow rate is 40-60 cc/min.

Before the gas enters the analyzer, it must be pressurized with a pump and passed through a suitable cleaning and drying system. In many cases, a small plug of glass wool is sufficient for cleaning and drying functions.

The entry of moisture or particulate matter into the analyzer will change instrument response characteristics. Therefore, the use of a suitable filter in the sample inlet line is recommended.

Any change in the temperature of a gas causes a corresponding change in its magnetic susceptibility. To hold this temperature of the gas in the analysis cell constant, the analyzer incorporates a thermostatically controlled heating circuit. Once the instrument reaches temperature equilibrium, the temperature inside the analysis cell is approximately 140°F. The sample should be admitted to the instrument at a temperature between 50 to 110°F. If the temperature is less than 50°F, the sample may not have time to reach temperature equilibrium before entering the analysis cell.

Calibration of the instrument consists of establishing two standarization points, i.e., a down scale and an upscale standarization point. These two points can be set by passing standard gases through the instrument at a fixed pressure, normally atmospheric pressure. First, a zero standard gas is admitted and the zero control is adjusted. Then the span gas is admitted and the span control is adjusted. Alternatively, the required practical pressures of oxygen are obtained by filling the analysis cell to the appropriate pressures with non-flowing oxygen or air. If the highest point is not greater than 21% oxygen, dry air is used to set the span point. If the point is greater than 21% oxygen, oxygen is used to set this point. The instrument should be handled carefully, as the fine quartz fiber supporting the test body may break.

MAGNETIC WIND INSTRUMENTS

Dyer (1947) describes a paramagnetic oxygen analyzer based on the phenomenon of *magnetic wind* which arises due to the motion of paramagnetic oxygen molecules into the non-uniform magnetic field. The flowing gas is made to pass over a heated

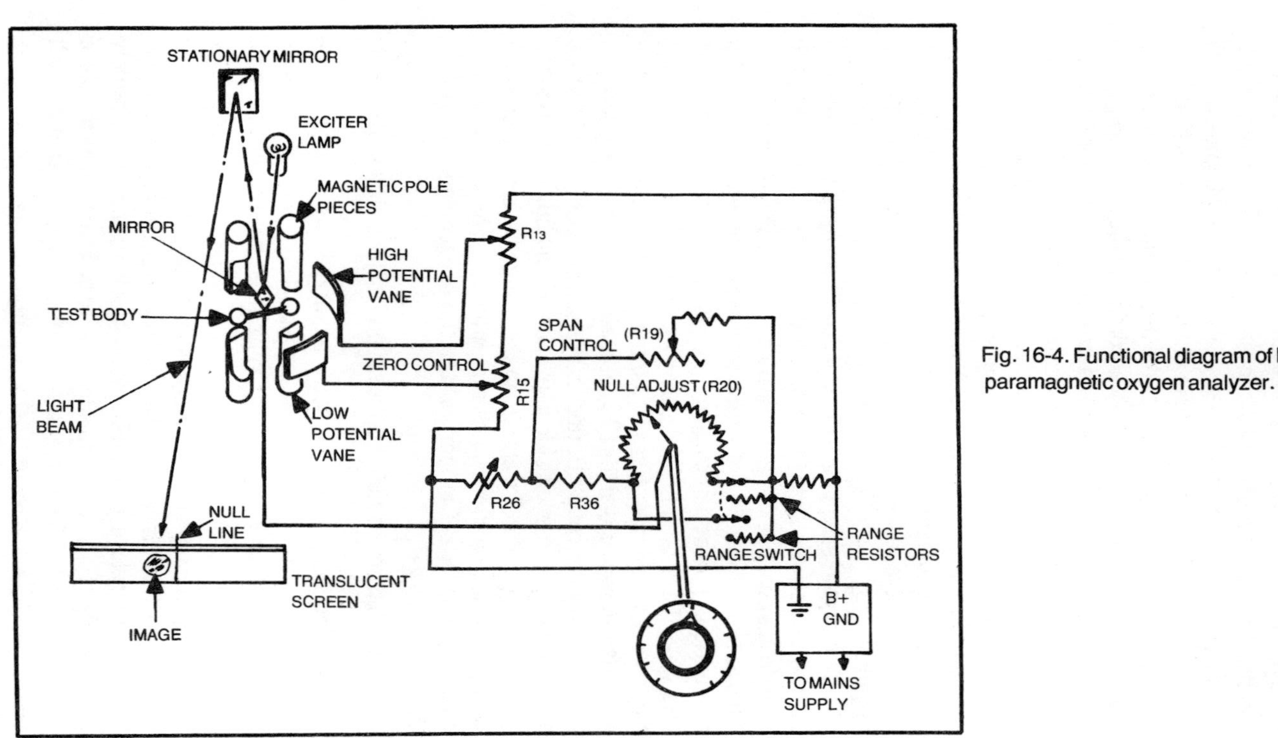

Fig. 16-4. Functional diagram of Beckman paramagnetic oxygen analyzer.

filament connected as one arm of a Wheatstone bridge circuit. The gas cools the filament and unbalances the bridge circuit. The output voltage from the bridge can be fed to a recorder to show a continuous indication of oxygen concentration in the sample.

Hartman and Braun Oxygen analyzer types Megnos 2T and Magnos 5T work on the principle of thermomagnetic action. The method consists in making use of the dependency of paramagnetic susceptibility on the temperature according to the *Curie* law.

Referring to Fig. 16-5, the analyzer consists of a ring chamber which has a metallic hollow ring arranged in a vertical plane, with a gas inlet below and a gas outlet above. The gas is conducted in the upward direction. The two halves of the ring chamber are connected by a glass tube. The transverse tube has taped wire winding on its outside. The winding forms a part of a Wheatstone bridge circuit whose one half is placed in the field of a permanent magnet.

In case the sample gas does not contain any oxygen, there is no flow in the horizontally placed transverse tube. If, however, the sample gas contains oxygen, it is attracted into the magnetic field. This gas flow is heated. Due to the increase of the temperature, the susceptibility of the paramagnetic oxygen molecules decreases. In this way, the heated gas is pushed away by the cold gas coming from the left hand side. A gas flow thus arises inside the transverse tube. This flow is called the magnetic wind whose velocity depends upon the oxygen concentration in the sample gas. The magnetic wind cools down the left side of the heated winding more than the right side. Therefore, a change in the bridge balance takes place which depends on the temperature gradient. The voltage difference of the bridge is proportional to the oxygen concentration in the measuring gas.

The instrument suffers from several major sources of error:

■ The filament temperature is affected by changes in the thermal conductivity of the carrier gas. Thus, the calibration is correct for only one gas mixture which must be specified for each analyzer.

■ Hydrocarbons and other combustible gases in the sample stream react on the heated filaments causing changes in temperature and therefore their resistance which results in extremely large errors. These hydrocarbons have to be removed by means of a cold trap, but if the percentage in the gas is high, an error will result due to the change in sample volume.

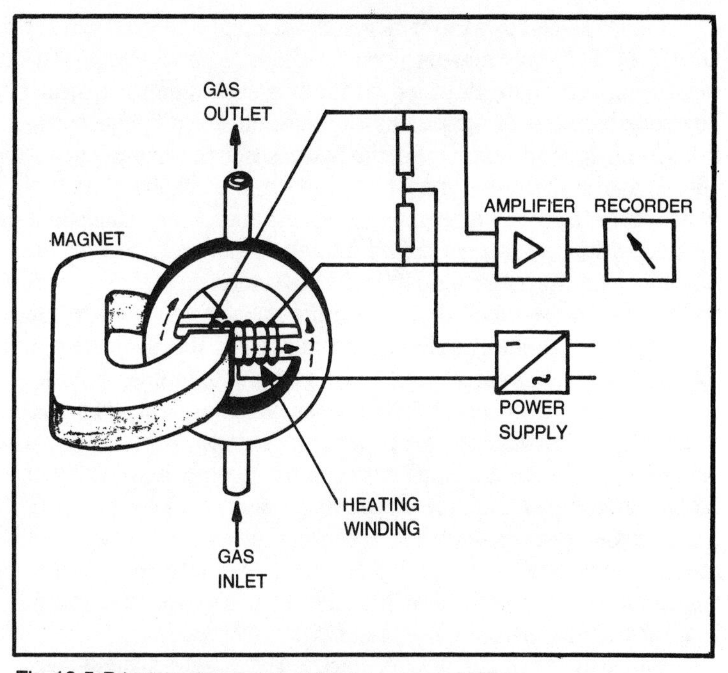

Fig. 16-5. Principle of Hartman and Braun oxygen analyzer.

■ The cross tube must be horizontal to avoid an error due to gravitational chimney flow effects.

It is important to note that magnetic wind and thermomagnetic analyzers are synonymous. The first name is generally used in Europe and the second in the United States.

In some instruments, the paramagnetic type of sensor works by passing the unknown gas between two powerful magnetic poles wherein a thin aluminum membrane is suspended. The magnetized oxygen is pulled into the space between one of the magnetic poles and the membrane, thereby displacing the membrane in the opposite direction. The displacement of the membrane can be conveniently measured by using LVDT or strain gauge transducer. The extremely low inertia of the thin membrane in the oxygen sensor allows it to come to equilibrium in 0.25 to 0.5 second. This enables the recording of rapid changes in oxygen concentration. This is particularly useful for making measurements during each breathing cycle while making measurements of respiratory gases.

Oxygen analyzers are employed in the areas of oxygen absorption studies on plants and tissues, respiratory studies, food processing, air pollution studies and anesthesiology.

Figure 16-6 shows the arrangement of a rapid oxygen analyzer OXYMAT-M from Siemens used for medical applications. The measuring procedure is based on the pressure difference which develops between two gases having different oxygen concentrations in a magnetic field. One of the two gases, usually ambient air, serves as the reference gas which flows in the channels, both of which open into a single measuring chamber. One channel is located within the magnetic field, the other in a position which is virtually free from magnetic influences. The respiratory air to be analyzed is sucked directly through the measuring chamber. On account of the paramagnetic property of oxygen, it is subjected, in the inhomogeneous parts of the magnetic field, to forces which act on the oxygen molecules in the direction of higher field strengths. If the oxygen concentrations of the two sample of gas differ, a pressure difference develops between the two points of entry of the reference gas into the measuring chamber. This pressure difference is compensated via a connecting channel in which a micro-flow sensor converts the stream of gas into an electrical signal which is proportional to the difference in oxygen concentration between the gas to be analyzed and the reference gas.

The instrument employs an electromagnet with changing flux intensity. As a result, an alternating pressure is created in the measuring chamber and an alternating electric voltage of 25 Hz is developed at the micro-flow sensor, which is rectified and amplified in the electronics part. The measurement of alternating pressure has the advantage of avoiding any unsymmetry and drift phenomena.

THE ELECTROCHEMICAL METHODS

Analyzers based on the electrochemical methods are mostly used for the determination of oxygen content of a gas. They utilize an electrolytic cell and can be broadly classified as galvanic, polarographic and conductometric methods.

Galvanic Methods

Galvanometric methods are based on the fact that the electrical current of a galvanic cell which is equipped with appropriate electrodes and an appropriate electrolyte would depend upon the oxygen concentration, it being related to rate of oxygen uptake by such a cell. It is the reverse of electrolysis in which oxygen is evolved at the anode and hydrogen at the cathode

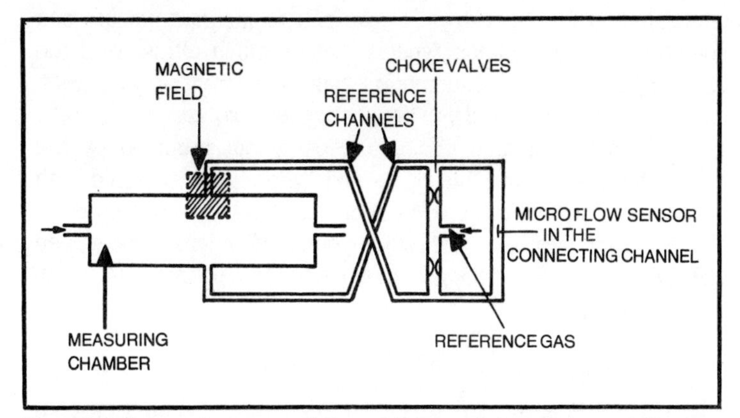

Fig. 16-6. Schematic representation of the measuring system of Siemens OXYMAT-M oxygen analyzer for medical applications.

when electrical current is made to pass through the cell. These analyzers are used for measurement of small oxygen concentrations.

The galvanic cell has two electrodes, one of which is made of noble metal such as silver and the other is made of a base metal such as lead. The oxygen contained in the sample gas is made to bubble through the electrolyte. A magnetically driven stirring system helps to ensure a quick and efficient mixing of the reaction liquid and the gas.

Analyzers based on this principle are used to measure the content of dissolved oxygen in boiler feed water. For this purpose, the boiler feed water flows through the cell and acts as an electrolyte and the cell is used for continuous monitoring.

There is a need to control the oxygen uptake at the cathode. This is generally done by having a porous carbon cathode and semipermeable membranes. Since the electrons are supplied by the dissolving anode, the life of the cell is limited. The cell is affected with a very high temperature gradient approximately 4% per °C. This is compensated using a combination of NTC and PTC resistors.

Speed of response and sensitivity is improved by using silver gauze instead of a smooth electrode and reducing the volume of electrolyte. The cell could be cylinderical in shape with a central anode of porous material like lead saturated with electrolyte and dipped in a reservior over which the sample gas flows. The cathode is formed by a gauze which surrounds the anode. This arrangement is known as a *Hersch* cell. Several improvements have been suggested in the basic form of the cell to increase its life and sensitivity. One common type is that in which diffusion of the oxygen through a Teflon membrane causes a current flow between two electrodes separated by a liquid or gel electrolyte.

The noble metal cathode if used is not attacked by the electroyte. Therefore, the drifts and instability associated with porous cathodes are eliminated. Also, the cell has no output in the absence of oxygen. A definite zero is obtained which does not need calibrating. Current in the galvanic cell obeys Faraday's law which is given by the relation:

$$I = 0.263\ CFP\ (298/T)$$

where I is the expected current in micro amps when a gaseous sample containing C ppm of oxygen by volume passes through the cell at a flow rate F cm^3/minute measured at P atmosphere and T°K. The expression assumes that the perfect gas laws apply. With a sample flow rate of 100 cm^3/min. at 1 atmosphere pressure and 298°K., the theoretical sensitivity is 26.3 micro amps/ppm.

These instruments are generally slow in operation and the sample gas must be scrubbed to remove CO_2, SO_2, H_2S or any acidic gas, but one attraction is that they can be used to measure dissolved oxygen in liquids.

Another type of electrochemical analyzer employs the high temperature galvanic cell. This cell is manufactured by a number of companies like Westinghouse, Thermolab, Kent and Engelhard.

All of these cells consist of a calcium stabilized zirconium oxide electrolyte with platinum electrodes. At the operating temperature, oxygen molecules on the side of the cell exposed to a high partial pressure of oxygen (the anode) gain electrons. Simultaneously, oxygen molecules are formed by the reverse action at the other electrode (the cathode).

For a cell operating at 850°C, the standard Nernst equation for an oxygen cell is

$$\text{EMF (open circuit)} = 55.7 \log_{10} (P_a/P_b)$$

where P_a is the partial pressure of oxygen within the cell, and P_b is the partial pressure of oxygen outside the cell. Since this effect is specific for oxygen, the instrument output is not affected by the presence of water or CO_2. However, hydrocarbons, hydrogen and other combustible gases will burn at the operating temperature and result in an indication of less oxygen than is actually present. The response of analyzers using such cells is very fast.

Westinghouse claims that this probe can be used for direct determination of oxygen in the flue provided that the flue temperature is less than 800°C. The instrument can be calibrated only by removing the probe from the flue and inserting it into an enclosed container which can be filled with calibration gases.

The instrument manufactured by Thermolab is mounted outside the flue. The sample gas is drawn through a short sample tube to the measuring cell by an air ejector pump. This analyzer provides a means of introducing calibration gases for setting zero and span. It is important to note that both these types of instruments require a supply of clean, dry air at the reference side of the zirconium electrolyte.

Polarographic Cells

Polarographic cells are generally used to measure the partial pressure or percentage of oxygen from injected samples, continuous streams or in static gas monitoring. They find maximum utility in the respiratory and metabolic laboratories. Polarographic cells are based on the redox reactions in a cell having both the electrodes of noble metals. When a potential is applied, oxygen is reduced at the cathode in the presence of KCl as the electrolyte and a current will flow. The cathode is protected by an oxygen permeable membrane. The rate at which oxygen reaches the cathode will be controlled by diffusion through the membrane. The voltage current curve will be a typical polarogram (Fig. 16-7). A residual current flows in the cell at low voltages. The current rises with the increase in voltage until it reaches a plateau where it is limited by the diffusion rate of oxygen through the membrane. For a given membrane and at a constant temperature, this would be proportional to the partial pressure of oxygen across the membrane. When the voltage is applied in the plateau region, the current in the cell is proportional to oxygen concentration.

Polarographic cells are temperature sensitive as the diffusion coefficient changes with temperature. The temperature coefficient is usually 2-4% per degree centigrade. Therefore, temperature compensation circuits are used to overcome this problem.

Polarographic oxygen cells are used mainly for portable gas detectors, where simplicity, low cost and light weight are important. They are preferably used for measuring oxygen in liquids, especially in water pollution and medical work.

The Beckmen OM-1 oxygen analyzer incorporates an oxygen sensor which contains a gold cathode, silver anode, potassium chloride electrolyte gel and a thin membrane. The membrane is precisely retained across the exposed face of the gold cathode, compressing the electrolyte gel beneath into a thin film. The membrane, permeable to oxygen, prevents airborne solid or liquid contaminants from reaching the electrolyte gel. The sensor is insensitive to other common gases. A small electrical potential (750 mV) is applied across the anode and cathode.

Although the composition of the atmosphere is remarkably constant from sea level to the highest mountain, i.e., oxygen 21% and nitrogen 79%, there is a great difference in the partial pressure of oxygen at different altitudes. The polarographic sensor which actually senses partial pressure would therefore require some adjustment to read approximately correctly the percentage oxygen at the altitude at which it is used. Humidity can also affect oxygen readings, but to a lesser degree. Water vapor in air creates a water vapor partial pressure that slightly lowers the oxygen partial pressure. Therefore, for precision work, it is often desirable to use a drying tube on in inlet sample line. Also, care should be taken to calibrate and sample under the same flow conditions as required for the gas to be analyzed. The range of the instrument is 0-1000 mm Hg O_2 and the response time is 10 seconds for 90%, 35 seconds for 99% and 70 seconds for 99.9%. The instrument can measure oxygen against a background of nitrogen, helium, neon, argon, etc., with no difficulty. The sensor is very slightly sensitive to carbon dioxide and nitrous oxide with typical error less than 0.1% oxygen for 10% carbon dioxide and 4% oxygen for 100% nitrous oxide.

Conductometric Method

The *conductometric* method is convenient and the most widely used method for trace gas analysis. In practice, the sample gas is passed through a cell containing a liquid reagent which can react with the gas of interest. The conductivity of the liquid is measured before and after the reaction with the gas. The difference in conductivity is proportional to the gas concentration.

To obtain reproducible results, the flow of gas and reagent must be kept constant. Therefore, the measuring gas must enter the analyzer at a constant velocity, which is generally adjusted by a pneumatic bypass and indicated by a capillary flowmeter. A slow stream of reaction solution enters the reaction cell via a second capillary and its flow rate is also kept constant. In the Hartman and Braun analyzer, the chemical reaction between the measuring gas

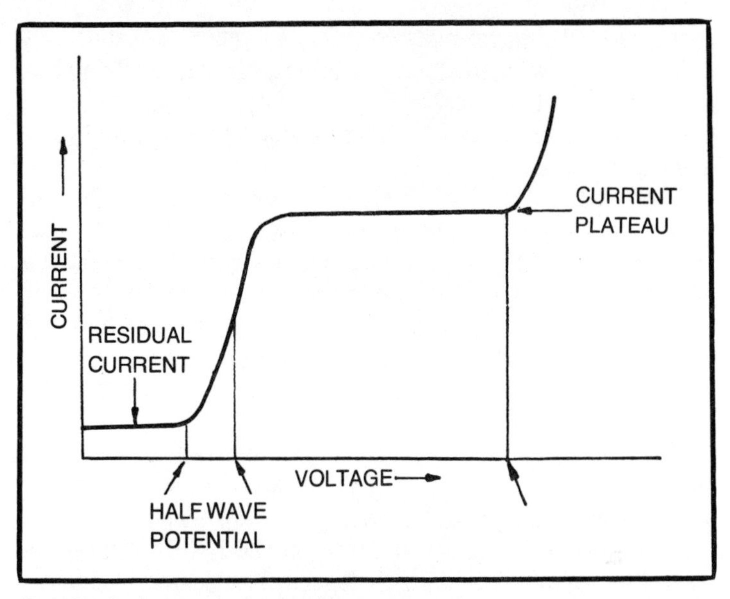

Fig. 16-7. Response of polarographic sensor.

and the reagent takes place quantitatively in a spiral reaction cell where the gas is separated from the liquid. The liquid is then passed through the conductivity measuring cell, while the gas is vented from the analyzer.

To minimize errors due to changes in ambient temperature, the measuring cell is kept in a precisely controlled temperature environment. By proper selection of the reagent, analyzers based on this principle can be made specific for various gas components. The method is especially suitable for measuring traces of H_2S, SO_2, NH_3 and H_2O in the ppb range.

INFRARED GAS ANALYZERS

Infrared gas analyzers depend for their operation upon the fact that some gases and vapors absorb specific wavelengths of infrared radiation. One of the most commonly measured gases using the infrared radiation absorption method is carbon dioxide. The technique used for this purpose is the conventional double-beam infrared spectrometer system having a pair of matched gas cells in the two beams. One cell is filled with a reference gas which is a non-absorbing gas like nitrogen whereas the measuring cell contains the sample. The difference in optical absorption detected between the two cells is a measure of the absorption of the sample at the particular wavelength. Since the vibration excitation only occurs if we have heteroatomic molecules, the infrared absorption principle is not applicable for the analysis of gases whose molecules are formed by two identical atoms like oxygen, hydrogen and nitrogen.

Infrared analyzers are used for the determination of a large number of components including CO, CO_2, SO_2, NH_3, H_2O and nitric oxide as well as most gaseous hydrocarbons. The selectivity is however restricted by the fact that the absorption bands partly overlap mutually. This can be eliminated by providing a filter cell which is filled with the interfering component. The selectivity can also be enhanced by the negative filtering so that it is possible to distinguish interfering components.

A simple method of using the infrared technique for gas analysis is shown in Fig. 16-8. This method does not require the use of a wavelength dispersing device. Two identical infrared sources in the pickup or sensing head emit beams of radiation that are pulsed by a motor driven chopper. The source of infrared radiation is the hot-wire spiral. The rotating chopping disc occludes each beam twice per rotation. For industrial analyzers, the chopping frequency is 2-10 Hz, whereas for medical applications it is 20-50 Hz. One beam passes through the sample cell, the other beam through a reference cell, and both beams enter opposite ends of the detection chamber. The detection chamber is a permanently sealed unit divided into two compartments by a thin, metal diaphragm. Both compartments are charged to the same pressure with the gas being measured. For example, the detector of a CO_2 analyzer would contain pure CO_2, and that of a *halothane* analyzer would contain pure halothane vapor.

When the gas being measured enters the sample cell, it absorbs infrared radiation at the same wavelengths as the gas in the detection chamber. This reduces the amount of radiation reaching the gas in the sample side of the detection chamber and produces a lower pressure in that side. The diaphragm bends toward the side of lower pressure, and this movement is converted into electrical impulses. The diaphragm thus vibrates at the chopping frequency and periodically bends towards the sample half of the detector. The metal diaphragm usually forms one plate of a capacitor. The movement of the diaphragm thus results in a variable capacitor, which in turn forms part of the tuning circuit of a radio frequency oscillator used in an amplitude modulation arrangement. In an alternative arrangement, the capacitor is supplied with a constant charge and

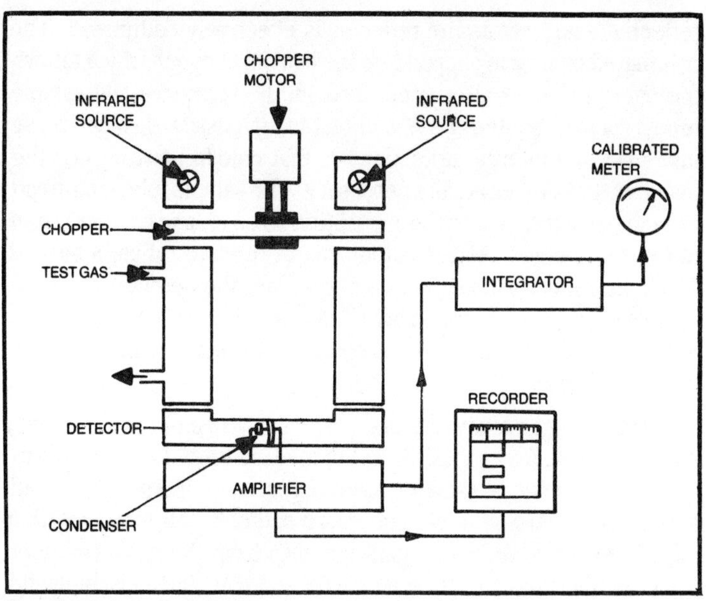

Fig. 16-8. Infrared gas analyzer.

the resulting voltage changes at the chopping frequency are amplified in a three-stage tuned amplifier with an electrometer tube in the input stage. The signal is rectified and smoothed and the output signal is displayed on a meter or recorder.

Comventional infrared analyzers of the type described above have the following drawbacks:

—The source must be balanced either electrically or mechanically.

—An unbalanced chopper causes failure of a chopper motor bearing.

—A Luft detector cell using gas as a detection mechanism is vibration sensitive and subject to cross talk with other gases. The cell is difficult to maintain and expensive to replace.

To overcome these difficulties, Infrared Industries, Cal., USA, have introduced a range of infrared gas analyzers employing a single source in an arrangement shown in Fig. 16-9. The solid state detector is PbSe. The chopper has a high speed of 3000 rpm and provides response times up to 100 m sec for 90% reading.

The infrared source operates at a temperature of about 1500°F where it emits infrared energy optimized for the spectral bands of interest and long life. The infrared energy source is located at the front focal plane of a parabolic reflector so that the reflected energy from the reflector is effectively collimated. The collimated energy is chopped by the coaxial chopper which allows the energy to pass alternately through the reference and sample tubes. Since the energy is collimated, it passes through these tubes without internal reflections so that gold foil coatings on the inside of these tubes are not necessary. The sample tube length can be selected according to the absorption strength and concentration of the sample gas. At the output end of the two tubes, a second parabolic reflector images the energy onto the detector filter assembly. The filter is a narrow band-pass interference filter with band-pass characteristics matched to the absorption spectra of the gas of interest.

Infrared gas analyzers are particularly useful for measuring carbon dioxide in respired air in the medical field. In these instruments, two types of samples are employed—a *micro-catheter* cell and a *breathe-through* cell. The micro-catheter cell is used with a vacuum pump to draw off small volumes from the nasal cavity or *trachea*. It's typical volume is 0.1 ml and it is particularly useful when larger volumes could cause patient distress. The breathe-through cell accepts the entire tidal volume of breath with no vacuum assistance. It can be connected directly into the circuit of an anesthesia machine. These instruments have a typical response time of 0.1 sec and a sensitivity range of 0 to 12% CO_2.

The calibration of the CO_2 analyzer is done in the following way. To establish zero calibration, an inert gas is sampled and the meter adjusted to zero. Zero calibration is generally made with room air. The upper end of the meter scale is calibrated with a cylinder of calibration gas. The frequency of calibration depends primarily upon the accuracy desired.

Jones et.al. (1971) explain a simple infrared gas analyzer developed especially for detection of hydrocarbons. This analyzer employs a partially selective source and a partially selective detector so chosen that their combined characteristics limit the sensitivity of the combination to a narrow spectral region with the absorption band centered at 3.4 μm. The indium arsenide photovoltaic detector operating at ambient temperature has maximum detectivity at about 3.4 μm, a rapid decrease in sensitivity at longer wavelengths and no response to radiation of wavelength greater than 4 μm. It is used in conjunction with a source of radiation consisting simply of a bead (about 3 mm diameter) of borosilicate glass encapsulating a platinum rhodium heating coil. The optimum

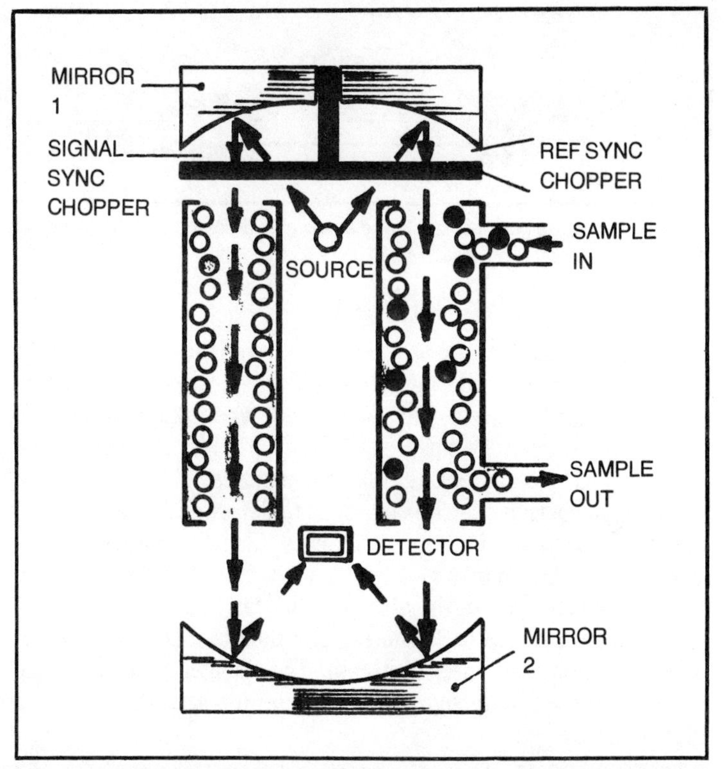

Fig. 16-9. Improved version of infrared gas analyzer (courtesy of Infrared Industries, USA).

temperature of operation was found to be in the region 400-500°C by these workers.

The instrument is very simple in construction. It consists of a single absorption cell (Fig. 16-10) the inner wall of which is silvered, with the source at one end and the detector at the other. The gas is made to diffuse into the cell through a fine stainless steel sinter. The change is the dc signal from the detector is amplified and can be displayed either on a suitably calibrated meter or chart recorder. No chopper arrangement is necessary.

A similar system comprising a heated quartz source and an indium antimonide detector which would cover the spectral region from 4-6 μm, could be used for the measurement of high concentrations of carbon monoxide and carbon dioxide. To determine either gas unambiguously a filter cell would be required; this would be filled with carbon dioxide to eliminate the unwanted radiation when detecting carbon monoxide and vice-versa.

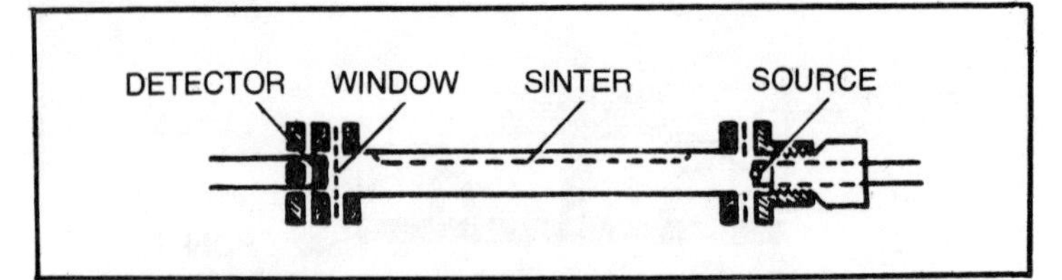

Fig. 16-10. Infrared gas analyzer for detection of hydrocarbons (after Jones et. al. 1971).

THERMAL CONDUCTIVITY ANALYZERS

Thermal conductivity of a gas is defined as the quantity of heat (in calories) transferred in unit time (seconds) in a gas between two surfaces 1 cm^2 in area, and 1 cm apart when the temperature difference between the surfaces is 1°C. The ability to conduct heat is possessed by all gases but in varying degrees. This difference in thermal conductivity can be employed to determine quantitatively the composition of complex gas mixtures. Changes in the composition of a gas stream may give rise to a significant alteration in the thermal conductivity of the stream. This can be conveniently detected from the rise or fall in temperature of a heated filament placed in the path of the gas stream. The changes in temperature can be detected by using either platinum filament (hot wire) or thermistors.

Figure 16-11 shows the relative thermal conductivity of a series of gases of interest for analysis. A gas analysis based on the thermal conductivity procedure presupposes binary gas mixtures or such gas mixtures, respectively, which include a measuring component whose thermal conductivity differs sufficiently from the thermal conductivity to the carrier gas. Typical examples of application are the measurement of hydrogen in blast furnace gases, the determination of argon in oxygen in the process of air decomposition, and of sulfur dioxide in roasting gases in the production of sulfuric acid.

In a typical hot-wire cell thermal conductivity analyzer, four platinum filaments (Fig. 16-12) are employed as heat-sensing elements. They are arranged in a constant current bridge circuit (Jesop, 1966) and each of them is placed in a separate cavity in a brass or stainless steel block. The block acts as a heat sink. The material used for construction of filaments must have a high temperature-coefficient of resistance. The materials generally used for the purpose are tungsten, *Kovar* (alloy of Co, Ni and Fe) or platinum.

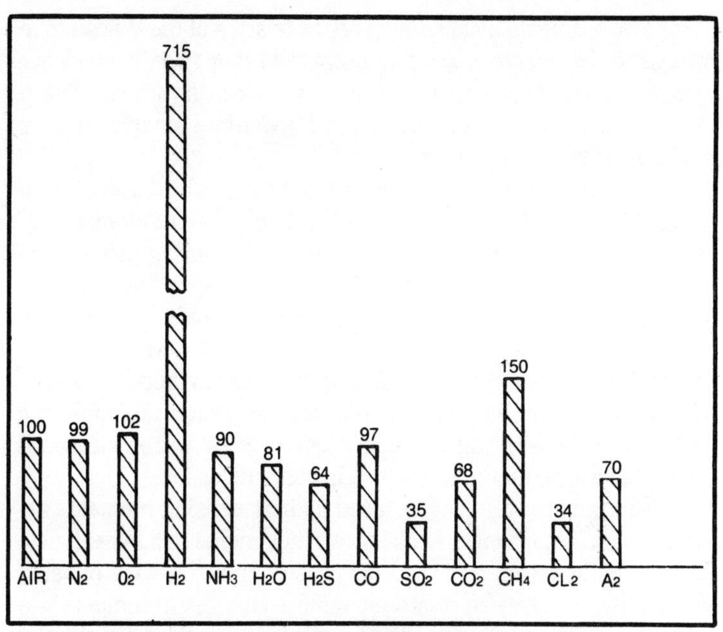

Fig. 16-11. Relative thermal conductivity of different gases.

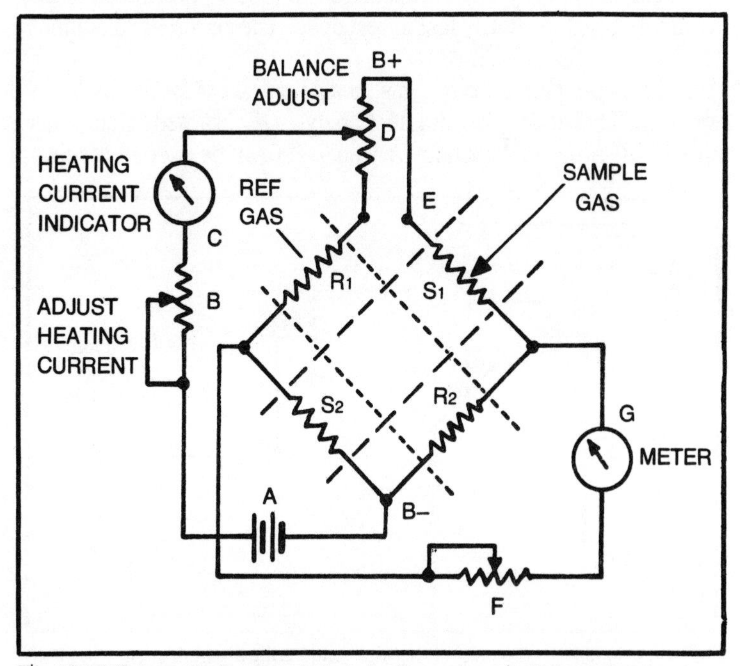

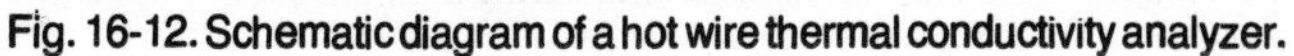
Fig. 16-12. Schematic diagram of a hot wire thermal conductivity analyzer.

Two filaments connected in opposite arms of the Wheatstone bridge act as reference arms whereas the other two filaments are connected in the gas stream which act as measuring arms. The use of a four cell arrangement serves to compensate for temperature and power supply variations.

Initially, reference gas is made to flow through all the cells and the bridge is balanced precisely with the help of potentiometer D. When the gas stream passes through the measuring pair of filaments, the wires are cooled and there is a corresponding change in resistance of the filaments. The higher the thermal conductivity of the gas, the lower would be the resistance of the wire and vice-versa. Consequently, the greater the difference in thermal conductivities of the reference and sample gas, the greater the unbalance of the Wheatstone bridge. The unbalance current can be measured on an indicating meter or on a strip chart recorder.

Thermistors can also be used as heat sensing elements arranged in a similar manner as hot wire elements in a Wheatstone bridge configuration. Thermistors possess the advantage of being extremely sensitive to relatively minute changes in temperature and have a high negative temperature coefficient. When used in the gas analyzers, they are encapsulated in glass. Thermistors are available which are fairly fast in response. The circuit arrangement is shown in Fig. 16-13.

Thermal conductivity gas analyzers are inherently non-specific. Therefore, the simplest analysis occurs with binary gas mixtures. A thermal conductivity analyzer can be used in respira-

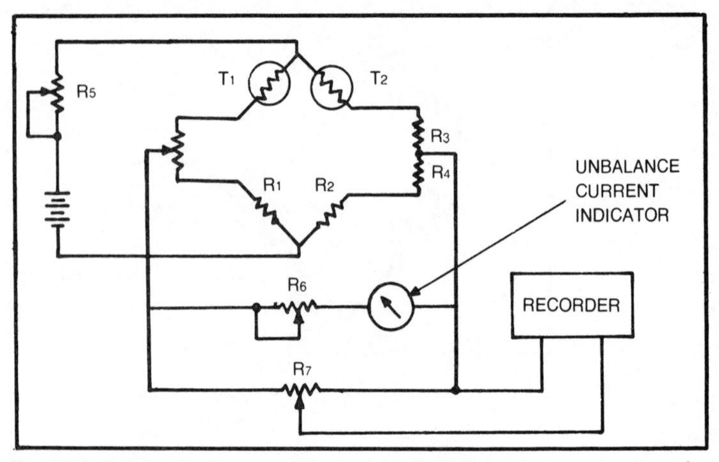

Fig. 16-3. Schematic diagram of thermal conductivity analyzer using thermistors.

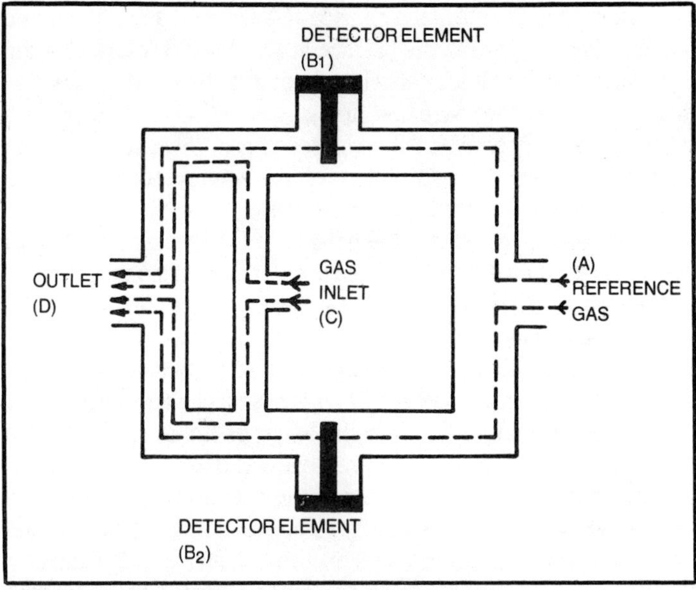

Fig. 16-14. Principle of gas analyzer based on gas density.

tory physiology studies to follow CO_2 concentration changes in the individual breaths of a patient. A high speed of response necessary for this purpose can be obtained by reducing the pressure of the gas surrounding the filaments to a few millimeters of mercury absolute. The variations in the proportions of oxygen and nitrogen in the sample stream will have little effect, since they both have almost the same thermal conductivity. The effect of changes in water vapor content can be minimized by arranging to saturate the gas fed to both the sample and reference filaments.

An analysis of a multicomponent mixture is possible if all components but one have almost thermal conductivity so that it can be treated as a binary mixture. Similarly, analysis is also possible if all components of the mixture other than the one being measured vary in the same ratio from each other.

ANALYZERS BASED ON GAS DENSITY

It is known that the density of an ideal gas has a direct linear relation with the molecular weight of that gas. Fortunately, all real gases behave as ideal gases at room temperature and normal atmospheric pressure. Instruments based on the principle of gas-density balance are commercially available.

Figure 16-14 illustrates the principle of operation of a gas-density balance based on the design suggested by Nerheim (1963).

The reference gas enters the balance at A where it splits itself into two streams and leaves the balance at D. Two detectors (B_1 and B_2) which may be either hot-wires or thermistors and mounted in the path of the two streams are connected as two arms of a Wheatstone bridge. When the reference gas flows such that the flow is balanced, the two detectors are equally cooled and the recorder would indicate a zero baseline.

The sample gas enters the balance at C and it also splits into two streams. It mixes with reference gas in the horizontal conduits and leaves at D. If the sample gas has the same density as the reference gas, there will be no unbalance of reference streams or of the detector elements. If the sample carries a gas having a higher density than the reference gas, it will cause a net downward flow, partially obstructing the flow in the lower path like A-B_2-D. This would result in raising the temperature of the detector element B2. This, in turn, increases the flow in the path A-B_1-D and causes more cooling of the element B_1. This temperature differential causes an unbalance in the bridge, the unbalance being linearly proportional to the gas-density difference between the reference and the sample gas. If the detectors used are hot-wires, it may require some factor of amplification before the signal can be given to a recorder. The use of a thermistor generally eliminates the requirement of amplification. The effective sample volume is 5 ml in the Nerheim design.

METHOD BASED ON IONIZATION OF GASES

The spectral regions for maximum radiation absorption for different gases are of different wavelengths. For example:

N_2	less than 900 A (for ultraviolet)
O_2	1450 A (ultraviolet)
CO_2	2.73, 4.25 and 14.93 μ.
Water Vapor	2.6, 20 and 52 μ (infrared)

Neither nitrogen nor oxygen analyses are routinely done using these absorption bands. However, with sufficient electrical excitation and at suitable pressures, gases emit radiation in different ways like spark, arc and glow discharge in different parts of the radiation spectrum. Measurement of the emitted radiation can help in the determination of unknown concentration of a gas in a mix-

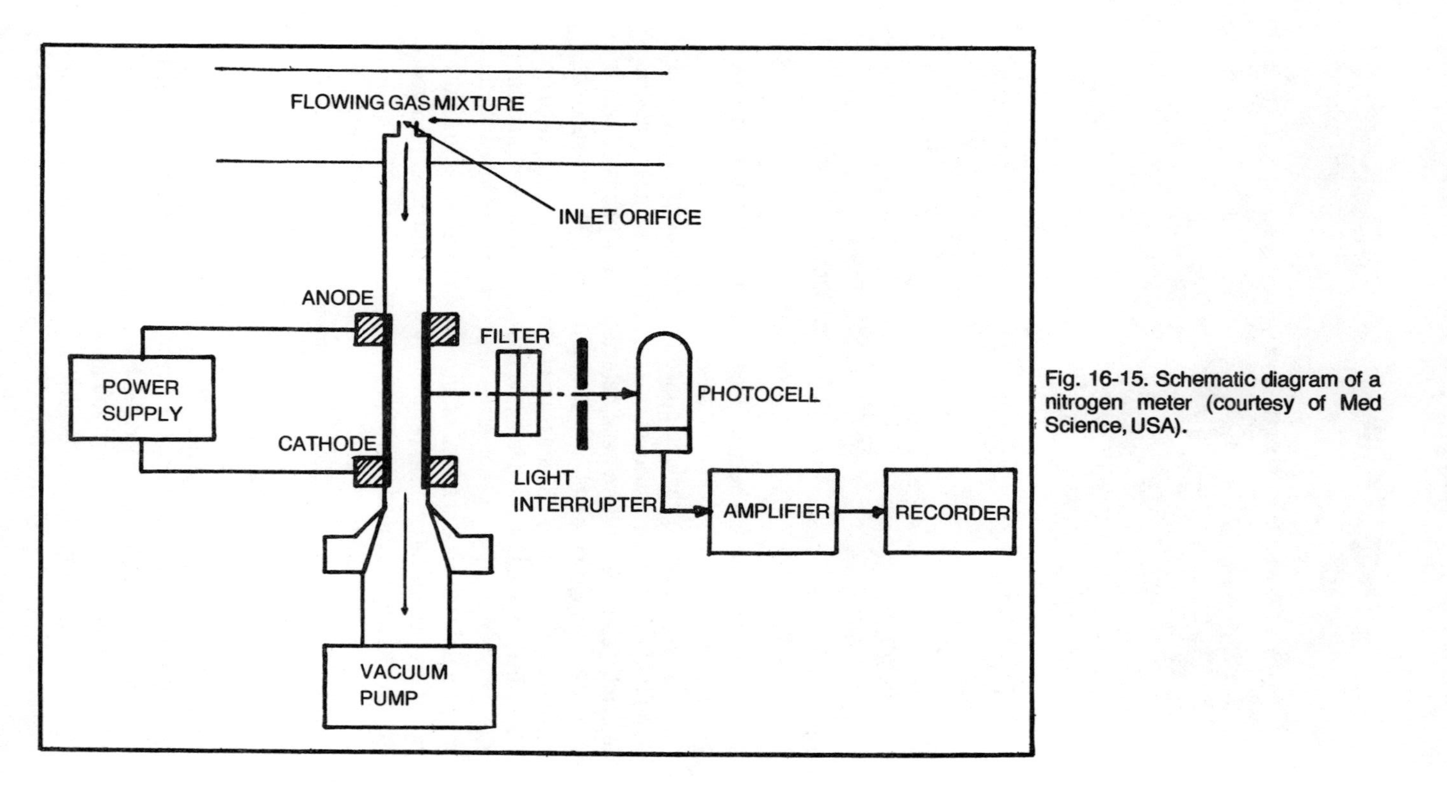

Fig. 16-15. Schematic diagram of a nitrogen meter (courtesy of Med Science, USA).

Fig. 16-16. Nitrogen analyzer (courtesy of Med Science, USA).

ture. This technique has been utilized for measurement of nitrogen gas particularly in respiratory gases.

The measuring technique utilized for measuring nitrogen is essentially that of a photospectrometer, wherein a gas sample is ionized, selectively filtered and detected with a photocell which provides an appropriate electrical output signal. The presence of nitrogen is detected by the emission of a characteristic purple color when discharge takes place in a low pressure chamber containing the gas sample. Nitrogen meters are usually employed in the medical field for measurement of nitrogen concentration to follow breath-by-breath variation in respiratory gases and other nitrogen gas analysis applications.

The instrument is generally in two parts. The sampling head contains the ionizing chamber, filter and the detector. The other part contains the power supply, amplifier and display system. The ionizing chamber or the discharge tube is maintained at an absolute pressure of a few torr. A rotary oil vacuum pump draws a sample and feeds it to the discharge tube. The voltage required for striking the discharge in the presence of nitrogen is of the order of 1500 volts dc. This voltage is generated by using a dc-dc converter or rectifying the output of a high voltage transformer.

The light output from the discharge tube is interrupted by means of a rotating slotted disc (Fig. 16-15) so that a chopped output is obtained. This light is then passed through optical filters to the wavelength corresponding to the purple color. The intensity of light is measured with a photocell and an amplifier specifically tuned to the chopping frequency. The light intensity is proportional to the nitrogen concentration.

The Med Science 505 Nitralyzer measures and displays digitally (Fig. 16-16) the concentration of nitrogen. The sampling rate is adjusted with the help of a needle valve which is normally set at 3 ml/min. The vacuum system provides 600-1200 microns Hg. The instrument is calibrated for water saturated mixtures of nitrogen and oxygen as a reading error of up to 2% can be expected with dry gases. Compensation for this error can be simply made by adjusting the sampling head needle valve, if it is desired to monitor dry gases.

Chapter 17

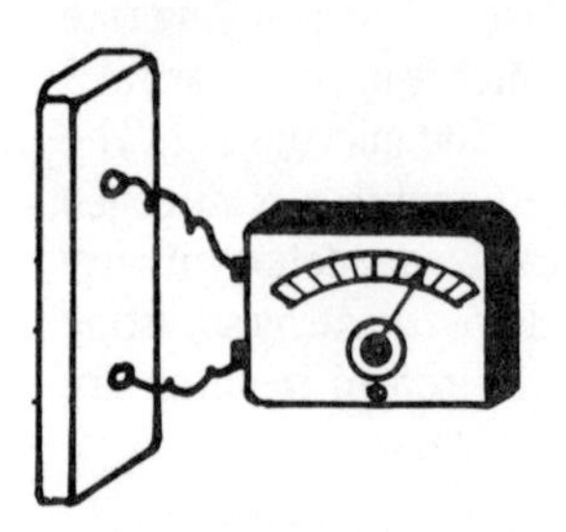

Radiochemical Instruments

The use of radioactive *isotopes* has lead to the development of radiochemical methods for analyses and has made possible the examination of phenomena, the measurement of which was formerly complicated or even impossible. These operations are based on the fact that radioisotopes (isotopes of elements with unstable atomic nuclei) emit radiation which can be detected by suitably located detectors. The proportion of radioactive atoms in the volume of material perceived by the detector can thus be determined by the measurement of the intensity of such radiation. Radiochemical methods offer the advantage of elimination of chemical preparation that usually precedes the measurement. These methods are both sensitive and specific and are often characterized by good accuracy.

In one method, a pure but radioactive form of the substance to be determined is mixed with the sample in known amount. After equilibrium, a fraction of the component of interest is isolated and the analysis is then based upon the activity of this isolated fraction. Alternatively, activity is induced in one or more elements of the sample by irradiation with suitable particles. The measurement of this activity gives information about the element of interest.

Before the advent of the cyclotron and more recently the chain-reacting pile, most of the work with radioisotopes was done with naturally occurring radioactive elements. However, it is now possible to obtain artificially produced radioisotopes of most of the elements. Also, it is possible to obtain these in large quantities and with extremely high activity.

TIME DECAY OF RADIOACTIVE ISOTOPES

Each radioactive isotope is characterized not only by type and energy of radiations emitted, but also by the characteristic life time of the isotope. This is most conveniently designated by half-life or half-period of the isotope. The half-period of a radioactive isotope is the time required for half of the initial stock of atoms to decay. Thus, after one half period has elapsed the total activity of any single radioactive isotope will have fallen to one-half its initial value; after two-half periods the activity will be one-fourth its initial value and so on. After 6.6 half-periods the activity will be 1% of the initial activity.

The half-life of a radioactive isotope is given by

$$t_{1/2} = \frac{0.693}{\lambda}$$

where λ is the decay constant for a particular radioisotope. In practice, disintegration rates are determined by counting the number of disintegrations over a certain time t_m and finding the ratio of the number of disintegrations to the time t_m.

UNITS OF RADIOACTIVITY

The unit of radioactivity is *curie*. This was originally defined to represent the disintegration rate of one gram of radium, but is now used as the standard unit of measurement for the activity of any substance, regardless of whether the emission is *alpha* or *beta* particles, or *X* or *gamma* radiation. When used in this way, the curie is defined as an activity of 3.7×10^{10} disintegrations. The curie represents a very high activity. Therefore, smaller units such as millicurie or microcurie are generally used.

TYPES AND PROPERTIES OF PARTICLES EMITTED IN RADIOACTIVE DECAY

Several different types of particles are emitted by radioactive atoms. These particles have definite energies related to energy levels in the parent nucleus. In most of the radioactive processes, the products of a radioactive decay event would consist of an altered nucleus, energy and possibly an elementary particle. The primary action is the emission of an alpha or beta particle or a neutron. If the resulting or new atom is in an excited nuclear state, it may be followed immediately by gamma radiation until the new nucleus is in its ground state. Different types of particles are distinguished by their penetration power, particles with the greatest mass and charge being the least penetrating.

The energies of alpha and beta particles and gamma radiations are expressed in terms of the *electron volt*. One electron volt is the energy that an electron would acquire if it were accelerated through a potential difference of one volt. Radioactive emissions have energies of the order of thousands or millions of electron volts.

Alpha emission is characteristic of the heavier radioactive elements such as thorium, uranium, etc. The energy of alpha particles is generally high and lies in the range 2 to 10 MeV (million electron volt). Their penetrating power is low and are completely stopped by foils, solid materials like aluminum. Due to the larger ionizing power of alpha particles, they can be distinguished from beta and gamma radiations on the basis of the pulse amplitude they produce on a detector.

Beta emission consists of a very energetic electron or positron (beta particles that carry a unit positive charge). Their penetration power is substantially greater than alpha particles, with an energy range of 0-3 MeV. Gamma rays are high energy photons having high penetrating and low ionizing power.

INTERACTION OF RADIATIONS WITH MATTER

Beta particles interact primarily with the electrons in the material through which they pass. The absorption depends mainly upon the number of electrons in their path. The molecules of the matter may be dissociated, excited or ionized. However, it is the ionization which is of primary importance in the detection of beta particles.

Alpha particles have relatively large mass and higher charge. The specific ionization produced by them is much larger than for beta particles.

Upon interaction with matter, gamma rays lose energy by three modes. The *photoelectric effect* transfers all the energy of the gamma ray to an electron in an inner orbit of an atom of the absorber. This involves ejection of a single electron from the target atom. This effect predominates at low gamma energies and with target atoms having a high atomic number. The *Compton effect* occurs when a gamma ray and an electron make an elastic collision.

The gamma energy is shared with the electron and another gamma ray of lower energy is produced. This travels in a different direction. The Compton effect is responsible for the absorption of relatively energetic gamma rays. When a high energy gamma ray is annihilated following interaction with the nucleus of a heavy atom, *pair production* of a positron and an electron results. Pair production becomes predominant at the higher gamma-ray energies and in absorbers with a high atomic number. The number of ion pairs per centimeter of travel is called specific ionization.

RADIATION DETECTORS

Several methods are available for detection and measurement of radiation from *radionuclides*. The choice of a particular method depends upon the nature of the radiation and the energy of the particle involved.

If the radiation falls on a photographic plate, it would cause darkening when developed after exposure. The photographic method is useful for measuring the total exposure of workers to radiation who are provided with film badges. Better methods are available for an exact measurement of the activity.

Ionization Chamber

The fact that the interaction of radioactivity with matter gives rise to ionization makes it possible to detect and measure the radiation. When an atom is ionized, it forms an ion pair. If the electrons are attracted towards a positively charged electrode and the positive ions to a negatively charged electrode, a current would flow in an external circuit. The magnitude of the current would be proportional to the amount of radioactivity present between the electrodes. This is the principle of the ionizing chamber.

An ionization chamber consists of a chamber which is gas filled and is provided with two electrodes. A material having a very high insulation resistance such as polytetrafluoreoethylene is used as the insulation between the inner and outer electrodes of the ion chamber. A potential difference of a few hundred volts is applied between the two electrodes. The radioactive source is placed inside or very near to the chamber. The charged particles moving through the gas undergo inelastic collisions to form ion pairs. The voltage placed across the electrodes is sufficiently high to collect all the ion pairs. The chamber current will then be proportional to the amount of radioactivity in the sample. Ionization chambers are operated either in the counting mode in which they respond separately to each ionizing current or in an integrating mode involving collection of ionization current over a relatively long period.

Figure 17-1 shows an arrangement for measuring the ionizing current. The current is usually of the order of 10^{-10} A or less. It is measured using a very high input impedance vacuum tube voltmeter which has an electrometer tube in the input stage. The current is indicated on a moving coil type ammeter. Alternatively, the null method can also be used. In this method, the change in voltage produced across a capacitor by the ionizing current is counterbalanced by an equal and opposite voltage supplied from a potentiometer. A potentiometric recorder of the self-balancing type can be used to record the signal.

The magnitude of the voltage signal produced can be estimated from the fact that the charge associated with the 100,000 ion pairs produced by a single alpha particle transversing approximately 1 cm in air would be around 3×10^{-14} coulomb. If this average charge is made to pass through a resistance of 3×10^{10} ohm in 1 second, a difference of potential of approximately 1 mV would develop across the high resistance. This voltage is a function of the rate of ionization in the chamber.

Liquid samples are usually counted by putting them in *ampoules* and placing the ampoules inside the chamber. Gaseous compounds containing radioactive sources may be introduced directly into the chamber. Portable ionization chambers are also used to monitor personnel radiation doses.

Geiger-Muller Counter

The *Geiger counter* is commonly called the G-M tube. This tube consists of a metal cylinder (Fig. 17-2) which acts as a cathode

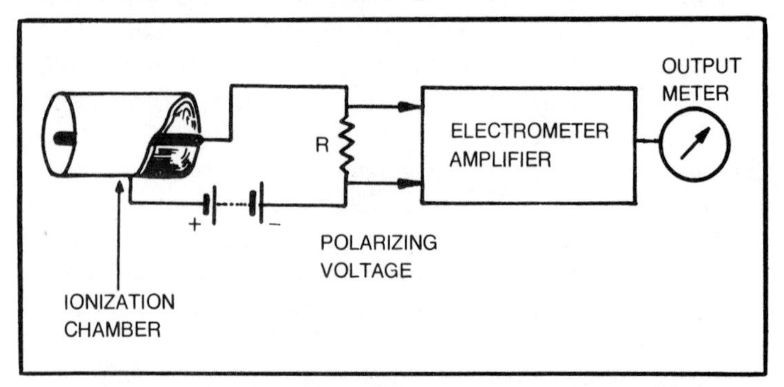

Fig. 17-1. Schematic diagram for measuring ionizing current using dc ionizing chamber.

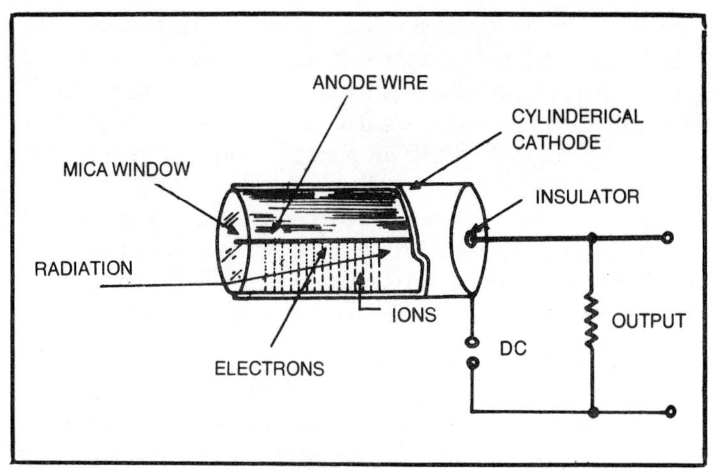

Fig. 17-2. Construction of a Geiger-Muller tube.

and is about 1 to 2 centimeters in diameter. It has an axial insulated wire working as an anode and is capable of being maintained at a high positive potential of the order of 800 to 2500 volts. This assembly is placed in a tubular glass envelope containing a gas or mixture of gases which is easily ionizable. The envelope is gas-tight and is typically filled to a pressure of 80 mm of argon gas and 20 mm of alcohol. Alcohol, butane or bromine acts as a quenching gas and argon as the ionizing gas. The tube contains a window of thin mica or other suitable material which permits effective passage of beta and gamma radiation but not of alpha radiation.

Principle. As the gas is ionized in the counting tube, migration of ions takes place toward the appropriate electrodes under the voltage gradient. They soon acquire sufficient velocity, cause further ionization and give rise to an avalanche of electrons traveling toward the central anode. As a result, ion multiplication spreads to a complete sheath around the anode, and the same pulse size is observed for each primary ionizing event. The process, in fact, produces a continuous discharge which fills the whole active volume of the counter in less than a microsecond. Each discharge builds up to a constant pulse of 1-10 volts. This pulse amplitude is sufficient to operate a ratemeter or scaler without using any amplifier. Kemp (1969) describes a Geiger counter ratemeter of this type.

Operation. Figure 17-3 shows the variation of the count rate recorded by a typical Geiger counter when the polarizing voltage is altered. The tube works in the voltage range exhibited by the plateau. Below the starting voltae, no counts are recorded. Between the starting voltage and beginning of the plateau, the voltage is too low to produce constant pulse sizes. Also, beyond the plateau the count rate increases because of breakdown and spurious discharges through the tube. The plateau is observed between 800 to 1400 volts for commercial tubes. The slope of the plateau is generally expressed in terms of a percentage of the count rate per volt. It should not be more than 0.1% per volt for a counter in satisfactory condition. In order to minimize the counts due to background, the counter is normally placed inside a lead shield.

Dead Time. Positive ions produced by ionization, being much heavier, have much lower mobilities than electrons. Therefore, they move only a very short distance towards the cathode in the time interval required for the electrons to cross the space between electrodes. This travel time may be several hundred microseconds and can vary from counter to counter. During this period, the positive ions form a sheath around the anode wire, which effectively lowers the potential gradient to a point where the counter becomes insensitive to the entry of fresh ionizing particles. This is called *dead time* of the counter.

Attempts to reduce the dead time of a self-quenched G-M counter by rapidly dropping the counter voltage below the starting potential have been made by several research workers. Use of an external quenching circuit offers two advantages: reduction of the counter dead time and prolongation of life time of the counter by reducing the number of organic molecules dissociated in each discharge. Eckey (1969) describes a simple quenching circuit which reduces the dead time of G-M counters by lowering the anode voltage immediately after the beginning of the discharge. The circuit delivers negative pulses with 300 V amplitude and 28 nS rise time. The dead time of the counter which was originally 320 μ sec reduced to 45 μ sec with this circuit.

Life of G-M Tubes. In a G-M tube containing ethyl alcohol as a quenching agent, some of the vapors are dissociated each time a count is recorded so that the counter has a limited life. The life of these tube is limited to perhaps 10^8 to 10^{10} counts. The tubes containing halogens have a much longer life because halogens simply dissociate during ionization and recombine afterwards. The counting tubes containing halogens (bromine) can be used at low temperatures. The halogens, however, may be consumed by reaction with electrodes and other metallic parts of the tube.

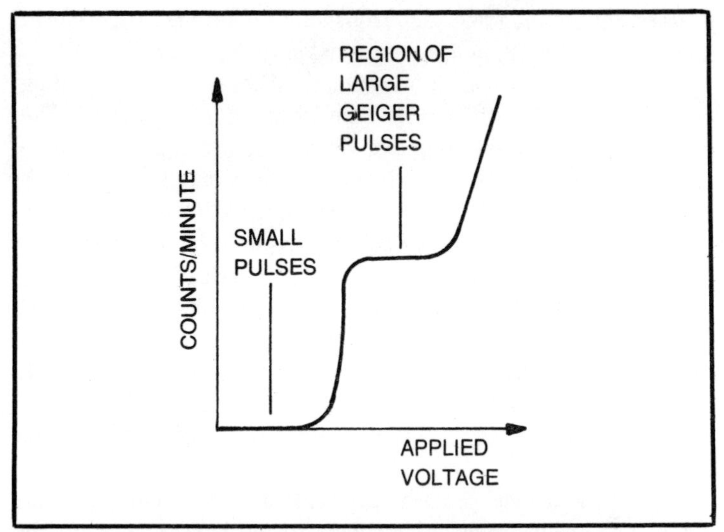

Fig. 17-3. Variation of the count rate with voltage using Geiger-Muller counter.

Geiger counters are utilized more often than any other counter. They have also been made in the miniature form having dimensions less than 1 inch and a diameter of 2 mm for biomedical applications.

Proportional Counter

The *proportional counter* is an ionization chamber that is operated at voltages beyond the ordinary ionization chamber region but below that of the Geiger region. These counters are called proportional counters because the output pulse from the chamber starts to increase with the increase in the electric field strength at the central electrode, but is still proportional to the initial ionization. In these counters, the pulse produced is amplified by a factor of 1,000 or more. The design of the counter and the value of the applied voltages are so chosen that a high voltage gradient exists close to the anode. The radius of cathode is about 1 cm and that of anode 0.001 cm, with a polarizing voltage of the order of 1000 volts. The output pulse is generally of a few millivolts and, therefore, requires amplification before the signal can be given to a scaler for counting.

The radioactive source may be placed inside the counter or outside the counter. In the former case, it avoids window absorption. In the continuous gas flow type counters, an argon-methane mixture flows at atmospheric pressure from a compressed gas tank at a rate of 200 ml/min. Counter life is, therefore, unlimited as the filling gas gets constantly replenished.

Figure 17-4 shows the schematic diagram of the counting equipment used with a flow type proportional counter. The preamplifier is a voltage follower which provides high input impedance. This is followed by a low-noise linear amplifier having a very stable gain in the range of 500 to 1000. The amplifier is required to be of a non-overloading type, since large pulses from cosmic rays or gamma ray background may overload a conventional amplifier for an appreciable time causing counts to be missed (Fairstein and Hahn, 1965). In order to avoid attenuation and distortion of pulses due to capacitance of the long connection cables, the preamplifier should be placed very near the detector. Figure 17-5 shows the range of operation of different types of counters.

In a proportional counter, different particles would yield output pulses of amplitude proportional to the isotope energy. By employing a pulse-height analyzer, which counts a pulse only if its amplitude falls within certain specific limits, a proportional counter can be made to respond to beta rays or X-ray frequencies, etc. In single channel pulse height analyzers, provision is made for lower and upper energy discrimination so that only pulses having amplitudes between the levels are passed. The voltage between the discriminator settings is called window, gate or width. The scaling unit counts down the pulses from the analyzer so that they are digitally displayed. A decade system of counting is employed which displays units, tens, hundreds, thousands and ten thousands of counts. Most of the scalers incorporate a counter/timer which displays time taken to record a definite number of counts or number of counts which occur within a definite time interval.

Scalers can often be replaced by ratemeters which continuously indicate or record the mean value of the rate at which pulses from the analyzers are applied to it. The incoming pulses are shaped into fixed width pulses and are used to charge a capacitor. The mean value of the voltage to which the capacitor is charged gives the mean count rate. Ratemeters may be connected to a potentiometric recorder to have record of the changes in count-rate with time. Also, a loudspeaker is often fitted to give an audible indication of the count-rate or changes in rate. Ratemeters or scalers also incorporate high voltage power supply of 500-1000 volts for operating counters. Stabilized EHT units may be separately used if the requirements are for higher voltage.

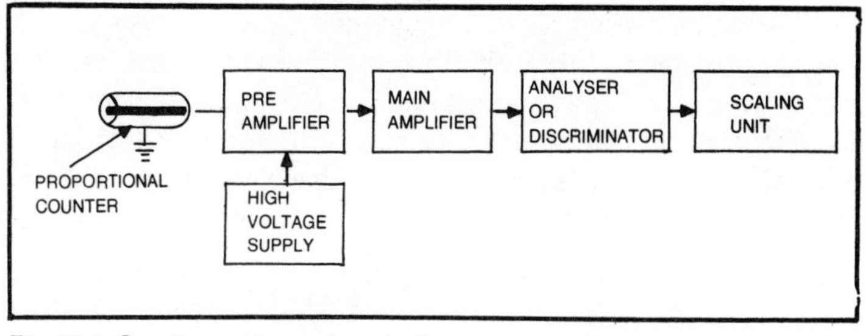

Fig. 17-4. Counting equipment used with a proportional counter.

Scintillation Counter

A *scintillator* is a crystalline substance which produces minute flashes of light in the visible or near ultraviolet range when it absorbs ionizing radiation. In such cases, the number of fluorescent photons is proportional to the energy of the radioactive particle. The flashes occur due to the recombination and de-excitation of ions and excited atoms produced along the path of the radiation. The light flashes are of very short duration and are detected by using a photomultiplier tube which produces a pulse for each particle. A scintillator along with the photomultiplier tube is known as a scintillation counter.

Gamma radiations cannot be detected directly in a scintillating material because gamma rays possess no charge or mass. The gamma-ray energy must be converted into kinetic energy of electrons present in the scintillating material. Thus, the conversion power of the scintillating material will be proportional to the number of electrons (electron density) available for interaction with the gamma rays. Because of its high electron density, high atomic number and high scintillating yield, the scintillating material which is generally used as a gamma ray detector is a crystal of sodium iodide activated with about 0.5% of thalium iodide. For counting beta particles, scintillator crystals of anthracene are employed. The crystal, being hygroscopic in nature, is usually mounted in a sealed aluminum container having a glass window on the side which is in contact with the face of the photomultiplier.

For in vitro counting, the geometry of the scintillation counters plays an important role. When it is convenient, the sample is placed in a well within the crystal. The crystal is coated on all sides with reflecting material except for the side which is bonded to the face of the photomultiplier tube. In order to reduce the background counts, the crystal-photomultiplier tube assembly is mounted inside a cylindrical lead shield having a lead lid.

Instruments used for counting gamma particles are called *gamma counting spectrometers.* They may include an oscilloscope display which is called an *energy scope*. The energy scope provides a visual indication of an isotope spectrum. The *Beckman biogamma counting system* provides a means of selecting the counting window by adjusting the variable discriminators and aids in selecting the proper high voltage.

The vertical gain can be set in two positions to determine the display height of the scope. The low position is used when measuring high-energy isotopes such as ^{59}Fe and ^{22}Na. The high position is used with low energy isotopes such as ^{125}I and ^{57}Co. The activity indicator indicates the amount of activity in terms of flashes of light. If the light glows constantly, a highly active source is in the counting chamber. If the light flashes, a less active source is in the chamber. Two variable discriminators permit adjustment from 0 to 1000 divisions to cover any part of an isotope spectrum. They are in fact two ten-turn potentiometers and are marked as upper and lower. The time selector switch is used to select one of 10 time intervals (in minutes) that determines the length of time each sample is counted. A low count selector switch selects the minimum number of counts that must be accumulated in the first 0.1 minute. If the minimum is not reached, counting is terminated and the system moves to the next sample. Gamma counting systems generally include automatic sample changers which may hold 20 vial trays with 10 vials each. The vials are molded from polypropylene to reduce gamma absorption in the vial walls. A teletype writer can also be included to have a sample printout of the count.

Samples containing weak beta emitters such as H^3 and C^{14} can be counted more efficiently by mixing the sample with a liquid scintillator so that the scintillator is in intimate contact with the short-range beta rays. Counting is carried out with a 1-inch diameter photomultiplier. The compound containing the radioactive source is dissolved in *toluene* or *xylene* to which is added a primary scintillator and a secondary scintillator to increase the pulse height by acting as a wavelength shifter. The photomultiplier tube is dipped directly into the scintillator solvent and the counts are made.

Normally five adjustments are made in a scintillation counter for proper counting of the ionizing particles. They are: high voltage setting, pulse height-analyzer threshold voltage, analyzer channel width, amplifier gain and time constant of the ratemeter.

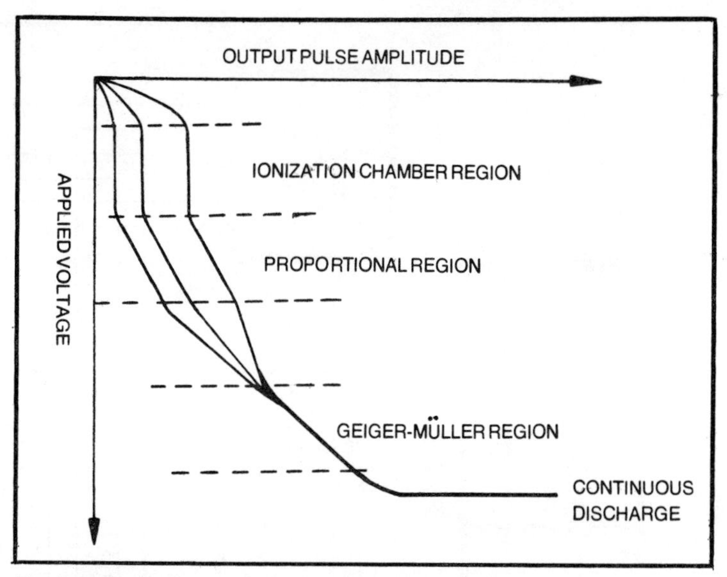

Fig. 17-5. Range of operation of different types of counters.

Figure 17-6 shows a block diagram of the automatic gamma counting system MR 1032 from Roche Medical Electronics. The maximum capacity of the sample changer in this system is 86 racks, each holding 12 sample tubes. The bidirectional sample changer has facilities for advance and reverse travel and up-down movement for single measurements. The sample changing cycle takes approximately 7 seconds.

Sample and rack identification are sensed by an optical reader. Thus, each sample is identified and the rack number can be matched to a laboratory or to a specific experiment. The detector is a well-type sodium iodine crystal activated with a trace of thallium and coupled to a photomultiplier tube. The crystal offers 4 pi geometry. The crystal is hermetically sealed in a thin can of aluminum and optically coupled to the sensitive face of an end-on PMT. Crystal and tube are covered with lead shielding for low and constant background.

The low energy models incorporate a pulse height analyzer with four preset windows which discrimate between the following low-energy isotopes: ^{125}I, ^{57}Co, ^{75}Se or ^{131}I. The high energy models incorporate a 7 isotope pulse height analyzer. The capacity of the scaler is 1,000,000 counts. There is one scaler associated with each pulse height analyzer. An increase in the number of counts accumulated will lead to a reduction in the percentage of statistical

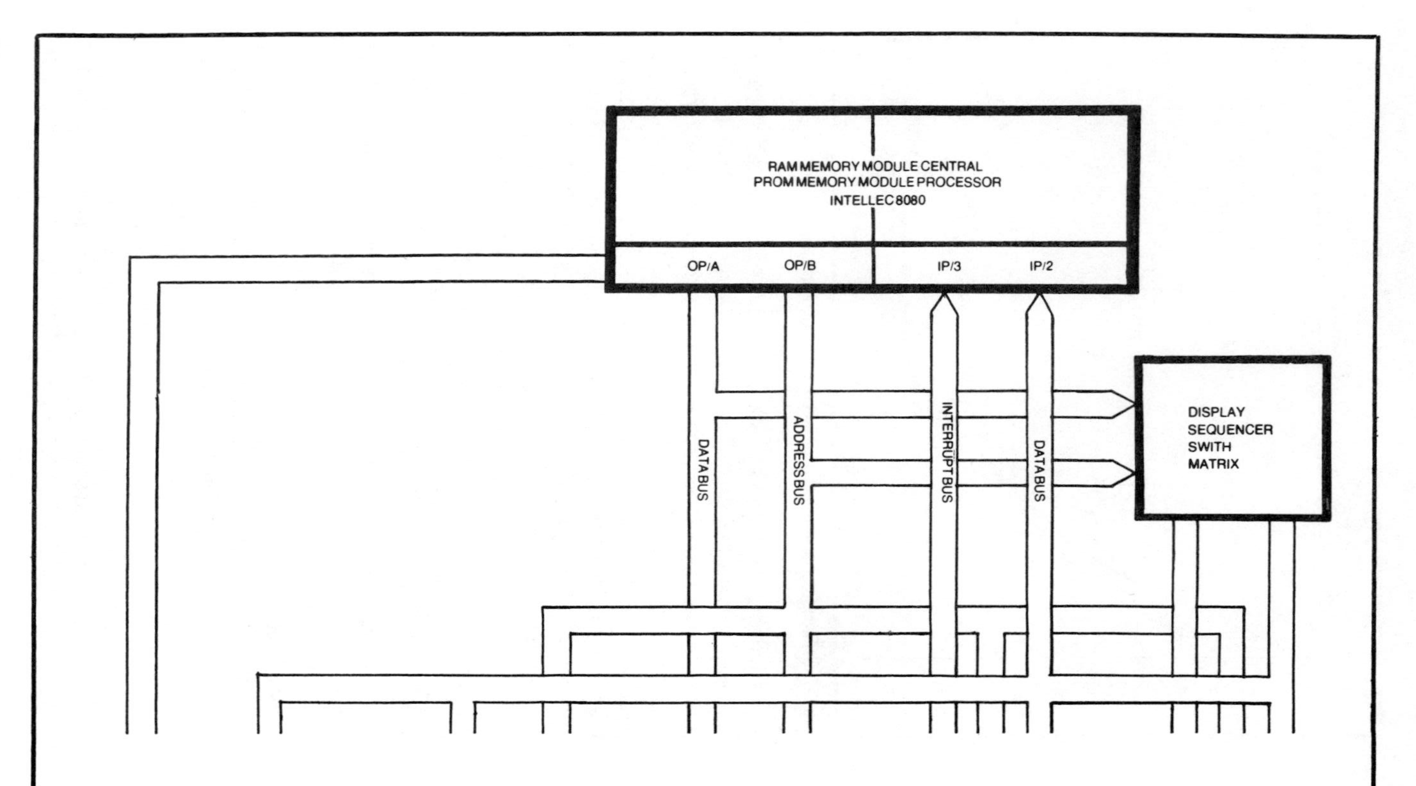

RAM MEMORY MODULE CENTRAL
PROM MEMORY MODULE PROCESSOR
INTELLEC 8080
OP/A
OP/B
IP/3
IP/2
DATA BUS
ADDRESS BUS
INTERRUPT BUS
DATA BUS
DISPLAY
SEQUENCER
SWITH
MATRIX

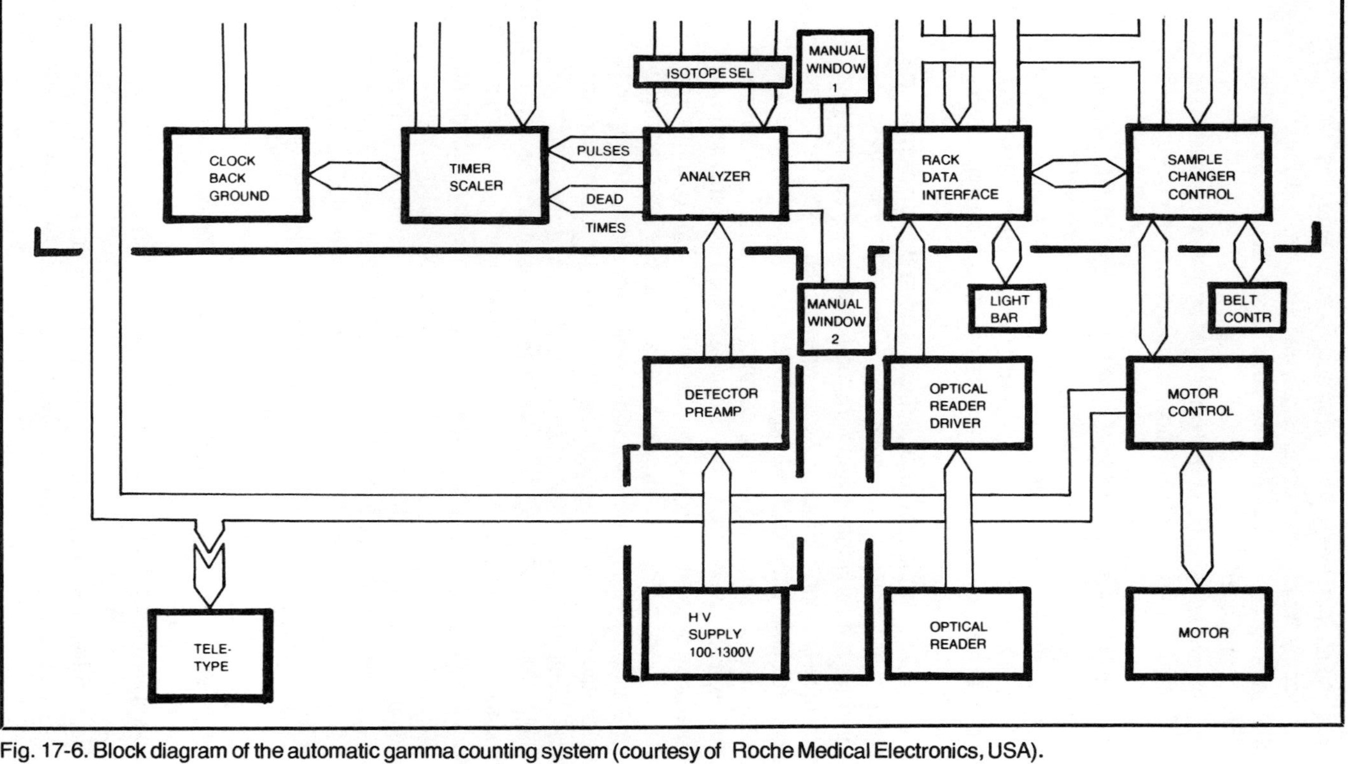

Fig. 17-6. Block diagram of the automatic gamma counting system (courtesy of Roche Medical Electronics, USA).

error and an increase in measuring accuracy. The display has a 7 digit display which shows counts per minute, elapsed time of the count in progress or identification number and the number of the sample being counted.

The system is directed and controlled by a microcomputer set MCS-8080. This parallel 8-bit modular microcomputer has up to 65 K bytes memory capacity in PROM and RAM. The 8080 incorporates a central processor module, a RAM module, PROM module and an input-output module which is designed for interface to a teletype writer. The results are printed out in numbered A-4 formats. The maximum printing speed is 10 characters per second. The tape punch can be used as the data transfer to a data processing system.

It is known that photomultipliers used in liquid scintillation counters are a source of instability. The sources of these instabilities are variation of gain with temperature, variation of gain with tube current, variation of gain with sudden large changes in tube current and effects due to aging. Stabilization systems based on the use of radioisotopes are common in scintillation counters used in nuclear spectroscopy. However, such systems are not practical in a liquid scintillation counter intended for counting low activity beta radiation sources. Several workers have reported stabilization systems which are not based on radioisotopes and which, therefore, are potentially useful for liquid scintillation counters. These include systems based on the use of gas discharge tubes (Haun and Kamke, 1960) and those based on the use of modulated and other light sources (Ageno and Fellici, 1963). Light emitting diodes have also been employed for the stabilization of photomultipliers. This is done by the use of LEDs for stabilization of gain in the photomultiplier by means of adjustments of high voltage or preamplifier amplification.

Gilland and Ried (1969) report on the use of GaP light-emitting diodes for photomultiplier stabilization, using also a negative temperature coefficient of the photomultiplier to less than 0.2% per °C. Soini (1975) describes a special stabilization system using GaAsP light emitting diodes in the measurement chamber to produce separate reference light pulses for each photomultiplier tube. The operation of the stabilizer is controlled by the reference pulse generator which feeds pulses to both detector channels. These pulses are used to trigger a pulse from the light emitting diode and to turn off the pulse height analyzer and turn on the feedback loop of the high voltage supply. The electronic circuitry is so arranged that when the light pulse signal from the preamplifier arrives at the peak detector, the difference between it and a reference voltage is passed through to the integrator. This signal is integrated in the integrator which controls the high voltage. The arrangement provides a significant improvement in both short and long term stability and in operational reliability

Semiconductor Detectors

There has been a great deal of development work on *semiconductor radiation detectors*. These detectors can be made very small and robust. Silicon and germanium crystals have been employed mainly for counting alpha and beta particles. They function in a manner similar to that of the gas ionization chamber. On absorption of radiation in the crystal, electrons and positive holes are formed which move towards opposite electrodes under the influence of applied potential. The resulting current is proportional to the energy of the ionizing radiation.

Pulse Height Analyzer

In radioactivity measurements, the individual particles are detected as single electrical impulses in the detectors. Also, various types of detectors can be set up to operate in a region in which the particular particle produces an electrical impulse having height proportional to the energy of the particle. The measurement of pulse height is thus a useful tool for energy determination. In order to sort out the pulses of different amplitudes and to count them, electronic circuits are employed. The instrument which accomplishes this is called *pulse height analyzer*. These analyzers are either single or multiple-channel instruments.

Figure 17-7 shows a block diagram of a single channel pulse height analyzer. The output pulses from the photomultiplier are amplified in a high input impedance low noise preamplifier. Amplified pulses are fed into a linear amplifier of sufficient gain to produce output pulses in the amplitude ranges of 0-100 volts. These pulses are then given to two discriminator circuits. A discriminator is nothing but a schmitt trigger circuit which can be set to reject any signal below a certain voltage. This is required for excluding scattered radiation and amplifier noise. As shown in Fig. 17-7, the upper discriminator circuit rejects all but signal 3 and the lower discriminator rejects signal 1 only and transmits signals 2 and 3. The two discriminator circuits give out pulses of constant amplitude. The pulses having amplitudes between the two triggering levels are counted. This difference in two levels in called the window width, the channel width or the acceptance slit and is analogous to monochromators in optical spectrometry.

Schmitt trigger circuits are followed by an *anti-coincidence circuit*. This circuit gives an output pulse when there is an impulse in only one of the input channels. It cancels all the pulses which trigger both the schmitt triggers. This is accomplished by so arranging the upper discriminator circuit that its output signal is reversed in polarity and thus cancels out signal 3 in the anti-coincidence circuit. As a consequence, the only signal reaching the counter is one lying in the window of the pulse height analyzer. The window can be manually or automatically adjusted to cover the entire voltage range with width of 5 to 10 volts. The scaler and counter follow the anti-coincidence circuit.

Multichannel pulse height analyzers are often used to measure a spectrum of nuclear energies and may contain several separate channels, each of which acts as a single channel instrument for a different voltage span or window width. The schmitt trigger discriminators are adjusted to be triggered by pulses of successfully longer amplitude. This arrangement permits simultaneous counting and recording of an entire spectrum.

A parallel array of discriminators is generally used, provided the number of channels is 10 or less. If the number of channels is more than 10, the problems of stability of discrimination voltages and adequate differential non-linearity arise. Agarwal et. al. (1976) describe a simple 100 channel pulse height analyzer which incorporates an A-D converter, a channel pulse height analyzer which incorporates on A-D converter, a channel sorter and preset timer circuits. A channel stability of 0.03% and selectivity of 0.1% is achieved. The instrument is based on the principle of conversion of channel threshold into time threshold.

Radioisotope Scanners for Medical Applications

Radioisotopes are used in medicine both for therapeutic and diagnostic applications. In diagnostic practice, radioisotope tracers are principally used either as sealed sources implanted in tissue are principally used either as sealed sources implanted in tissue or as unsealed sources through injection into the circulatory system. The distribution of radioactive material within an organ or part of the body is studied by using radioisotope scanners involving mechanical and electronic systems.

The heart of the system is the detector-collimator. The detector is usually a 3 or 5-inch diameter NaI crystal situated behind a

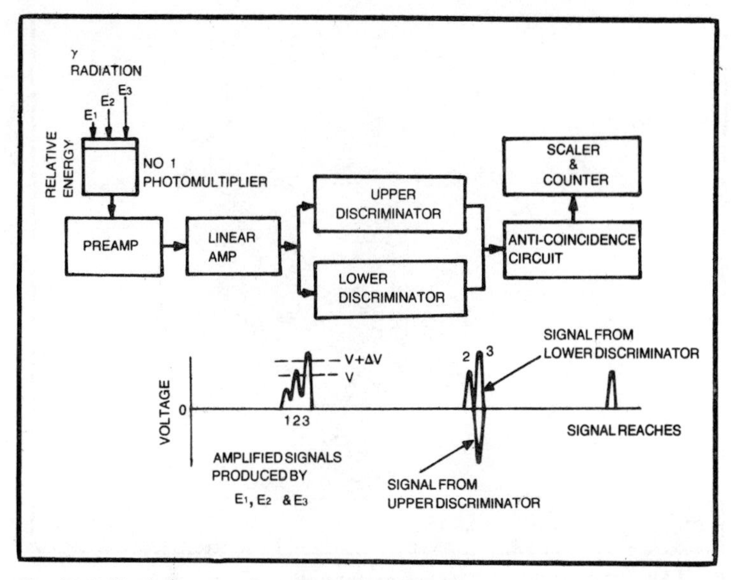

Fig. 17-7. Block diagram of a pulse-height analyzer.

focusing collimator. This is also called a probe and is so mounted that it can travel in a regular scanning pattern back and forth across the area of interest so that detected and amplified signals can be plotted to give a picture or contour map of radioactivity within the organ. Usually the detector-collimator assembly, the photomultiplier and the preamplifier are housed in a single unit which is attached to a motor driven device. This device defines the lateral and longitudinal limits of the scan.

The recording may be done either by a photographic recorder or by dot recorders. In a photographic recorder, the light flashes can be photographed on a film from the face of a cathode ray tube.

The *dot recorder* is most commonly used. It produces a map (Fig. 17-8) of the distribution of activity within the area of interest by recording dots or slit-like marks on paper. The dot recording mechanism consists of an electrically heated stylus to burn a small spot on a sheet of electrically conducting paper each time a pulse, passes through the stylus. The pulses to the stylus are delivered from the pulse height analyzer after scaling down the counts by an adjustable scaling factor from 1 to 256. A scaling factor of 16, for example, would mean that for every 16 counts arriving at the input of the scaling circuit from the pulse height analyzer, one dot appears on the paper. This reduction in counting rate is necessary because extremely high counting rates will drive the stylus wild. A

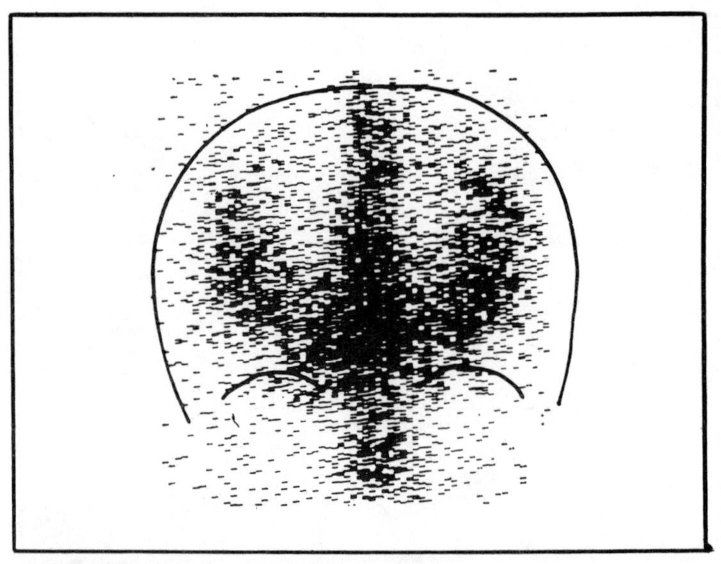

Fig. 17-8. Typical scintogram using a dot recorder.

count ratemeter is also incorporated to display or record average count rate.

Some scanners make use of color printing. In this technique, the maximum count rate is first established by moving the detector on the patient's body. This is then divided into 6 ranges each being associated with a different color print. As each count rate is recorded, it is allocated to one of the groups and the corresponding color is printed. In this way, a colored map showing the distribution of the isotope is built up.

THE GAMMA CAMERA

The *gamma camera* or the scintillation camera produces a position-intensity picture of a radioactive area by using multiple scintillation detectors. The radioactive contour of the scanned area is generally displayed on an oscilloscope. The arrangement consists of a camera head containing a 12-inch diameter and ½-inch thick crystal of sodium iodide. Various collimators are fixed on the patient's side of the head. On the other side of the crystal a matrix of 19 photomultipliers are coupled by a Perspex light guide. The camera is initially set using a collimated source of an isotope such as ^{203}Hg which has a single photopeak. This source is placed over each photomultiplier and the high voltage adjusted to equalize the pulse heights produced from each tube.

When a gamma ray interaction produces light in the crystal, each phototube will respond by producing an electrical signal proportional to the intensity of the light incident on its photocathode. Each of the 19 phototubes produces a signal that is inversely proportional to the square of its distance from the site of the interaction. The pulses from the tubes are summed up by a resistor network and are divided into four components X^+, X^-, Y^+ and Y^-. These signals are then added vectorially by means of a capacitor network to produce four positioning signals. Thus, an X-pulse and Y-pulse corresponds to the coordinates of the scintillation. The origin (X=0, Y=0), is taken as the center of the crystal face. The X and Y signals are applied to the X and Y deflecting plates of the cathode ray tube. The energy of the Z pulse from the photomultipliers is passed through an analyzer and caused to unblank the cathode ray tube beam. The Z signal is obtained by adding the four positioning signals and analyzing according to height. Therefore, only those pulses that fall within the window of the pulse height analyzer cause light flashes to appear on the face of the cathode ray tube.

In another arrangement, the difference circuits have been replaced by ratio circuits in which the resultant X and Y signals are divided by the Z signal. This makes the positioning more independent of pulse height and permits the use of wider windows. Figure 17-9 shows a block diagram of scintillation camera electronics.

Two types of collimators are generally used with a scintillation camera. A pinhole collimator is used specifically for thyroid work. In this arrangement, gamma rays from a radioactive source concentrated in the thyroid gland pass through a pinhole in a lead collimator and interact with sodium iodide crystals. The straight

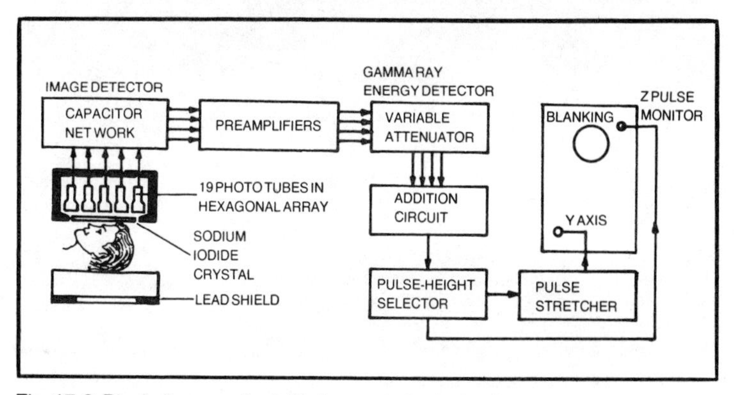

Fig. 17-9. Block diagram of scintillation camera electronics.

bore collimators are used to visualize larger organs such as the liver, brain, etc. They have perhaps 1000 holes and are compatible with high energy gamma emitters.

A Polaroid camera is mounted on the oscilloscope for photographing the buildup of perhaps 50,000 dots on the screen. In this way, a map is available of the distribution of activity.

Computerized Multicrystal Gamma Camera

A major limitation in using a scintillation camera for rapid dynamic studies is the counting losses that occur at the high count rates required for statistical studies. The count rate seen by the single channel analyzer is always greater than the registered count rate that passes the window of the analyzer. For each scintillation the output of all photomultipliers is summed to give a Z signal that is proportional to the total amount of light emitted during the scintillation. The Z pulse analyzer accepts only those events which fall within the spectrometer window. Two pulses occurring within a short time of each other are piled up and the resulting summation pulse is rejected because its amplitude exceeds the upper window level. This results in the loss of two valid pulses.

With a multicrystal data accumulation matrix, every gamma event coming from the patient, interacting in any crystal, is detected as a separate event at a unique location. If two or more

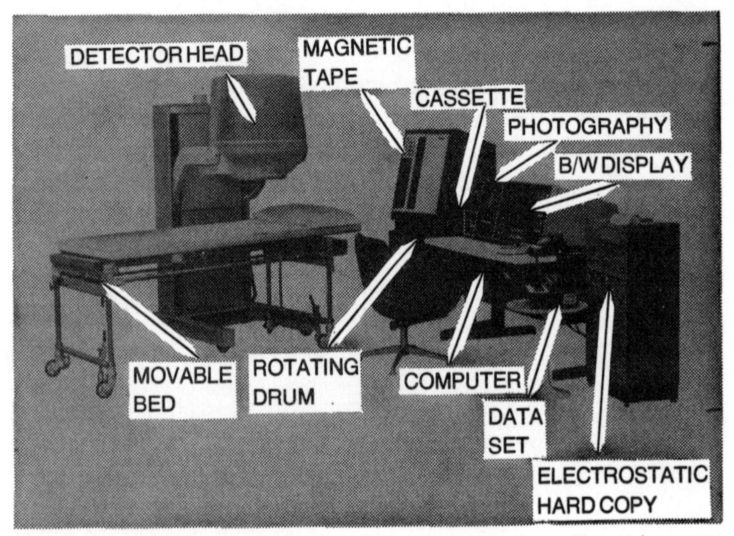

Fig. 17-10. Different sub-systems of the computerized multicrystal gamma camera (courtesy of Baird Atomic, USA).

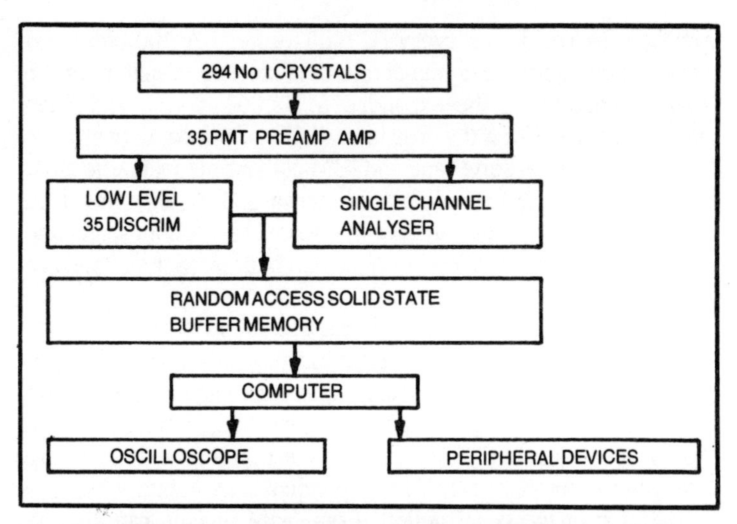

Fig. 17-11. Block diagram of the system 70 (courtesy of Baird Atomic, USA).

events interact at the same time in different crystals, both events are discarded. This is not possible with a monocrystal system, in which events which occur within the scintillation decay time appear to the positioning phototubes as a single scintillation event and are, therefore, erroneously positioned.

Baird Atomic, System 77 is a computerized multicrystal gamma camera system (Fig. 17-10) which can accumulate high count rates (200,000 Hz) at rapid time intervals (20 per second), making possible both clinical and research applications. Figure 17-11 the block diagram of this system.

The detector system of this camera consists of 294 discrete crystals arranged in 14 rows and 21 columns. Each column and row is optically coupled by a lucite light guide to a photomultiplier tube. Thus, there are 35 photomultiplier tubes. Collimation is achieved by limiting the field of view of each crystal with a single tapered hole case in lead collimators. Pulses from 35 preamplifiers and amplifiers serving photomultiplier tubes pass to 35 low-discriminators which eliminate events observed simultaneously in adjacent crystals. Pulses that are generated simultaneously in any row and column phototube are uniquely identified as to location in which the interaction has taken place. Anti-coincidence logic rejects all pulses arising simultaneously in more than one crystal.

Valid events are stored in a random-access, solid-state buffer memory. Information in binary form representing data from the entire detector passes into memory of an 8000 16-bit-word Nova computer as rapidly as every 50 milliseconds. A 100,000 word magnetic drum provides immediate data storage. Capacity of the Nova computer may be expanded to 32,000 words, and drum memory with 1,500,000 words is available. Optional data storage devices include cassette tape and IBM compatible magnetic tape. A variety of peripheral devices process and display the data. Computer programs for data accumulation, correction, manipulation and display are retained on a portion of drum memory and are executed by simple commands.

The movable bed which performs a scanning motion is used to improve spatial resoultion in static imaging measurements. Bed motion permits each detector crystal to scan a square area within 1.11 cm sides in 16 programmed movements of 2.78 mm each. Data observed by each of the 294 detectors at each of the 16 sites are arranged in memory as 4,704 independent data points which are utilized for image construction. Composite images can also be constructed from more than one detector position by combining information from fields of observation centered 15.4 cm apart. Bed motion cannot be used during a dynamic study and spatial resolution may be somewhat less in these studies.

Three types of devices are available with this camera for data output. The visual display is provided by a 525-line black and white TV monitor with a long-peristence phosphor to eliminate flicker. The display is refreshed at TV frame rates of 60 or 50 interlaced fields per second. A color TV system is also available for visual display.

Hard copy is provided by photography or by an electrostatic printer. A special TV monitor with a fast-fluorescent pure white phosphor is dedicated to photography to optimize the display of film characteristics.

Permanent data storage is provided by the nine-track magnetic tape or by a single-track digital cassette. The nine track tape is a fast, large capacity storage medium, about 10,000 frames per 10-inch reel of tape, that is also compatible with large IBM and other computer systems. The single-track digital cassette is a slower storage medium with a capacity of 400 frames. The digital cassette is also useful for entering new computational programs into the system at any time, as may be desirable.

Each crystal of the detector is canned individually, has a square cross section with 5/16 inch to a side, and is 1.5 inches long. The center-to-center spacing of the crystals in the array of 7/16 inch. The dimensions of the whole array are 6 × 9 inches.

The energy resolution obtainable with these crystals depends on the number of quanta from a scintillation event that are transmitted through the optical window of the crystal. The mosaic is assembled only from crystals with an energy resolution of 10% or better for 662 KeV-rays when each crystal is coupled directly to a multiplier phototube.

The light pipe array that is required to localize each event must transmit as much light as possible to optimize the energy resolution of the system. The light pipe array is shown schematically in Fig. 17-12. The address of each crystal is obtained by

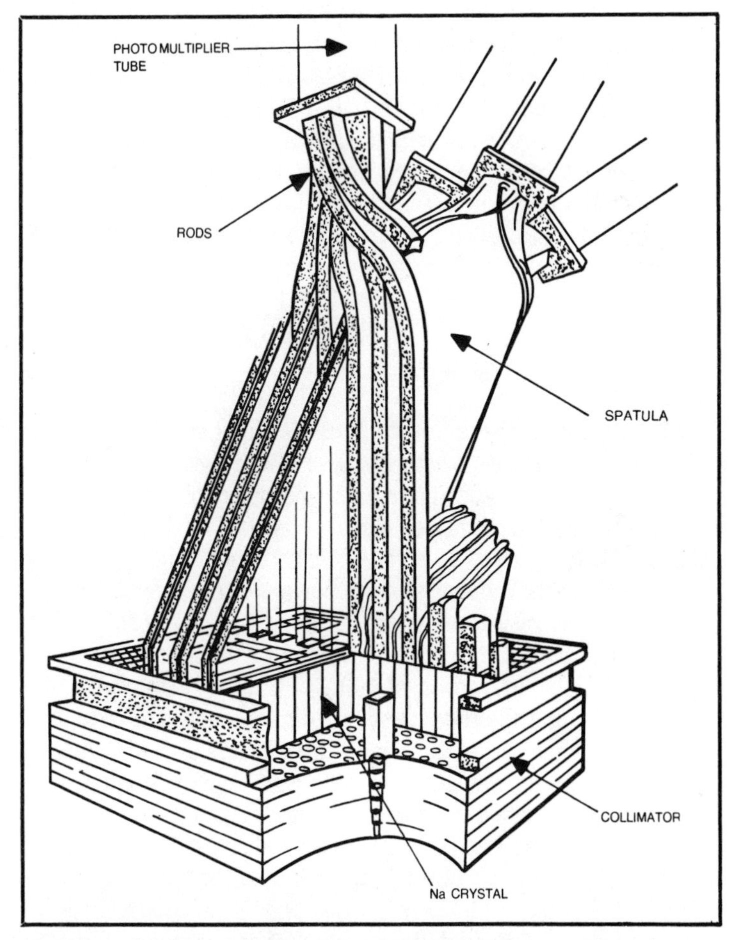

Fig. 17-12. Perspective drawing of parallel hole collimator, multicrystal array, light pipe array and phototube multiplier interconnections of the multicrystal camera (courtesy of Baird Atomic, USA).

placing two light pipes on each crystal in such a way that one-half the light from each scintillation event is guided down each pipe. The average length of the light pipes in the array is approximately 20 inches. The Y-coordinates are obtained by gathering the 11 rods to a 2-inch diameter multiplier phototube. The Y coordinates for the entire array require 14 groups of rods gathered in this manner. The X coordinates are derived from 21 spatulas that are shaped and bent to fit on to the end of 2-inch diameter phototubes.

The main advantage of the addressing scheme is that detection and positioning are made independent. The chief disadvantage of the light-pipe scheme is that light is attenuated in the light guides, leading to a degrading of the gamma ray energy resolution.

The light pipe array is fabricated from highly polished Plexiglas light guides. Maximum light transmission is achieved by total internal reflection of the light photons by maintaining an angle of incidence less than the critical angle. The latter is a function of the index of refraction of the material, which is about 1.5 for Plexiglas. The critical angle, ϕ, for light transmission in Plexiglas is given by

$$\phi = \sin^{-1}\left(\frac{1}{1.5}\right)$$

$$\text{or} \quad \phi = 26°$$

Therefore, photons that enter the Plexiglas light guide at an angle of 64° or less with respect to the side of the light guide can be transmitted to the photocathode of the multiplier phototube at the other end of the light guide. No light is lost by total internal reflection if the surface is highly polished; all light losses are caused by the bulk transmission properties of the light guides.

The display of images uses a scale of 16 gray shades. The computer searches for the highest count in any measurement to be displayed. This count represent 100% of the grey scale, i.e., the darkest shade, and all other counts are digitally normalized to this 100% shade. Each of the 16 shades corresponds to 6.25% increments in counts in the image. Front panel controls allow any percentage of the data to be displayed with 16 shades. In this way, any count percentage may be selected for display. For example, data from 30 to 70% may be selected for display on the TV monitor with all 16 shades. In this case, 30% is exhibited with the first or lightest shade and 70% with the sixteenth shade where each shade now represents 2.5% count increments. Any portion of the image may, therefore, be displayed with the full range of contrast. The color system works on the same basis using a color TV monitor, except that the 16 shades of grey are replaced with 16 separate colors.

Theory of Operation

The principle of operation of the multicrystal camera is illustrated in Fig. 17-13 by a simplified 2 × 2 crystal array. The detectors are arranged in a matrix to form orthogonal rows, A and B, and columns, 1 and 2. The light pipe array couples the detector array to row and column phototubes that uniquely define the coordinates of the detector. If an event occurs in detector B1, the light pulse from the scintillation is directed to phototubes B and 1 exclusively, and their outputs are processed in amplifiers B and 1, respectively. The outputs from these two amplifiers are summed and and analyzed in the single channel pulse-height analyzer, and they also trigger the corresponding lower level discriminators. The discriminator outputs are led to the row and column memory address registers and these set their corresponding flip-flops, which in turn provide an output if only one row and one column are addressed simultaneously. An AND gate then starts the Read/Write memory cycle, if an output from the pulse-height analyzer has been obtained.

The memory address registers are compared after pulse-height analysis and reset immediately with no storage in memory if any of the following conditions occur: if more than one row or more than one column memory address register flip-flop are set, if a single row memory address register flip-flop is set but no column flip-flop is set, or vice versa, and if no output from the pulse-height analyzer is obtained.

The above conditions guarantee that only events that occur in one and only one detector at the correct energy range will be stored in the memory and, also, the numbers stored at each memory location correspond to the number of events detected at the corresponding detector location.

The 294-crystal array can be regarded as an assembly of 294 rectilinear scanners, fixed in space with respect to each other. Whenever any one of these scanners is moved, the other 293 follow exactly the same motion. Because the detectors are spaced 11 mm apart, it is necessary for each detector to scan only an 11 mm × 11 mm area to cover the entire field of view as all other detectors scan a corresponding area simultaneously.

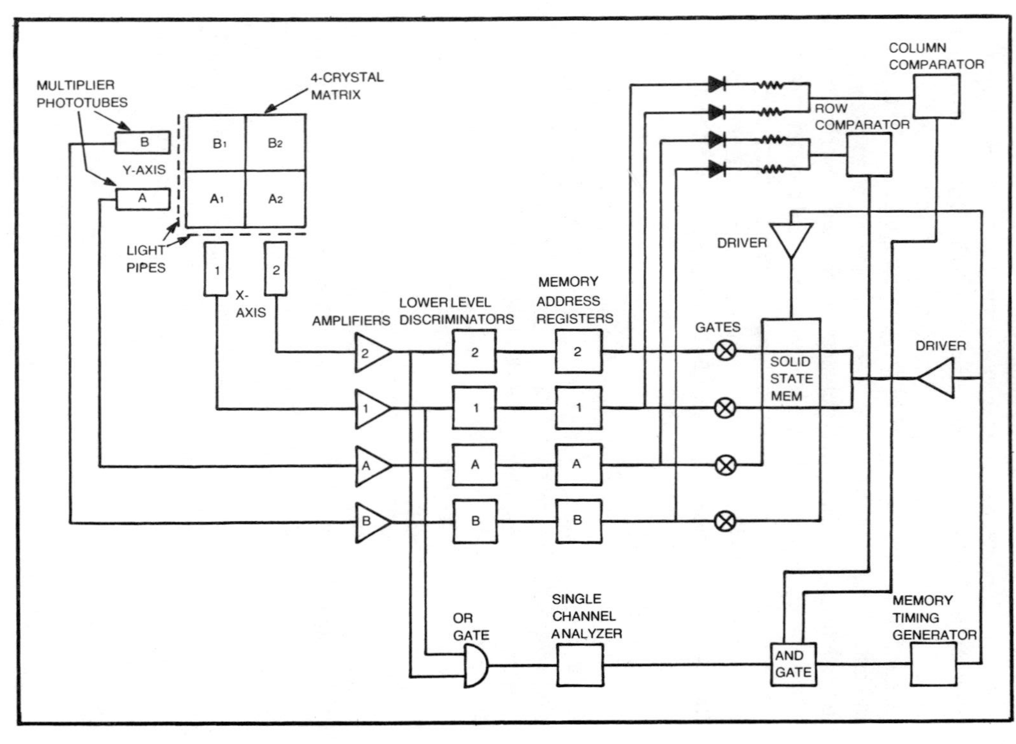

Fig. 17-13. Simplified block diagram showing data accumulation principle of the multicrystal camera illustrated by a 2 × 2 array.

The scan would be performed by a continuous motion over the 11 × 11 mm area just as a rectilinear scanner. In practice, however, the array is displaced in 16 discrete steps of 2.78 mm in both the X and Y directions. The total number of independent image elements generated during a 16 position static study is 294 × 16 or 4704 from an area of 378 cm^2.

The multicrystal camera requires a data accumulation system for the full range from multiposition statics to rapid dynamic function measurements. The data evaluation system was designed primarily to facilitate the interpretation of the data both quantitatively and qualititively. The data handling system allows direct user control of two important areas of data management—namely data accumulation and data evaluation.

Chapter 18

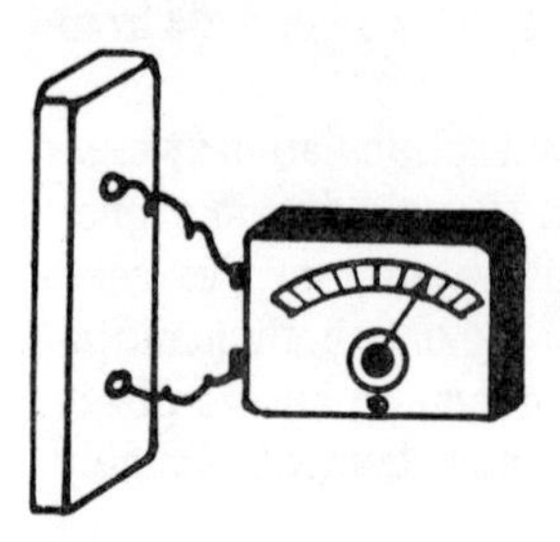

Automated Biochemical Analysis Systems

The chemical analysis of blood and other body fluids is one of the earliest forms of diagnostic criteria leading to the investigation of the diseases. In the early years of clinical biochemistry, most of the required analyses were performed on urine as it may be made available easily and in large quantities. However, with the development of semi-micro and microanalytical techniques, analysis can be carried out with minute quantities of samples. Therefore, the analysis of the blood serum or plasma is becoming more frequent. The great increase in the number of determinations coupled with heavy workload in many laboratories and the development of optical and electronic techniques have stimulated attempts to introduce a high degree of automation in the clinical biochemistry laboratories. A majority of the blood analysis can now be performed by an automated system.

The developments in new concepts and more advanced techniques in analytical methodology have resulted in estimations of blood constituents as a group whose metabolic roles are related and which collectively provide more meaningful information than the individual analyses. For instance, the group of important anions and cations of the blood plasma (electrolytes) like sodium, potassium, chloride and bicarbonate which together with serum urea form a related set of tests that are useful to perform on patients with electrolyte distrubances. Another group consists of the several analyses—protein, *bilirubin, alkaline phosphatase* and *SGOT* which together assess liver function. The effect of this trend is in the replacement of single isolated analysis by groups of analyses, all of which are carried out routinely on each sample with highly reproducible and accurate results. With this object in view, automatic analysis equipment has been designed and put to use. Automated analysis systems are available in multichannel versions. A full description of the detailed working instructions and details of techniques for individual substances are given in literature supplied by the manufacturers to the purchaser of their equipment.

SYSTEM CONCEPTS

The automated system is usually a continuous flow system in which individual operations are performed on the flowing stream as it moves through the system. The end product passes through the colorimeter where a balance ratio system is applied to measure concentrations of various constituents of interest. The final results are recorded on a strip chart recorder along with a calibration curve so that the concentration of the unknowns can be calculated. The output may also be connected to a digital computer to have a digital record along with the graphic record.

The automated system consists of a group of modular instruments (Fig. 18-1) interconnected together by a manifold system and electrical system. The various sub-systems are: sampling unit, proportionating pump, manifold, dialyzer, heating baths or constant temperature bath, colorimeter/flame photometer/ fluorometer, recorder and function monitor.

The sample to be analyzed is introduced into a stream of diluting liquid flowing in a narrow bore of flexible plastic tube. The stages of the analytical reaction are completed by the successive combination of other flowing streams of liquids with the sample stream by means of suitably shaped glass functions. Bubbles of air are injected into each stream so that the liquid in the tubes is segmented into short lengths separated by air bubbles. This segmentation reduces the tendency for a stationary liquid film to form on the inner walls of the tubes and decreases interaction between a sample and the one which follows it. The diluted samples and reagents are pumped through a number of modules in which the reaction takes place, giving a corresponding sequence of colored solutions which then pass into a flow-through colorimeter. The corresponding extinctions are plotted on a graphic recorder in the order of their arrival into the colorimeter cell. The air bubbles are removed before the liquid enters the colorimetric cell or flame emission.

TYPICAL FLOW SYSTEM OF INDIVIDUAL CHEMISTRIES

Figure 8-2 shows the flow system of individual chemistries used in SMA 12/60 Technicon Autoanalyzer. After the sample is aspirated, it is split six ways.

Stream 1 (SGOT). This is incubated with substrate. The mixture is dialyzed and diazo dye is added to the recipient stream. After incubation for color development, the mixture goes to the colorimeter.

Stream 2 (Alkaline Phosphatase). This is incubated with paranitrophenyl phosphate substrate, dialyzed into a recipient stream of aminomethylopropanol buffer and phased to the colorimeter.

Stream 3 (Uric Acid). This is treated with sodium hydroxide and dialyzed into a recipient stream of sodium tungstate and hydroxylamine. Phosphotungstic is then added for color development and the stream is phased to the colorimeter.

Stream 4 (Inorganic Phosphate). This is diluted with sulfuric acid and dialyzed into sulfuric acid. After dialysis, ammonium molybdate and stannous chloride hydrazine are added, and the stream is phased to the colorimeter.

Stream 5 (Cholesterol). This is mixed with Lieberman-Burchard reagent and phased to the colorimeter.

Stream 6 (T.P., Albumin, Calcium, Glucose, BUN, LDH and Bilirubin). The stream is diluted with water, and then the "main" sample stream is further split into 11 sub-streams.

(Glucose) Stream. The stream is diluted with saline and dialyzed into and treated with sodium carbonate. Copper neocuproine is added. The mixture is heated to 90°C and phased to the colorimeter.

(BUN) Stream. This is diluted with water and dialyzed into a recipient stream of diacetyl monoxime and thiosemicarbazide. It is then treated with acid ferric chloride, heated to 90°C and phased to the colorimeter.

(LDH) Stream. This is mixed with substrate, DPN diaphorase and then a tetrazolium dye. After incubation, and the addition of HCl to stop the reaction, the stream is phased to the colorimeter.

(LDH Blank). Same as assay except that a blank solution is used in place of DPN diaphorase.

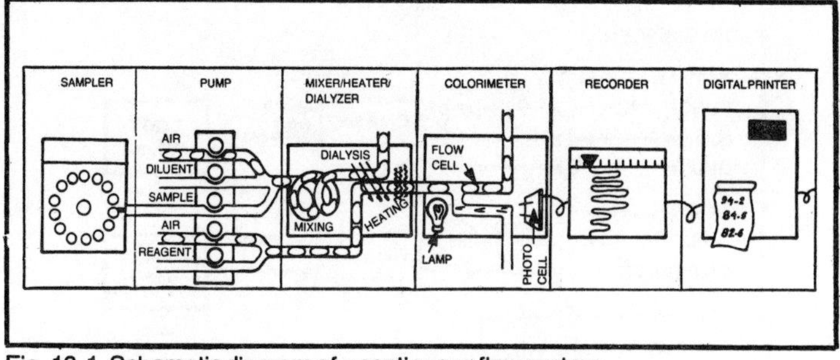

Fig. 18-1. Schematic diagram of a continuous flow system.

(Calcium) Stream. This is mixed with HCl and 8-hydroxyquinoline and dialyzed into a solution of HCl, 8-hydroxyquinoline and cresolphthalein complexone. After dialysis, a base is added and the stream is phased to the colorimeter.

(Total Bilirubin) Stream. This is mixed with caffeine, diazo, and tartrate and phased to the colorimeter.

(Total Bilirubin Blank) Stream. This stream is mixed with caffeine, sulfanilic acid, and tartrate and phased to the colorimeter.

(Total Protein) Stream. This is treated with biuret and phased to the colorimeter.

(Total Protein Blank) Stream. This is treated with alkaline iodide solution and phased to the colorimeter.

(Albumin) Stream. This stream is treated with HABA reagent and phased to the colorimeter.

(Albumin Blank) Stream. This is treated with phosphate buffer and phased to the colorimeter.

SYSTEM COMPONENTS

The sampling unit enables an operator to introduce unmeasured samples and standards into the autoanalyzer system. The unit in its earlier form consists of a circular turntable (Fig. 18-3) carrying around its rim 40 disposable polystyrene cups of 2 ml capacity. The sample plate carrying these cups rotates at a predetermined speed. The movement of the turntable is synchronized with the movement of a sampling crook. The hinged tubular crook is fitted at a corner of the base. The brook carries a thin flexible polythene tube which can dip into a cup and allow the contents—water, standard or test solution to be aspirated. At regular intervals, the crook is raised so that the end of the sample tube is lifted clear of the cup.

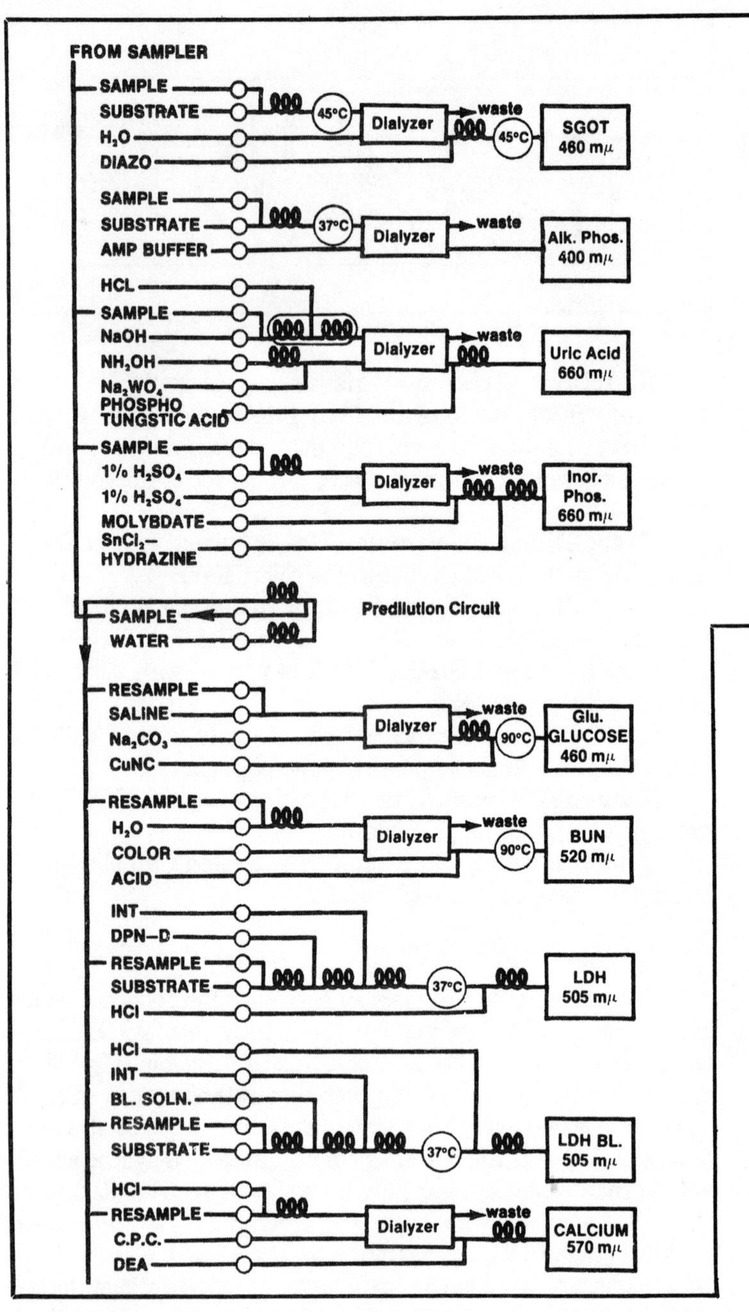

Fig. 18-2. SMA 12/60 Flow diagram for different tests (courtesy of Technicon Corp., USA).

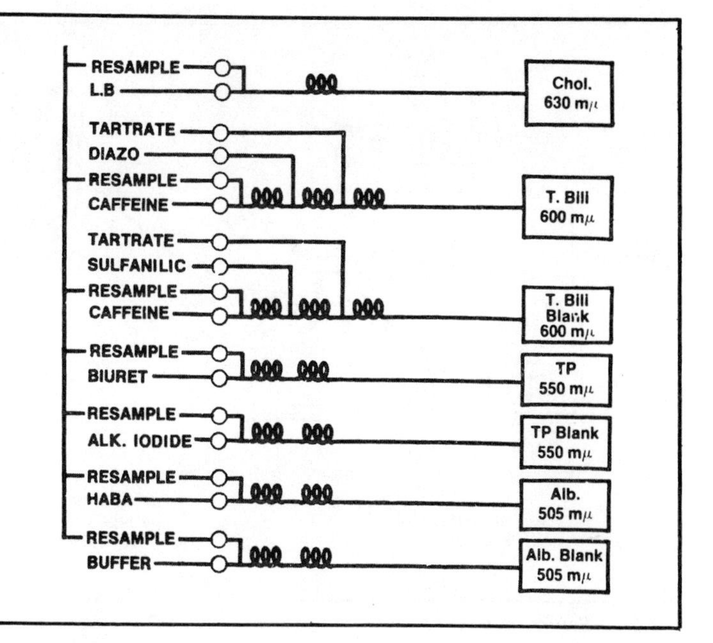

Sampling Unit

Between each sampling, the crook enters a receptacle of water or other suitable wash fluid to reduce cross contamination of one sample with another. The ratio of sampling time to wash time is normally 2:1. The plate then rotates a distance sufficient to allow the tube, when it next moves down, to dip into the next cup. One complete rotation of the plate thus presents 40 samples. As the sample plate completes a cycle, a switch is operated which stops the rotation action of the plate and the sampling action of the sample probe. The sampling rate can be adjusted to 20, 40 or 60 per hour giving, with the above ratio, times during which the sample is being drawn in, of two minutes, one minute or 40 seconds respectively. The volume of liquid taken up in most cases ranges from about 0.2 to 1.0 ml. This depends upon the rate at which the plate is run and the diameter of the pump tube.

The earlier version has been replaced by a more versatile from of the sampler in which during the time the sample tube is out of the specimen, the crook quickly comes down into water, and thus successive samples are separated by a column of water instead of air. This provides a better separation between them. With this sampler, the sample size may range from 0.1 to 8.5 ml. It utilizes a cup of sizes 0.5, 2.3 and 10 ml. The sample plate is kept

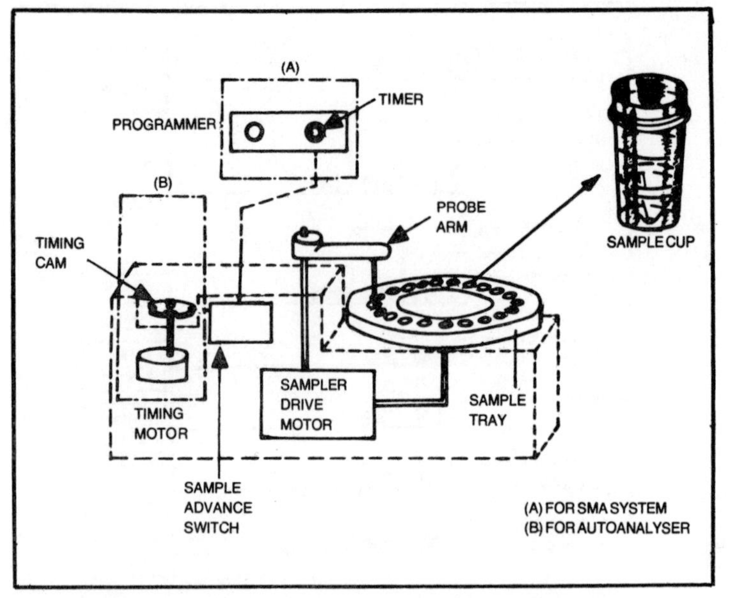

Fig. 18-3. Sampler controls (courtesy of Technicon Corp., USA).

covered to prevent evaporation which may sometimes lead to errors up to 5%. Sampling and washing periods are controlled by a programming cam. The sample speed and sample wash cycles are selected by the markings on the cam such as 40 and 2:1. This implies the speed is 40 per hour at a sample wash ratio of 2:1.

The samplers are fitted with a sample mixer which enables the sample to be mixed before and during aspiration. This is important in cases like when whole blood is used for analysis, the mixer prevents separation of the plasma from the cells. The mixer can be raised up by means of a clip. It will remain there until manually released. The sample base plate is rotated by means of a motor drive gear. When this gear drives the second drive gear for the sample probe assembly, it also rotates a gear chain of internal gears to the Geneva cam. The *Geneva cam* is an index type cam that may be designed for any acceleration, deceleration or dwell period. During the complete cycle of the sample wash probe, the cam makes one complete turn. As the probe moves back into the wash reservoir, the second point enters the next hole in the drive assembly plate. This rotates the drive assembly, sampler housing and sample plate at a controlled speed. In SMA-12 sequential multiple analyzer, the sample speed is controlled by the sampling rate cam on the programmer and not by a sample cam. In this system, a new sample is aspirated every two minutes and the sample wash ratio of 95% sample and 5% wash.

The *Chemlab automatic sampler* (Fig. 18-4) employs a different probe washing action between samples. Whereas most automatic samplers operate by dipping the probe alternately into the sample cup and then into a wash pot, the Chemlab sampler has a probe which simply moves up and down.

The probe washing device consists of a washing chamber through which the probe moves vertically. When the probe is in the sampling position, an O ring at the lower end of the internal wash reservoir seals the outlet. Water is pumped through the reservoir by the peristaltic pump used in the analytical system. At this stage the water flows upwards through the reservoir and then through the outlet at the top of the device. After sampling, the probe moves to the top of the reservoir and the O ring seals the outlet. The probe then aspirates water and the excess runs down the probe exit hole where it is sucked away through the annular channel which surrounds the channel. There is a subatmospheric pressure at the

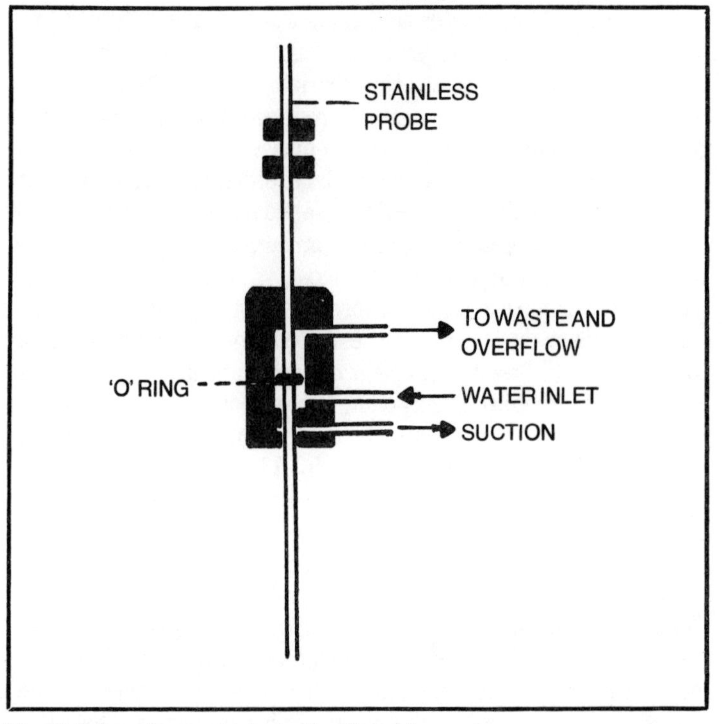

Fig. 18-4. Sampling mechanism of the Chemlab sampler.

probe exit and entry hole which causes any liquid adhering to the probe to be sucked away as it passes through the chamber. The sample and wash times can be adjusted from 0-120 seconds.

The Proportionating Pump

The function of the *proportionating pump* is to continuously and simultaneously push fluids, air and gases through the analytical chain. In fact, it is the heart of the automatic analysis system. Here, all the sample and reagent streams, in any particular analysis, are driven by a single peristaltic pump which consists of two parallel stainless steel roller chains with finely spaced roller thwarts.

A series of flexible plastic tubes, one from the sampler and the others from reagent bottles or simply drawing in air, are placed lengthways along the platen spring-loaded platform. The roller head assembly is driven by a constant-speed gear motor. When the rollers are pressed down the motor switched on, they compress the tubes containing the liquid streams (sample, standard and reagents) against the platen. As the rollers advance across the platen, they drive the liquid before them.

The roller head rotates at a constant speed. The different flow rates required in the several streams (0.15 to 4 ml/min) are achieved by selecting tubes of appropriate internal diameter but of constant wall thickness. Since the proportions of the various reagents are fixed by the tube sizes, no measurements are needed.

Proportionating pumps are available either for single speed or for two-speed operation. The single speed pump has the capacitor synchronous gear head utilizing 10 rpm output shaft at 50 Hz. The two-speed pump has a non-synchronous 45 rpm motor. The slow speed in this pump is used for the ordinary working during a run and a much quicker one for filling the system with reagents before a run, and for rapid washing to clear out reagents after the run. It is also utilized for rapid cleaning of the heating bath or of the complete system when fibrin problems are evident and are disturbing the run. Fast speed is not used for analysis. A heavy duty pump is also available which enables 23 pump tubes to be utilized simultaneously.

The plastic tubes are held taut between two plastic blocks having locating holes which fit onto pegs at each end of the platen. Before beginning a run, the tubes are stretched. With use and time, the tubes lose elasticity and pumping efficiency is reduced. Therefore, each block has three sets of holes so that the tubes can be increasingly stretched and the tension thus maintained. The tubes are replaced at the first sign of aging. In fact, they should be replaced at regular intervals to forestall failure. When not in use, one of the blocks is removed so that the tubes are not kept in tension.

Actually the sample or reaction stream is separated by air bubbles into a large number of distinct segments. The air bubbles completely fill the lumen of the tubing conducting the flow, thereby maintaining the integrity of each individual aliquot. In addition, the pressure of the air bubble against the inner wall of the tubing wipes the surface free of droplets which might contaminate the samples which follow. The proportionating includes an *air bar* device (Fig. 18-5) which adds air bubbles to the flowing streams in a precise and timed sequence. The air bar is actually a pinch valve connected to the pump rollers that occludes or opens the air pump tubes at a timed interval. Every time a roller leaves the pump platen, and this occurs every two seconds, the air bar rises and lets a measured quantity of air through. The release of air into the system is carefully controlled, thereby insuring exactly reproducible proportionating by the peristaltic pump.

Manifold

A *manifold* mainly consists of a platter, pump tubes, coils, transmission tubing, fittings and connections. A separate manifold is required for each determination and the change can be effected within a few minutes.

The pump tubing and the connected coils are placed on a manifold platter, which keeps them in proper order for each test. The pump tubing is specially made. It is of premeasured length and

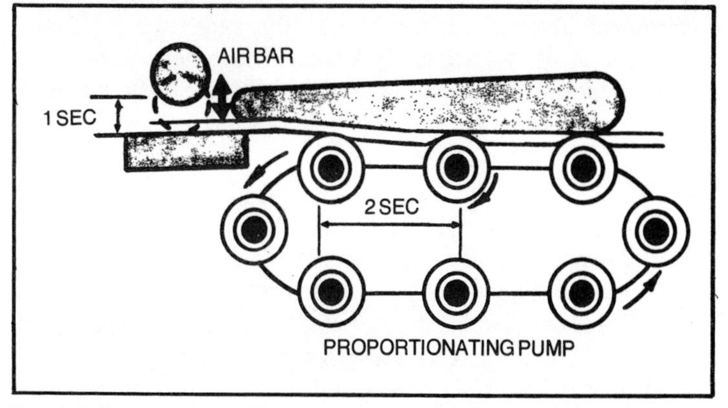

Fig. 18-5. Principle of air segmentation in the continuous flow system.

is meant to introduce all constituents of an analysis into the system. The physical and chemical properties of the tubing are extremely important in the correct functioning of the pump. It must not be so flexible as to expand beyond its normal internal dimensions on release of pressure which may lead to variation in the flow, thereby affecting reproducibility and accuracy of the system. The tubes should be chemically inert for the constituents which are expected to flow through the tube. The constant and correct tension also provides the continual delivery of a constant volume. The inside diameter of the pump tubing determines the flow rate per minute.

Several other tubes are required to introduce reagents and to transport the specimen from one module to another. There are five types of such tubings. They are of varying sizes and are to be selected according to the requirements. These are: standard transmission tubing (Tygon), solvaflex tubing, acidflex tubing, polyethylene tubing and glass tubing. Table 18-1 shows the pump tube sizes and delivery volumes for standard and solvaflex tubing.

Two types of coils are employed in the system—*mixing* coils and *delay* coils. Coils are glass spirals (Fig. 18-6) of critical dimensions in which the mixing liquids are inverted several times so that complete mixing can result.

Mixing coils are used to mix the sample and/or reagents. As the mixture rotates through a coil, the air bubble along with the rise and fall motion produces a completely homogeneous mixture. The mixing coils are placed in a horizontal position to permit proper mixing. Delay coils are employed when a specimen must be delayed for completion of a chemical reacting before reaching the colorimeter. These coils are selected in length according to the requirement. The standard delay coil is 40 ft. long, 1.6 mm I.D. and has a volume of approximately 28 ml. The time delay can be calculated by dividing the volume of the coil by the flow rate of specimen plus bubbles.

Phasing Coils

With 12 tests to be recorded on each sample and a sampling rate of 60 samples per hour, it follows that 5 seconds are allowed to record each steady state plateau. The reaction streams in the 12 channels and up to four blank channels must, therefore, be phased to arrive at the colorimeter in waves 5 seconds apart. For example, if the cholesterol stream arrives at X time, calcium muust arrive at X+5 seconds, total protein at X+10 seconds, albumin at X+15

Table 18-1. Pump Tube Sizes and Nominal Delivery Volumes.

Tube ID (inch)	Shoulder Colors	Clear Standard Delivery (ml/min)	Solvaflex Delivery (ml/min)
0.005	Orange black	0.015	0.015
0.0075	Orange red	0.03	0.03
0.010	Orange blue	0.05	0.05
0.015	Orange green	0.10	0.10
0.020	Orange yellow	0.16	0.16
0.025	Orange white	0.23	0.23
0.030	Black	0.32	0.32
0.035	Orange	0.42	0.42
0.040	White	0.60	0.56
0.045	Red	0.80	0.70
0.051	Gray	1.00	-
0.056	Yellow	1.20	1.06
0.065	Blue	1.60	1.37
0.073	Green	2.00	1.69
0.081	Purple	2.50	2.02
0.090	Purple black	2.90	2.42
0.100	Purple orange	3.40	2.89
0.110	Purple white	3.90	3.39

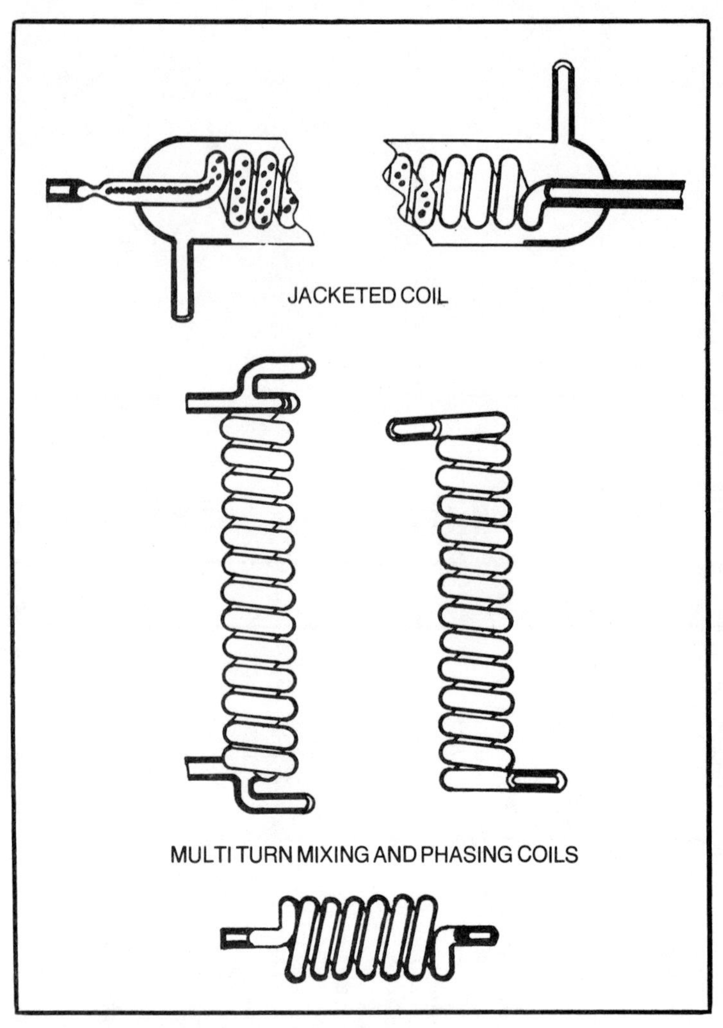

Fig. 18-6. Typical phasing and mixing coils.

seconds, etc. In order to ensure proper sequencing for presentation of the results, a number of devices have been provided to make this adjustment an extremely simple operation. Phasing coils are used in the SMA-12 system to permit the channels to enter the colorimeter in the proper sequence.

In the earlier models, sleeving was employed for interconnection of manifold tubes and cyclohexamone was applied to seal the joints. Polyethylene nipples are now used for these connections. Various types of fittings (Fig. 18-7) are employed to join streams of liquids, to split a stream or to introduce air segmentation to the stream.

Mixing with other reagents begins on leaving the pump. The first reagent with which the test specimen is mixed is usually a simple diluent. The reagent lines are segmented by introducing air through one or more additional tubes into the manifold. This produces a series of bubbles at regular intervals in the liquid stream. This is designated as *bubble pattern*. A uniform bubble pattern is very essential for accurate analysis. However, it may be noted that it is not necessary that every bubble be absolutely identical in length, but a firm consistent flowing segmented stream is required.

Dialyzer

In analytical chemistry, it is often necessary to remove proteins or cells to obtain an interference free analysis. This is ac-

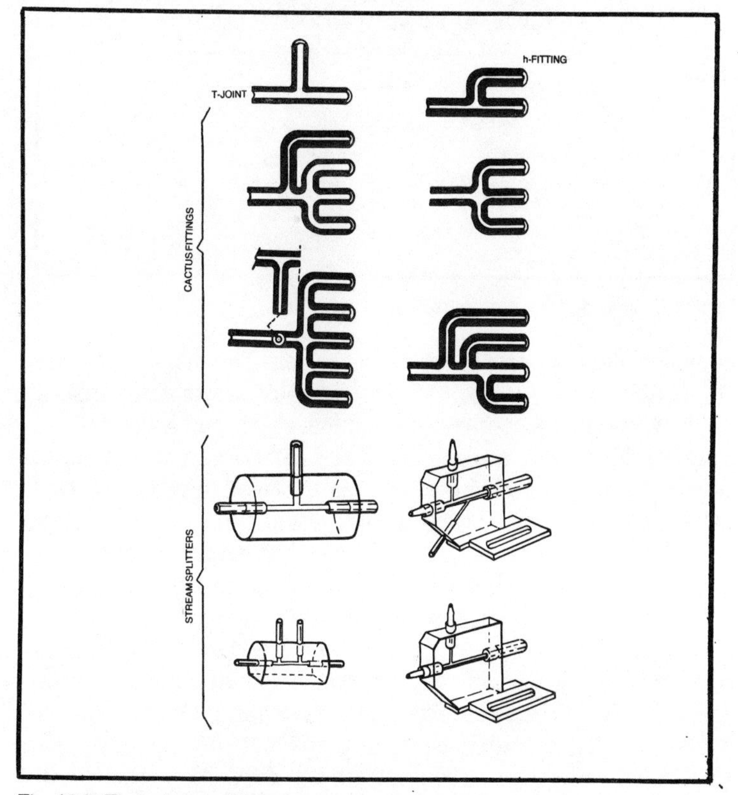

Fig. 18-7. Typical glass fittings in an automated analysis system.

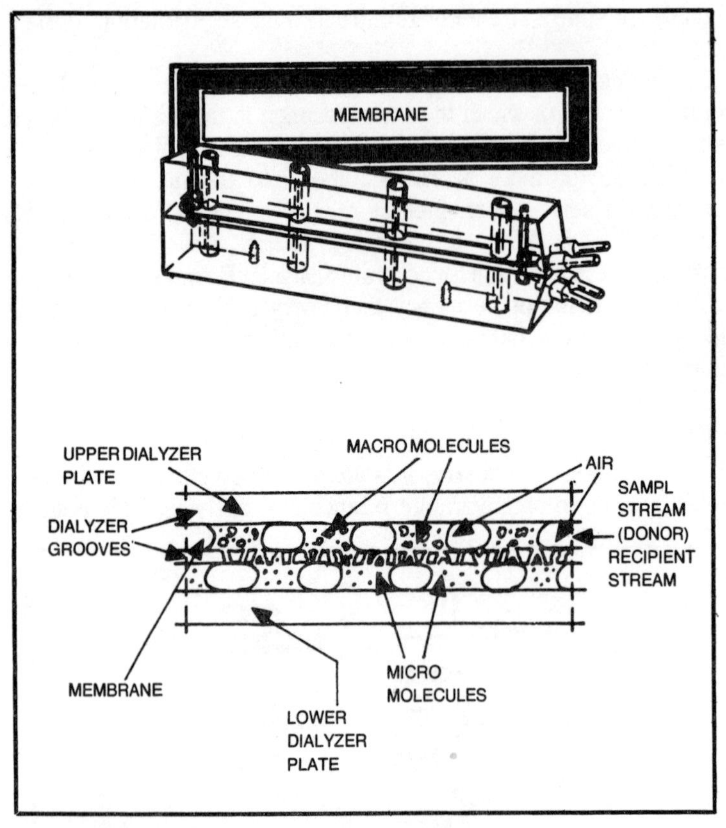

Fig. 18-8. Simplified diagram of the dialysis process.

complished by dialysis in the autoanalyzer. The dialyzer module (Fig. 18-8) consists of a pair of Perspex plates, the mating surfaces of which are mirror grooved in a continuous channel which goes in towards the center, turns on itself and returns to the outside. A semipermeable cellophane membrane is placed between the two plates and the assembly is clamped together similar to the kidney dialyzer. The continuous groove channel thus gets divided into two halves and the dialysis occurs across the membrane. A solution containing the substance to be analyzed passes along one half, usually the upper one, of the channel while the solvent that is respective to the substance to be removed enters the other half. The substance to be separated from the sample diluent stream will diffuse through the semipermeable membrance by osmotic pressure into the recipient stream and the non-diffusible particles will be left behind.

The cellophane membrane usually used in the dialyzer has a pore size of 40-60 A°. The rate of dialysis is stated to be dependent upon temperature, area and concentration gradient. For this reason, the dialyzer unit is usually immersed in a water bath maintained at a constant temperature (37°C ± 0.1°C). The temperature is kept constant with a thermostatically controlled heater and a motorized stirrer. Both streams pass through preheating coils before entering the dialyzer unit. The channel path is 87 inches long which provides a large surface presentation to the dialyzing membrane. The plates of the dialyzer must be a matched set. If the plates are not matched set, the channels may be slightly off, causing leakage, poor bubble patterns and loss of dialyzing area which would ultimately result in loss of sensitivity.

Heating Bath

On leaving the dialyzer, the stream may be combined by one or more additional reagents. It is then passed to a heating bath. This module is not used in all the tests performed by the autoanalyzer. The heating bath is a double walled insulated vessel in which a glass heating coil or helix is immersed in mineral oil. A thermostatically controlled immersion heater maintains a constant temperature within ± 0.1°C which can be read on a thermometer. Inside the bath the stream passes along a helical glass coil about 40 feet long and 1.6 mm I.D. immersed in oil which is constantly stirred. The heating bath may have a fixed temperature as 95° or 37°C or an adjustable value. Passage through the heating coil takes about five minutes, but it would obviously vary with the rate at which the liquid is moving which, in turn, depends on the diameter of the tubes in the manifold.

Detectors

After the sample is processed, the end product is quantitated by a suitable sensing device which may be a colorimeter, flame photometer or fluorimeter. However, most of the clinical analysis ends with the measurement of color intensity which is accomplished in a photoelectric colorimeter. Usually, a double beam system with wavelength selection by interference filters in the sample and reference beams is employed.

Colorimeter

The colorimeters used in the automated systems continuously monitor the amount of light transmitted through the sample. They employ flow-through cuvettes. In the earlier designs of flow cells the arrangement was such that as the incoming stream entered the cell, the air bubbles escaped upwards through an open vent so that a continuous stream of liquid could fill the cell before going to waste. The flow cell size varied from 6 to 15 mm. The later designs of flow cells are all of tubular construction. This requires a much smaller volume of fluid so that a smaller volume of sample can be used. Being completely closed, it does not require separate cleaning. Before the stream enters the flow cell, it is pumped to a debubbler where the air bubbles are removed. The stream is then pulled through the flow cell under the action of another pump.

Figure 18-9 shows the optical system of a dual beam type colorimeter used with the autoanalyzer system. The instrument has a tungsten filament lamp which provides a common light source for reference and sample beams. The reference beam passes through a collimating lens, an aperture and a filter before reaching the photocell. The sample beam passes through a set of focusing mirrors, a filter and the sample flow cell before striking at the sample photocell. The focusing mirrors are so made that their chromatic aberration is much less than that which occurs with lenses. The filters used are of Fabry-Perot interference type in the spectral range of 340 mμ to 900 mμ with peak transmittance in the range of 25% to 50% and half widths as narrow as 10 μ. Filters normally supplied with Technicon instruments are in pairs—a reference and a sample filter. These two filters are identical.

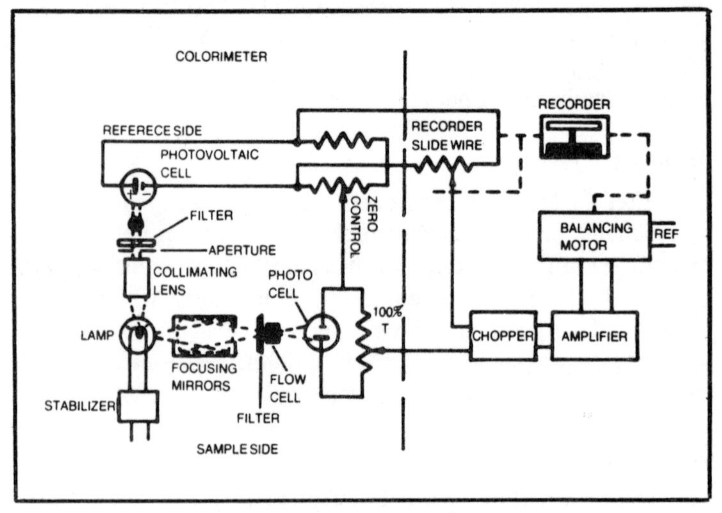

Fig. 18-9. Block diagram of the Technicon autoanalyzer colorimeter.

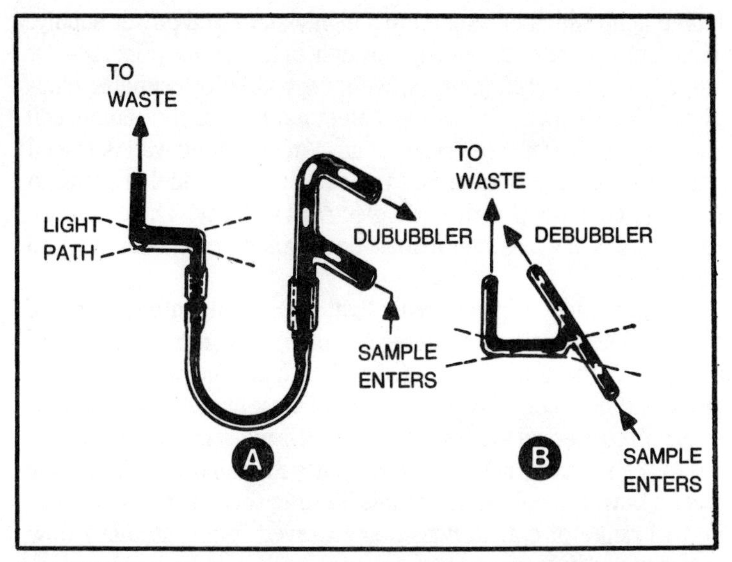

Fig. 18-10. Construction details of the flow cells (A) Flow cells used in the autoanalyzer colorimeter. (B) Cell of one piece construction (courtesy of Technicon Corp., USA).

Selenium photovoltic cells with a wavelength range of 400 to 660 mμ are the standard detectors for this range. Blue sensitive tubes for spectral range of 340 to 660 mμ are also used. They have a cathode of antimony-cesium-oxide alloy. Red sensitive phototubes are used in the wavelength range of 660 to 1000 mμ.

The two photoelectric cells are arranged in a null balancing circuit. The ratio of the voltages of the sample and reference photocells are recorded by a null-balance strip chart recorder. When the light intensity reaching the sample photocell is exactly equal to the light intensity reaching the reference photocell, the ratio is one or 100% T. Under these conditions, the recorder servo-system is said to be balanced.

Figure 18-10A shows the construction of the two types of flow cells used in the autoanalyzer colorimeter. They are designed to exhibit optimum wash (clean out) characteristics. A debubbler is coupled with the flow cell so that the entrapped air bubble may leak out. The cell shown in Fig. 18-10B is of one piece construction. It has a smaller volume while it maintains the same optical path length.

The Chemlab automatic continuous flow chemical analysis system incorporates a multichannel colorimeter. The colorimeter employs a fiber optics system using a single high intensity quartz iodine lamp which is coupled to a highly stabilized power supply. This lamp passes its energy through light guides (Fig. 18-11), which are connected to up to five independent colorimeter modules. Each colorimeter has two detectors, one for the sample cell and the other for the reference beam. An insulating wall is placed between the light source and the colorimeter modules so as to maintain good temperature and electrical stability. The colorimeter contains a glass continuous flow sample cell with integral debubbler.

Light guides from the main light source plug into one side of the cell block and the photodetectors are placed on the other side of the block. The filters are inserted into a slot between the light guide and flow cell. The detectors are closely matched and their output is connected to a highly stabilized bridge circuit. The output fed to the recorder is logarithmic, hence for solutions which obey Beer's Law, the recorder output is linear in concentration over the normal range of optical densities observed in a continuous flow automatic analysis system.

Flame Photometer

The measurement of sodium and potassium is carried out by a *flame photometer*. This consists of an oxygen propane burner into which samples for analysis are pumped and atomized. The light output from the flame passes through and falls on three photocells. One filter transmits the wavelength of sodium flame, one the potassium wavelength, and the third transmits the light emitted by lithium. An acid lithium nitrate solution is used as a sample diluent, and the output of the lithium sensitive photocell is used as a reference. The ratios of the outputs of sodium and potassium detectors to this reference signal are recorded on a strip chart recorder. This minimizes any minor signal changes caused by small variations in gas pressure, air or oxygen pressure, dialysis rate and sample flow rate. This is because these variations would affect the signal output of both the sample and reference photocells to the same degree.

Fluorimeter

Fluorimetric analysis permits measurements to be made at concentrations as low as 0.01 part per billion. The fluorimeters used for automated work, like colorimeters, have a flow-through system. The continuous flow cuvette is made of Pyrex glass which transmits light from the visible region to approximately 340 mμ.

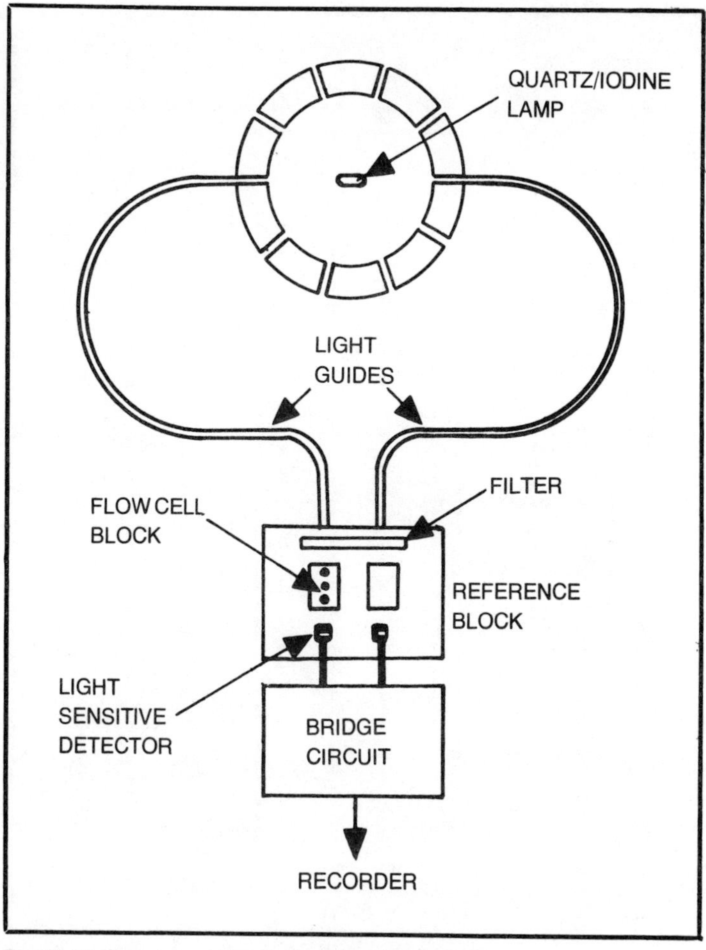

Fig. 18-11. Chemlab colorimeter for automated analysis.

For the ultraviolet region below 320 mμ, quartz cuvettes are available. The fluorimeter used with a Technicon autoanalyzer is a self-balancing, double beam type instrument.

Recorders

All 12 tests for each sample are reported in directly readable concentration units on a single strip of precalibrated chart paper. Since the normal ranges for each parameter are printed as shaded areas, the physician does not have to remember the normal values. Thus each abnormality stands out clearly. Figure 18-12 shows a typical record obtained on a blood sample. The most common type

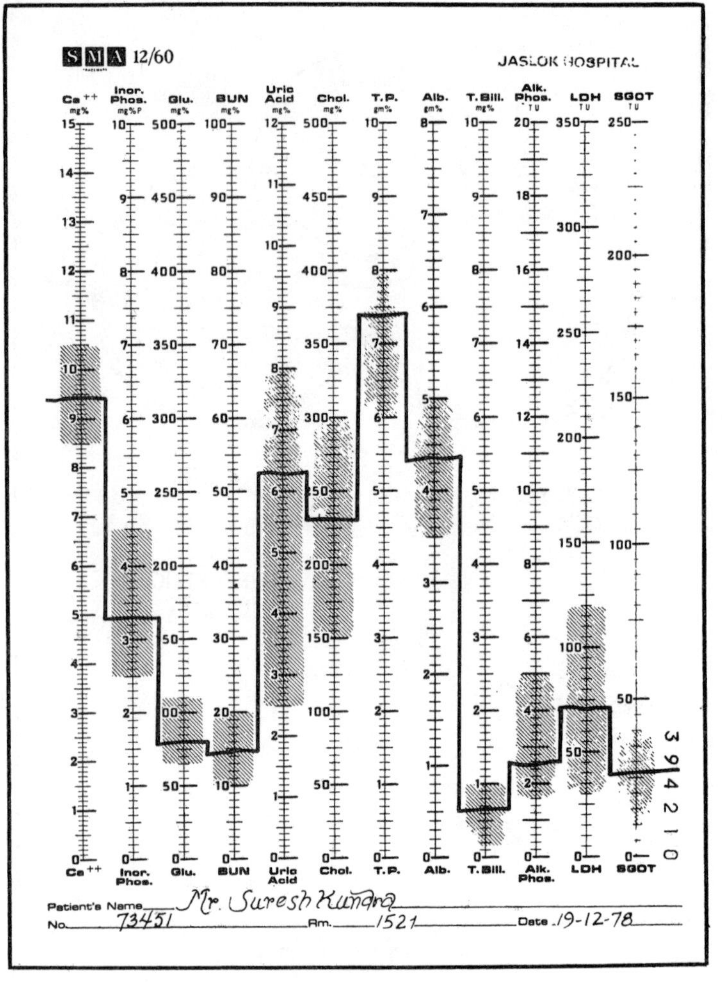

Fig. 18-12. Typical record obtained on a blood sample.

of recorder used with automated systems is the dc voltage null-balance potentiometric recorder of the type described in Chapter 1.

Initially, the setting of the recorder pen is made to almost 100% transmission, with a reagent blank which is kept running for some time until a smooth baseline is established on the chart. Ideally, this should be a smooth line parallel to the 100%T line. Significant fluctuations in the flow pattern may result in irregularities of the baseline. A good bubble pattern with air bubbles of equal size and equal spacing between them is one of the important factors in producing a smooth line with good peaks. The tubes and reagents must be clean. Irregularities are also produced due to drift in electronic circuit. Pulse suppressors are used to smoothen out fluctuations due to pump action and are specially required when microsample tubing is used.

After obtaining a good baseline, the run can be begun with a series of standards followed by test samples. It is usual to introduce standards at the end of the run and at intervals during the course of runs to ensure whether any change has occurred.

Since concentration of a substance in a sample is related logarithmically to the percent transmission, when this is plotted on a graph, the curve between them will not be linear. Therefore, a linearizing recorder with a logarithmic slide wire is employed which plots chart readings and yields a straight line if Beer's Law is strictly applicable. Alternatively, sufficient standards are run to give an accurate standard curve and the test results are then read from the same.

Actual measurements are made only after the analytical curve (Fig. 18-13) reaches its steady-state plateau (equilibrium condition in the system at which there are no changes in concentration with time). At this steady-state plateau, all effects of possible sample interaction have been eliminated, and the recorded signal gives a true reflection of the concentration of the constituent being measured. Herein lies the importance of segmenting each of the sample and reagent streams with air bubbles. In effect, the air bubbles act as barriers to divide each sample and reagent stream into a large number of discrete liquid segments. Equally important, the air

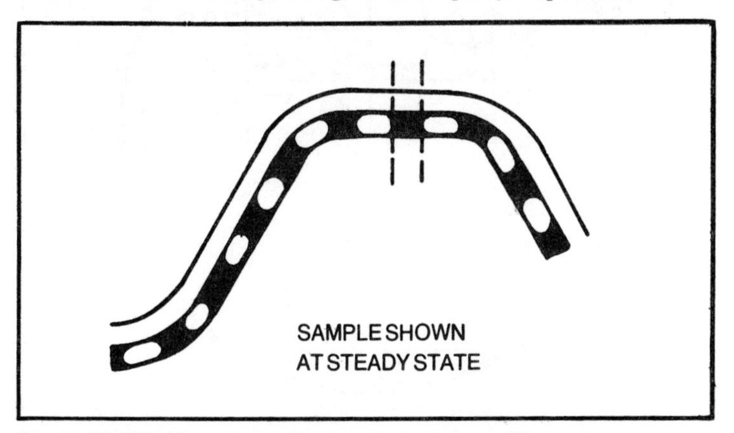

Fig. 18-13. Typical curve showing the steady state conditions when measurements are made in a continuous flow system.

bubbles continually scrub the walls of the tubing. This sequential wiping of the walls diminishes the possibility of contamination in succeeding segments of the same sample. Thus, should there be any interaction between two samples, it can easily be seen that the effects of this interaction will occur only in the first few segments of the second sample. In the middle segments, the air bubbles immediately preceding will have effectively cleansed the system and prevented further interaction. It is these middle and final segments, free from interaction, which are recorded as the steady-state plateau and appear as flat lines on the *Serum Chemistry Graph.*

Function Monitor

In order to see whether the system is in phase and recording at steady state levels, the electrical output of the phototubes is given to an oscilloscope called the *function monitor.* It consists of a direct view storage tube. The tube screen is divided into two columns of eight channels each. Each sample curve is recorded in its entirety, thus enabling the operator to see all 16 curves for each sample at all times. While these curves are still on the oscil-

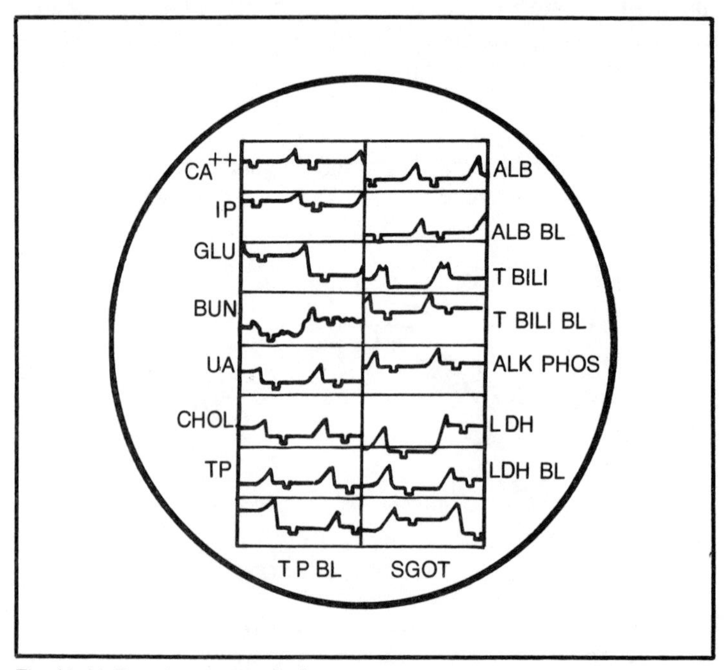

Fig. 18-14. Function monitor display screen.

loscope, the 16 curves for the next sample are traced alongside, so that one has a complete visual presentation of the relationship between two consecutive curves. If desired, these curves may be held for study. Further, a raised portion of the steady state plateau indicates where the Serum Chemistry Graph (SCG) was recorded. Figure 18-14 demonstrates the value of the function monitor.

On the SCG, all tests have been recorded at their steady-state plateaus with the exception of cholesterol and alkaline phosphatase. A quick glance at the tracings for these procedures on the function monitor immediately indicates the malfunction. In the case of cholesterol, the *blip* on the tracing occurs between samples and not on the steady state portion of the curve. This is indicative of an improperly phased system.

While the alkaline phosphatase tracing in the SCG is slanted very much like cholesterol, the function monitor reveals the problem to be one of noise rather than phasing. With this type of monitoring phasing and detection of incipient irregularities can be accomplished while the test is being performed, rather than after the run is ended.

Chapter 19

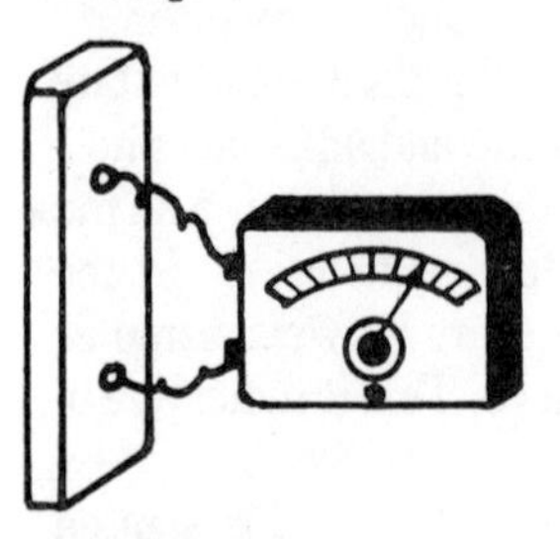

Environmental Pollution Monitoring Instruments

The awareness of and concern about the deteriorating environment is increasing the world over. It is necessary to monitor changes taking place in the quality of of the environment for initiating efforts to control it. There are a large number of instruments required to accomplish the environmental pollution monitoring program. Representative sampling of pollutant concentrations at the point of discharge or in the environment require an understanding of the pollutant characteristics as well as specific objectives in consideration like monitoring or control. Only then can appropriate measurement techniques be identified and employed. Pollution monitoring is thus a very complex task which involves systematic collection and evaluation of physical, chemical, biological and related information pertaining to environmental quality and effluent discharges.

AIR POLLUTION MONITORING INSTRUMENTS

Rapid industrialization and steadily increasing vehicular traffic on the roads have generated an acute problem of air pollution. At places, gaseous and dust pollutants in the atmosphere reach such a magnitude that suitable measures to limit the emission of pollutants become imperative. One of the basic measures to environmental quality management lies in the ability to continually monitor environmental characteristics and provide reliable, accurate and automatically recorded data for timely interpolation. Pollutants are required to be monitored not only in the environment but also as they are discharged from the multitude of stacks, exhaust pipes from industrial establishments and vehicles. Today, the pollutants that must be measured are many in number and they are emitted from a still larger number of sources.

The most reliable and useful information on the degree of pollutant concentrations is obtained by continuous sampling using officially accepted methods. The data about the peak concentrations obtained from such sampling is indispensable when monitoring the air for potentially hazardous pollutants. For some specific purpose, however, samples collected intermittently may be useful as a basis towards a more meaningful monitoring program. Regardless of whether one is involved with continuous monitoring or monitoring on an intermittent basis, the analysis of the air we breathe has become a necessity for the survival of mankind.

REPRESENTATION OF CONCENTRATION OF GASES

Gas concentrations in the atmosphere are generally represented as parts per million by volume, i.e., ppm/v or simply ppm or parts per hundred million (pphm) i.e., parts per billion (ppb). On the other hand, toxicological data is generally represented on a gravimetric basis, e.g., micrograms per cubic meter or milligrams per liter.

Conversion from volumetric to gravimetric concentration can be obtained by applying gas laws, the general equation for this being

$$\mu g/m^3 = (ppm) \times \frac{PM}{RT} \times 10^3$$

where P = total pressure (atm)
M = Molecular weight of gas of interest
R = Gas constant = 0.0821 1-atm/(mole) (°K)
T = absolute temperature °K.

TYPES AND CONCENTRATION OF VARIOUS GAS POLLUTANTS

Many types of pollutants are present in the atmosphere. However, for many reasons, it is not simple to make definite statements about pollutant concentrations. This is because of the extreme variability of the pollutant concentrations themselves which vary constantly with air turbulence and the strength of emissions. The major gas pollutants are: carbon monoxide, sulfur oxides, hydrocarbons, nitrogen oxides and particulates.

Carbon Monoxide. *Carbon Monoxide* is an especially hazardous pollutant. It is a by-product of combustion processes in which incomplete oxidation of fossil fuels takes place. It is also associated with automotive exhaust and deep mining operations. The average concentrations of this gas are found to be much below 200 ppm.

Hydrocarbons. *Hydrocarbons* enter the atmosphere from a wide variety of sources like petroleum-refining processes, incomplete combustion of organic fuels, and evaporation of fuels and solvents. Gasoline is the major source of their emission from internal combustion engines since they exhaust unburned and partially burnt hydrocarbons. These are also important because of their reaction in the atmosphere in the formation of ozone with nitrogen oxides and sunlight and also to form photochemical smog. Methane constitutes the major component of the total hydrocarbon emission.

Sulfur Oxides. Sulfur dioxide is the most common and the most abundant pollutant which is emitted into the atmosphere from heating and from industrial plants using high-sulfur coal and other sulfur-containing fossils. It is also associated with the most serious urban pollution disasters. Sulfur trioxide, which is estimated to be one-hundredth the concentration of sulfur dioxide, is formed in a secondary reaction and becomes sulfuric acid aerosol. It is reported that SO_2 is present, at an average level of 0.024 ppm for urban areas.

Nitrogen Oxides. The emissions of nitrogen oxides are chiefly from the products of fuel combustion in furnaces and engines. It has been observed that the distribution of nitrogen oxides closely follows the population concentrations. Its level ranges from 0.5 to 0.12 ppm on an annual average basis.

Oxidant (Ozone). The presence of oxidants in air can have a significant effect on ambient air quality. The major component of total oxidants is the ozone that has damaging effect on plants, animals and material if present in higher concentrations. Other pollutants which may be present are hydrogen sulfide, ammonia, halides (chlorine, fluorine and bromine) and carbon dioxide. See Table 19-1.

CARBON MONOXIDE

Non-dispersive infrared analysis depends on the characteristic energy of absorption of a CO molecule at a wavelength of 4.6 μ. Infrared energy is also absorbed by other gases like CO_2, H_2O, SO_2

Table 19-1. Instrumental Techniques and Measurement Ranges for Some Gases.

Gas	Full Scale Range	Measurement Technique
Carbon monoxide	0-50 ppm 0-200 ppm	Infrared absorption Gas chromatography (Flame ionization detector)
Hydrocarbons	0 - 80 ppm	Ultraviolet absorption Gas Chromatography (Flame ionization detector) Mass spectrometry
Sulfur Oxides	0 - 2 ppm	Ultraviolet abosrption Infrared absorption Gas Chromatography (Flame photometric detector) Flame photometric detector Colorimetric method Conductimetric method Coulometric method Electrochemical transducers
Nitrogen Oxides	0 - 1 ppm	Colorimetric method Coulometric method Chemiluminescence method Electrochemical transducers Infrared spectroscopy Lasers
Oxidants	0 - 500 ppb	Chemiluminescence method Coulometric method Colorimetric method Ultraviolet absorption method

and NO_2. The differentiation of CO from such types of gases depends on the difference in energy absorbed as infrared radiation is passed through a sample cell containing CO and a reference cell containing a fixed quantity of nitrogen, CO and water vapor.

Non-Dispersive Infrared Analyzer

Figure 16-8 shows a block diagram of the instrument. The reference cell contains a fixed quantity of the gases whereas an air sample is made to flow at about 150 ml/min through the sample cell.

CO analyzers based on infrared absorption would give greater sensitivity with larger cell path lengths. Instruments with 1-meter cell length would measure from 1 to 50 ppm. However, some of the latest instruments are capable of measuring 1 to 25 ppm even with cell paths of 10 cm. The response time of such instruments varies from 1 to 5 minutes. They are calibrated by passing a known part per million concentration of CO in nitrogen into the sample. The effects of interfering gases like CO_2 and water vapor can be further minimized by placing optical filters ahead of the sample cell so that IR radiation window is limited to a range where radiation absorption by these does not significantly take place.

Gas Chromatography

When an air sample containing CO is passed through a stripper column, the heavy hydrocarbons are retained and CO and methane are passed into a chromatographic column and then into a catalytic reducing chamber. The methane would pass through the reducing chamber unaffected while CO is reduced to methane. By using a hydrogen flame ionization detector, both methane peaks can be detected. The first peak is due to methane while the second peak would correspond to CO. The accuracy is about ± 2%. Peak heights of CO and CH_4 would give sensitivity of about 50 ppb. Ortman (1966) describes details of the method for semicontinuous monitoring of atmospheric carbon monoxide and methane.

SULFUR DIOXIDE

With colorimetry a known volume of air is passed through an aqueous solution which contains reagents that absorb SO_2 and produce a colored substance. The amount of colored substance is proportional to the component of interest (SO_2). This is determined by measuring a solution's optical absorbance spectrophotometrically. Within limits, the absorbance is linearly proportional to the concentration of the colored species in accordance with Beer's Law.

The most widely used method involving colorimetry for determination of SO_2 is that the West-Gaeke (1956) which is applicable to concentrations in the range of 0.005-5 ppm. The air to be analyzed is passed through an aqueous solution of 0.1 M sodium tetrachloromercurate. The sulfur dioxide reacts with the mercuric salt to form sulfatomercuric compound. The solution is further treated with acid-bleached para-rosaniline and formaldehyde which would result in a red-purple color. The amount of color is determined photometrically. The color must be developed within a week after sample collection.

Sulfur dioxide may also be determined by measuring the decolorization of an aqueous starch-iodine solution, which normally has a deep blue color. When SO_2 is passed through the solution containing starch-iodine solution, the iodine is reduced by the sulfur dioxide which, in turn results in decrease of the intensity of starch-iodine solution, the iodine is reduced by the sulfur dioxide which, in turn results in decrease of the intensity of starch-iodine color.

The advantages of colorimetry are simplcity, high sensitivity and good specificity. Color intensity is sensitive to temperature, purity of reagents, pH, development time and age of solutions. The method is free from interference from H_2, SO_4, SO_3, NH_3, etc.

Conductimetry

Analyzers based on measuring the change in conductivity of a solution, when a sample of air containing sulfur dioxide is bubbled through it, are the oldest and mst commonly utilized instruments for ambient air monitoring. The solution consists of sulfuric acid and hydrogen peroxide. The change in electrical conductivity takes place due to the formation of sulfuric acid by oxidation of the sulfur dioxide.

$$H_2O_2 + SO_2 \rightarrow H_2SO_4 \rightarrow H^+ + HSO^-_4$$

These instruments are characterized by fast response and high sensitivity. However, their performance gets affected due to interference by non-SO_2 gases that produce or remove ions in solution. Therefore, these analyzers are employed only if interfering gases are not in high concentrations or if they can be effectively removed from the air sample.

Killick (1969) describes the construction of a conductivity cell (Fig. 19-1) for the continuous measurement of sulfur dioxide in air. The cell (C) is made of glass, 1 cm inside diameter with a jet (J), orifice 0.5 mm diameter, located 1 cm above the reagent surface.

Two electrodes (E) made of 18 SWG stainless steel wire are inserted through a Perspex cap (P). The cap is sealed to the base of the cell with araldite. Reagent enters the cell from a central tube inserted in the Perspex cap. A small glass bead (B) in the cell acts as a non-return valve on the entry of the feed tube and prevents sulfuric acid diffusing from the cell. The end of the jet is made from a piece of capillary tube. A filter is placed before the jet to prevent blocking due to solid materials. The flow of air through the jet is maintained at approximately 200 ml/min. The cell is of small size (1.5 ml) and its capacity to absorb SO_2 is thus limited. It is therefore operated intermittently, the electrolyte being discharged and replaced at regular intervals of 15 minutes. The resulting output is recorded as a sawtooth waveform.

To measure the conductivity of the cell, 5 volt alterating current is applied across the electrodes. Alternating current avoids polarization. When normal urban concentrations of SO_2 are measured, the current through the cell increases from its zero value of 20 μA to up to 40 μA at the end of a 15 minute sampling period. In conditions of heavy pollution, it may go up to 2 mA. Because the current is recorded every 15 minutes, the concentration of sulfur-dioxide at any instant is proportional to the slope of the sawtooth at that instant. The calibration is carried out by using known concentrations of sulfur dioxide in air.

Gas Chromatography

There are two major problems when a gas chromatographic technique is employed for measuring pollutants in air. First, most of the pollutants are extremely reactive materials. Therefore, they may not pass through the column and appear at the detector. This necessitates the use of special column and support materials. The amount of pollutants of interest is generally so small that when a reasonably sized gas sample is injected into the carrier gas stream, it is not detected by ordinary GC detectors. Hence, extremely sensitive and specific detectors must be employed.

Stevens et. al. (1971) devised a chromatographic column support and stationary partitioning liquid that would separate a number of sulfur compounds present in the atmosphere. A special Teflon column was employed in order to minimize reaction of the constituents of interest on the walls. They employed the flame photometric detector specifically for the detection and measurement of the sulfur containing effluents from the column. These detectors measure the emissions from sulfur compounds introduced into a hydrogen rich flame. A narrow band optical filter selects the 394 nm emission band. These detectors are not susceptible to interference from non-sulfur pollutants. However, they are subject to interference from other sulfur compounds. This type of interference can be minimized by selective filters.

Coulometry

The ability of sulfur dioxide to reduce iodine can be employed to monitor SO_2 by using iodine coulometry in aqueous solution. The mass of I_2 reacted per unit time during any given interval of time would indicate concentration of SO_2 in air.

Coulometric arrangement would require two electrodes made of platinum which act as anode and cathode. These electrodes

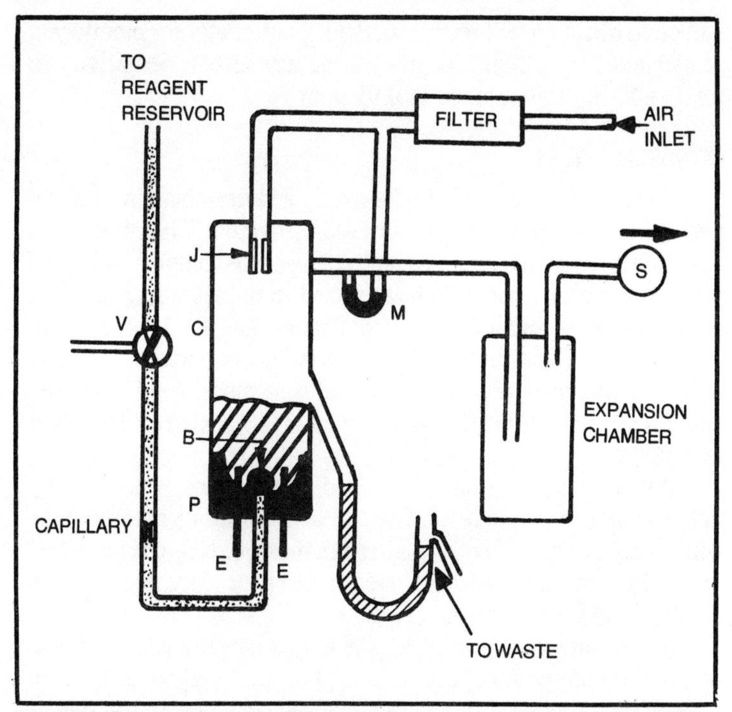

Fig. 19-1. Conductivity method for the measurement of SO2 in air (after Killick, 1969).

maintain a trace of I_2 in equilibrium with potassium bromide to maintain conductance. The shift in the anode-cathode potential potential is sensed by a third reference electrode when SO_2 from sample air is bubbled through the solution in a detector cell. The coulometric method provides a detection limit of 0.01 ppm. The response is not instantaneous and it may take about 4 minutes for 90% of the signal to appear for any concentration of SO_2.

Flame-Photometric Detector

Flame photometers, in which sample air is introduced into a hydrogen-rich air flame, and by using a narrow band interference filter that shields the photomultiplier tube detector from all but 394 nm emission energy of flame-excited sulfur atoms, offer another method for monitoring combined-sulfur. Research work has been continuing to increase the specificity of the method to actual SO_2 rather than total sulfur measurements. Suitable filters have been developed to remove interferences from hydrogen sulfide and some other organic sulfur. It is expected that improvements in selective filtration will result in a flame photometer for monitoring of ambient SO_2 within acceptable accuracy levels. Sensitivity to total sulfur by this method is 0.01 ppm.

NITROGEN OXIDES

Colorimetric method can be used for determination of nitrogen dioxide using the Saltzman (1954) method. This method is based on a reaction in which a pink colored dye complex is formed when the air containing NO_2 is passed in an absorbing solution consisting of the sulfanilic acid and diamine dissolved in the acetic acid medium. This method is sensitive in the parts per million range. A modified reagent that results in more rapid color development and a more efficient absorber of NO_2 has been described by Lyshkow (1965).

Nitric oxide may be determined by the same procedure by first passing the sample through an acid permanganate bubbler which oxidizes nitric oxide to nitrogen dioxide. NO_x can be determined by using Jacobs-Hochhesier (1958) procedure and its modifications by Purdue et. al. (1972).

When monitoring NO_x in stack effluents in concentration range of 5 ppm to several thousand ppm, the method consists in passing the sample into an evacuated flask containing a solution of H_2O_2 in sulfuric acid. The oxides of nitrogen are converted to nitric acid, and the nitrate ions react with phenol-disulfonic acid to produce a yellow color which is measured colorimetrically.

Chemiluminescence

The phenomenon of emission of radiation from chemi-excited species is known as *chemiluminescence*. It results due to the formation of new chemical bonds. The species in the excited state possess higher energy levels than the ground state and usually have a very short life. This phenomenon is very useful for measurement of air pollutants, particularly NO and NO_2. Instruments based on the measurement of chemiluminescent emission have been developed. Examine the following reaction.

$$NO + O_3 \longrightarrow NO_2 + O_2$$

$$NO_2 \longrightarrow NO_2 + h\nu \ (\lambda \max = 6300°A)$$

Since NO_2 reacts only slowly with ozone and the reaction which produces NO_3 is not accompanied by chemiluminescence, it is necessary to reduce NO_2 to NO before admission into the reactor.

$$NO_2 \xrightarrow{energy} NO + O$$

Nitric oxide and ozone containing gas stream are mixed in a vessel at sub-atmospheric pressure of about 2 mm Hg. Light emission is measured with a photomultiplier. With the use of high gain, low dark current photomultiplier tubes, extremely low levels of radiation can be measured. The response of the instruments based on chemiluminescence is linear from 1 ppb to 1000 ppm of NO. The technique is extremely useful for measurement of NO_x in automotive exhaust gases.

Use of CO Laser

Bonczyk (1975) describes an apparatus (Fig. 19-2) for detecting nitric oxide in 0.25 ppm concentration. The apparatus consists of a CO laser, which emits radiation that is absorbed by the NO in the mixture, the amount of absorption being proportional to the concentration of NO present. The wavelength match between laser and NO is made exact, and hence the absorption is enhanced by placing the NO in a magnetic field of the few KG intensity. The field shifts the absorption wavelength of the NO into coincidence with the fixed laser wavelength (Dousamanis et. al. 1955).

The CO laser used is a dc excited continuous working laser which operates at a single wavelength of 5.307 μ and at liquid

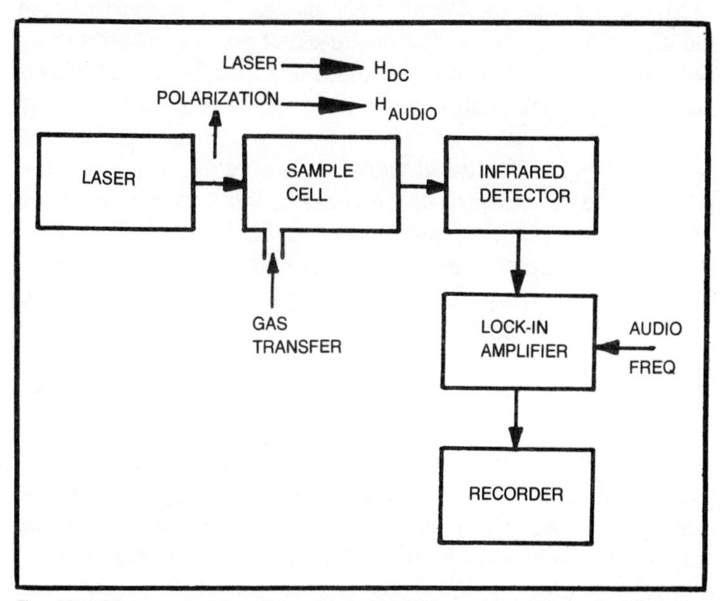

Fig. 19-2. Measurement of nitric oxide using CO laser (after Bonczyk, 1975).

nitrogen temperature. A diffraction grating is used at one end of the cavity as a line selector. The laser yields 5-30 mW of single line power.

The absorption cell is made of Pyrex and is 15 mm dia and 90 cm long. It is evacuated and pressure in the empty cell is 10^{-6}-10^{-5} torr. In order to produce modulating audio frequency magnetic field along the axis of the cell, insulated wire is closely wound around the outside of the cell over about half its length. The coil is excited with a current of 1A in the frequency range of 50-150 kHz. This produces a varying magnetic field of about 50 G peak to peak intensity. The dc magnetic field is produced by a solenoid which produces a field up to 2.5 KG.

The detector is a liquid nitrogen cooled Ge-Au element. The signal is amplified in a lock-in amplifier before it is given to the recorder. The signal amplitude varies linearly with the concentration on NO in the sample.

Laser Optoacoustic Spectroscopy

Optoacoustic detectors, in conjunction with thermal IR sources, have been widely used in gas detection and measurement systems. These have been developed with the intention of their application to air pollution measurement. Burkhardt et. al. (1975) successfully used this technique to measure trace amounts of nitrogen oxides in the stratosphere. The optoacoustic effect was demonstrated by Bell. He showed that the absorption of an amplitude modulated beam of IR by a gas could result in the generation of sound. Energy absorbed by the gas molecules from the IR beam excites the molecules to the rotational-vibrational energy levels above the ground state. The main path for decay of these excited states is collisional de-excitation which results in the transfer of absorbed energy into heat and raises the gas temperature. The temperature rise causes a corresponding pressure rise in the gas. When the beam intensity is modulated, the gas temperature and pressure change accordingly. The periodic pressure variations in the gas results in the generation of sound.

Kreuzer (1978) describes a laser optoacoustic detector, a block diagram of which is shown in Fig. 19-3. The arrangement makes use of a CO_2 laser that is tuned by rotating a diffraction grating at one end of the laser cavity. It is turnable to 64 different emission lines in the range from 927 to 1085 cm^{-1}. The laser beam is brought to a focus at the chopping wheel and then refocused into the detector. The chopping frequency is selected to optimize the

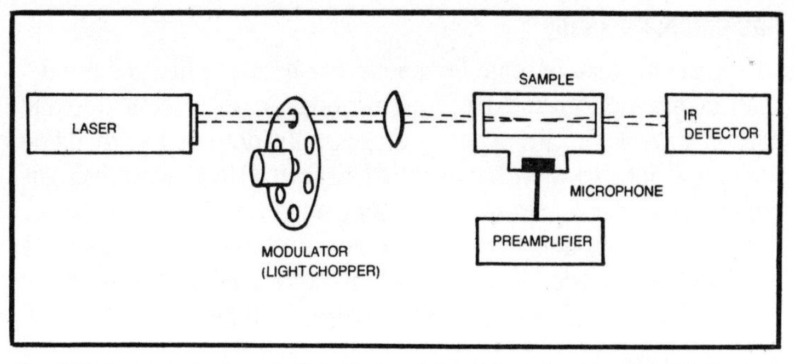

Fig. 19-3. Laser optoacoustic detector for nitric oxides (after Kreuzer, 1978)

signal to noise ratio. The microphone is commercially available (model 4144 manufactured by Bruel and Kjaer Instruments Inc., Cleveland, Ohio). The acoustic signal is amplified in a preamplifier and displayed. The detector is highly sensitive and sample amounts as small as 20 pg may be detected.

HYDROCARBONS

Organic compounds easily pyrolyze when introduced into an air-hydrogen flame. The pyrolysis produces ions that can be collected by having a cylindrical grid surrounding the flame. The detector response would be in proportion to the number of carbon atoms in the chain. For example, propane would roughly give three times the intensity of response as compared to methane. The ions collected on the positively charged grid are amplified in a high input impedance amplifier whose output is given to a chart recorder. The variation in the ion intensity resulting from flame ionization of any organic compound in the air sample is recorded on the chart. Generally, 0-20 ppm range is adequate for atmospheric sampling.

Flame Ionization Detector

Frostling and Brantte (1972) describe an apparatus for continuous measurements of hydrocarbons in air. The sample, vapors or aerosols, is analyzed in a flame ionization detector. The sensitivity for most of the gaseous hydrocarbons and aerosols is about 0.05 mg m^{-3}. The detector works at a reduced pressure achieved by using a carbon dioxide driven aspirator as a suction pump. The detector is thermostatically controlled by combustion of hydrogen gas in a catalytic heater. The whole apparatus, including transistor amplifier and a recorder, is driven by six 1.5 volt dry cells which run for 15 hours.

Gas Chromatography

Gas chromatographic technique has been applied for the detection and measurement of hydrocarbons, carbon monoxide and carbon dioxide in air. They are generally detected with flame ionization detectors. Villalobos and Chapman (1071) describe a gas chromatographic apparatus which they employed for this purpose. Flame ionization detectors are not sensitive to carbon monoxide and carbon dioxide. These are first converted to methane by hydrogenation and then measured with this detector.

Use of Lasers

Fixed frequency, infrared emitting lasers have demonstrated potential in atmospheric monitoring and research. This is based on absorption of CO_2 laser frequencies by some gases. Kanstad (1977) describes a tunable dual-line CO_2 laser for atmospheric spectroscopy and pollution monitoring. The arrangement enable you to measure 0.5 and 2 ppb densities of ethylene in a calm and turbulent atmosphere respectively. Ethylene carries considerable interest as a pollutant from the petrochemical industry and automobile exhausts.

OZONE

Total oxidants are usually measured by liberation of iodine from potassium iodine under buffered conditions. Saltzman and Gilbert (1959) have studied both the neutral and alkaline methods.

Colorimetry

The gas is passed through a solution of neutral, buffered potassium iodide (1% KI buffered at pH 6.8). The iodide is oxidized to iodine in the presence of an oxidizing agent and gives the solution a pale yellow color whose intensity is measured colorimetrically.

Chemiluminescence

The *chemiluminescent* method is based on the emission of light from solid organic dye samples due to the passage of ozone over the surface. Very low ozone concentrations can be detected. However, periodic calibrations are necessary both for the instrument and for the ozone source used.

Other chemiluminescent reactions due to ozone have also been used to determine ozone concentrations. Warren and Babcock (1970) described a portable ozone meter which utilizes the reaction between O_3 and ethylene.

The instrument depends on the reaction that occurs when air and ethylene are drawn into a Pyrex container and mixed directly in front of a Pyrex window. A photomultiplier tube is mounted behind this window which produces an electrical signal due to photons of chemiluminescene energy and is proportional to the total ozone present in the sample. Concentrations as low as 0.001 ppm are detectable.

Absorptiometry

The optical technique consists in measuring the absorption of light by ozone in the ultraviolet part of the spectrum. This method is the standard procedure for total ozone determination in the atmosphere. This method has the advantage that the absorption is specific to ozone. However, a long optical path is necessary for appreciable light extinction to occur expected O_3 concentrations. Also, several absorptions at different wavelengths must be taken since comparison with ozone-free air is clearly impossible and data reduction is somewhat complicated.

Conductimetry

A wet chemical method which uses the oxidizing properties of O_3 can be employed to construct a sensitive automatic meter for continuous sampling of contaminating oxidants in the atmosphere. The ozone containing air is bubbled into a potassium iodide solution and the resulting iodine is determined by measuring the current through the cell. The current is related to ambient O_3 levels by previous calibration with a known ozone source. This technique was employed by Brewer and Milford (1960) to construct an air ozone meter which measures and records instantaneous ozone concentrations.

Steinberger and Goldwater (1972) improved the accuracy of this method by determining the concentration of iodine and hence ozone by titrating the solution with *sodium thiosulfate*. The apparatus (Fig. 19-4) consists of a hermetically sealed glass jar containing 150 cm^3 buffered 10% KI solution and about 0.5 cm^3 sodium thiosulfate of known concentration. Two spiral platinum electrodes dip into the solution and a bias voltage of 30 mV is applied across them. The air above the solution is evacuated whereas the outside is let in through a Tygon tube which is inert to ozone.

When ozone enters the solution, the following reaction takes place:

$$O_3 + 2I^- + H_2O \longrightarrow I_2 + O_2 + 2OH^-$$

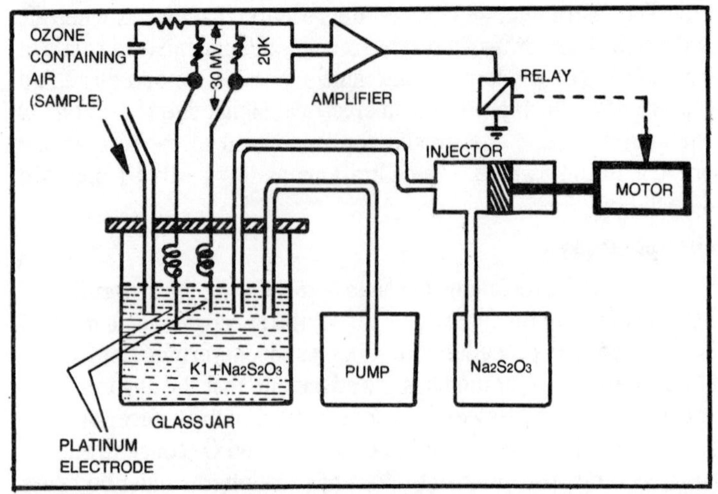

Fig. 19-4. Arrangement for measuring ozone in air using the oxidizing properties of ozone (after Steinberger and Goldwater, 1972).

The iodine then reacts with the thiosulfate:

$$I_2 + 2(S_2O_3)^- \longrightarrow 2(I^-) + S_4O_6^-$$

The reaction continues so long as there is thiosulfate in the solution. When all the thiosulfate has been reacted, free iodine appears and reacts at the electrodes.

The electrical resistance is high so long as there is an excess of thiosulfate. The resistance decreases when it is used up. This change is used to control the operation of the instrument. The voltage drop across 20 K ohm resistance which is in series with the electrode is used to operate the recorder as well as a relay which controls a motorized injector which injects 0.5 cm^3 thiosulfate in each operation. The recorder serves mainly to indicate as to when the injection was made, and thus the average ozone concentration between any two injections can be calculated. Since the pumping rate is known, knowledge of the time intervals gives the total volume of air sampled.

AUTOMATED WET-CHEMICAL AIR ANALYSIS SYSTEM

Wet-Chemical methods are the accepted means for obtaining data in the majority of environmental studies. Specific chemistries are recommended for most of the common pollutants in the atmosphere. Basically, an automated wet-chemical analysis system would be as shown in Fig. 19-5.

In this scheme, the sample is automatically introduced and prepared. Reagents are then added in proper quantities and sequence. A chemical reaction will take place. The presence of a particular constituent is then detected, displayed and recorded.

In any air analysis system, the equipment must be capable of obtaining a proper sample. This is essential in order to obtain correct results. In automated wet-chemical methods, the accuracy of the sampling system is dependent upon maintaining a constant ratio between the amounts of sample ratio and absorbing solution. The rate of flow of the absorbing solution is maintained constant by a proportionating pump.

Sampling System

Figure 19-6 shows, schematically, the principle of operation of a sampling system due to Zehnder and Belew (1970). The air sample control system utilizes the fluid dynamic principle whereby a constant pressure drop is maintained across a fixed orifice thus resulting in a constant rate of flow. In order to avoid dependence upon critically dimensioned orifices, a precision needle valve is used to establish the constant orifice. Once set, the orifice is maintained constant by a 0.45 micron filter which is placed in front of the needle valve. The relatively small flow rates, combined with the large surface area of the filter, afford a long filter life with negligible change of filter pressure drop. This arrangement permits the flow rates to be adjusted between 300 and 800 ml/min depending upon the chemical parameter being analyzed. A carbon

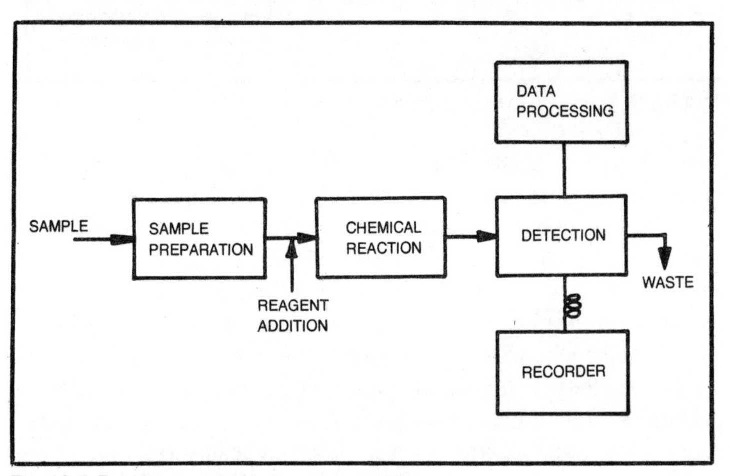

Fig. 19-5. Basic automated air analysis system.

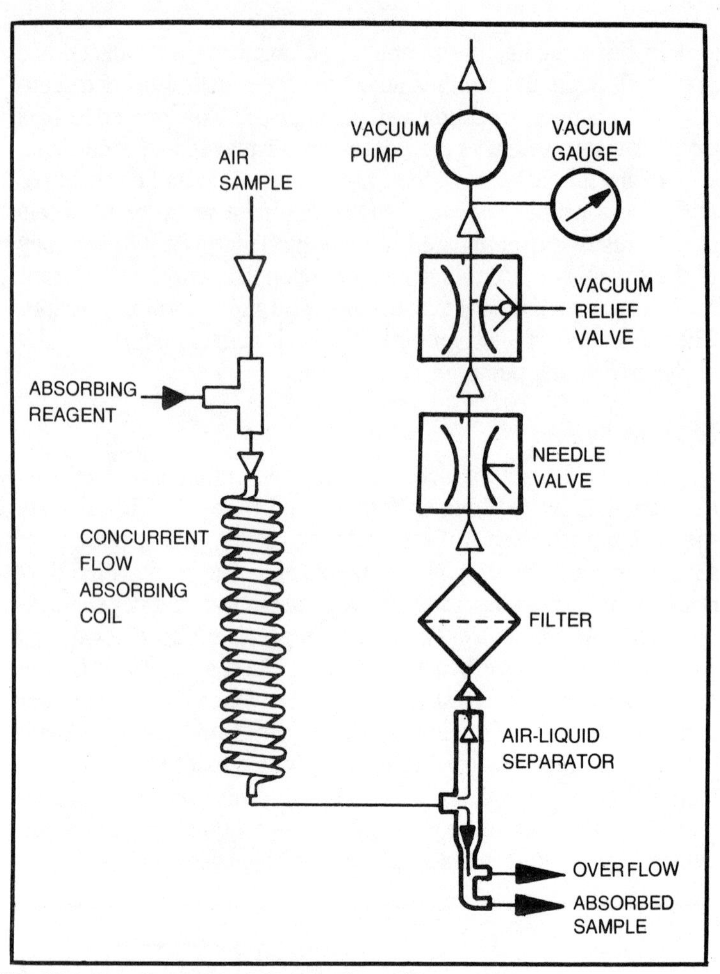

Fig. 19-6. Sampling system (after Zehnder and Belew, 1970).

vane pump equipped with a vacuum relief valve provides the necessary vacuum for air to flow through the system.

Thus, with the pressure losses upstream of the needle valve constant, the orifice area maintained constant by a protective filter, and the pressure vacuum downstream of the orifice held constant by the vacuum relief valve, a constant flow sytem is established. This arrangement provides a constant air flow within better than ± 5% per week of continuous operation. Air is then made to enter a bubbler through a glass fiber plug. A series of samples are thus collected for fixed durations at a predetermined flow rate. Automatic programs control the duration of sampling and the interval between samples, and switch automatically from one collector to the other.

Total Oxidants

Wet-chemical methods have been used consistently by regulatory agencies. They have the advantage over other methodologies in that they are specific, sensitive, accurate and reproducible. The measurement of total oxidants and SO_2 are two perfect examples of how automated wet-chemical analysis fulfills the requirements just outlined.

Figure 19-7 shows the flow-diagram for measurement of total oxidants in air. It also incorporates an ozone generator to enable dynamic calibration and stabilization of the system.

The measurement system makes use of the well known neutral-buffered potassium iodide reagent method. The sample air

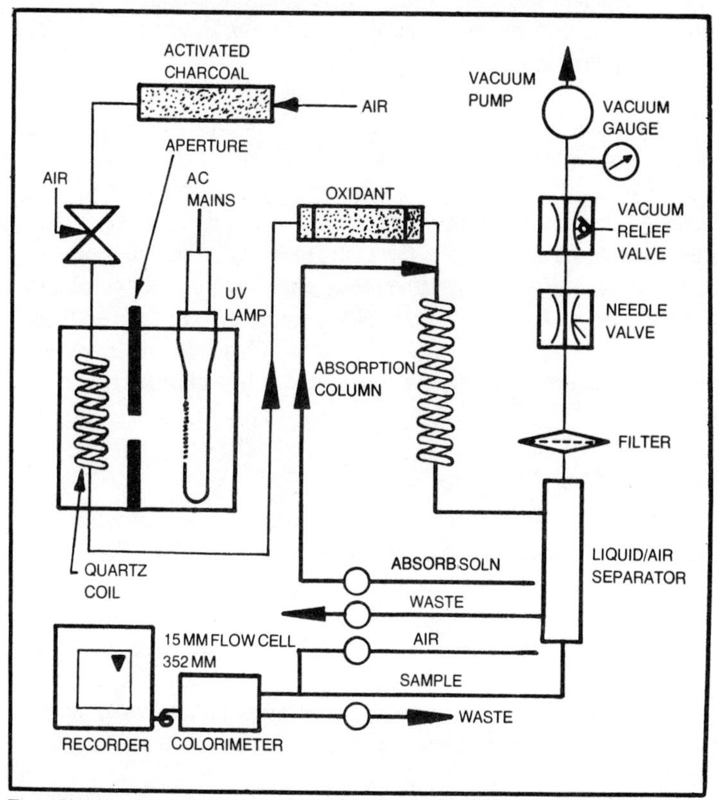

Fig. 19-7. Flow diagram for the measurement of total oxidants in air (technicon Air Monitor IV).

first is purified by means of an activated charcoal filter. Upon leaving the filter, the purified air passes through a quartz helix where part of the oxygen is converted to ozone by a constant current UV source. The sample then passes through a chromium trioxide-firebrick column which eliminates negative interference of SO_2. It then enters the absorption column where the ozone reacts with a solution of 1% potassium iodide. After separating the spent air, the liberated iodine is measured colorimetrically at 352 nm. The photoelectric colorimeter comprises of a dual beam optical assembly, having usual provisions for filters. flow cells and phototubes. Phototubes range from 340 to 900 nm. Interchangeable flow cells are provided for 15 mm and 50 mm.

Ozone in fixed amounts is generated in the quartz helix. The amount generated is dependent upon the intensity of the incident UV radiation and is adjustable by altering the aperture between the quartz helix and the UV lamp. By inserting the proper size apertures, known amounts of ozone, e.g., 0.05 ppm, 0.10 ppm and 0.20 ppm are obtained.

In an alternative method of monitoring total oxidants, the oxidation tube is first removed and the ozone generator aperture adjusted to produce a constant concentration of approximately 0.10 ppm of ozone, to be read as a baseline set at midscale of the recorder. Depending upon either oxidative or reductive conditions prevailing in the ambient air, the instrument will read either above or below the midscale baseline, thus enabling the user to obtain quantitative measurement both of oxidizing as well as reducing substances. This mode of operation is advantageous in areas where, for example, excessive emissions of SO_2 and/or H_2S occur periodically.

Measurement of Sulfur Dioxide

The *West-Gaeke procedure* was described for the estimation of SO_2 in air. The same procedure can be automated (Adelman 1970) by having a scheme as shown in Fig. 19-8.

Here, the specific absorbent, sodium tetrachloromercurate, is used to form a stable, non-volatile complex with sulfur dioxide. Subsequent reaction with acid-bleached prosaniline and formaldehyde gives a sensitive, specific and temperature-independent color reaction. It is measured with a colorimeter at a wavelength of 560 nm.

It is found that the method is not prone to interference by sulfur trioxide, sulfuric acid, ammonia or carbon monoxide which

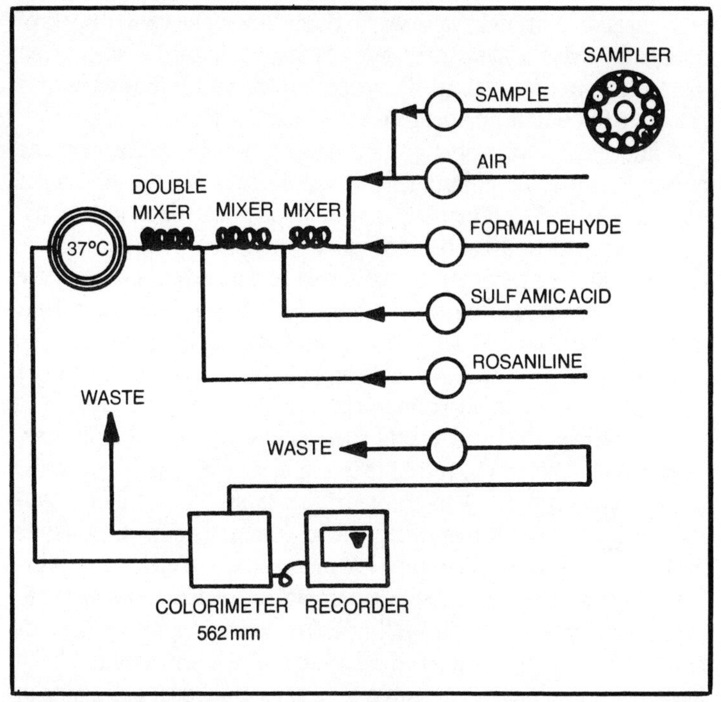

Fig. 19-8. Automated wet analysis system for SO_2 based on West-Gaeke procedure.

can be present in the atmosphere. Nitrogen dioxide, known to interfere at equivalent concentrations to that of SO_2, is eliminated by adding sulfamic acid (wests and Ordoveza, 1962). Where any influence of heavy metals is known to be present, it is removed by complexing with EDTA.

Air Monitor IV of the Technicon provides for the totally automated simultaneous measurement of sulfur dioxide, nitrogen dioxide, oxides of nitrogen, total oxidants, aldehydes and hydrogen sulfide with a built-in feature of flexibility to change parameters as the need arises for a system in a permanent location, as well as in air mobile laboratories.

WATER POLLUTION MONITORING INSTRUMENTS

Pollution of water sources can occur from a very large number of compounds which have varying degrees of potential pollution. Their access to the water resources can adversely affect the quality of water and may render it unsuitable for drinking purposes, industrial use, production of fish and aquatic foods, irrigation and even transport through waterways. Sometimes, the constituents of industrial waste waters may present health hazards even when present in minute amounts. The problem of water pollution is thus one of considerable importance for the society.

The quality of water can be determined by measuring the quantity of specific pollutants present in it. The measurements generally include elemental analyses, physical determinations, microbiological and bacteriological examinations, radiochemical analyses, etc. Water analyses were earlier carried out either at the field site or in the laboratory on samples gathered from the site. It is obvious that taste, smell and visual inspection are inadequate to determine the chemical, physical and biological characteristics of water and, therefore, instrumental methods have been adopted to perform this work. Instruments such as a pH meter, colorimeter, conductivity meter and turbidimeter are routinely used in water analysis laboratories. Modern instruments such as an atomic absorption spectrophotometer, gas chromatograph, thin layer chromatograph and polarograph are increasingly used for detecting trace metals, pesticides and toxic organics. Instruments are preferred over wet chemical methods due to their rapidity of estimates, greater precision and accuracy of measurements. With these instruments, it is possible to estimate several pollutants at microgram and nanogram level.

TYPES OF POLLUTANTS AND TECHNIQUES

Different types of investigations would be required for monitoring pollutants depending on the purpose of the monitoring program. The following standard parameters are determined for assessment of normal pollution characteristics of waste waters: alkalinity, biological oxygen demand, C.O.D., ammonia, nitrates, kjeldahl nitrogen, total phosporous and total dissolved suspended and volatile solids. Several other parameters are monitored in case of industrial effluents and similarly for water used for drinking or domestic purposes.

Conductivity

Conductivity is the measure of dissolved ionized solids in water. Conductivity is temperature dependent. It is non-specific and measures to some degree all ions present in water. A sudden increase in conductivity is an indication of pollution by strong acids, bases or other highly ionized substances. Conductivity is expressed in micromhos per centimeter. For ultra-pure water, the conductivity is 0.055 micromhos per cm at 25°C. A conductivity cell made up of two platinized platinum electrodes can be used for the measurement of electrical conductivity of the sample solution by the null method.

Dissolved Oxygen

Dissolved oxygen is a measure of the ability of water to sustain aquatic life. The solubility of oxygen in water decreases with an increase in temperature. Also, almost all other types of waste consume oxygen from the receiving stream. Dissolved oxygen is expressed in mg per liters (ppm). The dissolved oxygen can be measured by a special sensor kept in an electrochemical cell by the amperometric method. The cell consists of a sensing electrode, a reference electrode and a supporting electrolyte, a semipermeable membrane which serves a dual function. It separates the water sample from the electrolyte and at the same time permits only the dissolved oxygen to diffuse from the water sample through the membrane into the supporting electrolyte. The dissolved gas may subsequently react at the sensing electrode, thus causing a current flow. The current will be proportional to the dissolved oxygen. The range for this measurement is 0-20 ppm with an accuracy of ± 1 ppm. Biological oxygen demand of water could be determined from dissolved oxygen measurements.

pH

pH is the logarithm of active hydrogen ion concentration in moles per liter. The pH of a neutral solution is 7. Natural water shows only slight fluctuations in pH. Rapid changes in pH are generally due to pollution such as dumps of acids and alkalines. The pH of the water sample is measured with the combined glass-calomel electrode immersed in the water sample.

Oxidation-Reduction Potential

Oxidation-reduction potential is an electrochemical potential developed by oxidizing or reducing materials in water. Industrial wastes usually contain strong oxidizing and reducing agents. Oxidants and reductants present in neutral waters tend to balance out to give zero ORP. A platinum electrode is used in conjunction with a calomel electrode for oxidation-reduction potential measurements.

Temperature

Temperature affects the solubility of oxygen and other chemical contents of water. The changes in water are normally slow. They are seasonal unless pollution is experienced. Temperature of water can be measured by using a forward biased silicon diode as a sensor. The diode forms part of a Wheatstone bridge, and the unbalanced voltage is amplified and given to the meter for display. The meter is calibrated in terms of temperature in °C. The range of 0-50°C with a reading accuracy of ± 0.5°C.

Turbidity

Turbidity is a commonly accepted criterion of water quality in water treatment for industrial or potable purpose. It is an expression of optical property of a sample which causes light to be scattered and absorbed rather than transmitted in a straight line through the sample. It is a measure of the undissolved solids in water including silt, clay, algae, rust, bacteria and other microorganisms. Turbidity is expressed in Jackson turbidity units.

Size, shape and refractive index of the particles affect the transparency of water. It is, therefore, not possible to correlate turbidity and quantity of these particles by weight. Instruments for measuring turbidity work on the principle of measuring the intensity of scattered light at an angle from a strong light beam. Two methods are commonly used.

Turbidimetric analysis is based upon measurement of the diminution in power of a collimated beam due to suspension of particles. *Nephelometry* is based upon the measurement of light scattered by a suspension. This is usually done by measuring the scattered radiation at right angles to the collimated beam.

The choice between nephelometric and turbidimetric measurement depends upon the fraction of light scattered. When scattering is extensive, owing to the presence of many particles, the turbidimetric measurement is more satisfactory. If the suspension is less dense, and diminution in power of the incident beam is small, the nephelometric method is the more satisfactory method. In dilute suspensions, the attenuation of a parallel beam of radiation by scattering is given by

$$P = P_0 \, e^{-Jb}$$

where P_0 and P are the power of the beam before and after passing through the length b of the turbid medium. The quantity J is called the turbidity coefficient and is linearly related to the concentration c of the scattering particles. So,

$$\log_{10} \frac{P_0}{P} = kbc$$

where k = 2.3 J/c.

The relationship between $\log_{10} P_0/P$ and c is established with standard samples, the solvent being used as a reference to determine P_0. The resulting calibration curve is used to determine the concentration of samples from turbidimetric measurements.

Turbidimetric measurements are carried out with a filter photometer and a typical optical path is shown in Fig. 19-9. Highly monochromatic light is not required.

The concentration of a variety of ions can be determined by this technique by the use of suitable precipitating agents, so as to form a stable colloidal suspension. Surface active agents such as gelatin are frequently added to the sample to prevent coagulation of the colloid.

Anions and Cations

Pollution of water is also determined to a large extent by the presence of *anions* such as halides, nitrate, sulfate, cyanide, car-

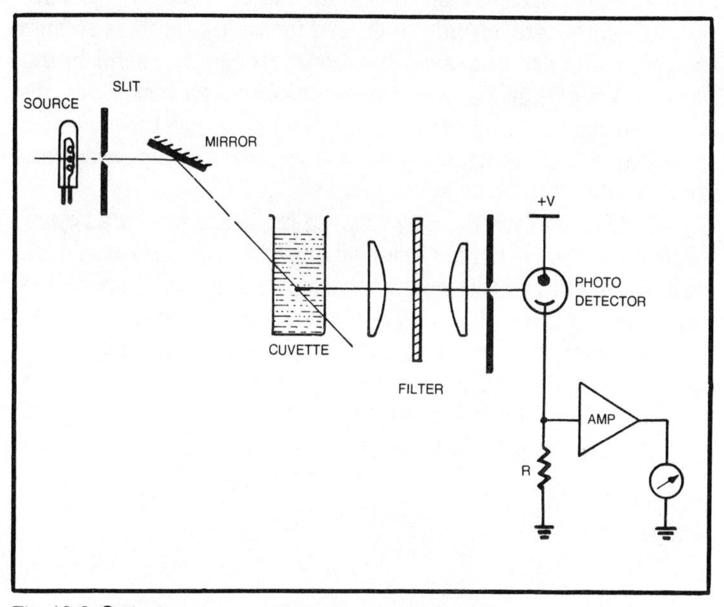

Fig. 19-9. Optical arrangement for measurement turbidity of water.

Table 19-2. Performance Requirements for a Typical Automatic Water Pollution Monitoring Instrument.

Parameter	Transducer	Range	Full Scale Accuracy
Conductivity	Potentiometric	0-60,000 mhos	1%
pH	Glass and calomel electrode	2-12	1%
Dissolved oxygen	Polarographic	0-24 mg/1	1%
Temperature	Silicon diode Thermistor or Thermocouple	0-50°C	0.5°C
Chloride	Ion-selective electrode	0-240 mg/1 0-2400 mg/1	5%
Turbidity	Optical	0-120 JTu 0-1200 JTu 0-2400 JTu	2%

bonate, etc. and *cations* such as lead, zinc, chromium, arsenic, copper ,etc. Inorganic anions are generally estimated by spectrophotometric and electrochemical techniques. With the availability of ion-selective electrodes, the task of estimating both anions and cations has become easy and, at the same time, accurate and reliable. Ion-selective electrodes for anions such as fluoride, chloride, carbonate cyanide, etc., and for cations such as sodium and potassium are now available. Polarography is useful in the analysis of metal ions in water because of its high sensitivity, its ability to analyze mixtures and to tolerate large quantities of dissolved solids. Bacteriological and radioactive measurements are limited only to very special cases.

Table 19-2 gives the performance requirements for a typical automatic water pollution monitoring instrument. The instrument must have automatic temperature compensation and should be stable at least up to two weeks. It is not always possible to fix up the water sample from the site thoroughly and expect the results not to vary in the laboratory measurements. Therefore, portable instruments have been designed which can be conveniently taken to site and the results obtained almost immediately.

Chapter 20

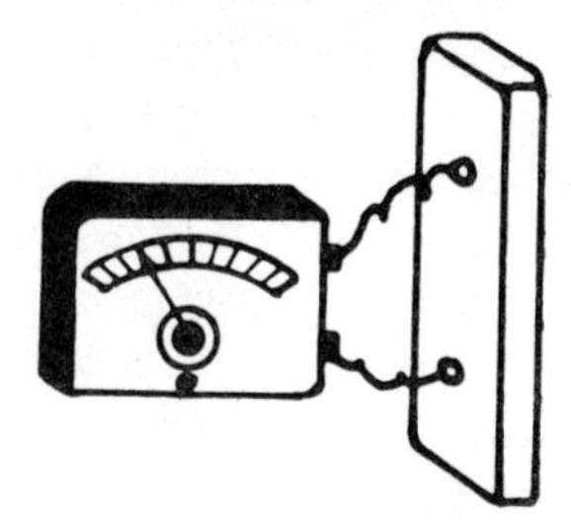

Computer Directed Analyzers

It is well recognized that computers have much to offer in the analytical field both in routine applications as well as in research. This can be in the way of automatic calculation of the calibration curve, automatic drift correction, processing the data from the interfaced instruments into directly readable signals and increased productivity. The primary benefit, however, is in reduction of time and labor of the technician which is usually required for conversion of raw results to proper functional scales. This reduces the associated errors which inadvertently creep in with manual methods of computation.

A computer is a combination of devices which serves to make mathematical calculations and thus find solutions to various types of problems which are otherwise difficult and time consuming. Computers may depend for their function upon mechanical devices, electrical devices or electronic devices and circuits. Electronic computers are the ones which are more versatile and find multiple applications in the analytical field. Therefore, only this type shall be considered in this chapter. Electronic computers utilize a number of electronic devices and circuits for their action. There are two main classes of electronic computers, analog and digital. In general, digital computers have a much higher degree of accuracy than the analog computers.

ANALOG COMPUTERS

Analog computers may be used to perform basic mathematical operations such as to add, subtract, multiply, divide, integrate and differentiate. They are most useful for solving problems involving differential equations.

Problems are solved on the analog computers by simulating the magnitude of the physical quantities in terms of analogous voltages, hence the name analog computer. The physical variables represented by the voltages are added, subtracted, multiplied, and so forth, through the various components of the computer. The heart of the analog computer is the operational amplifier. Several operational amplifiers are required to construct an analog computer. They are fabricated in the integrated form and are used just like other passive components. Interconnections of various identical circuit components are done by using plug-ended cords in the proper jacks.

Inputs and outputs of analog computers are generally in the form of voltages which vary with time. Performance curves are plotted for variables which can be read by special devices to convert into corresponding voltage variations. In a similar way the output is in the form of a curve plotted by a computer controlled recorder. This output curve represents the solution to the problem.

Analog computers are used to study systems or equations for which solutions by standard computing methods are difficult and for systems where non-linearity plays an important role.

DIGITAL COMPUTERS

A *digital computer*, as the name implies is based for its operations on the handling of digits and numbers. The operation is basically arithmetic and covers addition, subtraction, multiplication and division. Digital computers also incorporate additional facilities to perform calculus type of solutions.

All of us are accustomed to work with the decimal system of dealing with numbers. But in digital computers, the system used for solutions of problems is the binary system, which means that we take 2 as the base instead of 10 in the decimal system. Any number when expressed in the binary system contains a series of only ones and zeros. Representation of numbers in this way very much facilitates the transmission and storage of numbers by two simple signals for one and zero. Such signals may merely be positive pulses transmitted at regular intervals. The occurrence of the pulses at any time may mean a one and its absence at another time interval a zero. These time generated pulses are employed to perform the desired computing operations.

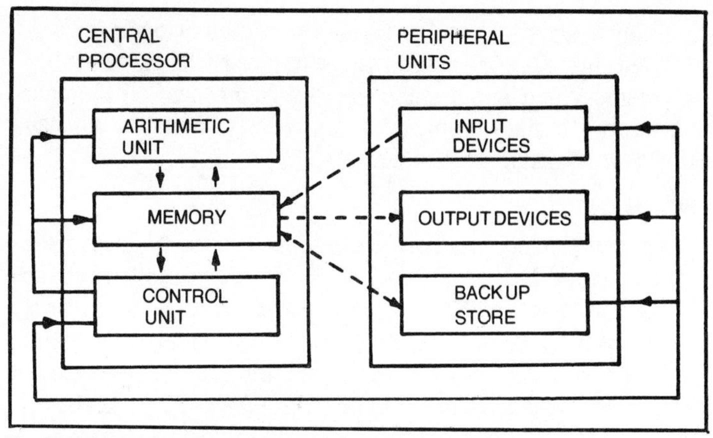

Fig. 20-1. Block diagram of a digital computer.

The block diagram (Fig. 20-1) shows a general picture of the components and functions of a typical digital computer. It consists of two major parts, namely the *central processor* and *peripheral* units. The central processor includes: internal storage or memory, arithmetic unit and control unit. The peripheral unit includes: input devices, output devices and auxiliary memory or backup store.

For obtaining a solution of any problem, all essential information is fed into the input. This information consists of the numerical data involved in the problem and the coded instructions of what should be done with the data. This combined data are passed on to the memory unit in the central processor, where each part of the information is stored in separate locations in the form of electrostatic or magnetic storage.

The heart of the digital computer is the memory unit. It stores the information received from the input and from other functional parts and delivers this information at an appropriate time during the computing process. The control unit gets instructions from numerical codes which are stored in the memory and interprets these codes in the appropriate sequence and directs the operations of the entire machine. The function of the control unit can be illustrated by considering a simple problem of summation. The control unit shall send a signal to the memory unit to pass the stored numbers to the arithmetic unit where they are temporarily stored in the shift registers. As a next step, it signals the arithmetic unit to add these numbers and send the output back to the memory unit from where it is passed to the output under instructions from the control unit. In practice, however, the problems are not as simple and they involve a series of arithmetic operations to be carried out before the results can be passed to the output. If the usual memory unit is not sufficient to accommodate all the data involved in a certain problem, it can be supplemented with an auxiliary or a backup storage unit.

Input Devices

The function of an *input device* is to present the basic symbols 0, 1, 2, . . . 9; A, B, C . . . Z; +, −, ÷, %, ×; etc. in a coded form so that they can be held within the computer's internal storage. This is achieved by using devices capable of reading punched cards, paper tape or magnetic tape which carry information in a coded form as intermediate media. It may also be done by using a keyboard similar to that of a typewriter.

Punched cards have been used as a means of input from the earliest days of computers. A punched card carries over it, in the printed form, a number of vertical columns and a number of horizontal punched positions. The data is recorded in the card by punching holes in it which may be read by a photosensitive system.

Punched paper tape is a continuous strip of paper on which the data is recorded by punching holes to represent a character across the tape. Paper tape can be prepared by several methods. However, the most commonly employed method is either by a manually operated keyboard device or an on line computer controlled paper tape puncher. A paper tape reader consists of a number of photoelectric cells and corresponding exciter lamps.

Specially designed typewriters (teletypewriters) can also be used to input data to a computer. By depressing keys on the typewriter keyboard, a character code is generated for each character and this code is directly fed to the computer's internal store. This avoids the use of intermediate coded media.

In the case of large digital computers, the reading of data from punched cards and manual keyboards is usually too slow and represents an uneconomical use of the computer. The computer can be kept fully occupied through the use of auxiliary equipment which records the input information from keyboards or punched cards, etc., on magnetic tape. When all necessary information is recorded in this manner, it may be transferred to the computer memory at a very high speed.

Output Devices

The solution of a problem by a digital computer is ultimately transferred to the output section. The output may be on the punched cards, paper tape or magnetic tape. The solution of the problem is usually in binary form and hence requires conversion into decimal system. This can be effected by auxiliary electronic equipment.

The output can be in the form of a display on the cathode ray tube. A graph plotter incorporated in the system can also be operated with programmed instructions so that graphs can be drawn from a pattern of discrete digits held in internal storage. The output data may also be obtained in the printed form from on line printers.

Internal Storage or Memory Unit

Internal storage is the nerve center of a digital computer. Its function is to store a huge number of bits at a high speed. Besides, it must be capable of returning this stored information to other parts of the computer again at a high speed.

The system most usually employed for storage in the memory is the magnetic storage, which can be done in two ways: storage on ferromagnetic surfaces (e.g. magnetic tapes and drums) and static storage in ferrite magnetic cores. Most of the earlier commercial computers used a magnetic drum for internal storage. Storage on magnetic drums usually proves too slow to match with the very high speed operations of modern computers. So they are not used as the sole means of internal storage.

The storage of information on the magnetic tape follows the same principle as the recording of sound on tape. The tape has a large capacity of storage but only about 50% of the theoretical capacity can be realized in practice. This is because some space is to be left in the channels to allow for starting and stopping of reels for obtaining access to specific locations of storage. Also, space is used for coding and instructions. Obviously, the access time to reach information stores on tape is relatively large. Therefore, magnetic tapes are not preferred for memory units.

An alternative arrangement of internal storage is core storage. This is called static magnetic storage and can be produced in a matrix of wires with a toroid or core of ferrite at each intersection of the wires. Figure 20-2 shows the formation and circuits of such a matrix. Ferrite has the property of retaining magnetism. Data is stored by magnetizing a core by passing an electric current in a particular direction. A current applied in a particular direction represents one and a current passed in the opposite direction represents zero. When the current is removed, the core retains a certain amount of magnetism. The ferrite cores used in practice are about 2 mm in diameter and the windings which carry current consist of single wires threaded through them. A binary digit is stored in each tiny core.

To read the binary digit stored in a particular core (A), currents of one half magnitude are passed simultaneously through the horizontal wire (X) and vertical wire (Y). If these current forces are additive and in the direction to reverse the flux, an output pulse will be induced in the coil. No output signal shall arise from any other core except (A) because only one half current is made to flow.

This type of storage is used in most modern computers and is also known as *high speed memory* or *immediate access store*. This is because there is no delay as in a magnetic drum or tapes in accessing the contents of a required location in store upon the provision of an address. The size of the core store varies in size from 200 address locations to 0.5 million locations. The effective size of a computer's store will largely be determined by cost and the ability of an instruction to address all the locations.

Arithmetic Unit

The *arithmetic* or *computing unit* in the central processor performs arithmetic operations on data transferred from the memory unit. It also performs logical and data handling operations of shifting and transferring data. It also has the facility for temporarily storing the data in registers.

All these functions are performed by the arithmetic unit by employing a large number of simple electronic circuits. The basic circuits used for logical operations called gates are AND circuits, OR circuits and inverting circuits. For temporary storage of signals, the flip-flop circuit is most widely used.

Programming

Programming is a term which may be considered to consist of a series of instructions so organized that a required operation is performed by the computer. The basic stages in preparing a program are problem analysis, outline and detailed flow charting, coding, input preparation, testing and programming languages. This sequence of steps in carrying out the solution of the problem must be prepared and converted to numerical codes for storage in memories. When all this information has been fed from the input into memory, the computer is ready for a run.

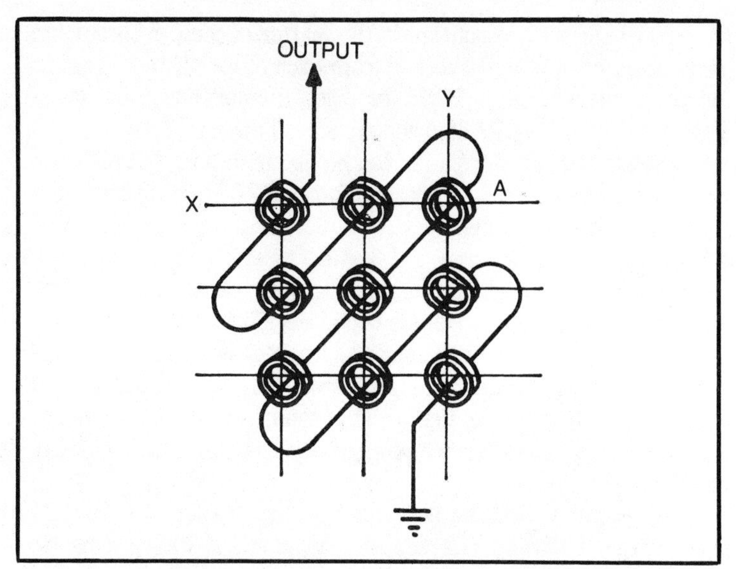

Fig. 20-2. Ferrite core matrix used in a digital computer for internal storage.

MICROPROCESSORS

Microprocessor is a high density semiconductor chip incorporating arithmetic logic and control units. Together with memory and peripheral circuitry, microprocessor chips form complete microcomputers. In complexity, they fall somewhere between conventional mini-computers and small hand-held calculators. They are compact and inexpensive as calculators but, like minicomputers, can be programmed for a wide range of tasks and work with similar peripheral computer devices.

The high density chip needed for microprocessors has been obtained by the use of MOS technology. At first PMOS was employed, but for increased speed the manufacturers turned to NMOS devices. More recently CMOS (Complimentary-MOS) chips consuming very low power have appeared. The latter form of MOS combines P and N channel transistors and features lower dissipation than either PMOS or NMOS devices.

Regardless of the technology used, microprocessor systems (Fig. 20-3) are organized in the same way as conventional computer systems. The major blocks are a central processing unit (CPU), memory and input/output (I/O) facilities. In their simplest form, each of these blocks can be a single chip.

Microprocessors are used extensively for dedicated control systems for performing a limited set of operations. They differ from processors of computers in the matter of their program and data, being completely divorced from each other. The program is located in one memory, a ROM (read only memory), and the data in a second memory, a RAM (random access memory).

From a ROM, information can be obtained, but that information cannot be altered during operation. A PROM (Programmable ROM) provides the same function, but internal bit patterns can be set by the user rather than the manufacturer. Data on the other hand reside in RAM. This kind of memory allows information to be written and modified as well as read. Because some scratch pad memory is usually needed, most microprocessors include some temporary storage called *index registers*. The microprocessor also requires output ports through which it communicates its results to a display or peripheral device, or provides control signals that may direct another system.

Input/output circuitry is available either as a separate IC or as a part of the ROM and RAM chips. Four, 8 and 16 bit systems are available with cycle times reduced from 20 microseconds for PMOS to 2 μs for NMOS devices. Because of their low cost and high performance, microprocessors can even be used for reasonably complex logic systems.

Microprocessors have been incorporated into the instruments to give full digital control, as is amply evident from modern automated clinical analytical and research equipment. For example, automatic liquid scintillation counters are available in which hundreds of glass vials or tubes are handled, monitored and positioned electronically. No user adjustable amplifier or discriminator controls are provided. The counting parameters are set depending on the sample's degree of quench, the stored standard pulse height spectrum data, and the detected activity and the disintegrations per minute occurring within the sample. The required data is printed out automatically.

With such microprocessor based systems all that is required of a user is that the samples are loaded, the program selected and the instrument started. The instrument would perform all measuring and computing functions.

OFF-LINE/ON-LINE COMPUTERS

An *off-line computer* accepts the experimental data from input devices such as magnetic tape, cards, punched paper tapes or typewriter. It does not take data directly from the analytical instrument; neither does it process in real time. On the other hand,

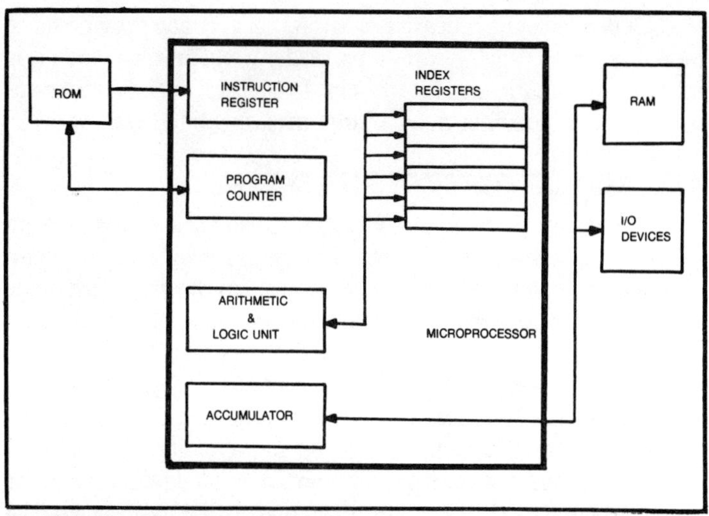

Fig. 20-3. Microprocessor system.

an on-line computer takes the data directly from the instruments or process and analyzes it to take decisions based on it. It is thus obvious that with the off-line computer, only the past history of an experiment can be studied, whereas with the on-line computer it is possible to determine results in sufficient time to investigate new and unusual occurrences during the course of an experiment.

ROLE OF DEDICATED COMPUTERS IN THE LABORATORY

The true small, general purpose computer dedicated to a specific function in a laboratory has been largely responsible for a revolution in the methodology, economy, quantity and quality of experiments performed and data analyzed. In combination with an analytical instrument, the dedicated small computer becomes an element in a total analytical instrumentation system and is capable of acquiring and analyzing data, and controlling the instrument and experiment based on the data received.

The development of on-line dedicated computers permits considerably more sophisticated instrumentation systems that were earlier not possible.

Laboratory systems built around the dedicated small computers are already in a variety of disciplines such as mass spectroscopy, gas chromatography, nuclear magnetic resonance and clinical chemistry. Programmed data processing systems and software systems are available as black box devices for a variety of applications.

Dedicated computers are of low initial cost and provide much more computer capability for a particular problem than when it is interfaced to a time-sharing system. In addition, a dedicated computer system does not need a large programming staff.

A TYPICAL COMPUTER CONTROLLED SYSTEM

The use of a computer to manipulate and organize data from analytical systems is well known, but the great advantage of computer control over systems is well known, but the great advantage of computer control over all the elements of the analytical process is only now becoming apparent. Such systems provide access to an almost infinite variety of procedures that can be rapidly called up and set up with minimal error. Multi-reagent procedures, any combination of time delay, sample handling, complex statistical calculation schemes and quality control procedures can all be accommodated by the computer that controls the entire analytical process.

In a typical computer controlled system, the system would consist of the following modules (Fig. 20-4):

- Sample handling module.
- Dispensers.
- Spectrophotometric unit.
- Printer unit for results.
- Minicomputer, programmed by a magnetic card which integrates the above modules into an analytical system according to the needs of the specific procedure and requirements.

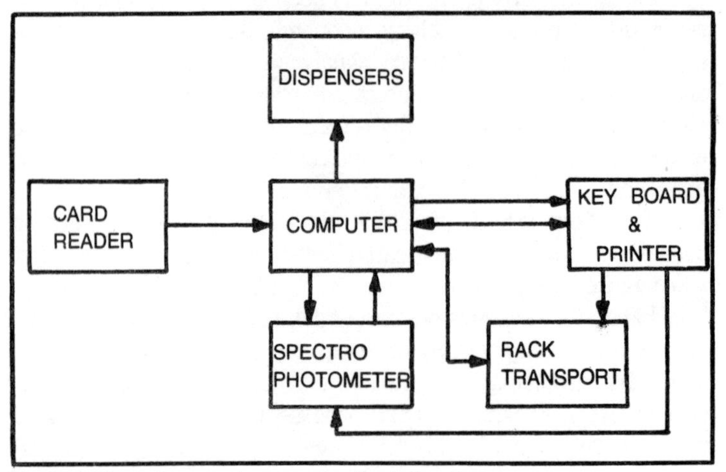

Fig. 20-4. Block diagram of a computer directed analyzer.

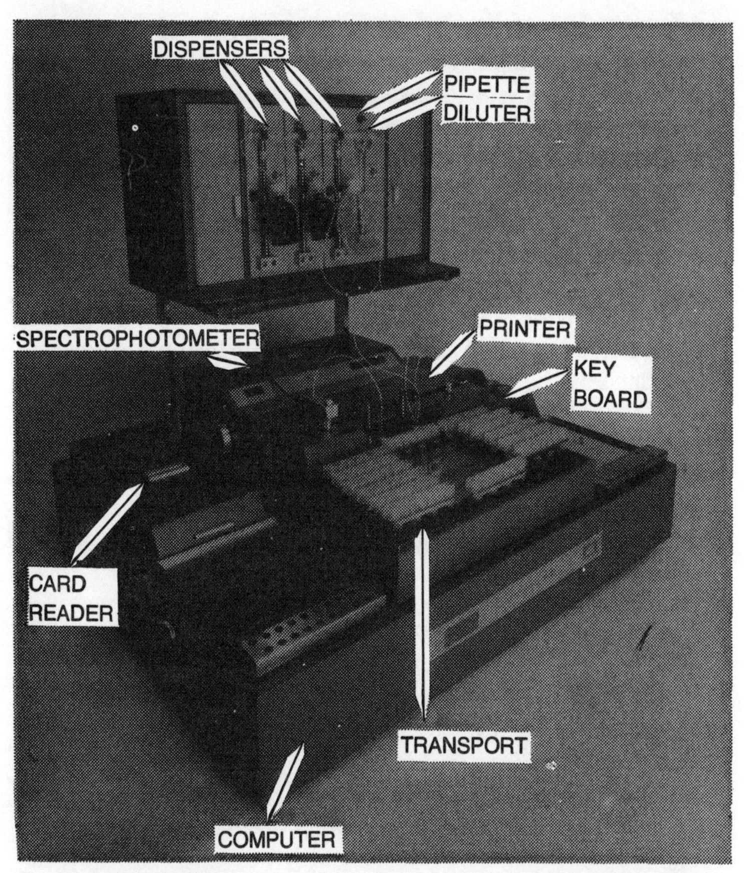

Fig. 20-5. System 3500 Computer Directed Analyzer (courtesy of Gilford, USA).

The movement of test samples from pickup to reagents dispensers and to the spectrophotometer, the final analysis of the data by an appropriate computation scheme, and finally the printout of results with positive identification are all under control of the computer.

Figure 20-5 shows the system 3500 computer directed analyzer from Gilford, USA. This is an automated analytical system which uses preprogrammed magnetic cards to program its functions. By selecting and inserting the desired magnetic card into the card reader, the system can be programmed for a variety of different procedures. The system can test up to 150 different samples per hour, depending upon the test being performed. Supplied with the system is a selection of magnetic cards to program the system to perform routine chemistries. Reagent preparation, setup procedures, instrument settings, expected values, error prevention messages and sample tape printout are included in the test procedures following each magnetic card.

When the program card is inserted into the card reader, the magnetic tape head in the card reader picks up the recorded data from the program card. The information contained on the program card is then stored in the memory of the computer. When the system is operated, the information is recalled from the memory and directs other components of the system to perform functions such as dispense, sample and transport. Calculation processes concerning the reaction are also recalled from the memory and are performed by the computer.

The spectrophotometer measures the absorbance of a sample in the spectral range of 340 mμ to 700 mμ. The sample takeup volume is of the order of 700μl. Temperature regulation circuitry for the thermocuvette is an integral part of the spectrophotometer. Peltier effect thermoelectric devices in the thermocuvette provide both heating and cooling.

The dispensers deliver reagent from an attached reservoir. They have a 2.5 ml full stroke volume and can dispense a range of different volumes.

The pipette-diluter is a precision device used for both sample pickup and sample dispense. It will pick up a selected amount of sample from a sample cup and dispense the sample along with the proper amount of reagent into a reaction cup. Like the dispensers, the pipette-diluter uses volume stops to determine the intake volume.

The card reader uses a magnetic read head. The card reader reads the pre-programmed magnetic card and programs the computer to perform the operation recorded on the card.

The computer controls the operation of the printer. The printer is an alphanumeric printer and employs a thermal print head.

The keyboard is used for entering standard values or other information a program may call for. Data from the keyboard is manually entered when called for by the preamble.

COMPUTERIZED DATA PROCESSING SYSTEM FOR ACQUISITION AND PROCESSING OF MEASURED DATA IN THE HOSPITAL LABORATORY

The operation of a laboratory computer system can be divided into two methods: operation of the *commercial process laboratory* and operation of the *hospital laboratory*. The essential difference in these two systems is the patient folder. In the hospital laboratory, the patient folder is critical to the organization and processing of multiple specimens relating to the same patient over extended periods of time. In the commercial process laboratory, quality control and handling of a more or less standard set of tests on a given sample are essential. In the commercial system, organization is centered around the specimen in the hospital system on the patient. In fact, one may consider the commercial laboratory to be a hospital laboratory system without a patient oriented record file.

The steadily increasing work load on clinical laboratories has encouraged the use of a substantial amount of automatic analytical equipment. A natural development has been the coupling of this equipment to a digital computer system. A well-designed arrangement will considerably relieve the staff of the tedium of measuring large numbers of recorder traces. At the same time, it would serve as the focal point for coordinating the flow of work through the laboratory.

Failure to associate results with the patient from whom the specimen is obtained has been the most serious problem confronting the pathology laboratory. The incidence of errors is indeed alarming. Introduction of computer systems has eliminated errors in the transcription and reporting stage and assured some degree of validation of identification data.

In a biochemical laboratory, the process for automated analysis would need the following information:

- Identification of the patient.
- Information about the patient.
- Information about the specimen.
- Work or test requested.
- Destination for the report.

In addition to this, the following is a list of information required for which a maximum number of characters must be defined:

Hospital Registration No.	For locating records and patient identification.
Surname, Initials	To remind the lab staff that they are dealing with human beings.
(a) Date of Birth *(b) Sex*	For interpretation of results and for statistical purposes.

Ward	Destination of the report.
Consultant	For determining workload and as a matter of courtesy.
Diagnosis	Provisional diagnosis for special handling of the specimen.
Urgency	For preference of emergency cases.
Laboratory Accession Number	For maintaining a link throughout the analytical process.
Date and Time of Collection.	
Test Code	To reduce the amount of data which has to be punched and inputted to the computer.

For feeding this information, the only real solution to the problem is the production of machine readable request forms which can be achieved with the aid of punched cards. On admission of a patient, a stack of, say, 10 cards is punched with the patient's identification data and placed in the patient's folder. The identification data can be later transferred to paper tape, one card at a time, and additional data added to the tape by manual key entry through the general purpose keyboard terminal.

In addition to punched cards and punched tape, data can also be entered in most systems through terminals. Types of terminals vary from special purpose entry stations to general purpose terminals such as a teletypewriter or a cathode ray tube. The terminal is also a station at which a program may be called.

In the majority of analytical methods, colorimeters or spectrophotometers are used to measure concentration of the substance of interest. As each processed specimen or standard passes through the detector, a peak is produced on the chart depending on whether the color is developed or removed by the reaction. Ideally a square wave is produced. Each peak is identified sequentially with its sample, having associated calibration peaks with the appropriate standards by the known pattern. Concentrations are determined by comparison of the sample peaks with the calibration and drift standards.

The input of data from the colorimeter or spectrophotometer, peak detection and interpolation of unknowns against standards, involves relatively trivial programming and negligible computer time, though the equivalent technician's effort is very substantial. For the majority of tests, signal levels in the range of one millivolt to 10 millivolts are obtained. For converting this data into digital form, it is required to be amplified at least 1000 times. The analog-to-digital converter converts an analog signal in the range 0-10 volts to a 12 bit binary equivalent. This provides 4096 numbers and a resolution, therefore, of 10 volts divided by 4095. An accuracy of 0.025% is thus available. Before any analog signal is given to the A-D converter, the required signal may be selected using a six bit word transmitted to the unit from the computer by the multiplexer, which is decoded to select any one of the 64 switches. On feeding the data to the computer, it carries out calculations on peak detection, validation tests and suggests dilutions if necessary as per the set program. Though the basic program may be the same for each test, it is possible to accommodate the wide variation in shape and quality of peaks by the suitable choice of parameters for each test, which are stored in the computer. Computers carry out calculations of concentrations from the peaks by incorporating drift correction and interpolation.

Maintaining laboratory records is sometimes a legal requirement. Good record keeping also helps in determining workloads and provides research data. Updating records by manual systems involves an enormous amount of effort, high copying cost and considerably valuable laboratory space. Even then records could be misfiled and transcription errors may creep in. Computers can effect a major contribution in this area. It is possible to store information in highly compact form on magnetic tape. Commercial systems are usually able to hold 800 alphanumeric characters per inch of tape and one reel may possibly be able to store a whole year's records. Even higher packing densities are possible but at a considerably increased cost and increased risk of error due to dust. Also, having stored all this information in numerical form on magnetic tape, it is really quite easy to carry out fairly advanced statistical research.

The end product of the laboratory is a report form, for which continuous fanfolded preprinted stationery is used. The report includes a patient identification section which prints out information such as registration number, surname, initials, date of birth, sex, consultant and destination. The standard test section gives the collection date of the specimen, laboratory number of the specimen and time of collection of the specimen. Test results are printed to a predefined precision. In the event of a result being outside the range of the analyzer and where no dilution is permitted, the result is printed as > or < a concentration. Reliable means and standard deviations as a function of age and sex have been established for the most common tests. If they are included in the program, it is possible to establish the degree of abnormality automatically. In addition to printing a report on a terminal, many systems provide line printer output of high-volume reports such as information summaries. Most terminals print at a rate of less than 30 characters per second; line printers typically print at speeds of greater than 300 lines per minute.

In the hospital laboratory, the blood and urine of the patient are examined with respect to chemical compositions and biological structures. Such laboratories are today provided with a large number of fully automatic, semi-automatic and also manually operated work places at which the analyses are carried. To illustrate, Fig. 20-6 shows the schematic of a system for acquisition and processing of measured data in the hospital laboratory.

The examination request card is filled out by the physician. This card is in directly legible form and is punched into the cap of the master examination vessel via a teleprinter in the ward. At the patient's bed,the blood after removal is put in the special examination vessel which is closed with the cap already prepared. It is then sent to the laboratory.

In the sample distributor, the specimen in the master vessel is first divided up among several individual examination vessels in accordance with the type of investigations to be carried out, which can be machine read on the cap of the master vessel. The code numbers of the patient and of the desired individual examination are punched into the edge of the individual examination vessel. The samples are then distributed to the different analysis stations. From the analysis stations, the measured values together with patient code numbers are passed on to the computer. The measured data constantly being fed into the computer are printed out, patient correlated and sent to the ward as a complete laboratory findings report.

CHROMATOGRAPHIC DATA PROCESSOR

Gas and liquid chromatography provides fast and convenient means for analyzing the chemical components of complicated mixtures. However, identifying and quantifying the raw chromatographic information obtained on a conventional strip chart recorder

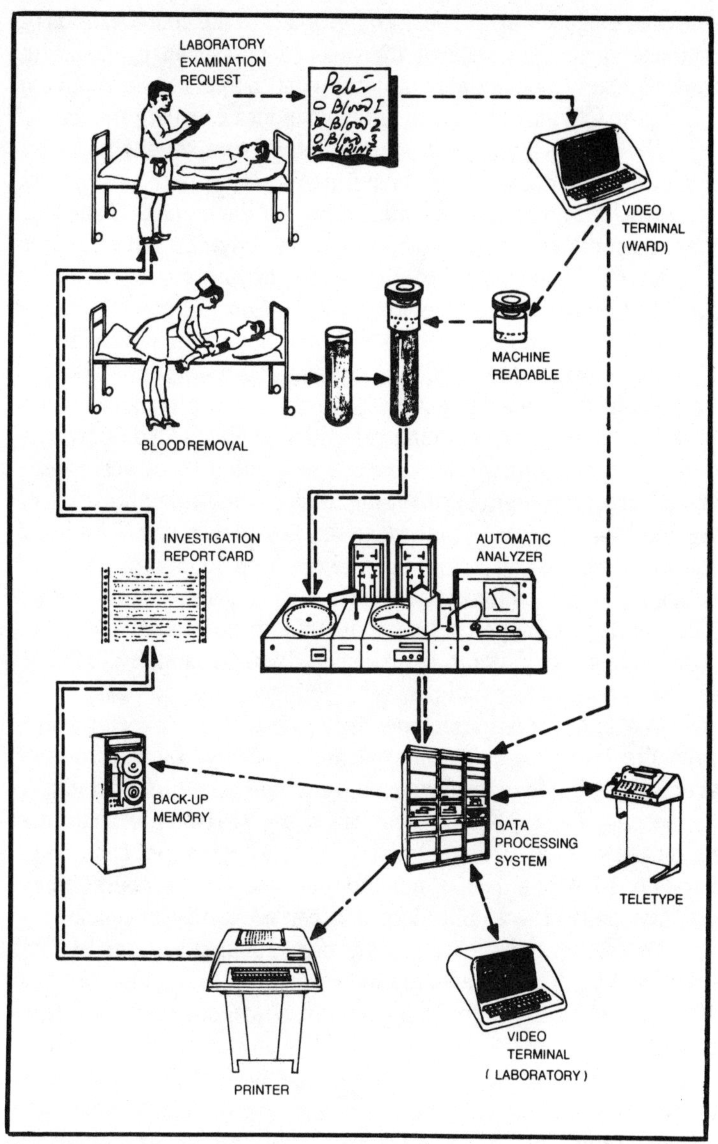

Fig. 20-6. Schematic of a system for acquisition and processing of measured data in a hospital laboratory.

requires a major effort. A typical chromatogram would consist of a series of peaks separated in the time domain, each peak corresponding to a chemical component detected. The time of occurrence of each peak corresponds to the travel time through the column and can be used to identify the corresponding chemical component.

The area enclosed by the peak corresponds to the concentration of that chemical. The area of the peak is measured by counting squares, using a planimeter or cutting out the peaks and weighing the paper. Obviously, a lot of effort goes into reducing the data.

The development of electronic integrators speeded up data reduction by automatically computing the areas under the peaks and print the areas and retention time for each peak. Basically, these integrators are voltage-to-frequency converters that monitor the output of the chromatograph detector and drive a counter activated by rather complex peak-recognition logic.

The next development was to derive final results with the aid of a computer working directly from an analog-to-digital converter. This could be possible only by time-sharing the computer with several chromatographs doing repetitive analyses. To overcome these problems, Hewlett Packard developed a large scale integrated circuit for digital processing, thus enabling simplification of the hardware design. This processor provided means of adding automatic calibration so that the integrator could identify the peak belonging to the calibrating signal and then scale results. It also incorporates means for reducing the effects of detector noise and for selecting the optimum slope sensitivity automatically so that it can be sensitive to small peaks while ignoring noise peaks.

The data processor includes analog-to-digital converter which uses the integrating digital voltmeter circuits to measure the average amplitude of the chromatograph detector output five times per second. The dual-slope technique is used to convert the detector output voltage to digital form. The voltmeter output consists of bursts of 10 MHz pulses, the number of pulses in each burst being proportional to the amplitude of the corresponding sample.

To smooth noisy chromatograms, a running average of consecutive samples is calculated by a weighted averaging method. This smoothes the high frequency noise without distorting true peaks.

The integrators universally select peaks on the basis of slope. If the slope threshold is set too low, noise on the baseline can trigger integration. If it is set too high, integration starts high on the peak and a significant part of the peak is lost. The digital processor can be set to start integration at the slightest hint of a peak but it discards the count if the peak presence is not confined. Thus, even with the slope threshold set high, total peak area is integrated. This is done as follows. The digital processor measures slope by continuously comparing each new averaged value to the previous value. If the difference is positive and exceeds a certain minimum for several successive samples, the processor judges that a peak is being detected. If the sample-to-sample comparison indicates that the slope reverses before the threshold criteria is reached, then it is assumed that a noise peak had been encountered and the total count is discarded. Once the processor has made the decision that a peak is being detected, counts continue to be totaled until the sample-to-sample difference indicates that the detector output has returned to the baseline. The processor memory is capable of holding counts obtained from 54 peaks in any one chromatograph run. The processor compares the beginning and end of each peak to detect baseline drift. It then adjusts the readings to account for drift, if present.

A difficult problem for designing integrators is finding the true areas of peaks that overlap or merge on the chromatogram. Two cases of merged peaks are shown in Fig. 20-7. The processor judges when the sample to sample comparison indicates that the slope of the chromatogram changes sense before it reaches the baseline. The accumulated count is stored up to that point and starts a new count. If the trace returns to the baseline on the next downslope, the two counts obtained are stored as the area counts for the two peaks. This is called the *dropline* method of merged peak separation.

When a small peak rides on the tail of a larger peak, the method known as the *tangent skim* is preferred. Here, the end of

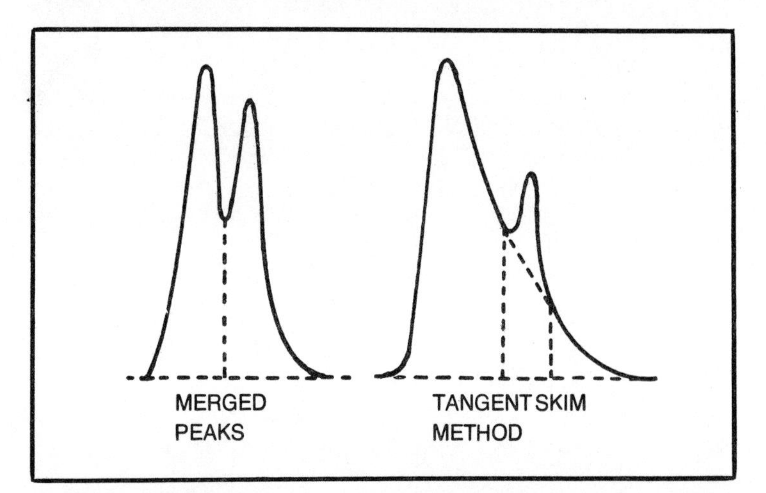

Fig. 20-7. Areas of merged peaks (left) are separated by a line dropped from the valley to the baseline. The area of a small peak riding on a tailing peak (right) is computed using the tangent line as a baseline (after Poole and Bilen, 1976).

the second peak is determined by continuously calculating the slope of a line drawn from the start of the second peak to the latest sample, and comparing the line's slope to the slope of the chromatograph curve. When the two slopes coincide, the end of the peak is indicated. The processor then calculates the area.

The parameters are entered in the processor through a calculator type keyboard. Some parameters may even be entered through slide switches. In fact, through the keyboard, all aspects of the analysis are controlled, for example, the column oven temperature program, the temperature of other heated zones, the integration parameters, the calibration and the type of computation. Operation of backflush valves, a change in recorder speed, a change in detector and other parameters can be programmed to occur at specific times following the start of a run. Poole and Bilen (1976) describe the details of the process.

Appendix

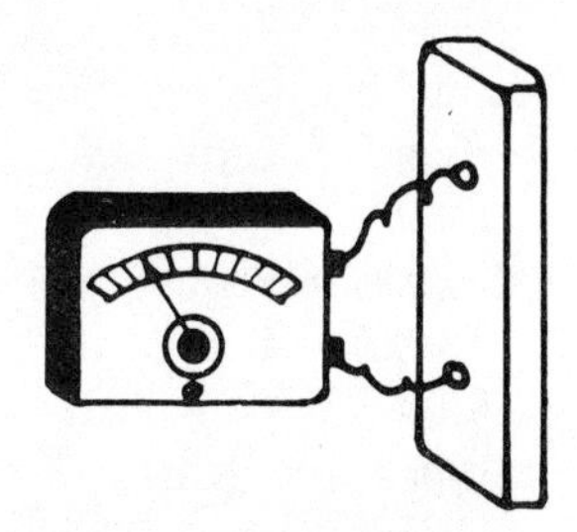

Additional Reading

Abragam, A. (1961)
The Principles of Nuclear Magnetism
Oxford, Clarenden.

Abramson, H.A.; Moyer, L.S. and Govin, M.H. (1942)
Electrophoresis of Proteins and the Chemistry of Cell Surfaces
New York, Reinhold Publ. Corpn.

Adelman, M.H. (1970)
Automated Wet Chemical Air Analysis: The Beginning to the Present.
Presented at the Technicon International Congress, New York.

Agarwal, G.P. et. al. (1976)
Jr. Phy. E (Sc. Instr.)
9, 626.

Ageno, M. and Felici, C. (1963)
Rev. Sc. Instr.
34,997.

Aldons, K.M. et. al. (1970)
Anal. Chem.
42,939.

Alexander, R.D.; Dudeney, A.W.L. and Irwing, R.J. (1974)
Jr. Phy. E (Sc. Instr.)
7,522.

Allendoerfer, R. et. al. (1975)
Anal. Chem.
47,890.

Anson, F.C. (1966)
Anal. Chem.
38,1141
Aoshima, R. and Sugita, T. (1974)
Jr. Phy. E (Sc. Instr.)
7,48.
Astrup, P. and Schroeder, S. (1956)
Scand. J. Clin. Lab. Invest.
8,30.
Astrup, P. (1956)
Scand. J. Clin Lab. Invest.
8,33.
Astrup, P.; Jorgensen, K.; Siggaard-Andersen, O. and Engel, K. (1960)
Lancet
1,1035.
Astrup, P. and Siggaard-Anderson, O. (1963)
Advances in Clinical Chemistry
(Ed. H. Sobotka and C.P. Stevens)
Academic Press, New York.
Aurich, F. (1969)
Jr. Phy. E. (Sc. Instr.)
2,109.
Ashley, J.W. and Reilley, C.N. (1965)
Anal. Chem.
37,626.

Balslev, I. (1966)
Phy. Rev.
143,636.
Bard, A.J. and Santhanam, K.S.V. (1970)
Electroanalytical Chemistry
Vol. 4 Ed. A.J. Bard
New York, Moscel Dekker.
Bartky, W. and Dempster, A.J. (1929)
Phy. Rev.
33,1019.
Bates, R.G. and Acree, S.F. (1945)
Res. Natn. Bur. Stand.
34,373
Bennett, W.H. (1950)
Jr. Appl. Phy.
21,143.

Beneken Kolmer, H.H. and Kreuzer, F. (1968)
Resep. Physiol. Neth.
4,109.
Berclaz, T. et. al. (1977)
Jr. Phy. E. (Sc. Instr.)
10,871.
Boling, E.A. (1965)
Anal. Chim Acta
37,482
Bonczyk, P.A. (1975)
Rev. Sc. Instr.
46,456.
Botten, D. (1975)
Instr. News
25,14.
Boutilier, G.D. et. al. (1977)
Applied Spectroscopy
31,307.
Bradley, A. et. al. (1976)
Jr. Am. Chem. Soc.
98,620.
Braun, W.H. et. al. (1976)
Anal. Chem.
48,2284.
Brewer, A.W. and Milford, J.R. (1960)
Proc. Royal Soc.
A 256,470.
Brooks, C.J.W. and Middleditch, B.S. (1977)
Mass Spectrometry, Vol. 4
Chemical Society, London.
Brosnan, S.J. et. al. (1977)
Appl. Phy. Lett.
30,330.
Brown, E.B. and Miller, F.A. (1952)
Am. Jr. Physiology
169,56.
Buckland, A.; Ramsbotham, J.; Rochester, C.H. and Scurrell, M.S. (1971)
Jr. Phy. E. (Sc. Instr.)
4,146.
Burge, D.E. (1970)
Jr. Chem. Edcn.
47,A81

Burkhardt, E.G.; Lambart, C.A. and Patel, C.K.N. (1975)
Science
188,1111.
Burlingame, A.L. et. al. (1976)
Anal. Chem.
48,368 R.
Burlingame, A.C. et. al. (1978)
Anal. Chem.
April 1978, 346 R.

Carr-Brion, K.G. and Gadsen, J.A. (1969)
Jr. Phy. E (Sc. Instr.)
2,155.
Chalzel, A. (1977)
Jr. Phy. E. (Sc. Instr.)
10,633.
Christie, J.H.; Laner; G. and Osteryoung, R.S. (1964)
Jr. Elect. Chem. Soc.
111,1420.
Clark, L.C. (1956)
Trans Am. Soc. Internal Organs.
2,41.
Clark, L.C.; Misrahy, G. and Fox, R.P. (1958)
Jr. Appl. Phys.
13,85.
Clark, L.C. and Becattini, F. (1967)
Ala. Jnl. Med. Sci.
4,337.
Claxton, G. (1959)
Jr. Chromatog.
2,136.
Coe, J.S. and Slaney, R.E. (1969)
Jr. Phy. E. (Sc. Instr.)
2,98.
Collins, A.T. (1975)
Jr. Phy. E. (Sc. Instr.)
8,1021.

Dankleman, W. and Daemen, J.M.H. (1976)
Anal. Chem.
48,401.

Davies, P.W. and Brink, F. (1942)
Rev. Sc. Instr.
130,524.
Davis, H.M. and Seaborn, Joyce, E. (1953)
Electron, Engg.
25,314.
Deave, N. and Smith, H.W. (1957)
Jr. Biol. Chem.
227,101.
Dempster, A.J. (1918)
Phy. Rev.
11,316.
Denton, M.B. and Swartz, D.B. (1974)
Rev. Sc. Instr.
45,81.
Desty, D.H. (1975)
Chromatographia
8,452.
Dill, D.B.; Daly, C. and Forbes, W.H. (1937)
Jr. Biol. Chem.
117,569.
Dousmanis, G.C.; Sander, T.M. and Townes, C.H. (1955)
Phy. Rev.
100,1735.
Dresner, S. (1975)
Pop. Sc.
1,45.
Dubowski, K.M. (1976)
Clin. Chem.
22,863.
Dyer, C.A. (1947)
Rev. Sc. Instr.
18,696

Eckey, H. (1969)
Jr. Phy. E. (Sc. Instr.)
2,1121.
Eisenman, G,; Rudkin, D.C. and Casby, J. U. (1957)
Science
126,831
English, T.H. (1976)
Jr. Phy. E. (Sc. Instr.)
3,69.

Erickson, J.O. and Surles, T. (1976)
Am. Lab.
8,41.
Ernst, R.R. and Anderson, W.A. (1966)
Rev. Sc. Instr.
37,93.
Erwine, Barbara (1975)
Sensitivity Enhancement for Analytical NMR Spectroscopy App. Note No. NMR-75-1
Varian Instruments Division, Palo, Alto, Cal. USA.
Evans, B.L. and Thompson, K.T. (1969)
Jr. Phy. E. (Sc. Instr.)
2,327.

Fairstein, E. and Hahn, J. (1965)
Nucleonics
7,56.
Frant, M.S. and Ross, J.W. (1966)
Science
154,1553.
Frostling, H. and Brantte, A.L. (1972)
Jr. Phy. E. (Sc. Instr.)
5,251.

Geary, Stephen (1975)
Instr. News.
25,5
Geddes, L.A. and Baker, L.E. (1968)
Principles of Applied Biomedical Instrumentation
John Wiley & Sons Inc. New York.
German, A.V. and Heyman, H.W.G. (1972)
Jr. Phy. E. (Sc. Instr.)
5,413.
Giese, A.T. and French, C.S. (1955)
Appl. Spectrosc.
9,78.
Gilland, J.R. and Ried, L. (1969)
IEE Trans. Nucl. Sci.
16,277.
Golay, M.J.E. (1947)
Rev. Sc. Instr.
18,347.

Goldberg, I. and Crowe, H. (1975)
Jr. Mag. Reson,
18,497.
Goodrich, G.W. and Wiley, W.C. (1961)
Rev. Sc. Instr.
32,846.
Gore, W.G. and Smith, G.W. (1974)
Jr. Phy. E. (Sc. Instr.)
7,644.
Gooley, C.; Gross, P.K. and Keyzer, U. (1969)
Jr. Phy. E. (Sc. Instr.)
2,531.
Gough, W. (1977)
Jr. Phy. E. (Sc. Instr.)
10,867.
Grob, K. (1975)
Chromatographia
8,423.
Grunesis, F.; Schneider, S. and Dorr, F. (1975)
Jr. Phy. E. (Sc. Instr.)
8,402.

Haagen-Smit, J.W.; Ting, P.; Johns, T. and Berry, E.A. (1971)
The Beckman EC News
1,1.
Hahn, C.E.W. (1969)
Jr. Phy. E. (Sc. Instr.)
2,48.
Halliday, J.D.; Hill, H.D.W. and Richards, R.E. (1969)
Jr. Phy. E. (Sc. Instr.)
2,29.
Haraguchi, H. et. al. (1976)
Chem. Abst.
84,159
Haun, S. and Kamke, D. (1960)
Nuc. Instr. Methods
8,331.
Hazebroek, H.F. (1972)
Jr. Phy. E. (Sc. Instr.)
5,180.
Hazen, G.; Grinvald, A.,; Maytal, M. and Steinberg, I.Z. (1974)
Rev. Sc. Instr.
45,1602.

Hell, A. and Ramiraz-Munoz (1970)
Jr. Anal. Chem. Acta
51,141.
Henson, A.F. (1950)
Jr. Appl. Phy.
21,1063
Hertz, C.H. and Siesjo, B. (1959)
Acta Physiol, Scand.
47,115.
Hieftye, G.M. and Copeland, T.R. (1978)
Anal. Chem.
50,300 R.
Hogg, A.M. (1969)
Jr. Phy. E. (Sc. Instr.)
2,289.
Honkawa, T. (1976)
Anal. Lett.
9,839.
Horlick, G. (1976)
Appl. Spectroscopy
30,113.
Howard, W.E.; Selzle, H.L. and Schlag, W.E. (1975)
Jr. Phy. E. (Sc. Instr.)
8,783.
Hupe, K.P. and Bayer, E.J. (1967)
Jr. Gas Chromatogr.
5,197.

Itoh, T. et. al. (1976)
Jr. Appl. Phy.
15(7), 1281.

Jacobs, M.B. and Hochheiser, S. (1958)
Anal. Chem.
30,426.
Janate, J. and Moss, S.D. (1976)
Biomed. Engg.
11,241.
Jesop, G. (1966)
Jr. Phy. E. (Sc. Instr.)
11,777.

Johansen, K. and Krog, J. (1959)
Acta, Physiol. Scand.
46,106.
Jones, T.A.; Firth, J.G. and Jones, A. (1971)
Jr. Phy. E. (Sc. Instr.)
4,792.

Kabanova, M.A. and Sautina, E.N. (1977)
Chem. Abstr.
87,77793q.
Kanstad, S.O. et. al. (1977)
Jr. Phy. E. (Sc. Instr.)
10,998.
Karasek, F. et. al. (1976)
Jr. Chromatogr.
124,179.
Killick, C.M. (1969)
Jr. Phy. E. (Sc. Instr.)
2,1017.
Kim, S.S. and Weissman, S. (1977)
Jr. Mag. Reson.
24,167.
Knipe, A.C.; Mclean, D. and Tranter, R.L. (1974)
Jr. Phy. E. (Sc. Instr.)
7,586.
Koizumi, H. and Yasuda, K. (1976)
Spectrochim Acta, Part B,
31,237.
Kreuzer, L.B. (1978)
Anal. Chem.
50,597A.

Labhart, H. and Staub, H. (1947)
Helv. Chemi. Acta
30,1954.
Langelaar, J.; Vries, G.A. de and Bebelar, D. (1969)
Jr. Phy. E. (Sc. Instr.)
40,149.
Lenfant, C. (1961)
Jr. Appl. Physiol.
16,909

Longsworth, L.G. (1939)
Jr. Am. Chem. Soc.
61,529.
Lotmar, W. (1953)
Plasma
2,209.
Lovelock, J.E. et. al. (1964)
Anal. Chem.
36,1410.
Ludbrook, J. (1959)
Symposium on pH and Blood Gas Measurement
(Ed. R.F. Woolmen) Churchill, London.
Page 34.
Lundsgaard, J.S. and Petersen, H.A. (1974)
Jr. Phy. E. (Sc. Instr.)
7,524.
Lyshkow, N.A. (1965)
Jr. Air Poll. Control Assn.
15,481.

Mailer, C. et. al. (1977)
Jr. Mag. Res.
25,205.
Martin, A.A. and Sinclair, Y. (1976)
Jr. Bioengg.
1,55.
Mattauch, J. and Herzog, R. (1936)
Phy. Rev.
50,617.
Mattock, G. (1961)
pH Measurement and Titration
Heywood, London.
Page 82.
McConn, R. and Robinson, J.S. (1963)
Br. Jr. Anaesth.
35,679.
McIvor, Mc. (1969)
Jr. Phy. E. (Sc. Instr.)
2,292.
Meigh, D.R. and Oetzmann, E.H. (1971)
Jr. Phy. E. (Sc. Instr.)
4,66.

Meili, J. et. al. (1977)
25th Annual Conference on Mass Spectrometry and Applied Topics
Washington, D.C.
Page 175.
Melhuish, W.H. (1975)
Jr. Phy. E. (Sc. Instr.)
8,815.
Mitchell, J. Jr. and Smith, D.M. (1948)
Aquametry,
Interscience Publishers Inc. New York.
Montgomery, H. (1957)
Am. Jr. Med.
23,697.
Moore, D.H. and White, J.U. (1948)
Rev. Sc. Instr.
19,700
Moss, D.G. (1977)
Jr. Phy. E. (Sc. Instr.)
10,1170.

Nakajima, F. and Sakai, K. (1976)
Bunseki Kagaku,
25,378.
Nerheim, A.G. (1963)
Anal. Chem.
35,1640
Newmann, S. and Ewald, H. (1962)
Z. Phy.
169,224.
Nicholas, D.J. (1971)
Jr. Phy. E. (Sc. Instr.)
4,68.
Nier, A.O. (1940)
Rev. Sc. Instr.
11,212.
Nishita, N.; Farmer, R. and Peterson, S. (1972)
Anal. Chem. Acta
58, 1.
Norris, M.O. and Strange, J.H. (1969)
Jr. Phy. E. (Sc. Instr.)
2, 1106.

Novak, J and Janak, J. (1977)
Jr. Chromatogr.
138,1.
Nunn, J.F. (1958)
Br. Jr. Anaes.
30,264.
Nunn, J.F. (1964)
Br. Jr. Anaes.
36, 366.

Ortman, G.C. (1966)
Anal. Chem.
38, 644.
Otvos, J.W. and Stevenson, D.P. (1956)
Jr. Am. Chem. Soc.
78, 546.

Pailthrope, M.T. (1975)
Jr. Phy. E. (Sc. Instr.)
8, 194.
Palmer, D.A. (1971)
Jr. Phy. E. (Sc. Instr.)
4, 41.
Parker, C.A. and Hatchard, C.G. (1978)
Rep. Admty Mater Lab. Am.
R7602, 10.
Passenheim, B.C.; Kitterer, B.D.; Flangan, T.M. and Denson, R. (1974)
Rev. Sc. Instr.
45, 1365.
Pauling, L; Wood, R. and Sturdevant, C.O. (1946)
Science
103, 338.
Pellizzari E.D. et. al. (1976)
Ana. Lett.
9, 579.
Perley, G.A. (1948)
Jr. Electro. Chem. Soc.
92, 485.
Philpot, J.S. (1938)
Nature
141, 283.

Pierce, J.R. (1945)
Proc. IRE
33, 112.
Poole, J.S. and Bilen, L. (1976)
Hewlett Packard Journal.
April 1976, Page 13.
Poppelwell, R.J.L. (1972)
Jr. Phy. E. (Sc. Instr.)
5, 307.
Purdue, L.J. et. al (1972)
Environ. Sc. Technol.
6, 152.
Pursall, B.R. and Dugar, A.C. (1972)
Jr. Phy. E. (Sc. Instr.)
5, 862.

Reyes, R.J. and Neville, J.R. (1967)
USAF School Aerospace Med.
Tech. Rept SAM-IR-67-23.
Rigby, W.; Whyman, R. and Wilding, K. (1970)
Jr. Phy. E. (Sc. Instr.)
3, 572.
Riggs, W.A. (1971)
Anal. Chem.
43, 976.
Rosenthal, T.B. (1948)
Jr. Biol. Chem.
173, 25.
Ross, J.W. (1967)
Science
156, 1378.

Saltzman, B.E. (1954)
Anal. Chem.
26, 1949.
Saltzman, B.E. and Gilbert, N. (1959)
Anal. Chem
31, 1914.
Sanz, M.C. (1957)
Clin. Chem.
3, 406.
Sathanapati, J. et. al. (1977)
Jr. Phy. E. (Sc. Instr.)
10, 1.

Sathanapati, J. et. al. (1977)
Jr. Phy. E. (Sc. Instr.)
10, 221.
Schomburg, G. and Husmann, H. (1975)
Chromatographia
8, 517.
Schrenker, H. (1975)
Hewlett Packard Journal
Oct. 1975, Page 17.
Sevensson, H. (1946)
Ark. for Kemi,
229, No. 10.
Sevensson, H. (1945)
Ark. for Kemi.
22A, 10.
Sevensson, H. (1951)
Act. Chem. Skand.
5, 1001.
Severinghaus, J.W.; Stupfel, M. and Bradley, A.F. (1956)
Jr. App. Physiol.
9, 189.
Severinghaus, J.W. and Bradley, A.F. (1958)
Jr. Appl. Physiol.
13, 515.
Severinghaus, J.W. (1962)
Acta Anaes. Scand.
6 (Suppl XI) 207.
Shepherd, J.R. and Hedgpeth, H.R. (1973)
Rev. Sc. Instr.
44, 338.
Siddiqui, A.S. and Stewart, D. (1974)
Jr. Phy. E. (Sc. Instr.)
7, 318.
Slevin, P.J. and Harrison, W.W. (1975)
Appl. Spectrosc. Rev.
10, 201.
Smith, G.E.; Blankdership, R.E. and Klein, M.P. (1976)
Jr. Mag. Reso.
48, 282.
Soini, Erkki (1975)
Rev. Sc. Instr.
46, 980.

Sole, M.J. and Walker, P.J. (1970)
Jr. Phy. E. (Sc. Instr.)
3, 394.
Spackman, Stein and Moore (1958)
Anal. Chem.
30, 1185.
Staeudner, R. (1976)
Proc. Anal. Div. Chem. Soc.
13, 212.
Staynov, D. and Stamov, G. (1969)
Jr. Phy. E. (Sc. Instr.)
2, 1114.
Steinberger, E.H. and Goldwater, F. (1972)
Jr. Phy. E. (Sc. Instr.)
5, 373.
Stephens, W.E. (1946)
Phy. Rev.
69, 691.
Stevens, R.K. et. al. (1971)
Anal. Chem.
4, 827.
Sthanapati, J. et. al. (1977)
Jr. Phy. E. (Sc. Instr.)
10, 26.
Stow, R.W.; Baer, R.R. and Randall, B.F. (1957)
Arch. Phy. Med. Rehabil.
38, 646.
Sturgeon, R.E. (1977)
Anal. Chem.
49, 1255A.
Subramanian, M.S.; Srinivasan, P.S. and Mithapara, P.D. (1973)
Jr. Phy. E. (Sc. Instr.)
6, 43.

Talmi, Yair; Baker, D.C.; Jadamec, J.R. and Saner, W.A. (1978)
Anal. Chem.
50, 936A.
Thomas, J.R.D. (1977)
Laboratory Practice
26, 313.
Tiselius, A. (1938)
Kolloid Z.
85, 129.

Tobias, J.M. (1948)
Rev. Sc. Instr.
20, 519.

Villalobos, R. and Chapman, R.L. (1971)
ISA Trans.
10, 356.

Wade, C.H. (1975)
Rev. Sc. Instr.
45, 987.

Warren, G.J. and Babcock, G. (1970)
Rev. Sc. Instr.
41, 280.

Wasson, J.R. and Corvan, P.J. (1978)
Anal. Chem.
4, 92R.

West, P.W. and Gaeke, C.C. (1956)
Anal. Chem.
28, 1816.

West, P.W. and Ordoveza, F. (1962)
Anal. Chem.
34, 1324.

Wetzel, R. et. al. (1969)
Jr. Phy. E. (Sc. Instr.)
2, 841.

White, J.U. (1976)
Anal. Chem.
48, 2089.

Woolmer, R.F. (1956)
Br. Jr. Anaes.
28, 118.

Yoshimura, H. (1937)
Bull. Chem. Soc. Japan
12, 443.

Yotsuyanagi, T. et. al. (1976)
Anal. Chem. Acta
82, 431.

Zehnder, H. and Belew, W. (1970)
The Air Monitor IV, A New Approach in Air Monitoring.
Presented at the Technicon International Congress, New York.

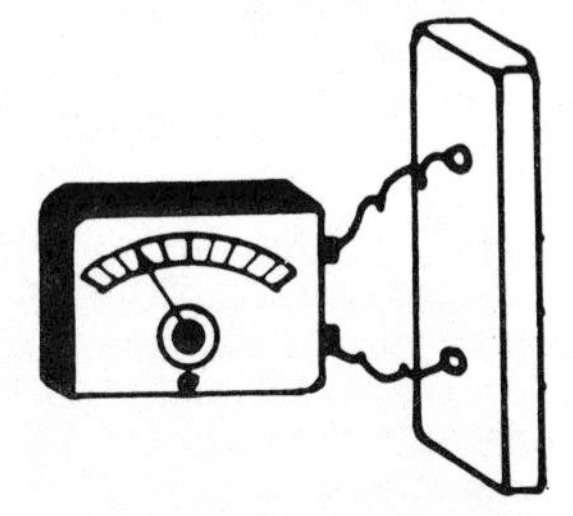

Index